DATE DUE

DEC 21 2001			
APR 07 2011			

Wetland Ecology
Principles and Conservation

Wetlands are among the most productive and biologically diverse ecosystems on Earth. Their very diversity has confounded ecologists and produced a fragmented area of study where each wetland type has tended to be considered in isolation. The discipline has been further compartmentalized by narrow focus on specific organisms and geographic regions. This book, in contrast, provides a synthesis of the existing field of wetland ecology, using a few central themes. These themes include basic characteristics of wetlands, key environmental factors that produce wetland community types and some unifying problems such as assembly rules, restoration and conservation. The volume draws upon a complete range of wetland habitats and geographic regions, including examples from Africa, Asia, Europe, Australia and New Zealand, as well as from North and South America. No other book provides up-to-date ecological syntheses over the entire geographical and habitat range of wetlands. As such, *Wetland Ecology* is essential reading for anyone planning research or management in wetland habitats.

PAUL KEDDY is the Edward G. Schlieder Professor for Environmental Studies at Southeastern Louisiana University. He is author of *Competition* (1989) which won both the Lawson Medal and the Gleason Prize, and co-editor of *Assembly Rules: Perspectives, Advances, Retreats* (1999).

CAMBRIDGE STUDIES IN ECOLOGY

This series presents balanced, comprehensive, up-to-date, and critical reviews of selected topics within ecology, both botanical and zoological. The Series is aimed at advanced final-year undergraduates, graduate students, researchers, and university teachers, as well as ecologists in industry and government research.

It encompasses a wide range of approaches and spatial, temporal, and taxonomic scales in ecology, experimental, behavioural and evolutionary studies. The emphasis throughout is on ecology related to the real world of plants and animals in the field rather than on purely theoretical abstractions and mathematical models. Some books in the Series attempt to challenge existing ecological paradigms and present new concepts, empirical or theoretical models, and testable hypotheses. Others attempt to explore new approaches and present syntheses on topics of considerable importance ecologically which cut across the conventional but artificial boundaries within the science of ecology.

Wetland Ecology Principles and Conservation

PAUL A. KEDDY
Southeastern Louisiana University

PUBLISHED BY THE PRESS SYNDICATE OF THE UNIVERSITY OF CAMBRIDGE
The Pitt Building, Trumpington Street, Cambridge, United Kingdom

CAMBRIDGE UNIVERSITY PRESS
The Edinburgh Building, Cambridge CB2 2RU, UK
40 West 20th Street, New York, NY 10011-4211, USA
10 Stamford Road, Oakleigh, VIC 3166, Australia
Ruiz de Alarcón 13, 28014 Madrid, Spain
Dock House, The Waterfront, Cape Town 8001, South Africa
http://www.cambridge.org

First published 2000

Printed in the United Kingdom at the University Press, Cambridge

Typeface Monotype Bembo 11/13pt. *System* QuarkXPress™ [SE]

A catalogue record for this book is available from the British Library

Library of Congress Cataloguing in Publication data

Keddy, Paul A., 1953–
Wetland ecology principles and conservation / Paul Keddy.
p. cm. – (Cambridge studies in ecology)
ISBN 0 521 78001 2 (hb) – ISBN 0 521 78367 4 (pb)
1. Wetland ecology. 2. Wetland conservation. I. Title. II. Series.
QH541.5.M3 K44 2000
577.68–dc21 99-045443

ISBN 0 521 78001 2 hardback
ISBN 0 521 78367 4 paperback

Contents

PART II. FACTORS CONTROLLING PROPERTIES OF WETLANDS

Preface

According to Bernard Shaw, writer of many a lengthy preface, the lesson intended by an author is hardly ever the lesson the world chooses to learn from his book. If Shaw is right (and who would risk disagreeing with him), why would anyone trouble to write a book? And why a book on wetlands?

In answer to the first question, the motivation of all writers includes a healthy dose of inspiration, frustration, and ego gratification. Events can conspire to feed these forces, with unfortunate consequences for both writers and the public. When Dr Birks first asked me to write this book, I therefore declined. In part, I was not convinced that a need existed. Bringing a new book into the world requires the shouldering of parental obligations. (While one can keep one's rowdy children at home, a book is always on public display.) The world certainly has too many children, and only the most devoted reader, deep-pocketed publisher, or hardened bibliophile could believe that every author who is inclined towards writing should do so. The self-restraint that is a virtue in biological procreation, may be equally so for aspiring authors.

Events can, however, over-ride caution. Illness, like the threat of a hanging, tends to concentrate and clarify one's mind. Moreover, during the days chained to the wall before the hanging, one is inclined to dwell on shortcomings, particularly those of one's associates. But I digress. This is, after all, a preface to a book on wetlands.

The principal objective of this book is to try to provide some unity and coherence in the study of wetland ecology. To do so, I have organized this book into three sections. The first section (Chapters 1–3) emphasizes the properties of wetlands, or, in statistical terms, the dependent variables in our inquiry. The next section (Chapters 4–9) addresses the environmental factors that control these properties: in statistical terms, the independent variables. In these chapters, I freely range across wetland types and geographic regions. My self-assigned task is to illustrate the relationships among properties and environmental factors, whether they occur in an Amazonian floodplain, prairie pothole, boreal peatland or tidal marsh. The book may still tend to emphasize the types of wetlands with which I am most familiar, but this should not distract a reader from principles and scientific generalities. The final section (Chapters 10–12) illustrates some larger frameworks for studying the relationships between ecological properties and conservation biology. Assembly rules, functional groups, and restoration ecology receive particular attention.

The book has been prepared with several audiences in mind. It is intended as a text book for senior undergraduates, an introduction to key factors controlling wetlands for busy managers, and as general reading for any scientist intending to work in wet habitats. Further, the first chapter, will I hope, introduce the essential features of wetlands to a general reader; while superficial in places, it is less so than many popular treatises, and it will simultaneously remind more experienced readers about the salient features that make wetlands of particular interest to humanity. The main body of the book presents a general framework for the study of wetland communities. For practising wetland scientists I had an expanded purpose. The discipline of wetland ecology is currently Balkanized by habitat types, geographic regions, and study organisms. Many of the studies of particular wet habitats that I have read over the past decade have seemed blissfully unaware of nearly identical work in other habitats, wet or dry. By combining all wetland types within one book, I have tried to restore some conceptual unity to the discipline by emphasizing the essential processes that all wetlands share, and then by illustrating the ways in which some of them differ. Hence the part of the title referring to principles. I hope that specialists will be stimulated by seeing the parallel advances in habitat types and geographic regions other than their own, and that this enriched context will assist them with further progress within their own areas of specialization. In exchange, I trust they will forgive the inevitable oversights that annoy a specialist.

The final part of the title mentions conservation. Sound science is the essential foundation of good ecosystem management. Ecosystem management emphasizes ecological processes and their interconnections. This book takes exactly such a perspective: it begins with patterns present in wetlands, and then proceeds to the processes and interconnections that produce the patterns. The focus is upon communities and ecosystems themselves; implications for global biogeochemical cycles are mentioned from time to time, but they are not a primary focus. Rather, it is assumed that, in most cases, maintaining the normal processes within wetlands will ensure that their valued functions continue to occur. When, and if, it is necessary to manipulate wetlands in order to change some aspect of their global function, say, to increase wildlife production or to decrease methane production, this will always require knowledge of processes at the local community scale.

I first thought that such a book might be too personal a perspective on wetlands. Fields of enquiry are now so large that perhaps only multi-authored works are appropriate. However, my editor and advisor, Alan Crowden, has convinced me that many readers actually prefer a systematic and personal account of a field to a series of edited papers. Moreover, I have slowly convinced myself, too, that the existing literature is far too fragmented and diffuse and therefore confused. I have already written bluntly, perhaps too much so, of my views on symposium reports and festschrifts (Keddy 1991a, b, c). A number of recent symposium volumes on wetlands appear to be little more than expensive books with a haphazard collection of people giving a haphazard collection of papers with no unifying theme whatsoever except for the fact that all work in wet areas. Surely we can aspire to do better than this. While my own community-oriented perspective undoubtedly has its limitations, it at least compensates with continuity and consistency.

I have tried to emphasize several research strategies. These include (i) greater emphasis upon measurable properties of ecosystems and (ii) the relative importance of different environmental factors that produce pattern. Far too many studies in wetlands consist, it seems, of little more than drawings of transects through wetlands or autecological studies of small groups of species living in wet places. Neither of these latter styles will inspire bright young scientists to enter the field. In fact, wetland community ecology is exciting, challenging, socially significant, and worthy of our best minds.

At first I was going to include a chapter on applications. But then my continued resistance to the forced distinction between theoretical and applied ecology intervened. Throughout this book there is an interplay between theory and application. In combining them we can achieve maximum impact upon knowledge with a minimal expenditure of effort. An appeal to efficiency itself ought to be sufficient, but we are now faced, in addition, with the rapid loss of the very ecosystems we study. We must hasten if we are to solve some of the growing problems with management of wetlands. Throughout the book there are practical examples that show that wetland ecologists have a great many useful things to say to environmental managers. Altered hydrology, eutrophication, loss of species – these are fundamental environmental issues and conceptual axes in the study of wetlands. There is therefore no single chapter on conservation alone because the entire book is about conservation.

Were it not for the inseparability of theory and application, this book might be considered schizophrenic. It is written with both the basic researcher and the resource manager in mind. I hope that both bright, young graduate students and cynical, overworked managers can benefit from consulting it. I have made abundant use of subheadings and figures so that parts that, at least on first reading, appear of secondary importance can easily be skipped. Each chapter will, I hope, be able to stand alone. Those needing an immediate short course, or feeling too harried to deal with an entire book, can obtain an overview of essentials with Chapters 1, 4, 5, and 12. Chapters 10 and 11 are the most speculative, and can be safely omitted from a first reading since they deal more with future possibilities than established phenomena.

Some of the limitations of the book are deliberate. I have placed an emphasis upon communities and on the factors than influence them. Although nutrient cycling is an important topic, I have not dealt with it extensively except under the heading of eutrophication. Similarly, systems models are already well covered in works such as Good *et al.* (1978), Mitsch and Gosselink (1986), and Patten (1990). Apart from eutrophication, I have left the topic of toxic contaminants to other better-qualified authors. There are also two fine compendia which already describe wetland types by region (Gore 1983; Whigham *et al.* 1992). I have not tried to duplicate their efforts. The logical structure of this book is built upon similarities in process rather than geography.

Finally (restrictions on travel are inclined to make one long-winded) this is not *just* a book on wetlands. I have tried to present not only an overview of wetland ecology, but to illustrate the general procedures with which one can dissect an ecological community to search for patterns and the mechanisms that may cause them. In this way, I hope to not only contribute to our understanding of wetlands, but to illustrate practices that will be of use in other vegetation types and ecological communities.

Since we began with Shaw, let us also close with him too. A successful book, according to Shaw, will impress the strong, intimidate the weak and tickle the connoisseur.

Paul Keddy

Acknowledgements

I thank colleagues including Mark Bertness, Dan Brunton, Duncan Cameron, Paul Catling, Alan Crowden, Ian Davidson, Jim Grace, Phil Grime, Laura Gough, Mary Kentula, Doug Larson, Jorge Meave, Steward Pickett, Richard Reader, Curtis Richardson, Clayton Rubec, Gary Shaffer, Dan Simberloff, Evan Weiher, and Doug Wilcox for their advice, encouragement, criticisms, assistance or input. Several anonymous but tactful reviewers commented at length on early drafts; I hope this version leads to the conclusion that their efforts were worthwhile. I also thank my former students and co-workers, Scott Wilson, Bill Shipley, Céline Boutin, Connie Gaudet, Dwayne Moore, Nick Hill, Irene Wisheu, Lisa Twolan-Strutt and Maureen Toner for examples I have drawn upon in this volume. Since a book of this length and breadth is certain to have errors and omissions, I trust readers will be forthright in drawing them to my attention. Special thanks must go to Irene Wisheu, Evan Weiher and Lauchlan Fraser who ran my laboratory in Ottawa during 8 long years of adversity. Without their loyalty, attention to detail and sheer hard work, far less would have been possible. Rochelle Lawson and Teri Keogh did the constant day-to-day work. While it is a cliché to say that this project could not have been done without them, in this case it is also literally true. Finally, much of the funding was provided by the Natural Sciences and Engineering Research Council of Canada; their generous support over many years is gratefully noted.

The final product is a team effort. The staff at Cambridge University Press, in particular, Mary Sanders, Maria Murphy and Sue Tuck, deserve special mention for their care and patience during production. The cover draws upon an original 1976 line drawing of a snapping turtle by Howard Coneybeare (courtesy of Friends of Algonquin Park) in a design by Chris McLeod. Other line drawings were prepared by Rochelle Lawson. Michaelyn Broussard helped track down the last errant figures, and Kim

Fisher contributed some new ones. Cathy Keddy was relentless at proofing. Copyediting and proofing were done amidst the cypress swamps and pine forests of south Louisiana. Special recognition must therefore be given to my hosts there, including Sally Clausen, Randy Moffett, John Miller, Nick Norton, Bill Font, Bob Hastings, Mark Hester and Gary Shaffer. All were willing to take a leap of faith and offer me the new professional home where this project came to fruition.

Part I
Properties of wetlands

1 · *Wetlands: an overview*

All life contains water. From outer space, our planet appears as a mosaic of blue and green – blue for water, green for plants. This book is about the ecological communities that occur where green meets blue: wetlands.

Wetlands have always influenced humans. Early civilizations first arose along the edges of rivers in the fertile soils of floodplains (Figure 1.1). Wetlands are also the source of immense human suffering from diseases such as malaria (Figure 1.2). Philosophers and theologians may enquire how it is that one system can be both life giving and death dealing. Our more confined task as scientists is to explore the basic patterns that occur in wetlands, to offer some educated guesses as to the mechanisms that account for these patterns, and then, perhaps, to guide society in a wise co-existence with wetlands. I intend to make at least one step forward in this direction (i) providing a logical structure for this enquiry, by (ii) bringing together some existing information, and (iii) stressing the commonality in pattern and process among world wetlands.

Definitions and distributions

> A wetland is an ecosystem that arises when inundation by water produces soils dominated by anaerobic processes and forces the biota, particularly rooted plants, to exhibit adaptations to tolerate flooding.

This broad definition includes everything from tropical mangrove swamps to subarctic peatlands. Note that even this single sentence of definition has a complex structure: there is a cause (inundation by water), a proximate effect (reduction of oxygen levels in the soil) and a secondary effect (the biota must tolerate both the direct effects of flooding and the secondary effects of anaerobic conditions.)

Other definitions, particularly longer ones, are possible. We do not want to get tangled up in semantic arguments early in this enquiry, but

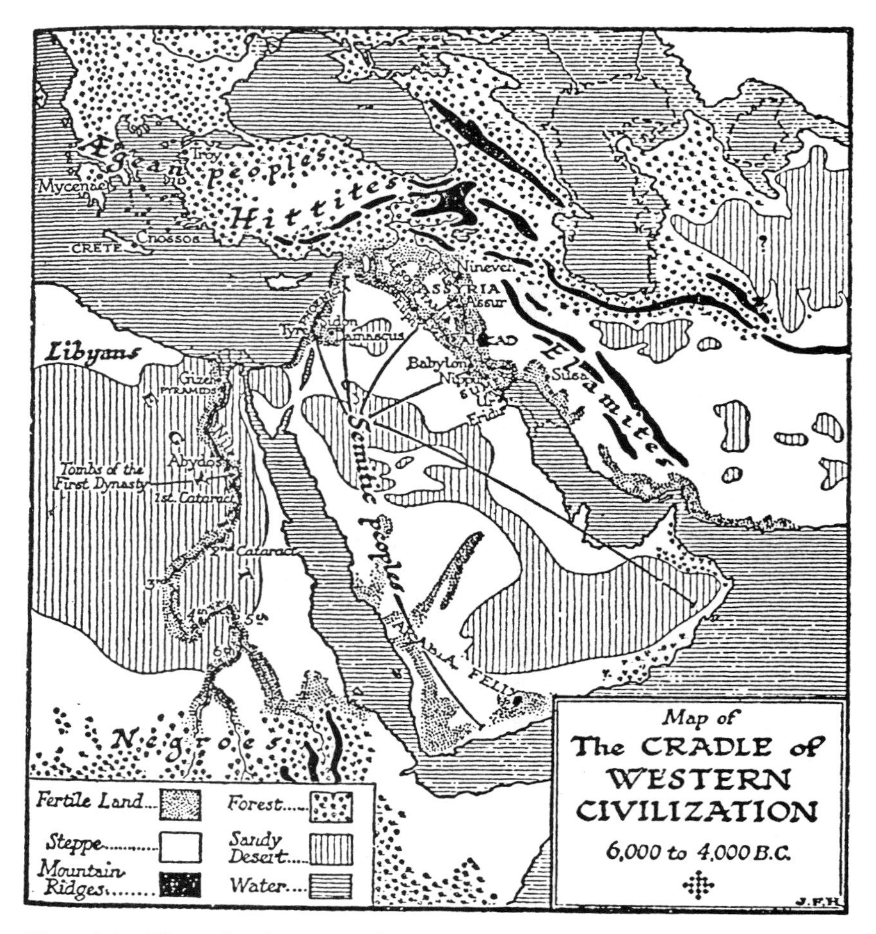

Figure 1.1 The earliest human civilizations are associated with floodplains (from Wells 1949).

perhaps something more needs to be said about definitions. If nothing else, we need to be able to defend ourselves from those annoying pedants who think that scholarly activity is primarily a form of word play. As Shakespeare observed, 'How every fool can play upon the word'. It seems that the point of definitions is widely misunderstood. There are at least two kinds of wetland definitions, with different purposes intended: the scientific and the legal, each with its own purpose.

A scientific definition is a tool for the analysis of nature. The above definition describes the domain of inquiry of wetland ecology. It also directs our attention to some salient features of wetlands. But a tool

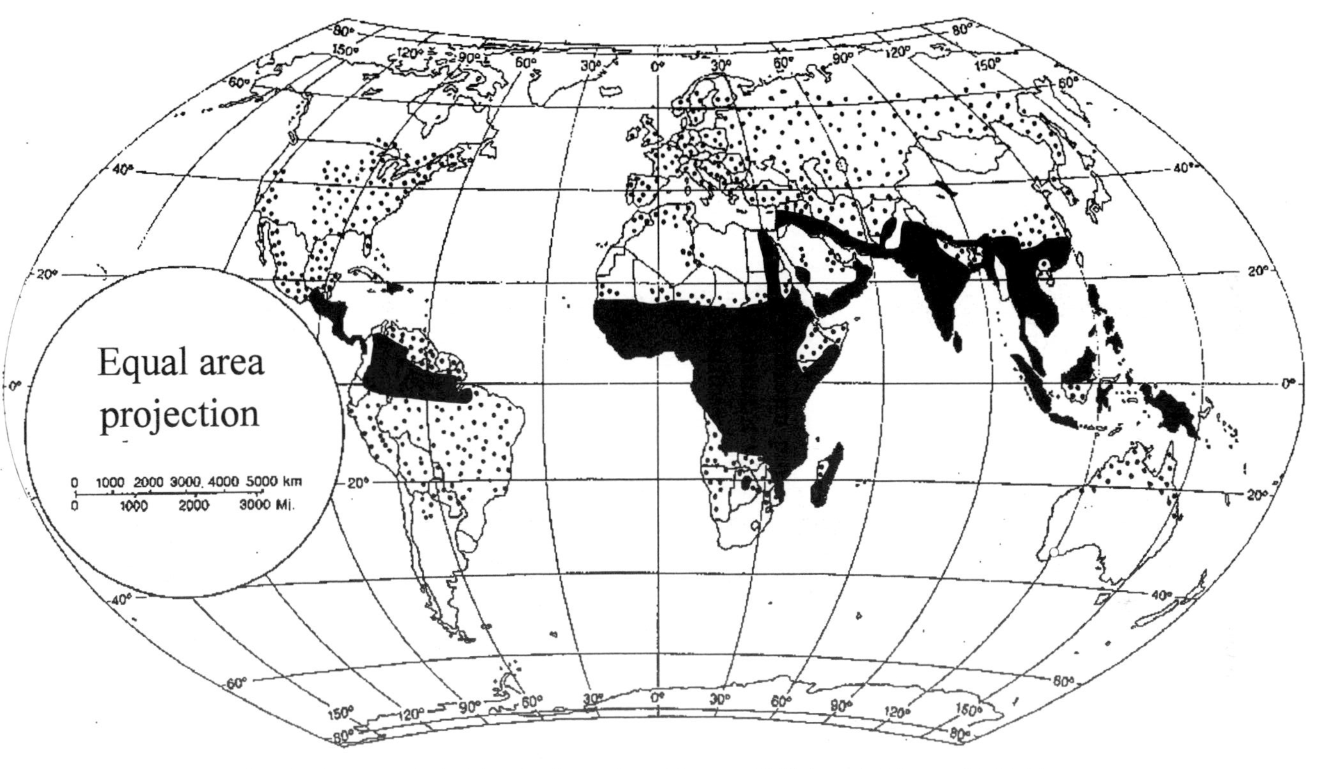

Figure 1.2 The present distribution of malaria (after Loffler and Malkhazova 1990) (stippled, low; solid, high).

should be retained only if it is useful. As our knowledge of a discipline grows, we may expect our definitions to slowly change too (Keddy 1989a). There is an evolutionary process at work here: definitions help us investigate nature, and our investigation of nature helps clarify definitions. Is a marine rocky intertidal zone a wetland? Are the contents of the leaf of a pitcher plant a wetland? How dry does a site have to become before it is no longer a wetland? Before becoming distracted by discursive definitional thoughts, we could relax: the above definition is quite satisfactory for launching a scientific enquiry.

A legal definition is another matter. Although legislative definitions can also evolve, the process is often much slower. Moreover, a clever lawyer can exploit any weakness in a definition, with the risk of serious social, economic and environmental consequences. The purpose of legal manoeuvring, after all, is not to refine scientific definitions, or to advance human understanding, but simply to win cases for a client. The proliferation of legislation designed to protect wetlands, and regulate their use, therefore requires a different kind of definition.

The Ramsar Convention is an international treaty for wetland conservation. Adopted in 1971 in Ramsar, Iran, the 'Convention on Wetlands of International Importance especially as Waterfowl Habitat' had 101 contracting parties by 1997. Altogether, 881 sites totalling some 62.7 million hectares have been designated. Participants accept the obligation to promote '. . . as far as is possible the wise use of wetlands in their territory'. Wise use is interpreted to mean the maintenance of the ecological character of wetlands (Navid 1988). The Convention uses a particularly broad definition of wetlands in Article 1:

> . . . areas of marsh, fen, peatland or water, whether natural, permanent or temporary, with water that is static or flowing, fresh, brackish or salt, including areas of marine waters the depth of which at low tide does not exceed six meters.

Moreover, the area of coverage is broadened by Article 2, which provides that wetlands

> . . . may incorporate riparian and coastal zones adjacent to the wetlands and islands or bodies of marine water deeper than six meters at low tide lying within the wetlands.

Navid (1988) observes that, as a consequence, the Convention includes habitat types including rivers, coastal areas and even coral reefs.

While such breadth may be admirable and necessary for international conservation conventions, from the perspective of this book it creates two problems. First, the definition uses terms including marsh, fen and peatland, which means that these in turn have to be individually defined. Secondly, the definition is so broad that it may lack the precision necessary for scientific enquiry. Although I have adopted a relatively broad coverage of wetland types, coral reefs and rocky marine shorelines are excluded, in part, because the ecological characteristics of such habitats appear to be sufficiently different to require a separate treatment.

A definition which circumvents these two issues is provided by the Committee on Characterization of Wetlands (1995), which has prepared an entire book on this topic for the United States! One ventures into terrain with 'national delineation manuals', 'interagency manuals', and 'revised manuals' that arise as different agencies and government bodies struggle to legally define wetlands. Flawed definitions, and flawed procedures for applying them, can lead to unnecessary constraints on land development, or to losses of important wetland habitats. The Committee on Characterization of Wetlands therefore developed a 'reference definition' to stand outside of any single agency, policy or regulation. Their reference definition is:

> A wetland is an ecosystem that depends on constant or recurrent, shallow inundation or saturation at or near the surface of the substrate. The minimum essential characteristics of a wetland are recurrent, sustained inundation or saturation at or near the surface and the presence of physical, chemical, and biological features reflective of recurrent, sustained inundation or saturation. Common diagnostic features of wetlands are hydric soils and hydrophytic vegetation. These features will be present except where specific physiochemical, biotic, or anthropogenic factors have removed them or prevented their development.

This longer definition still contains the essential elements of my shorter one: water, modified substrate, and distinct biota. We shall let Mitsch and Gosselink (1986) have the last word on this topic.

> Because wetland characteristics grade continuously from aquatic to terrestrial, any definition is to some extent arbitrary. As a result, there is no single, universally recognized definition of what a wetland is. This lack has caused confusion and inconsistencies in the management, classification, and inventorying of wetland systems, but considering the diversity of types, sizes, location, and conditions of wetlands . . . the inconsistency should be no surprise. (pp. 16–17)

Enough on definitions. Applying the simplest definition in a commonsense way opens the door to some preliminary enquires into the nature and distribution of wetlands. The Earth has about 5.6 million square kilometres

Table 1.1. *Mass of water in different forms on Earth*

Form	Mass ($\times 10^{17}$ kg)
Chemically bound in rocks[a]	
Crystalline rocks	250 000
Sedimentary rocks	2 100
Free water[b]	
Oceans	13 200
Icecaps and glaciers	292
Ground water to a depth of 4000 m	83.5
Freshwater lakes	1.25
Saline lakes and inland seas	1.04
Soil moisture	0.67
Atmospheric water vapour	0.13
Rivers	0.013

Notes:
[a] does not cycle
[b] part of hydrological cycle
Source: From Clapham (1973).

of wetlands (Dugan 1993), an area equivalent to roughly ten times the size of France or nearly four times the size of Alaska. This book will emphasize what such systems have in common. All the time, of course, we know that they will differ in detail. Until the similarities are described, it is difficult to decide which differences are the important ones. This book therefore deliberately takes a top down approach. I begin with commonalities among wetlands at the global scale. Then, gradually, patterns are dissected to uncover differences.

Wetlands require water, so the obvious place to begin this enquiry is the distribution of water. Table 1.1 shows that a majority of the Earth's available water is in the oceans; a much smaller amount is present as fresh water. Heat from the sun drives a distillery, removing water vapour from the oceans and returning it to the land as rain. Some wetlands form along the edges of oceans; these tend to be mangrove swamps in equatorial regions and salt marshes at higher latitudes. A majority of wetlands are, however, freshwater ecosystems. They occur where rainwater accumulates on its way back to the ocean.

Since life began in the oceans, most life, including freshwater life, has a chemical composition more like the ocean than fresh water (Table 1.2). It appears that most freshwater life did not originate in fresh water, but is

Table 1.2. *Concentration of some common ions in animals, sea water, and fresh water. Concentrations are given as mM/kg water*

Ions	Standard sea water	Fresh water (soft)	Fresh water (hard)	Crab (*Maia*) blood	Frog (*Rana esculenta*) blood	Crayfish (*Astacus fluviatilis*) blood (mM/l blood)	Rat (*Rattus rattus*) blood
Na^+	478.3	0.24	2.22	487.90	109	212	140
K^+	10.13	0.005	1.46	11.32	2.6	4.1	6.4
Ca^{2+}	10.48	0.067	3.98	13.61	2.1	15.8	3.4
Mg^{2+}	54.5	0.043	1.67	44.14	1.3	1.5	1.6
Cl^-	558.4	0.226	2.54	552.4	78	199	119
SO_4^{2-}	28.77	0.045	3.95	14.38	–	–	–
HCO_3^{2-}	–	–	2.02	–	26.6	15	24.3

Source: Modified from Wilson (1972).

secondarily adapted, having passed from ocean to land and then back again to fresh water. As implausible as this may seem, the bodily fluids of aquatic animals show a strong similarity to oceans, and indeed, most studies of ion balance in freshwater physiology document the complex regulatory mechanisms by which fish, amphibians and invertebrates attempt to maintain an inner ocean in spite of surrounding fresh water. It is these sorts of unexpected complexities and apparent contradictions that make ecology so interesting. The idea of a fish in a freshwater lake struggling to accumulate salts inside its body to mimic the ocean reminds one of the other great contradiction of the biosphere: plants are bathed in an atmosphere composed of roughly three-quarters nitrogen, yet their growth is frequently restricted by lack of nitrogen.

Wetlands have soil, so they are not truly aquatic like planktonic and pelagic communities. But they have standing water, so neither are they truly terrestrial. As a consequence, they are often overlooked. Terrestrial ecologists assume them to be within the territory of limnologists. Limnologists regularly assume them to be within the domain of terrestrial ecologists. The fact that wetlands intergrade with both terrestrial and aquatic environments illustrates the difficulties in delineating precise scientific and legal boundaries.

Another obvious place to start the study of wetlands is with distribution. This is more difficult than it may first seem. As a beginning, Figure 1.3 gives the proportion of wetland in each of four latitudinal zones. It is

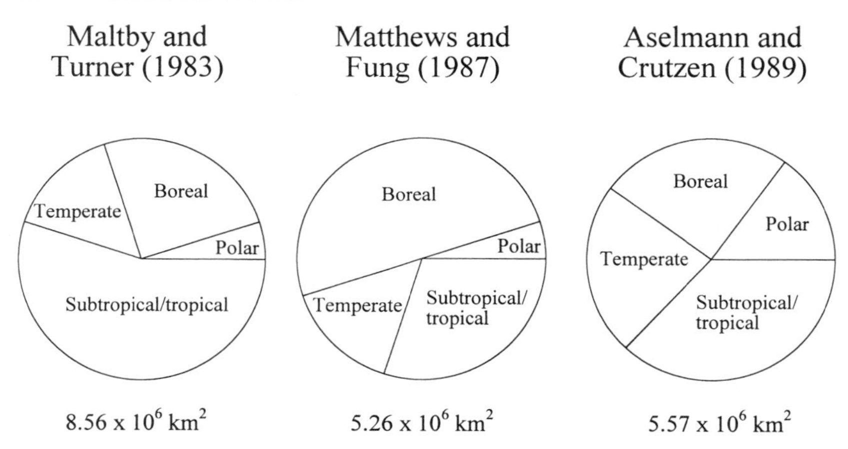

Figure 1.3 Area of wetlands in four latitudinal zones (*a*)Maltby and Turner (1983) based on Bazilevich *et al.* (1971); (*b*) Matthews and Fung (1987); (*c*) Aselmann and Crutzen (1989); drawing after Mitsch and Wu (1994).

difficult to map wetlands at the global scale for at least three reasons. First, wetlands are frequently a relatively small proportion of the landscape. Secondly, they are distributed in small patches or strips throughout biomes, and therefore cannot be mapped at a scale suitable for reproducing in a textbook. Thirdly, they are very variable within biomes, and one biome can therefore contain a wide array of wetland types. In spite of these limitations, Figure 1.4 presents a preliminary map of global wetland areas. Table 1.3 then lists the largest wetland areas in the world; these set an important priority list for research and conservation.

Causal factors in wetland ecology

Wetlands arise because there is water, but the particular kind of wetland, and the characteristics of its species or communities, will depend upon other environmental factors. Three broad principles will guide our investigation of these factors.

The first principle states that *any particular community or ecosystem is produced by multiple environmental factors acting simultaneously.* We can therefore picture any particular wetland (and that includes its species, communities and functions) as being a product of the pushing and pulling of opposing environmental factors (Figure 1.5). Any specific wetland encountered in the field has arisen as a temporary consequence of these multiple factors. It may be useful to consider this set of physical factors as a kind of *habitat*

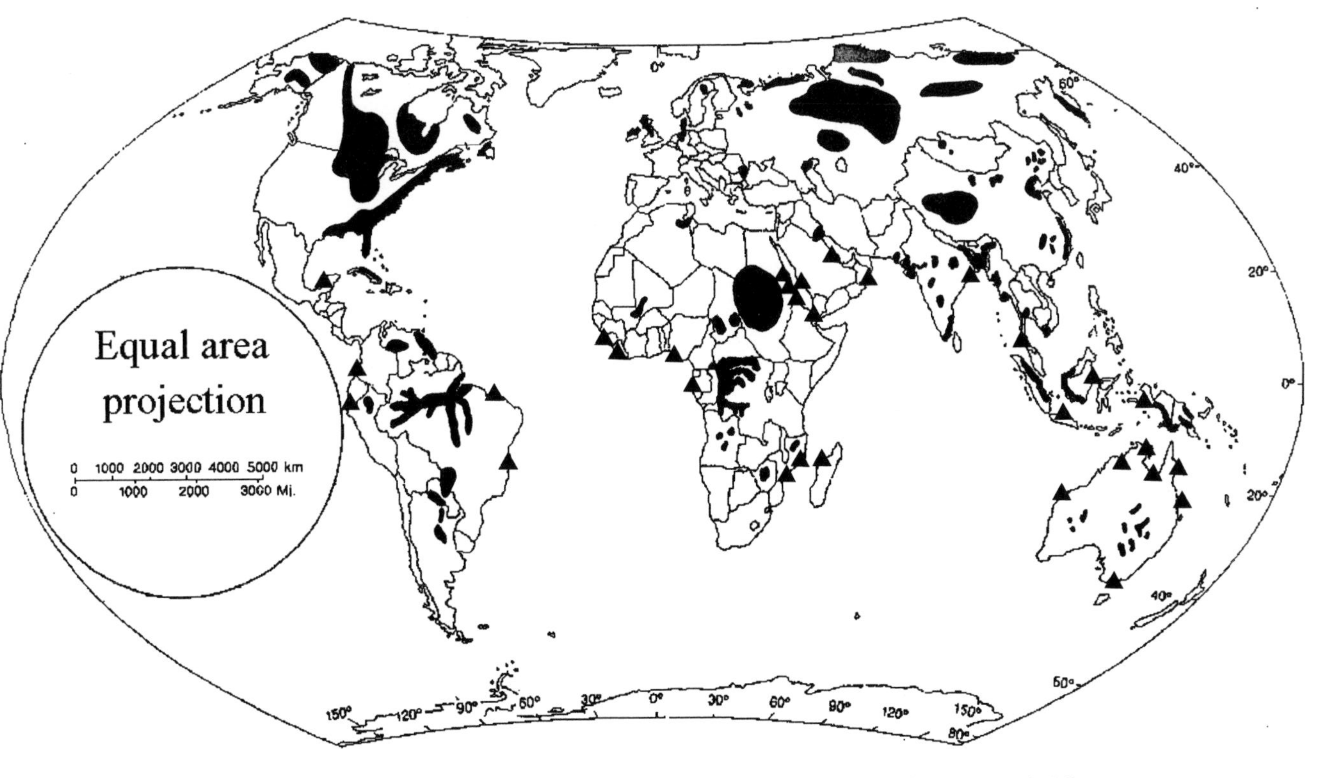

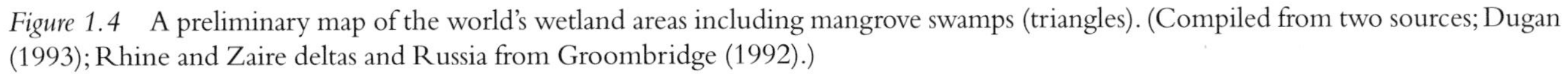

Figure 1.4 A preliminary map of the world's wetland areas including mangrove swamps (triangles). (Compiled from two sources; Dugan (1993); Rhine and Zaire deltas and Russia from Groombridge (1992).)

Table 1.3. *A preliminary list of the world's largest wetlands. At the lower limit of 30 000 km² a number of other candidate areas appear possible*

Rank	Continent	Wetland	Description	Area (km^2)	References
1/2	Eurasia	West Siberian Lowlands	Peat bogs, boggy forests, meadows	780 000 (up to 1 000 000)	Zhulidov *et al.* (1997) (p. 62)
1/2	South America	Amazon River	Large river floodplain 300 000	>800 000	Aselman & Crutzen (1989)
			Small river floodplain >500 000	> 90 000	Sippel *et al.* (1998) (Fig. 9)
3	North America	Hudson Bay Lowlands	Peatlands	>200 000	Wickware & Rubec (1989)
				320 000	Martini (1982), Riley (1982)
4	South America	Pantanal	Marsh, swamp, floodplain	120 000	Hamilton *et al.* (1996) (Fig. 9)
				140 000	Prance & Schaller (1982)
				200 000	Groombridge (1992)
5/6/7	Africa	Upper Nile Swamps (Sudd)	Swamps and floodplains	90 000+	Howard Williams & Walker (p. 206) in Denny (1985)
				50 000	Groombridge (1992)
5/6/7	Africa	Chari–Logone (drains into L. Chad)	Seasonal floodplain	90 000	Thompson & Hamilton (1983)
5/6/7	North America	Mississippi River floodplain	Bottomland hardwoods	86 000[a]	Llewellyn *et al.* (1996)
8/9	Eurasia	Papua-New Guinea	Swamp, bog	69 000	Aselman & Crutzen (1989)
8/9	Africa	Zaire–Congo system	Riverine swamps and floodplain	40 000+	Howard Williams & Walker (p. 206) in Denny (1985)
				80 000	Thompson & Hamilton (1983)

10	North America	Upper Mackenzie River	Marsh, fen, floodplain	60000	Fremlin (1974) (p. 39–40)
11	South America	Chilean Fjordlands	No published description available	55000	map in Groombridge (1992)
12	North America	Prairie potholes	Marsh	40000[b]	Leitch (1989)
13	South America	Orinoco River delta	Floodplain, swamp, marsh	30000	map in Groombridge (1992)

Notes:

[a] Only 20000 remain forested, one-third of which are in the Atchafalaya River basin in Louisiana.

[b] The Prairie pothole region is 300000 square miles containing 8 million ha before drainage; with one half drained this leaves 4 million ha (Leitch, 1989). The intermixing of land and potholes creates considerable problems in estimating wetland area; Kantrud *et al.* (1989) say that 40 to 60% of the area 'often was covered by wetlands' whereas the National Atlas of Canada (Fremlin 1974) shows extensive portions as <1% wetland.

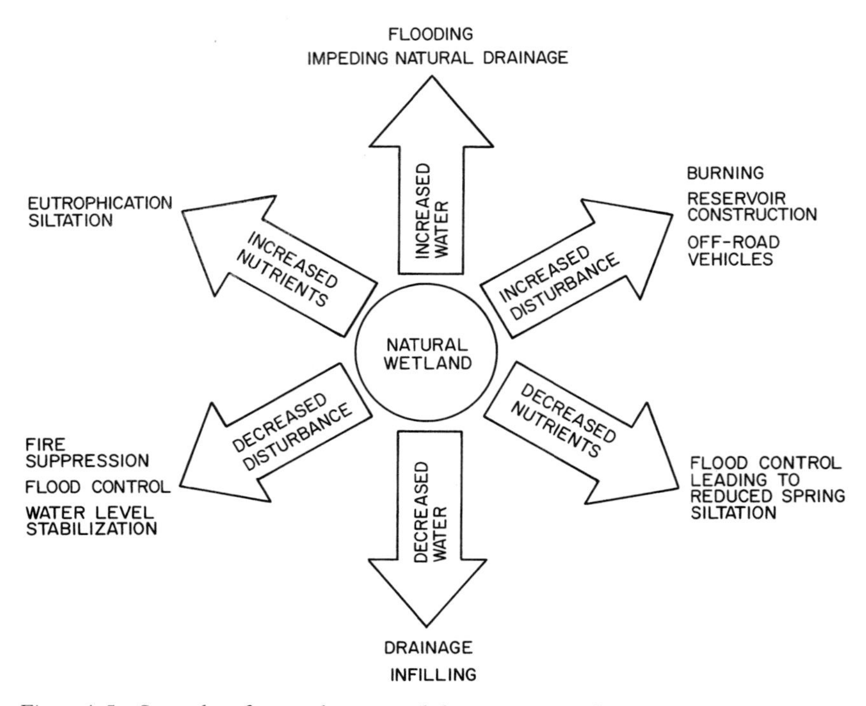

Figure 1.5 Some key factors that control the structure and species composition of wetlands.

template, which both guides and constrains the biological communities and ecological processes that occur. For example along most watercourses, the movement of water, and particularly the movement of floodwaters, creates gravel bars, eroding banks, sand bars, ox bows and deltas (Figure 1.6). Along such watercourses, the three most important processes producing the habitat template are (i) flooding, (ii) erosion and (iii) deposition. These are not entirely independent, of course, but for the purposes of this book I have tried to tease them apart somewhat. There are therefore separate chapters for water level fluctuations, disturbance and burial. Similarly, in heavily disturbed European peatlands, the most important factors in the habitat template are (i) fertility, (ii) water availability, (iii) frequency of disturbances such as fire and (iv) intensity of grazing. Again, these are not independent of one another, but there are separate chapters on flooding, fertility, disturbance and grazing.

Nested within the first principle are a series of more specific relationships between wetlands and environmental factors. The second princi-

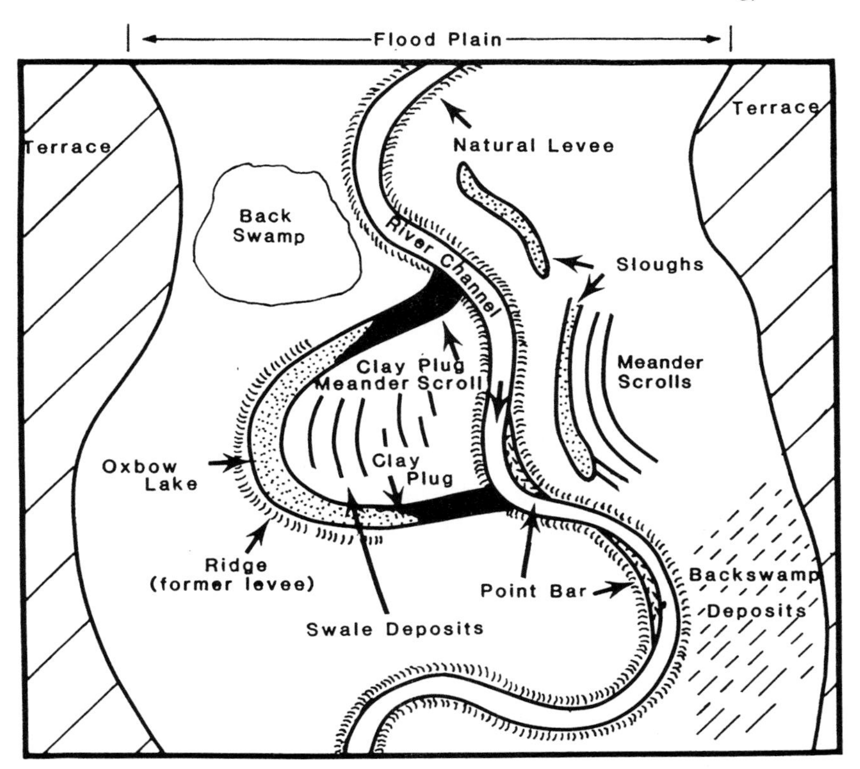

Figure 1.6 Dynamic processes create the wide array of wetland habitats along a water course. These habitats create the template upon which hydrological variation and ecological communities are superimposed (from Mitsch and Gosselink 1986).

ple therefore states that *to understand and manage wetlands, scientists must determine the quantitative relationships between environmental factors and the properties of wetlands.* The study of each factor in the habitat template provides an opportunity for exploration of such quantitative relationships. Examples might include (i) the production of fish as determined by the factor of floodplain area, (ii) the diversity of plants as controlled by substrate fertility, or (iii) the zonation of invertebrates produced by different rates of burial by sediment. These relationships summarize the state of human knowledge about the factors that create and control wetlands. The challenge for the scientist is to unravel these factors, discover their consequences for wetlands, and determine their relative importance. The challenge for managers and conservationists is to understand these relationships, and then, if necessary, manipulate or regulate one or more of them to maintain or produce the desired characteristics of a wetland.

These challenges are made difficult by the many kinds of wetlands, and the many factors at work in them. The difficulty is compounded by a third principle: *the multiple factors that produce a community or ecosystem will change through time.* Wetlands are no different from other ecosystems, where disturbance from fire, storms, landslides and floods controls the communities and species at a site. If humans change the factors acting on the wetland in Figure 1.5, say by decreasing spring flooding or increasing fertility, the balance of forces will shift and the wetland composition or function will begin to shift as well. The habitat template along water courses in Figure 1.6 is constantly changing as moving water reshapes the environment. It is far too easy, and therefore far too common, for humans to study small fragments of this habitat complex (say one species or one vegetation type in one oxbow lake), losing track of the fact that the particular species and community types are but transitory occurrences at any location. To understand wetlands, and to manage them wisely, it is essential to appreciate their multifactorial and dynamic nature right from the start of our inquiry. Thus the introduction of these principles is early in the book.

All of the chapters that follow can be considered to be elaborations of these principles, expanding the description of known relationships, and then discussing how these influence the study, conservation and management of wetlands. This approach begs the question of which environmental factors deserve priority. As the table of contents showed, the factors allocated separate chapters in this book are hydrology, fertility, disturbance, competition, grazing and burial, arranged (roughly) in declining order of importance. Three other factors might be added to this list: salinity, positive interactions, and time. I have not added salinity in as a separate factor because this would immediately create a ghetto of sorts for saline wetlands, leading readers to overlook the many ways that saline wetlands are similar to freshwater wetlands. If salinity is separated out, then pH could similarly be used to create a separate category for peatlands, leading to further loss of unity in the book. This sort of division by wetland type is something I actively sought to avoid. Further, there are also many excellent treatises on saline wetlands (e.g. Chapman 1974; 1977; Pomeroy and Wiegert 1981; Tomlinson 1986; Adam 1990) which a single chapter could not duplicate. Positive interactions are also important but excluded from chapter status. Positive interactions could include floodplain trees providing food for fish, beavers creating ponds for frogs, and *Sphagnum* moss creating peat for orchids. I have already commented on the manner in which textbooks overlook positive interactions (Keddy, 1990a), there are now several fine reviews on mutualism (Boucher 1985; Smith and Douglas 1987), and Bertness and his co-workers have increasingly documented the role of positive interactions in wetlands (Bertness and Yeh 1994; Bertness and Hacker 1994; Bertness and Leonard 1997). Rather than combine such positive interactions in one chapter, I have sorted them by their main effects: fish and plants are discussed under flooding, as are beavers, alligators digging is discussed under disturbance, and peat accumulation under succession and burial. Finally, time is a critical factor that could have been excised: pluvial lakes came and went from deserts over the past 20 000 years, peat accumulated as continental glaciers receded over thousands of years, beaver ponds come and go over centuries, and riparian wetlands are flooded each year. A synthetic approach to ecology requires us to be explicit about duration, rates, and the appropriate scale for measuring each ecological process (Delcourt and Delcourt 1988). I have not included time as a separate factor, deciding that time is really a surrogate for the specific factors that vary with time. Changes in hydrology during glaciation are therefore introduced in hydrology, changes in fertility with peat accumulation are discussed under fertility, fire cycles are discussed under disturbance, succession is discussed under zonation and burial, and so on. I do not want to argue or even imply that this is the only way to sort the many topics covered in this book. A book with separate chapters on salinity, positive interactions and time could also be written, and would undoubtedly have its merits.

Wetland classification

The six basic types

Given the many kinds of wetlands, an important first step in scientific study is to divide and sort them into similar types. Each type can be visualized as a particular set of plant and animal associations that recur. This recurrence probably means that the same causal factors, and similar ecological relationships occur within each type. Unfortunately, the terminology for describing wetlands varies both among human societies, and among their scientific communities. Thus one finds an abundance of words including bog, carr, fen, flark, hochmoor, lagg, marsh, mire, swamp, pocosin, pothole, quagmire, savannah, slough, swale, yazoo, etc. used in a contradictory way among English speaking people of the world. Many of these words can be traced back centuries to Old Norse, Old Teutonic or

Gaelic origins (Gorham 1953). Now add in modern foreign languages, and the problem is compounded.

One of the simplest classification systems recognizes only four types of wetland: swamps, marshes, fens and bogs. Let us begin with the four types and insist upon at least this much standardization.

Swamp

A wetland community that is dominated by trees that are rooted in hydric soils, but not in peat. Examples include tropical mangrove swamp (mangal) and bottom-land forests in floodplains.

Marsh

A wetland community that is dominated by herbaceous plants that are usually emergent through water and rooted in hydric soils, but not in peat. Examples include cattail (*Typha*) marshes around the Great Lakes and reed (*Phragmites*) beds around the Baltic Sea.

Bog

A wetland community dominated by *Sphagnum* moss, sedges, Ericaceous shrubs or evergreen trees rooted in deep peat. Examples include blanket bogs which carpet mountain sides in northern Europe, and floating bogs which cover the shores of many lakes in temperate and boreal regions.

Fen

A wetland community that is usually dominated by sedges and grasses rooted in shallow peat, often with considerable water movement through the peat. Examples include the extensive peatlands in northern Canada and Russia, as well as smaller seepage areas throughout the temperate zone.

Figure 1.7 shows each of these four types, and Table 1.4 presents some comparative data on their properties. Other wetland types could be added to these four. Two important ones are the following.

Wet meadow

A wetland community dominated by herbaceous plants rooted in occasionally flooded soils. Temporary flooding excludes terrestrial plants and swamp plants, but drier growing seasons then produce plant communities typical of moist soils. Examples include wet prairies along river floodplains, or herbaceous meadows on the shorelines of large lakes.

Figure 1.7 The four principal types of wetlands are swamp, marsh, bog and fen. (*a*) Silver Maple swamp, Ottawa River, Canada.

These wetlands are produced by periodic flooding and may be overlooked if visited during a dry period (Figure 1.8a, b).

Shallow water

A wetland community dominated by truly aquatic plants growing in and covered by at least 25 cm of water. Examples include the littoral zones of lakes, bays in rivers and the more permanently flooded areas of prairie potholes (Figure 1.8c, d).

Any attempt to sort the diversity of nature into four to six categories will have its limitations. With little effort it will be possible to find one or more situations that appear to defy categorization, or to recall a local community that the definition calls a bog but which has always been known locally as a mire. Rather than worry further about this, we should probably admit that wetlands show considerable variation, and agree to not get stalled or diverted by debates over terminology. As Cowardin and Golet (1995) observe 'no single system can accurately portray the diversity of wetland conditions world-wide. Some important ecological information inevitably will be lost through classification.'

Figure 1.7 (*cont.*) (*b*) Mangrove swamp, Caroni wetland, Trinidad.

(*c*) Temperate marsh, Ottawa River, Canada (courtesy of B. Shipley).
(*d*) Tropical marsh, Caroni wetland, Trinidad.

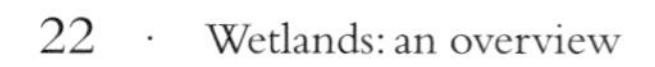

Figure 1.7 (cont.) (*e*) Lowland bog, Algonquin Park, Canada. (*f*) Upland coastal bog, Cape Breton, Canada.

(*g*) Patterned fen, northern Canada (courtesy of C. Rubec).

Other classification systems

Owing to the long history of interest in peatlands, there is a particularly diverse terminology here. Gore (1983) proposes that the term mire should be used to designate wetland types including bog, fen, moor, muskeg and peatland. Unfortunately, he also adds the word 'swamp', which some European ecologists use to describe eutrophic mires or marshes. Inconsistent use of the word 'swamp' causes problems even within England, says Burnett (1964), being defined virtually on a regional basis and differing again from usage on the continent. The term 'swamp' in British usage generally applies to wetlands in which the normal water level is above the soil surface, usually dominated by reeds (*Phragmites*), tall grasses, sedges or rushes; the commonest kind being a 'reed swamp'. Similarly, in Africa Thompson and Hamilton (1983) use the term swamp for grass, sedge and reed dominated herbaceous wetlands as well as for forested areas. In the definitions used in this book (p.18) herbaceous wetlands lacking peat are simply a kind of marsh; it may be further qualified with modifiers such as an 'emergent marsh', '*Phragmites* marsh', '*Papyrus* marsh'or 'lacustrine marsh' as the need arises.

Figure 1.7 (*cont.*) (*h*) Shoreline fen, Lake Ontario, Canada.

Table 1.4. *General characteristics of southern Ontario wetland types off the Canadian Shield*

	Bog ($n=23$)	Fen ($n=34$)	Mixed, hardwood, thicket swamp ($n=46$)	Marsh ($n=18$)
Surface water pH (mean)	4.3	6.0	6.1	6.5
cm depth-to-water (mean)	49	19	25	6
Number of vascular plant species (mean)	13	29	33	25
Peat depth (mean)	3.0	2.5	1.4	1.6

Source: After Riley (1989).

Russia has extensive areas of wetland, and yet no satisfactory classification system. Until recently, there was not even a word for the general concept of 'wetland'. More than 30 local names (from *alasy* to *zaymischa*) existed and could be understood differently in different parts of the country (Zhulidov *et al.* 1997).

When it took a month to sail across the Atlantic, cultural differences in classification were perhaps inescapable. But in an era of international flights, e-mail and global telephone linkages, scientific dialects are no longer acceptable. Let us hope that teachers will try to bequeath to their students the one standard terminology given above.

The distinction between, and gradations within, bog and fen also generate a great deal of terminology. The principal distinction is largely based upon differences in nutrient availability, which in turn, translate into differences in plant species composition (Gore 1983). Differences in nutrient availability are, in turn, a function of hydrology, with the infertile bogs often being dependent upon rainfall (hence *ombrotophic*), whereas the fens are connected to flowing groundwater (hence, *minerotrophic*). Within these general categories, there are many more specialized terms. In cold climates, the flowing water though mires creates alternating ridges and pools at right angles to the direction of water flow; these are known as patterned mire or string bog (Foster *et al.* 1983; Mark *et al.* 1995). Around raised bogs, where water flowing off the bog meets mineral soils, a trough called a lagg is often found (Godwin 1981; Damman and Dowhan 1981). When mires form in the valleys of slow-moving rivers, they may be called valley bogs or headwater bogs.

Figure 1.8 Wet meadow: (*a*) Long Point, Lake Erie, Canada (courtesy of A. Reznicek). (*b*) Tusket River, Nova Scotia, Canada (courtesy of A. Payne).

Shallow water: (*c*) *Nelumbo* covered bay, Lake Erie, Canada (courtesy of A. Reznicek). (*d*) Interdunal pond, Sable Island, Canada.

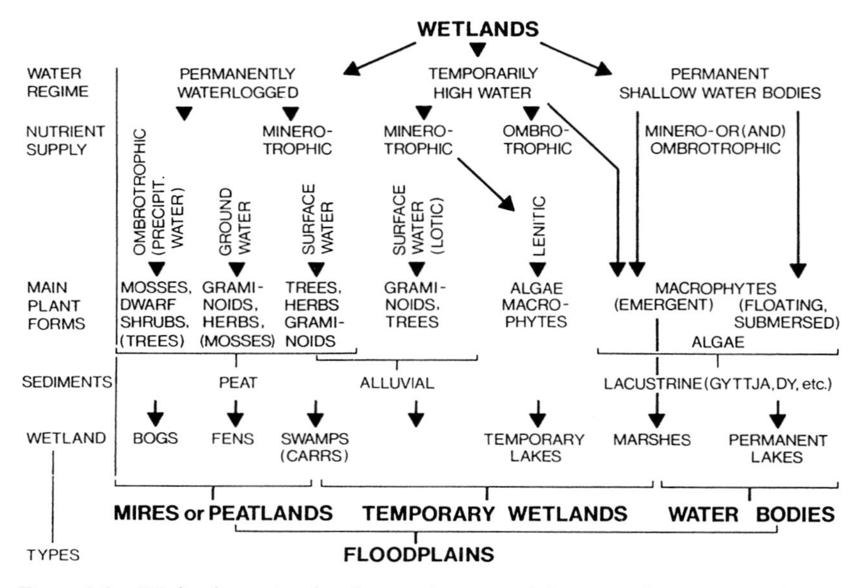

Figure 1.9 Wetland type is related to environmental factors such as water regime and nutrient supply (from Gopal *et al.* 1990).

There are many classifications of wetland types. Each wetland classification system tries to summarize the major types of wetland vegetation, and then to either implicitly or explicitly associate them with different sets of environmental conditions. Because so many are erected to cover small geographical areas with limited amounts of variation, they may add terminology and confusion to the literature. Figure 1.9 provides a summary that ties different schemes together. It begins with water regime, from permanently waterlogged on the left to permanent shallow water on the right, with temporary high water in the middle. Combining these three hydrological regimes with nutrient supply, one obtains peatlands on the left, swamps in the middle, and lakes on the right. Further, the scheme then goes on to address the main plant forms that will occur. Many systems can be superimposed upon this one by allowing for finer discriminations among vegetation types or processes. (One other system has been presented by Gopal *et al.* (1990) to summarize world wetland types. It again has two principal axes, hydrology and fertility, but it will be introduced in Chapter 5 after hydrology and fertility have been explored in more depth.) The widely used Cowardin classification system (Table 1.5) begins with the location of the wetlands rather than emphasizing functional properties; Cowardin and Golet (1995) have reviewed the performance of this

Table 1.5. *Cowardin classification of wetlands and deepwater habitats*

System	Subsystem	Class
Marine	Subtidal	Rock bottom
		Unconsolidated bottom
		Aquatic bed
		Reef
	Intertidal	Aquatic bed
		Reef
		Rocky shore
		Unconsolidated shore
Estuarine	Subtidal	Rock bottom
		Unconsolidated bottom
		Aquatic bed
		Reef
	Intertidal	Aquatic bed
		Reef
		Streambed
		Rocky shore
		Unconsolidated shore
		Emergent wetland
		Scrub–shrub wetland
		Forested wetland
Riverine	Tidal	Rock bottom
		Unconsolidated bottom
		Aquatic bed
		Rocky shore
		Unconsolidate shore
		Emergent wetland
	Lower perennial	Rock bottom
		Unconsolidated bottom
		Aquatic bed
		Rocky shore
		Unconsolidated shore
		Emergent wetland
	Upper perennial	Rock bottom
		Unconsolidated bottom
		Aquatic bed
		Rocky shore
		Unconsolidated shore
	Intermittent	Streambed
Lacustrine	Limnetic	Rock bottom
		Unconsolidated bottom
		Aquatic bed

Table 1.5 (*cont.*)

System	Subsystem	Class
Lacustrine (*cont.*)	Littoral	Rock bottom
		Unconsolidated bottom
		Aquatic bed
		Rocky shore
		Unconsolidated shore
		Emergent wetland
Palustrine		Rock bottom
		Unconsolidated bottom
		Aquatic bed
		Unconsolidated shore
		Moss-Lichen wetland
		Emergent wetland
		Scrub-Shrub wetland
		Forested wetland

Source: Cowardin *et al.* (1979).

classification, and have suggested some modifications to strengthen its utility. Other large countries have developed their own classification systems and procedures (e.g. Russia: Botch and Masing 1983; Zhulidov *et al.* 1997; China: Hou 1983; Lu 1995; Canada: Committee on Ecological Land Classification, 1988).

The situation in the tropics is more complicated. As a consequence, and perhaps counter-intuitively, Lugo and Brown (1988) present a slightly simpler classification system based upon their experience with tropical wetlands in the Caribbean Islands. They note that in Cuba alone, one can recognize 27 classes, 53 orders, 80 alliances and 186 plant associations, of which nearly one-quarter are wetlands, and more than half of these are forested wetland types; half the plant species on the island are endemic. Cuba also has the largest wetland complex in the Caribbean; on the south coast near the Peninsula de Zapata there are some half-million hectares of mangroves and freshwater marshes. The diversity of wetlands across the Caribbean as a whole probably results from a combination of factors including the excess of rainfall over evapotranspiration, the varied topography, and the complex geology. The island of Puerto Rico, for example, has volcanic, plutonic, limestone, serpentine, sedimentary and sandstone formations in an area of only 888 000 ha, with elevation exceeding 1000 m in the Cordillera Central, thereby producing 164 soil series. Given the rich flora of the Caribbean, a classification system that further emphasizes species names becomes inordinately complex. Lugo *et al.* (1988) therefore propose a classification system that recognizes three main kinds of wetlands based upon geology and hydrology: riverine, basin and fringe wetlands. Modifiers such as salinity, dominant life form and nutrient status can then be superimposed. Adding three salinity levels (freshwater, oligohaline and saline), three plant forms (herbaceous, scrub and forest) and three nutrient levels (oligotrophic, eutrophic, dystorphic) yields 81 basic wetland types, a system that is broad enough to include everything from mangal to montane seeps, yet simple enough to require only four main criteria for classification. The diversity of wetland types in Cuba, for example, is then reduced to 24. Further subdivision is possible by recognizing different types of wooded wetland based upon location (montane, lowland, coastal) or the dominant tree species (*Pterocarpus*, the palm *Prestoea monatana* or colorado (*Cyrilla racemiflora*)).

Thompson and Hamilton (1983) use an even simpler system of our basic categories for describing the wide array of African wetlands:

(i) Flushes, cushion-bog and tussock sedge mires;
(ii) Mangrove swamps and coastal peatlands;
(iii) Swamp forests;
(iv) Grass, sedge and reed swamps.

Some of these terms may be unfamiliar, but according to the classification scheme used in this book, (i) refers to fens and bogs whereas both (ii) and (iii) can be combined as swamps, differentiating further on the basis of saline as opposed to freshwater conditions. The last category includes freshwater marsh and perhaps, wet meadow. Further subdivision can be made based upon location (e.g. valley bogs) or dominant species (*Typha* marsh, *Papyrus* marsh). Compared with Lugo *et al.* (1988), formal rules and criteria may seem to be lacking, but this system none the less appears sufficient for a preliminary description of wetland types.

Those reading the European literature will find that classification continues to finer and finer scales where each community type is ultimately given a separate name (Shimwell 1971, Westhoff and van der Maarel 1973). For example, a *Phragmites communis* marsh in Poland would be placed in the Class Phragmitetea (R. Tx. et Prsg. 1942), Order Phragmitetalia eurosibirica (R. Tx. et Prsg. 1942), Alliance Phragmition (Koch, 1926) and Association Scirpo-Phragmitetum (Koch 1926). Palczynski (1984), for example, recognizes 7 classes, 10 orders, 14 alliances and 37 associations in the Biebrza valley in Poland. Beeftink

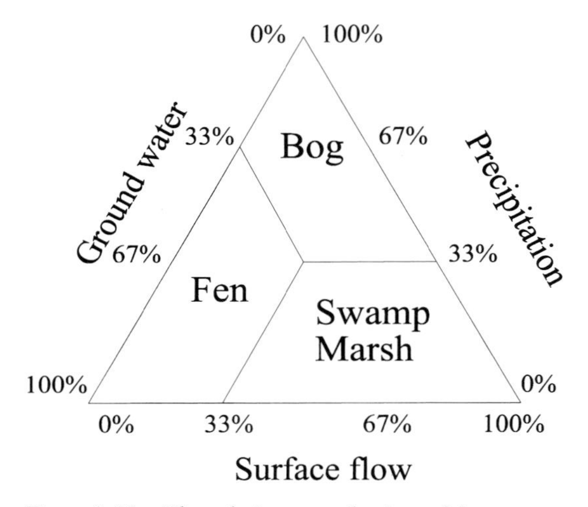

Figure 1.10 The relative contribution of three water sources: precipitation, groundwater discharge, and lateral surface flow determines three main wetland types (modified from Brinson 1993a, b).

(1977) uses the same approach to classify salt marshes in north western Europe. I have not used any such terminology in this book. At best, I think that this exercise detracts from more important work, and at worst, simply creates confusion by distorting plant names and making work in wetlands obscure to all but a narrow group of experts. In this book, I have tried to keep to relatively coarse scale classifications (p. 18) modified, where necessary, by the names of one or a few dominant species. Thus the Scirpo-Phragmitetum Association would simply be called a *Scirpus–Phragmites* floodplain or *Scirpus–Phragmites* marsh, depending upon the circumstances.

Another simpler classification system is based upon hydrology. There are three main sources of water for wetlands – precipitation, ground water, and water flows across the surface. Raised bogs are almost completely dependent upon the first, whereas flood plains are largely dependent upon the third. In practice, nutrient levels are often closely correlated with hydrology, since rainfall tends to be low in nutrients, whereas water flowing across the surface of the ground picks up dissolved nutrients and particulate matter. Wetlands can therefore be classified according to the relative proportions of three water sources (Figure 1.10). Apart from its importance in creating the main wetland types, the study of water flow through wetlands is a discipline in it own right. Water flow may be of interest for the study of plant growth, the transport of pollutants, the temporary storage of flood waters, or the recharge of aquifers.

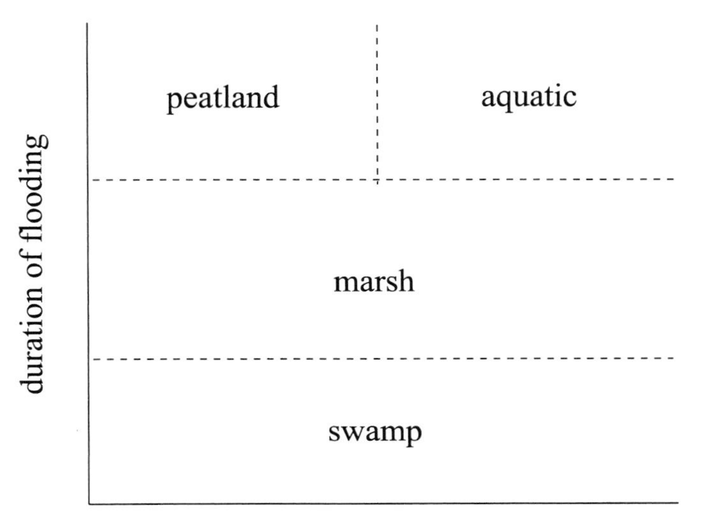

Figure 1.11 Four types of wetland in relation to two aspects of hydrology (modified from Brinson 1993a, b). These aspects of hydrology are the causes of secondary constraints given in Table 1.5.

The latter three topics will barely be discussed in this book, since I will be considering water primarily as a principal factor controlling the species composition of wetlands.

Figure 1.11 serves as a starting point for the study of causal factors by separating fens from bogs, and bogs from riverine wetlands. In the next section, we shall see that the figure can be further partitioned into marsh and swamp by the frequency and duration of the surface flows. Chapter 4 further elaborates upon the volume and variation of surface flow in producing different kinds of wetlands. Chapter 5 explores the effects of the nutrient composition of soil and water. The topic of hydrology is revisited in Chapter 9 where burial by peat and sediments are explored as controlling factors and Chapter 11 where hydrological types are examined. Those who wish to study more of the details of hydrology of wetlands can begin by consulting Ingram (1983), Mitsch and Gosselink (1986), Hook *et al.* (1988) Brinson (1993b) and Lent *et al.* (1997).

Plants, stress and wetland types

Returning to Figure 1.9, let us simplify it further to obtain five of the six vegetation types introduced on p. 18, yielding Figure 1.11. Fen and bog appear as peatland in the upper left, aquatic appears at the upper right, and marsh and swamp appear as broad zones across the bottom. In this approach only hydrology needs to be considered in order to derive the four vegetation types. Low water levels but steady duration produce peatlands, whereas high water levels with long duration produce aquatics. Depending upon the duration of flooding, one then has either marsh or swamp. Let us turn our attention first to the direct effects of flooding, and then consider the many secondary consequences.

Wetland soils

The presence of hydric soils, like the presence of wetland plants, may at first seem a circular way of defining wetlands. Some background will illustrate the merits of this kind of apparent circularity. The particular soil that arises in any environment is a function of parent material acted upon by organisms and climate and conditioned by relief over time (Faulkner and Richardson 1989). The reactivity of a soil is largely determined by surface area and surface charge of soil particles, and the surface charge in turn is determined by the fraction of clay-sized particles and organic matter. Cations which are ionically bound to the surface of soil particles or organic matter can exchange with other cations; metal ions (Ca^{2+}, Mg^{2+}, Na^{+}) dominate in mineral soils whereas H^{+} dominates at high organic contents. Oxidation, aerobic decomposition, leaching and dehydration are important processes that influence the properties of soils. All four of these processes are modified by flooding, principally because water displaces air from the pore spaces between the soil particles. Since oxygen and other gases diffuse in air about 10^3–10^4 times more rapidly than in water, oxygen is soon depleted from the flooded soils by the respiration of soil micro-organisms and plant roots. Wetland soils therefore tend to be deficient in oxygen (a condition of hypoxia) and in some cases there may be none (a condition of anoxia). Once molecular oxygen is depleted, oxidation of organic matter ceases and populations of micro-organisms begin to change the ionic composition of the rooting zone (Ponnamperuma 1972, 1984; Faulkner and Richardson 1989; Marschner 1995). These may seem to be largely technical details that can be left to professional soil scientists. Actually, they are of fundamental significance to wetland ecology. Organisms living in wetland soils have at least three problems to contend with: not only is there a shortage of oxygen, but there are accumulations of toxic gases, and atypical concentrations of ions. The significance of these conditions requires a brief digression to review several billion years of the Earth's prehistory.

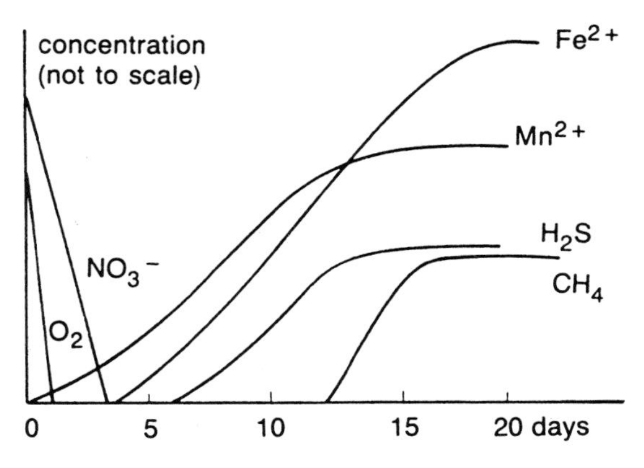

Figure 1.12 Some changes in wetland soils after saturation with water (from Brinkman and Van Diepen 1990 after Patrick and Reddy 1978).

Oxygen now makes up nearly one-quarter of the Earth's atmosphere. It is important to realize that this is a consequence of life, and more precisely, of photosynthesis. The early atmosphere was anoxic (Day 1984, Levin 1992); the evolution of photosynthesis completely changed planetary biogeochemistry. Wetland soils are therefore of particular interest because the hypoxic or anoxic conditions that result from waterlogging recreate conditions that once were widespread but now are relatively uncommon. Figure 1.12 shows the changes that occur in wetland soils after saturation with water. Not only do O_2 and NO_3^- disappear in only a few days, but gases such as methane (CH_4) hydrogen sulphide (H_2S) and ammonia (NH_3) begin to accumulate. As well, ions such as Fe_2^{2+} reappear. The word 'reappear' is appropriate because such ions were once common in the Earth's oceans until they were oxidized several billion years ago. Whereas oxygen was once a toxic waste product, it is now an element which is obligately required by respiring life forms.

Since a majority of the world's soils are oxidized, wetlands provide the major reducing system present in the biosphere. Thus wetlands function as transformers of nutrients and metals. In general a wetland can be divided into three components: standing water, the upper oxidized layer of soil, and the deeper reduced layer of soil. As compounds are transformed in one layer, diffusion gradients arise to transport ions across the boundaries of these layers. Since the degree of anoxia and intensity of reduction varies with flooding, water depth gradients and seasonal changes in water level further widen the array of conditions for microbial and chemical transformations. Whereas most aerated soils have a redox potential above 400 mV, wetland soils can range from −300 to +350 mV (Faulkner and Richardson

1989). Nitrogen, phosphorus, sulphur, iron and manganese cycles, for example, are all closely tied to the different layers of oxidation in wetlands. Flooded soils are therefore critical in biogeochemical cycles. Let us briefly consider changes in N, P and S in flooded soils. More details can be found in Good *et al.* (1978), Ponnamperuma (1972, 1984), Mitsch and Gosselink (1986), Faulkner and Richardson (1989), Gopal (1990), Armentano and Verhoeven (1990) and Marschner (1995).

In the case of nitrogen, the transformation of NH_4^+ to NO_3^- occurs in the shallow oxidized zone when chemoautotrophic bacteria oxidize the NH_4^+. This sets up a concentration gradient which drives upward diffusion of nitrogen as NH_4^+ from deeper anoxic regions to the upper oxidized layer. At the same time, nitrogen in the form of NO_3^- flows in the reverse direction. Most terrestrial ecosystems are sources of organic nitrogen and nitrate, whereas most wetlands appear to be sinks.

Phosphorus cycling in wetlands is very different from that of nitrogen. Nearly all of the phosphorus is stored in sediment and plant litter. Unlike nitrogen, there is no gaseous phase, and no valency changes occur during biotic assimilation or decomposition. The redox potential, pH, and concentrations of P, Fe, Al and Ca minerals are the principal controls on phosphorus retention. Wetlands again appear to act as sinks for phosphorus eroded from surrounding terrestrial ecosystems.

Sulphur cycling, like nitrogen cycling, is mediated by microorganisms. Decomposition of organic matter can release SO_4^{2-} and under reducing conditions the SO_4^{2-} is transformed to H_2S by anaerobic bacteria. This H_2S can diffuse into the atmosphere or react with organic matter. Most of the inputs to wetlands are atmospheric SO_4^{2-}. Whereas most terrestrial ecosystems are SO_4^{2-} sources, most wetlands appear to be SO_4^{2-} sinks.

The ability of wetlands to transform nutrients and metals, particularly the ability to store phosphorus and transform nitrogen and sulphur to gases, suggests that they play an important role in reducing nutrient concentrations in natural waters. Further, artificial wetlands may be used to treat nutrient-laden waters produced by human activities (Hammer 1989). We will return to nutrients on pp. 70–75 (the nitrogen cycle) and pp. 265–278 (eutrophication).

Flood tolerance: the primary constraint

The presence of hydric plants, like hydric soils, is another attribute used to define wetlands. Plants cannot move; this apparently trivial observation has great significance for their ecology and evolution. A plant, being non-motile, must possess the complete range of traits to tolerate the conditions a particular site is exposed to over many years. Unlike other organisms, it cannot migrate during an unsuitable period. Plant traits must therefore be capable of tolerating and integrating a broad suite of strong environmental pressures. In wetlands, an ability to withstand varying degrees of flooding is essential.

Flooding is one of many possible environmental constraints. A useful way to think about any response to an environmental constraint is to divide it into one of two categories: stress avoidance and stress tolerance (Levitt, 1980). In order to clarify the distinction, Levitt advocates the use of the words stress and strain, *stress* being the external pressures put on an organism by an environmental factor, and *strain* being the degree to which the physiology of the organism is damaged by that stress. An organism can therefore adapt to a stress by either (i) increasing the stress necessary to produce a specific strain or (ii) decreasing the strain produced by a specific stress. In the first case, while the organism cannot alter the external stress, it can prevent the penetration of the stress into its tissues; this is called *stress avoidance*. Some kind of barrier is generally used to prevent the stress from causing strain. Alternatively, an organism may allow the stress to enter its tissues, but all the same avoid the expected internal damage. Such *stress tolerance* implies the ability to prevent, decrease or repair the injurious strain induced by the stress. In the case of flooding, the strategy of stress avoidance supplies oxygen to the tissues; the strategy of stress tolerance modifies a metabolism to withstand low oxygen concentrations.

The effects of flooding in plants has been dealt with extensively (e.g. Crawford 1982, 1996; Kozlowski 1984a; Crawford and Braendle 1996) and particularly in books on aquatic plants (Sculthorpe 1967; Hutchinson 1975). The principal effect of flooding (we are here avoiding secondary effects) is hypoxic soil conditions. Since the roots of actively growing plants require oxygen to respire, there has been strong selective pressure to address the problems posed by hypoxia.

The principal mechanism of stress avoidance has been the evolution of air spaces or lacunae, which may extend from the leaf parenchyma through the petiole into the stem and into the buried rhizome or root. These spaces are formed either from splits between cells, or by the disintegration of cells. The continuity of the air spaces though the plant were, according to Hutchinson (1975), illustrated by Barthelemy in 1874, who found that, when a leaf was placed under reduced pressure, air could be drawn upwards from the rhizome. The system of lacunae is frequently called, in spite of Hutchinson's objections, aerenchyma. Further, in some cases it appears that as the roots elongate, they are able to oxidize their surroundings (Hook 1984; Moorhead and Reddy 1988) and even provide oxygen for the respiration of roots by neighbouring plants (Bertness and Ellison 1987; Bertness 1991; Callaway and King 1996). The lacunae also provide a route for methane produced in the soil to escape into the atmosphere; genera including *Scheuzeria*, *Carex*, *Peltandra* and *Typha* are known to transport methane (Cicerone and Ormland 1988). The presence of aerenchyma is one of the most obvious characteristics of wetland plants, being particularly well developed in marsh and aquatic plants.

In addition to the passive diffusion of oxygen, there can be bulk flow of air through aerenchyma if an internal pressure gradient exists. Such bulk flow (also called convective through flow (Armstrong *et al.* 1991) or pressurized ventilation (Grosse *et al.* 1991)), has now been reported in wetland plants including *Phragmites australis*, *Carex gracilis*, *Egeria elegans* (Armstrong *et al.* 1991) as well as *Hydrocleys nymphoides*, *Nelumbo nucifera*, *Victoria amazonica* and *Almus glutinosa* (Grosse *et al.* 1991). Typically, it was first documented by a German botanist in the late 1800s (Pfeffer 1897 in Grosse *et al.* 1991), but then overlooked for decades. The process is now best understood in water lilies, where pressurized gases in young leaves flow down through the petiole into the rhizome. The air then moves back up the old petioles and escapes through the pores in older leaves (Figure 1.13). As much as 22 litres of air a day can enter a single floating leaf and flow to the rhizome (Dacey 1980; Dacey in Salisbury and Ross 1988, pp. 68–70). Most trees do not show obvious morphological features to withstand flooding, but there is one conspicuous exception. Both major groups of woody plants, gymnosperms (e.g. *Taxodium*) and angiosperms (e.g. *Avicennia*) produce above-ground extensions of their roots called pneumatophores (Figure 1.14) that may allow roots direct access to atmospheric gases.

The processes of stress tolerance are less well understood, although the immediate responses to flooding have been documented. In well-drained soils, aerobic respiration of plant roots and soil micro-organisms typically consumes 5 to 24 g oxygen per square metre (Jackson and Drew 1984). When flooded, pore spaces become water filled, and gas exchange between the soil and the atmosphere is virtually eliminated. Toxic compounds such as ammonia, ethylene, hydrogen sulphide, acetone and acetic acid are formed from the anaerobic decomposition of organic matter (Ponnamperuma 1984). Flooding 'usually sets in motion a sequential and

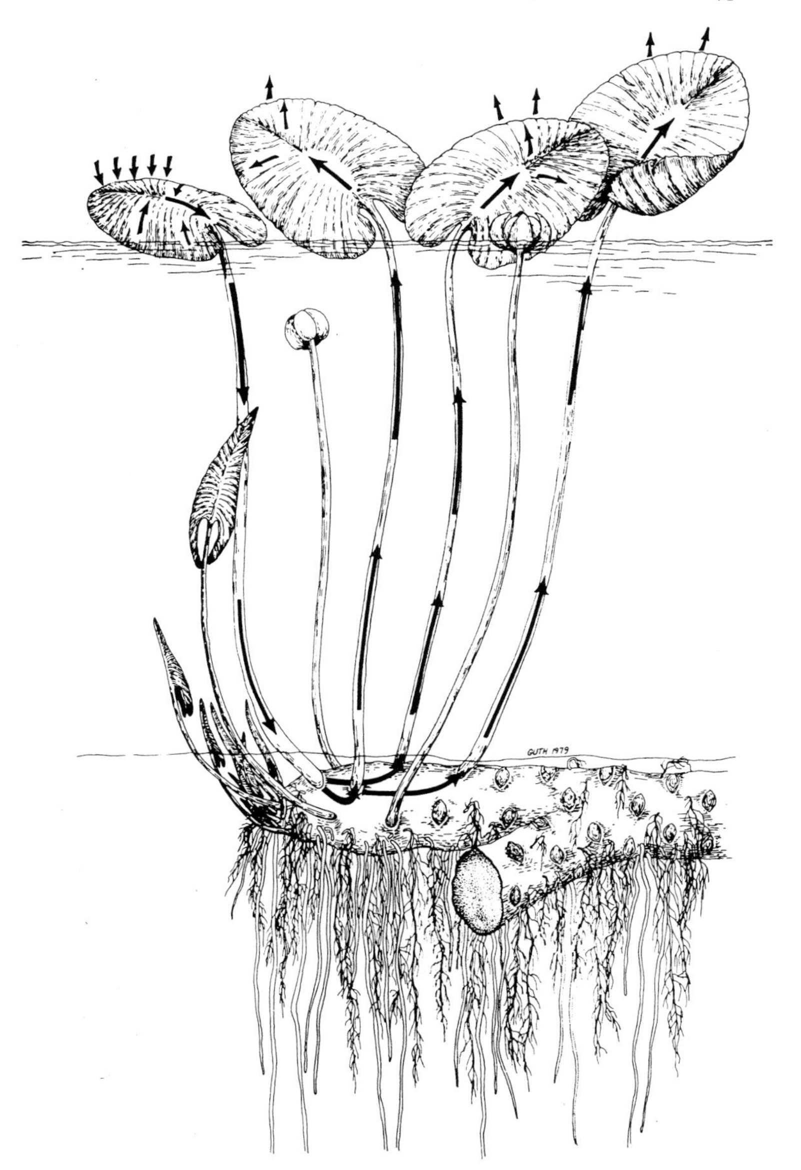

Figure 1.13 The movement of air through aerenchyma in water lilies (*Nuphar luteum*) (from Dacey 1981).

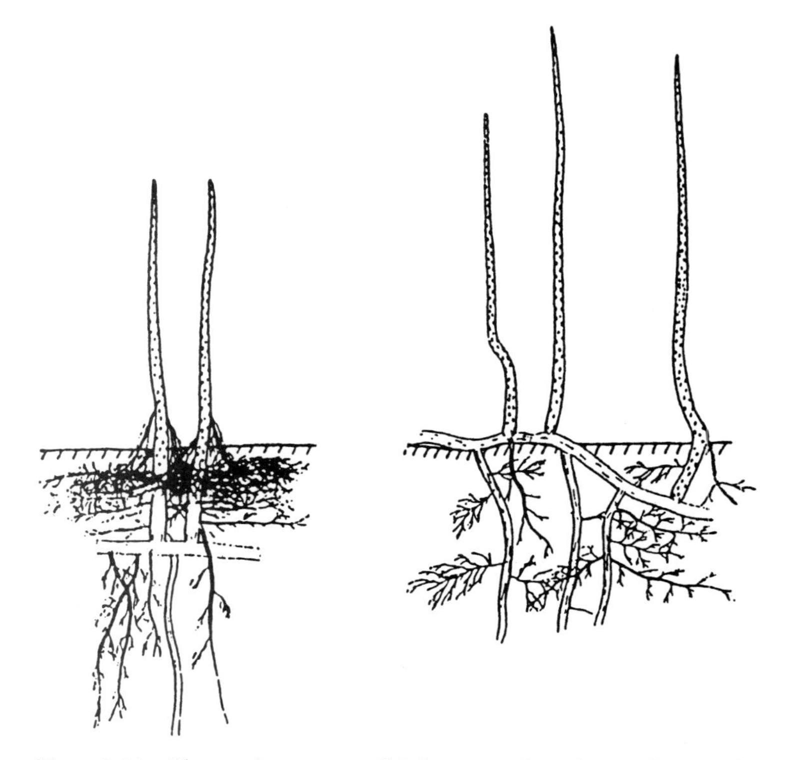

Figure 1.14 The rooting system of *Avicennia nitida* under conditions of peat formation (from Chapman 1940).

complicated series of metabolic disturbances' (Kozlowski and Pallardy 1984). In flood-intolerant plants, stomata close, the rates of transpiration and photosynthesis decline, and nitrogen and phosphorus contents of tissues decline. The roots cease extending, growth is reduced and apical meristems begin to degenerate. Ethylene levels in the tissues also increase (Reid and Bradford 1984), and these are implicated in stimulating the development of aerenchyma.

There is also evidence that some stress-tolerant plants have modified metabolic pathways, allowing plants to temporarily survive or even grow in the absence of oxygen; this mechanism may involve the capacity to avoid production of potentially toxic ethanol at the end of the glycolytic pathway. It has long been known, says Hutchinson (1975), that some water plants contain ethanol, and that rhizomes of water plants such as *Nymphaea tuberosa*, *Sagittaria latifolia* and *Typha latifolia* can live anaerobically for long periods of time. This remarkable tolerance of rhizomes to anaerobic conditions was demonstrated by Laing (1940, 1941) who grew rhizomes from genera including *Acorus*, *Nuphar*, *Peltandra* and *Scirpus* in water through which nitrogen was bubbled. The rhizomes were able to respire anaerobically, producing ethanol in 3% or less of oxygen. Both *Pontederia* and *Typha* showed long persistence even in pure nitrogen. More recent work has expanded the list of species with growth apparently unimpeded by lack of oxygen (Spencer and Ksander 1997). Crawford (1982) proposes that some flood-tolerant species can also produce malate rather than ethanol. Moreover, the malate is available for aerobic respiration when inundation ceases, leading Hutchinson to suggest that it may be more typical of frequently flooded rather than permanently immersed species. However, Jackson and Drew (1984) are of the opinion that some of the evidence for metabolic stress tolerance in these latter species is still equivocal. In any case, the minimum oxygen concentration needed to saturate respiration is a useful indicator of how sensitive metabolic activity is to hypoxia (Brändle 1991). The concentration of alcohol dehydrogenase (ADH) in roots has been proposed as a useful assay for anaerobic respiration; when *Spartina alterniflora* was transplanted from oxidized streamsides into sites with oxygen deficiencies, concentrations of ADH in the roots increased fivefold. Sulphides are, however, also associated with anoxia, and it is possible that sulphides themselves could be having direct toxic effects upon roots, or interfering with the uptake of ammonium (Mendelssohn and McKee 1988).

Hypoxic conditions can also occur in the water column itself, particularly after flooding, when shallow pools in floodplains are left filled with debris and exposed to the warming effects of sunlight (Junk 1984). There are many biochemical means by which animals can tolerate anoxic conditions (Hochachka *et al.* 1973); let us first consider the extreme case of Amazonian fish that occupy pools in drying floodplains (Kramer *et al.* 1978; Junk 1984; Junk *et al.* 1997). Using closed respiration chambers, Junk *et al.* (1997) found that lethal oxygen concentrations were normally less than 0.5 mg l^{-1} O_2, although some species from well-oxygenated bays were sensitive to levels twice this high. Some nine or ten families of Amazonian fish include species that can remove oxygen from the atmosphere; the swim bladder of *Arapaima gigas* can act like a lung, and catfish may use the stomach to remove oxygen from swallowed air. Other fish can, within a few hours of hypoxic conditions, enlarge the lower lip to better extract oxygen from water. In one small floodplain lake, Junk (1984) reports that 40 out of 120 species were regularly found under pronounced hypoxic conditions. Of these, ten could take oxygen form the air, and ten could use the lower lip like a gill; the adaptations of the others were unknown. Other fish avoid hypoxic conditions by diurnal migration from macrophyte stands during the day to open water at night. Many other animals avoid hypoxia by migrating into the river itself during dry periods; these include fish, manatees, turtles and invertebrates. Still others are killed, but survive the dry period with resting structures such as eggs or turions; these include clams, sponges, and cladocerans.

In parallel with observations on wetland plants, Junk *et al.* (1997) imply (without using the actual word) that fish of hypoxic waters are stress tolerators, able to escape predators or competition by occupying habitats unavailable to other species. Unfortunately, when life history strategies are explicitly discussed, Junk *et al.* (1997) refer only to the r–K dichotomy, and appear unaware of the third, stress tolerator strategy (Grime 1977; Southwood 1977; Levitt 1980) into which many of these fish seem naturally to fall. An important opportunity to find a generalization applying equally to fish and plants may thereby have been overlooked.

Tolerance to anoxia is equally an important factor in controlling the distribution of amphibians and reptiles (Goin and Goin 1971). Amphibians can exchange gases with the environment by means of gills, lungs, mucosa of the throat or the skin; the relative importance of these pathways varies with habitat. Reptiles have a skin that is relatively impermeable to water; while this has allowed then to colonize land and diversify in terrestrial environments, it has reduced respiration through the skin, placing greater emphasis upon lungs for oxygen uptake. Among the reptiles, turtles seem especially adapted to tolerate anoxia, and many of the aquatic turtles (e.g. in the genera *Trionyx* and *Sternothaerus*) can gain sufficient oxygen by pumping water in and out of the throat to allow long periods of submersion so long as the animal is not physically active. In contrast, snakes, lizards and even crocodiles are far less tolerant of anoxia (Table 1.6).

The importance of secondary constraints

The primary constraint caused by flooding is reduced availability of oxygen. Flooding imposes secondary constraints as well, and these secondary constraints determine the type of wetland that will develop (Table 1.7).

Peatlands

Sites at the top left of Figure 1.11 are flooded more or less permanently, but the water table is near the soil surface. Under these conditions

Table 1.6. *Tolerance of anoxia in various families of reptiles*

Order and family	Number of species tested	Mean time (minutes)
Testudinata		
Chelydridae	1	1050
Testudinidae	14	945
Pelomedusidae	2	980
Kinosternidae	5	876
Trionychidae	1	546
Chelidae	2	465
Cheloniidae	2	120
Squamata		
Lacertilia		
Iguanidae	6	57
Gekkonidae	1	31
Anguinidae	1	29
Scincidae	4	25
Teiidae	1	22
Serpentes		
Viperidae	3	95
Boidae	3	59
Elapidae	1	33
Colubridae	22	42
Crocodilia		
Crocodylidae	1	33

Source: Data from Belkin, *Science* (1963).

decomposition is reduced but since there are no waves, flowing water or tides to carry away debris, the organic matter accumulates. Once organic matter has accumulated to a depth of about 10 cm, roots are steadily isolated from access to the mineral soils beneath the peat. Plants therefore become increasingly dependent upon dilute nutrients deposited in rainwater (Gorham 1957; Godwin 1981; van Breemen 1995) and have distributions strongly related to nutrient concentrations in the ground water (e.g. Gore 1983; Glaser *et al.* 1990; Vitt and Chee 1990). Adaptation to such infertile conditions requires a variety of unusual plant traits, the most visible of which is the tendency toward leathery and evergreen (sclerophyllous) foliage. The cause of sclerophylly in wet habitats was at first unclear, and thought to be a consequence of physiological drought (Small 1972a,b; Richardson 1981), but it is now believed that deciduous leaves require conditions of relative fertility, since a plant must continually

Table 1.7. *Primary constraints create a series of secondary constraints that determine the ecological attributes of wetland communities*

Primary constraint	Nature of flooding: Mean level	Nature of flooding: Duration		Secondary constraints	Secondary characteristics	Key references to secondary constraints
Flooding	Low	High (continuous)	Peatland	Infertility	Evergreenness Mycorrhizae Carnivory	Grime 1979 Chapin 1980 Givnish 1988
Flooding	High	High (continuous)	Aquatic	Low CO_2 Low light Waves	Stress tolerance	Sculthorpe 1967 Hutchinson 1975
Flooding	Low (with seasonal highs)	Low (0.3 growing season)	Swamp	Shade Disturbance	Gap colonization Shade tolerance	Pickett and White 1985 Grime 1979
Flooding	Medium	Medium (0.5 growing season)	Marsh	Disturbance Herbivory Fire	Buried rhizomes Annual shoots Seed banks	White 1979

replace the nitrogen and phosphorus lost in deciduous foliage (Chapin 1980; Grime 1977, 1979; Vitousek 1982) Evergreen foliage is therefore typical of infertile sites, and the evergreen shrubs in the Ericaceae and evergreen trees in the Pinaceae come to dominate peatlands (Richardson 1991). A further consequence of sclerophyllous leaves may be recurring disturbance by fire (Christensen *et al.* 1981). Less obvious, but still present and linked to the theme of infertility, are mycorrhizae which assist plants in the uptake of nitrogen and phosphorus (Marschner 1995). Finally, in sites extremely deficient in nitrogen and phosphorus, carnivorous plants occur (Givnish 1988).

Peatlands also exhibit a distinctive abundance of bryophytes; in no other habitat of the world do bryophytes comprise such a large proportion of the biomass and primary production. One genus of bryophytes, *Sphagnum*, tends to be dominant in bogs, while brown mosses in the families *Amblystegiaceae* and *Brachytheciaceae* are found in fens (Vitt 1990; Gignac and Vitt 1990; Vitt *et al.* 1995). *Sphagnum* (Figure 1.15), is largely responsible for the long-term carbon storage of peatlands; indeed, there may be more carbon incorporated in *Sphagnum*, dead and alive, than in any other genus of plant (Clymo and Hayward 1982). The success of *Sphagnum* in dominating large areas has been attributed to at least three characteristics (Godwin 1981; Clymo and Hayward 1982; van Breemen 1995; Verhoeven and Liefveld 1997). Firstly, shoot morphology and anatomy ensures that *Sphagnum* carpets act like sponges to create permanently wet conditions; about 98% of a living *Sphagnum* carpet is pore space, 10–20% in the hyaline cells of the moss tissue, and the rest on the outside of the plant (van Breemen 1995). Secondly, the cell walls have a high cation exchange capacity, which both acidifies the local environment and retains nutrients. Thirdly, organochemical compounds such as phenolics and uronic acids may contribute to suppressing the growth of vascular plants. Fens are typified by mosses in the genera *Scorpidium*, *Drenpanocladus*, *Brachythecium* and *Calliergonella* (Malmer 1986; Vitt 1990; 1994). Infertile fens ('poor fens'), however, may also have *Sphagnum*, revealing their intermediate position along the bog/fen gradient). These differences in moss flora are associated with differences in essential nutrients and pH, the latter ranging from 4 in bogs to 8 in fens. Production of mosses generally lies in the range of 200 to 400 g $m^{-2}yr^{-1}$, with rates of decomposition at least 50% higher in fens than bogs. The lower decay in bogs in part explains the peat accumulations, although even in bogs, some 80% of annual production accumulates (Malmer 1986).

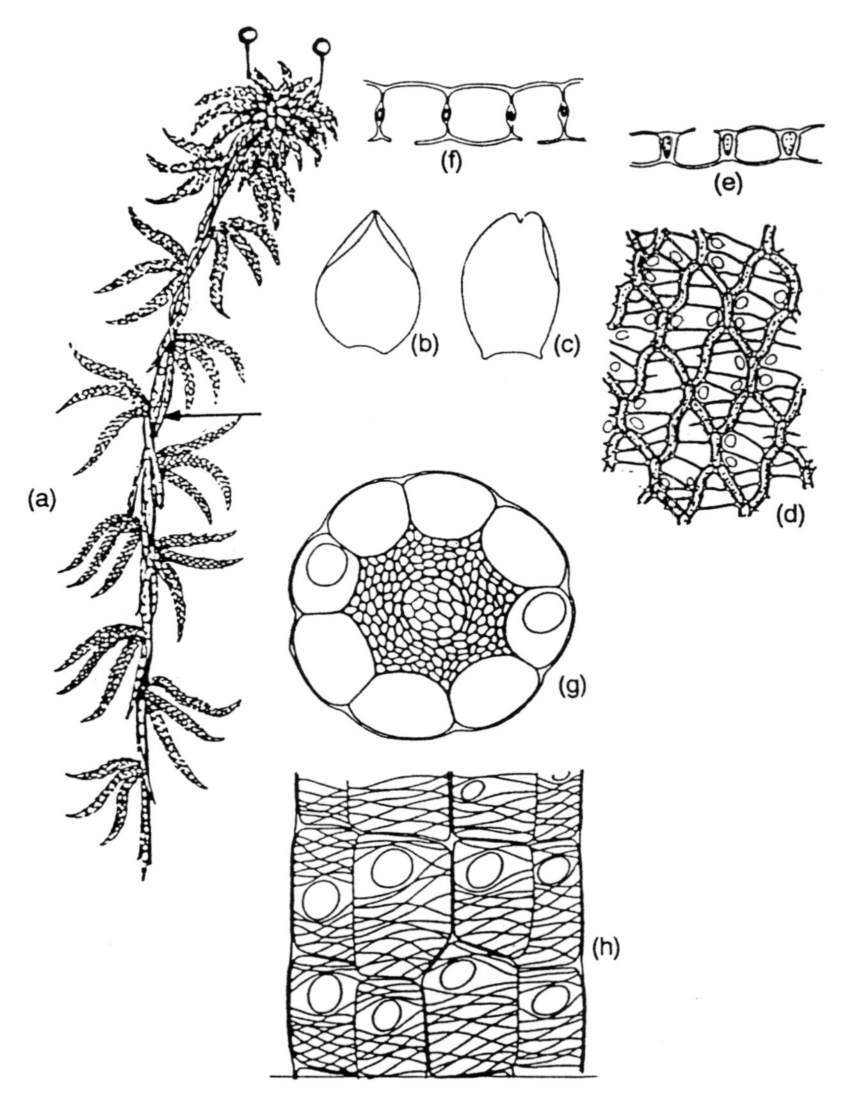

Figure 1.15 Morphology and anatomy of Sphagnum. (*a*)–(*e*) *S. papillosum*: (*a*) sporophyte-bearing shoot (with pendent branches; see arrow); (*b*) branch leaf; (*c*) stem leaf; (*d*) network of chlorophyllose (stippled) leaf cells, surrounded by porose hyaline cells; (*e*) cross-section of leaf; (*f*) same for *S. magellancium*; (*g*) and (*h*): cross section and external view of stem of *S. papillosum*, showing the large, porose hyaline cells, with fibril thickenings of cell walls, at the outside (after Scagel *et al.* 1966 in van Breeman 1995).

Dominant themes in the study of peatlands are the rates of peat accumulation, and the consequences of these for the distribution of other organisms. Measured accumulation rates range from 40 g m^{-2} in boreal bogs of Russia to 450 g m^{-2} in subtropical fens of Florida (Bakker *et al.* 1997). Over time, peat can accumulate to a depth of many metres, gradually transforming the landscape (Dansereau and Segadas-Vianna 1952, Gorham 1953), a topic to which we will return on pp. 87–93 and 396–401. As peat accumulates and absorbs water, the diminutive *Sphagnum* moss can even flood and kill forests, a process known as paludification (van Breeman 1995). Other plant species become distributed along gradients of peat depth (Slack *et al.* 1980; Glaser *et al.* 1990; Yabe and Onimaru 1997). Disturbances such as fire can reverse the process of peat accumulation (Kuhry 1994; White 1994). While the importance of peat accumulation may set peatlands in their own category, the remaining chapters of this book will show that, in other ways, the processes in peatlands are qualitatively similar to those of other wetland types.

Aquatic

Standing water, at the upper right (Figure 1.11), produces an different suite of conditions. Here, in addition to low oxygen concentrations, the environmental factors are constant submergence, disturbance by waves, reduced availability of carbon dioxide and the potential access to the atmosphere provided by floating leaves. The traits of aquatic plants therefore include well-developed aerenchyma, floating leaves, heavily dissected submersed leaves, and remarkably modified flowers (Figure 1.16). These adaptations are sufficiently dramatic that they are already thoroughly explored in two fine monographs (Sculthorpe 1967; Hutchinson 1975). From the perspective of this book, aquatic plants are a specialized subgroup of wetland plants that occupy permanently and deeply flooded sites; if readers focus too narrowly upon them, it will be easy to overlook the many other groups that are adapted to wetland conditions.

In aquatic wetlands, however, something more must be said about the problem of carbon acquisition, because this secondary constraint and its solutions are (almost) entirely restricted to aquatic macrophytes. Submersed aquatic plants are isolated from atmospheric supplies of carbon dioxide. In his monograph, *Limnological Botany*, Hutchinson (1975) points out that the concentrations of carbon dioxide in water are similar to those in air, and even a little greater at low temperatures, placing aquatic plants in a rather different situation from respiring aquatic animals. If CO_2 concentrations in water are roughly equal to those in the atmosphere, why

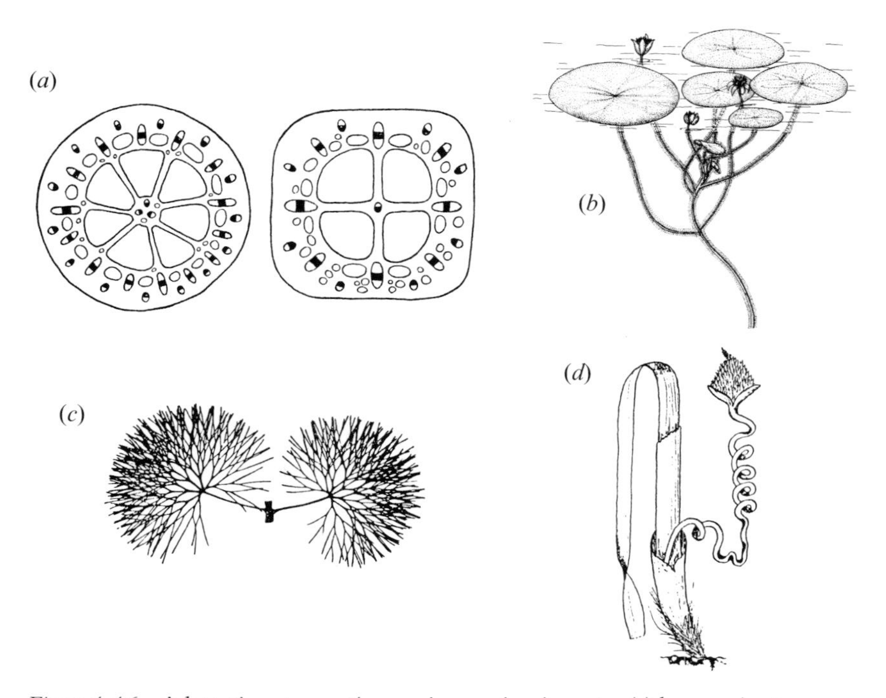

Figure 1.16 Adaptations to continuous immersion in water (*a*) lacunae in stems and rhizomes (of *Nymphaea* spp., Sculthorpe 1967, part Figure 9.5), (*b*) floating leaves of *Brasenbia schreberi* (from Hellquist and Crow 1984), (*c*) dissected submerged leaves of *Camboba caroliniana* (from Sculthorpe 1967, part Figure 5.8), (*d*) modified flowers that retract after pollination (*Enhalus acoroides*, from Sculthorpe 1967, part Figure 9.22).

should there be a problem at all? Hutchinson answers that 'the assimilating plant is still at a disadvantage in the water, owing to the much lower coefficient of molecular diffusion in a liquid than that in a gaseous medium. Submerging a land plant in water . . . may reduce its photosynthetic rate to a negligible value' (p. 145). There are at least three morphological or anatomical responses to increase access to dissolved CO_2: reduction in the waxy epidermis of leaves, reduced thickness of leaves, and most conspicuously, the elaboration of leaves to increase surface to volume ratios (Figure 1.16(*c*)). These are adaptations that would not be possible in terrestrial plants because of the need for support tissues, and because such changes would also greatly increase rates of water loss. The other main adaptation is a biochemical one, the uptake of bicarbonate ions rather than that of CO_2 directly. Hutchinson concludes that most higher aquatic plants have this capability. One set of exceptions, he says, includes *Lobelia dortmanna* and species of *Isoetes*; we shall return to them shortly. Another exception appears to be the carnivorous *Utricularia purpurea* (Moeller 1978). In such cases carbon uptake is directly related to the concentration of CO_2 in the water. All of these exceptions appear to be restricted to infertile soft water lakes, perhaps in part because these lakes lack the diurnal depletion of CO_2 found in some eutrophic lakes (Moeller 1978).

Those unusual plants that are unable to extract bicarbonate from water appear to be able to absorb carbon dioxide from their roots rather than from their leaves. Consider the bizarre group of plants in the genus *Isoetes*, an obscure group of herbaceous plants thought to be evolutionary relicts related to fern-like plants such as *Lycopodium* and *Selaginella*. *Isoetes* look rather like a small pincushion, and grow mostly in shallow water in oligotrophic lakes, although some species grow in temporary pools and a few are terrestrial. One member of this group grows at high altitudes (usually >4000 m) in the Peruvian Andes, and the following account comes from work by Keeley *et al.* (1994). At one time this plant was considered so distinct that it was put in its own genus (*Stylites*), but it is now considered to be more appropriately classified as *Isoetes andicola*. Over half this plant is composed of roots, and only the tips of the leaves emerge above ground and have chlorophyll (4% of total biomass). Most of the carbon for photosynthesis is obtained through the root system. There are no stomata in the leaves, and the plant has a CAM photosynthetic system, a system in which carbon dioxide produced at night by respiration is stored so that it can be recycled and used in photosynthesis. Studies of carbon isotopes were needed to work out the actual source of carbon being used by these plants. Here some background is necessary. In the early 1960s, atomic bomb testing contaminated the atmosphere with new carbon isotopes. Studies of isotope ratios in tree rings, for example, show dramatic increases in certain carbon isotopes after the 1960s. But Keeley *et al.* report that the $^{14}CO_2$ levels in this *Isoetes* are far below current levels, and conclude that the plant is growing by fixing carbon dioxide from decaying peat – peat that was formed from plants that grew long before atmosphere testing of nuclear weapons. There is an entire group of aquatic plants called, because of their superficial resemblance to *Isoetes*, 'isoetids', and these plants are similarly restricted to oligotrophic lakes where there are very low inorganic carbon levels in the water. Some of these (e.g. *Lobelia dortmanna*) are also known to use their roots to take up CO_2 from sediments (Wium-Anderson 1971), and some have CAM photosynthetic systems too (Boston 1986; Boston and Adams 1986).

Swamp

Moving to the bottom of Figure 1.11, trees that occur on flooded sites must be flood tolerant, and different species can be ranked in order of their tolerance to this stress (Kozlowski 1984b; Lugo *et al.* 1990). A defining property of swamps is that they are heavily forested, and as a consequence, the dense shade at the ground level inhibits the germination and growth of tree seedlings (Grubb 1977; Grime 1979). R. H. Jones *et al.* (1994) studied seedling regeneration over 4 years in four floodplain forests, cataloguing over 10 000 seedlings, and finding high rates of mortality during the first growing season. In all forests they examined, the composition of seedlings differed from the overstorey trees, suggesting that composition would change if the adult trees died. They concluded that flooding, shading and root competition were important environmental factors, but only in the years following establishment. The high rates of mortality and slow growth that they observed may be typical of densely shaded sites, but major disturbance did not occur during their study, so a period of rapid regeneration was not observed. Many tree species require gaps for their seedlings to establish (Grubb 1977; Pickett 1980; Duncan 1993). These gaps can be produced either by the death of individual trees, or when entire sections of forest are swept away by extreme floods or ice. The secondary constraint in this wetland type is therefore shading. Gap dynamics result. The processes of gap creation, seed dispersal to gaps, and establishment in gaps (Pickett and White 1985) therefore become prominent features of the ecology of swamp forests whether freshwater (Nanson and Beach 1977; Salo *et al.* 1986) or mangal (Lugo and Snedaker 1974).

Marsh

The final vegetation type in Figure 1.11 arises in the region where the above three vegetation types intersect. Here, plants are exposed to three sets of environmental factors: (i) Frequently inundated, plants here must be able to tolerate anoxia. (ii) Frequently exposed to the atmosphere, plants here can experience heavy herbivory or fire. During longer dry periods, dense canopies can develop, producing shade. (iii) Finally, when water levels are at intermediate levels, waves can break over the plants. In northern climates, waves can also grind ice onto the shore, and during cold weather, ice can freeze onto the shoreline, and when water levels rise, large pieces of vegetation can be torn away. These three combined factors create severe constraints: marshes therefore occur. Woody shoots are not suitable: they would be burned by fire, ripped out by ice, or torn away by floods. The above ground parts are therefore herbaceous. Rather like prairie plants, the herbaceous shoots arise from deeply buried rhizomes (Figure 1.17), these being horizontal stems that provide anchorage for shoots during light disturbance, storage during periods of unsuitable conditions, and regeneration after heavier disturbance (Archibold 1995). Raunkaier (1937) recognized the importance of protecting meristems from damage, and most of these plants would fall into his category of cryptophytes, plants that regenerate from buds, bulbs or rhizomes that are completely buried in the substrate.

Sculthorpe (1967) comments upon the fact that a majority of emergent and floating leaved plants must survive 'the unfavourable tropical dry season or temperate winter'. In the case of the tropical *Scirpus grossus*, he describes how

> As the water recedes at the approach of the dry season and the foliage dies back, the tip of each stolon swells . . . and forms a hard, dark, globose tuber. Laden with starch, each tuber survives until the next monsoon when it sprouts forth, forming a new plant either directly at its apex or at the end of a short stolon. (p. 343)

In some tropical climates, particularly dense aggregations of floating vegetation called sudd can form. 'The rhizomes and roots do not become anchored in the substrate; they form a stable raft floating at a depth of a few centimetres' producing floating islands, which can be a serious obstacle on navigable waterways. Floating sudd may illustrate another way rhizomatous plants can deal with the problems posed by fluctuating water levels.

In cases of severe disturbance, even deeply buried rhizomes may be killed. This may occur when a site is deeply flooded for several years running. Managers have established that, in genera like *Typha*, the dead shoots can carry oxygen to rhizomes, but if the shoots have first been destroyed by fire or grazing, the rhizomes are far more sensitive to flooding (Kaminski *et al.* 1985). Intense grazing by mammals is well documented to destroy extensive stands of marsh plants (van der Valk and Davis 1978; Fritzell 1989). In addition to buried rhizomes therefore, most of these species have long-lived seeds that persist buried in the soil (van der Valk and Davis 1978; Keddy and Reznicek 1986).

Finally, however, there is one further secondary factor, a factor alluded to above when shade was mentioned. Since disturbance is intermittent, and even infrequent (although it may be catastrophic when it does occur), there are also periods of intense competition. Deeply buried rhizome systems combined with dense arrays of shoots create strong

Figure 1.17 Marsh species have annual photosynthetic shoots emerging from deeply buried rhizomes (from Hutchinson 1975).

interference with neighbouring plants (e.g. Kershaw 1962). Large clones of *Typha* and *Phragmites* are able to dominate wetland communities to the detriment of other species with smaller growth forms and shorter shoots (Gaudet and Keddy 1988; Gopal and Goel 1993; Keddy *et al.* 1998). Further, many common genera of marsh plants appear to have the ability to interfere with neighbours through the production of toxins (Gopal and Goel 1993).

The need to tolerate damage to shoots and rhizomes, during some dry periods, and then tolerate competition in other periods, appears to have produced a distinctive growth form: dense stands of tall reed-like shoots arising from deeply buried rhizomes. *Papyrus* 'swamps' on the Nile, reed 'swamps' in Europe, cattail marshes in North America, wet meadows in Asia, and flooded savannas in South America, all show this same characteristic form. Most of these plants are also in the monocotyledonae, with the vast majority in the Cyperaceae (e.g. *Carex*, *Scirpus*, *Eleocharis*, *Cyperus*, *Rhynchospora*, *Cladium*) or the Poaceae (*Glyceria*, *Oryza*, *Phragmites*, *Phalaris*, *Panicum*). Other families include the Typhaceae (*Typha* spp.), Sparganiaceae (*Sparganium* spp.) and Juncaceae (*Juncus* spp.) The distinctive anatomy of the monocotyledonae appears to be closely associated with these conditions of recurring flooding and disturbance.

The shoots of some marsh plants are simplified to mere photosynthetic cylinders (e.g. *Equisetum*, *Eleocharis*, many *Scirpus*). The full significance of this reduction has not been explained, but it may have several functions. A cylinder has the minimum possible surface to volume ratio for a photosynthetic shoot and this may limit damage from flowing water by reducing the surface area exposed to mechanical damage. Alternatively, in the event that shoots are ripped off by flowing water, broken by waves, or consumed by herbivores, the investment of resources in each shoot has been minimized. The cost–benefit analyses of different shoot designs in these groups do not yet appear to have been conducted; certainly there are some species with more elaborate photosynthetic canopies such as *Papyrus* spp., *Scirpus maritimus*, and *Phragmites australis*.

Of the four main wetland types, marshes appear to have received the least thorough attention (Table 1.7) when compared with peatlands, aquatic plants and swamps. Rather, marshes are often treated as a peripheral part of mires (e.g. Gore 1983), a subset of the aquatic flora (Sculthorpe 1967), a shoreline zone (e.g. Bernatowiscz and Zachwieja 1966) or a habitat of wildlife impoundments (Weller 1994a). More in-depth treatments of marsh ecology are available, but always from a perspective that obscures the appreciation of their global similarities. All the same, some pieces of the story have emerged. Van der Valk (1981) has provided an account of the regeneration strategies of these species focusing upon seed banks and rhizomes, and their role in dealing with fluctuating water levels, grazing and competition; his study, however, emphasizes succession rather than the uniformity of process in marsh vegetation. Kantrud *et al.* (1989) discuss vegetation dynamics of some of these genera with changing water levels, but in the limited context of prairie pothole vegetation types. These processes have been well established in studies of New England salt marshes (Adam 1990; Bertness 1991; Bertness *et al.* 1992; Bertness and Yeh 1994; Bertness and Leonard 1997). White (1994) provides an excellent account of the role of fire and water level fluctuations in producing dense stands of *Typha* and *Cladium*, but the phenomena are presented in a volume on the management of the Everglades. Van der Toorn *et al.* (1990) were given the opportunity to treat marshes as a distinct global entity, but almost exclusively addressed commercial harvesting of thatch and reeds, and manipulation of marshes for wildlife; the natural dynamics of marshes are hardly addressed, perhaps reflecting the degree to which natural disturbance regimes have been eliminated from the European landscape. Mitsch and Gosselink (1986) treat marshes as a distinctive wetland vegetation type, but hardly address issues of disturbance or competition, and draw their examples solely from the continental United States of America.

In conclusion, while there are many studies that have been carried out in marsh habitats, like so much of wetland ecology, the synthesis and context has been lacking. I have therefore spent somewhat longer on marshes on this introductory section. Further, I have had to be somewhat speculative because of the lack of synthetic volumes that exist for other habitats. The secondary constraints producing marshes appear to be the costs imposed by living on the fringe: periods of flooding, periods of disturbance, periods of drought and periods of competition.

Wet meadows are the only one of the basic six wetland types that do not appear as a distinctive zone in Figure 1.11. Bogs and fens were combined into the peatland category, and similarly, wet meadows are probably best plotted in Figure 1.11 as a particular kind of marsh. The ecology of wet meadows appears to be related to that of marshes, except that most plants are smaller, and rather than persisting with deeply buried rhizomes, they appear more dependent upon frequent regeneration from reserves of buried seeds. Wet meadows may therefore represent an extreme type of marsh where rates of disturbance are so high, and the habitat so short lived, that even dense stands of rhizomatous marsh plants cannot survive.

Wetland functions

Historical context

We have reached the end of a great era in biology: the era of global exploration, map making and classification. While there remain new geographical discoveries to be made, particularly in poorly known groups such as the arthropods and fungi, and perhaps in regions under the oceans, the great period of explorers in sailing ships and steamers has passed. Morris (1973, vol. 1, pp. 232–49) recounts how only a little over 100 years ago, 16 September 1864, to be precise, the British Association for the Advancement of Science met in Bath, England, and among the celebrities were the two most controversial figures of African exploration, Richard Burton and John Hanning Speke. *The Times* called their impending formal confrontation a gladiatorial exhibition. The topic of debate? The source of the Nile.

There is little room here to recount their joint expedition to Africa in 1858, arrival at Lake Tanganyika with Speke nearly blind from trachoma and Burton half-paralysed by malaria, or Speke's solitary reconnaissance trip, which 25 days later brought him to the shore of Lake Victoria. On Livingstone's last and greatest adventure, still on the hunt for the Nile headwaters, he was 'delayed by tribal wars, constantly sick, losing his teeth one by one' when he reached the Arab slaver's village of Ujiji and languished near death. On 10 November, 1871 he was discovered by Henry Stanley of the *New York Herald*, and greeted with the now famous 'Dr Livingstone, I presume?' These and other stories of the era of exploration are delightfully recounted by James Morris (1973) in his three-volume history of the British Empire, *Pax Britannica.*

While the lack of detail in the global map (Figure 1.4), and the need for local wetland inventories, mean that some exploration will continue to be necessary, we are now entering a new era, an era where the essential challenges to the scientist require not the discovery of the headwaters of the Nile, but the discovery of the essential *factors* which control wetlands, their *relationships* with ecological processes and life forms, and the *functions* these provide to the biosphere.

Wetland Functions: the de Groot approach

Human societies are entirely dependent, both for their survival and well-being, upon the biosphere, the 20 km thick layer that provides all the necessities of life. One useful way of thinking about the linkages between humans and the biosphere is the concept of ecological function. Ecological function can be defined as 'the capacity of natural processes and components to provide goods and services that satisfy human needs' (de Groot 1992 p. 7). This concept challenges humans to recognize, inquire into, and quantify the benefits received from ecosystems. In his overview, *Functions of Nature*, de Groot lists 37 functions that natural environments perform for humans (Table 1.8). These range from the ozone layer's function of protecting humans from harmful cosmic influences to a landscape's function in artistic inspiration. Further, de Groot breaks these functions down into four categories:

(i) *Regulation functions* describe the capacity of ecosystems to regulate essential ecological processes and life support systems on Earth. Examples include regulation of the CO_2 and O_2 concentrations of the atmosphere.
(ii) *Carrier functions* describe the space or suitable substrate needed for the conduct of human activities such as living, cultivation and recreation. Examples include soil and rainfall for growing crops.
(iii) *Production functions* describe the resources provided by nature, including food, raw materials for industrial use and genetic raw material. Examples would include production of clean water for drinking and wood for building.
(iv) *Information functions* describe the role played by natural ecosystems in the maintenance of mental health by providing cognitive development, spiritual inspiration and scientific appreciation of the world. Examples would include areas for wilderness appreciation such as Parque Naçional Tortuguero in Latin America or historical landscapes such as the medieval castles of the Rhine River valley in Europe.

Christensen *et al.* (1996) argue for the benefits of ecosystem management using a slightly different terminology; they recognize three different categories of values provided by natural ecosystems: processes, goods, and services. Processes include hydrological storage, biological productivity, biogeochemical cycling and biological diversity. Goods include food, construction materials, medicinal plants and tourism. Services include regulating climate, cleansing air and water and detoxifying pollutants. Although the classification scheme is somewhat different, their proposal captures many of the same functions as are covered in de Groot.

Wetlands are certainly important for many, if not all of the functions listed in Table 1.8. Sather *et al.* (1990) and Larson (1990) examine some of these specific functions performed by wetlands (e.g. flood control,

Table 1.8. *Functions of natural environment*

Regulation functions

1. Protection against harmful cosmic influences
2. Regulation of the local and global energy balance
3. Regulation of the chemical composition of the atmosphere
4. Regulation of the chemical composition of the oceans
5. Regulation of the local and global climate (incl. the hydrological cycle)
6. Regulation of runoff and flood prevention (watershed protection)
7. Water-catchment and groundwater recharge
8. Prevention of soil erosion and sediment control
9. Formation of topsoil and maintenance of soil fertility
10. Fixation of solar energy and biomass production
11. Storage and recycling of organic matter
12. Storage and recycling of nutrients
13. Storage and recycling of human waste
14. Regulation of biological control mechanisms
15. Maintenance of migration and nursery habitats
16. Maintenance of biological (and genetic) diversity

Carrier functions

providing space and a suitable substrate for

1. Human habitation and (indigenous) settlements
2. Cultivation (crop growing, animal husbandry, aquaculture)
3. Energy conversion
4. Recreation and tourism
5. Nature protection

Production functions

1. Oxygen
2. Water (for drinking, irrigation, industry, etc.)
3. Food and nutritious drinks
4. Genetic resources
5. Medicinal resources
6. Raw materials for clothing and household fabrics
7. Raw materials for building, construction and industrial use
8. Biochemicals (other than fuel and medicines)
9. Fuel and energy
10. Fodder and fertilizer
11. Ornamental resources

Information functions

1. Aesthetic information
2. Spiritual and religious information
3. Historic information (heritage value)
4. Cultural and artistic inspiration
5. Scientific and educational information

Source: After de Groot (1992).

Table 1.9. *A list of some functions performed by wetlands in the biosphere*

Ground water recharge
Ground water discharge
Floodflow alteration
Sediment stabilization
Sediment/toxicant retention
Nutrient removal/transformation
Carbon transformation
Production export
Wildlife diversity/abundance
Wildlife breeding
Wildlife migration
Wildlife wintering
Aquatic diversity/abundance
Uniqueness/heritage
Recreation

Source: After Bardecki *et al.* (1989), Adamus *et al.* (1987) and Richardson (1995).

productivity) calling them 'wetland values'. I do not intend to exhaustively document each function carried out by wetlands (see Table 1.9); such a treatment might even become monotonous. In this first chapter, there is space only to address five important functions of wetlands: production (including wildlife production), regulation of atmospheric carbon dioxide and methane levels, maintenance of the global nitrogen cycle, storage of ecological records, and flood reduction. In the rest of the book the emphasis will be upon ecological properties, environmental factors, and wise management. This approach will lead naturally to an understanding of the processes that produce functions, and to the continuance of these functions. When humans manipulate wetlands, whether by draining for agriculture, or flooding to increase certain species, many properties, processes and functions are simultaneously changed, often with unknown consequences.

Production

The capture of solar energy by plants is the foundation of virtually all life on Earth. We now know that certain wetland types, particularly marshes and swamps, are highly productive, in part, because they occur where two

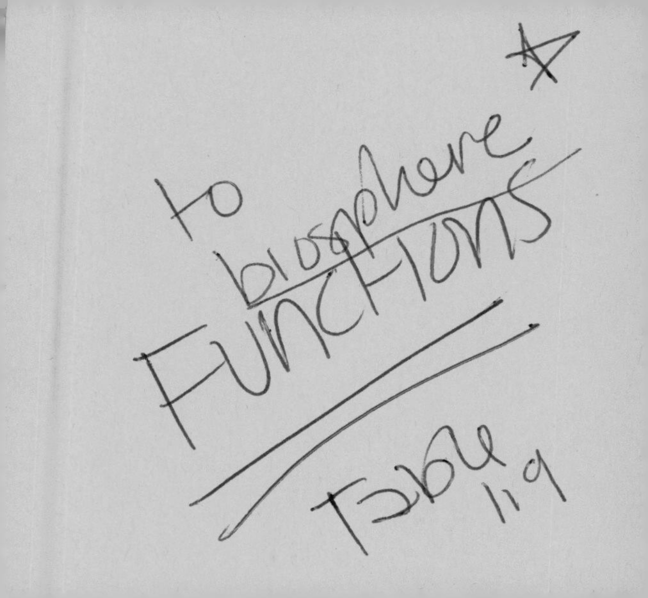
to biosphere
Functions
Table 1.9

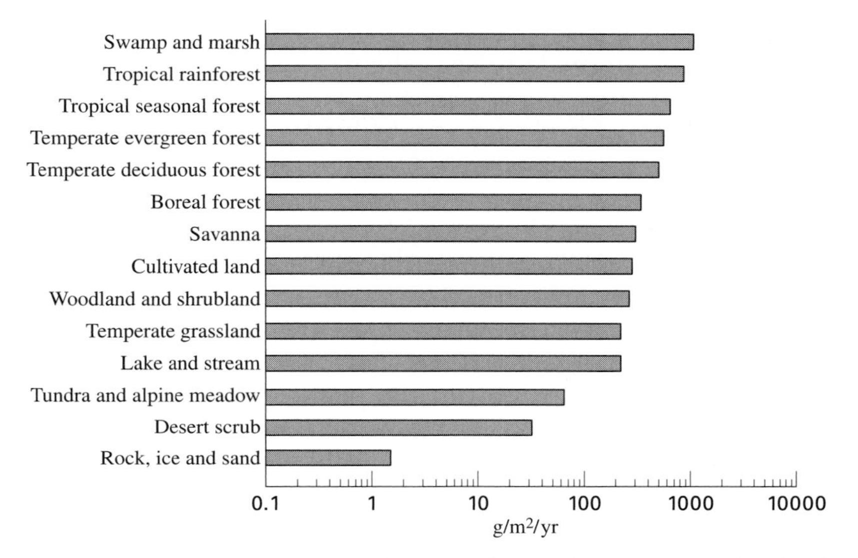

Figure 1.18 Mean net primary productivity of wetlands (top) compared with other ecosystems; note log scale (from data in Whittaker and Likens 1973).

common constraints on photosynthesis (water and nutrients) are simultaneously reduced. Wetlands are sites where nutrients tend to accumulate, and where the high water table ensures adequate supplies of water – in terrestrial ecosystems, in comparison, shortages of nutrients and water often limit plant growth. These basic patterns of primary production have only recently been determined. Leith (1975) recounts how photosynthesis itself was only discovered in the period from 1772 to 1779, and how in 1804 de Saussure gave the correct equation for photosynthesis. In 1919, Schroeder provided an estimate of dry matter production on land, 28×10^9 t. Future work required better mapping of world vegetation types, and better data on oceanic production. By 1960, Muller was able to estimate 10.3×10^9 t of carbon produced on land and 25×10^9 t in the sea. The creation of the International Biological Program (IBP) in the early 1960s co-ordinated attempts to estimate primary production better in different ecosystems, and to incorporate these data into large-scale ecosystem models. While the value of these systems models is doubted by some scientists (McIntosh 1985), it is still evident in many works on wetlands (e.g. Good *et al.* 1978, Patten 1990), although I will be taking a different approach in this book. Whatever the merits of large-scale ecosystem models, the actual measurement of primary production provides a useful empirical basis for the study of wetlands. Figure 1.18 shows that swamps and marshes are some of the most productive ecosystems on Earth; they rival both rain forest and cultivated land. But, unlike agricultural fields, this primary production occurs with no fossil fuel inputs in the form of petrol and fertilizer, no tending by humans, no artificial irrigation, and no heavy machinery. Wetlands can therefore be regarded as factories in the landscape that mass produce both organic matter and oxygen to support surrounding ecosystems. Draining such wetlands may therefore be compared to systematically smashing the factories that support life on Earth.

Aquatic plants, like peat bogs, do not appear to fit the above generalization: quantitative comparisons of primary production using carbon dioxide exchange rates show that the majority of aquatic ecosystems fall below 10 $\mu mol\ CO_2\ m^{-2}\ s^{-1}$. Aquatic macrophytes, phytoplankton and attached microalgae all exhibit similar low production figures relative to terrestrial plants. Sand-Jensen and Krause-Jensen (1997) attribute this to three causes: terrestrial plants, they say, have complex canopies with many leaf layers to intercept sunlight, their leaves can acclimate to high or low irradiance, and there is both rapid diffusion of gases and a large reservoir of carbon dioxide in the air. Regrettably, such broad comparative studies often distinguish only between aquatic communities and terrestrial communities. Wetlands are then overlooked, and we have no way of judging where they fall, nor how much difference there is among wetland types. This is not the only case we shall see where studies of 'aquatics' do not necessarily provide useful generalities about wetlands. Certainly, stands of marsh and swamp plants have the production advantages cited by Sand-Jensen and Krause-Jensen to explain the high production of terrestrial plants, yet marsh and swamp plants may simultaneously escape the constraints that drought and low nutrients often place upon terrestrial plants. This combination presumably explains the very high rates of production seen in Figure 1.18.

High rates of primary production provide raw materials for the construction of other life forms. The production of animal biomass in wetlands is some 9.0 $g\ m^{-2}\ yr^{-1}$, 3.5 times the value for terrestrial ecosystems (Turner 1982). This production has both direct economic values (e.g. fisheries, trapping, hunting) and values that are more difficult to measure (e.g. carbon flow, recreation, support of endangered species). Tables 1.10 and 1.11 show just two examples of species (turtles and water fowl, respectively) that consume plant material in wetlands, illustrating the dependence of animal production upon primary production. The actual pathways of energy flow are as variable as the species composition of individual wetlands. For example, wildlife species may not only feed

Table 1.10. *Contents of 22 turtle stomachs*

Food item	No. of stomachs	Percentage of samples
Vegetation	15	68.2
Potamogeton spp.	8	36.4
Algae	6	27.3
Polygonum spp.	4	18.2
Lemna spp.	9	40.9
Other		
Fish		
Carp (*Cyprinus carpio*)	16	72.8
Pike (*Esox lucius*)	6	27.3
Bass (*Micropterus* spp.)	4	18.2
Perch (*Perca flavescens*)	4	18.2
Molluscs		
Snail (*Physa, Planorbula, Gyraulus*)	21	95.4
Other	4	18.2
Insects	11	50.0
Birds	5	22.7
Turtles	1	4.5

Source: From Hammer (1969).

directly on plants, but also consume them indirectly by feeding on other secondary producers: the turtles in Table 1.10 feed not only upon plants, but also upon fish and molluscs. As well, in habitats such as salt marshes, some of the wildlife species are feeding not directly upon the plants, but rather are part of an elaborate food web that is based upon decomposers who have fed upon the plants. Further, the primary production of wetlands may support harvests of commercial species that do not even occur within wetlands – shrimp harvests in estuaries of the Gulf of Mexico are strikingly correlated with the area of salt marsh (Figure 1.19). Similarly, Welcomme (1976, 1979, 1986) has found that the area of floodplain in African rivers predicts the fish catch from these rivers. A production of 40–60 kg ha^{-1} of maximum flooded area is typical for tropical floodplains throughout the world. Further, on a world-wide basis, there is a quantitative relationship:

$$\text{catch (kg)} = 5.46 \times (\text{floodplain area (ha)})$$

In spite of such examples, little of the primary production of world ecosystems is directly consumed by wildlife. Kurihara and Kikkawa

Table 1.11. *Gizzard analysis of plant species in 1102 birds of 15 species of waterfowl in 58 locations in the eastern United States and Canada*

Scientific name	Common name	Volumetric percentage
Potamogeton spp.	Pondweed	13.29
Polygonum spp.	Smartweed	6.69
Zizania aquatica	Wild rice	5.10
Scirpus spp.	Bulrush	4.90
Najas flexilis	Northern Naiad	4.32
Lemna, *Spirodela*, etc.	Duckweed	2.97
Vallisneria spiralis	Wild celery	2.49
Leersia, chiefly *L. oryzoides*	Cutgrass	2.02
Setaria spp.	Pigeongrass	1.62
Echinochloa, chiefly *E. crusgalli*	Wild millet	1.59
Sparganium spp.	Burreed	1.33
Carex spp.	Sedge	1.21
Sagittaria spp.	Arrowhead	1.00
Brasenia schreberi	Watershield	0.95
Nymphaea spp.	Waterlily	0.77
Ceratophyllum demersum	Coontail	0.77
Bidens spp.	Beggar's tick	0.65
Cyperus spp.	Sedge	0.57
Pontederia cordata	Pickerel weed	0.48
Zea mays	Corn	2.30
Fagopyrum exculentum	Buckwheat	1.40
Sorghum vulgare	Sorghum	0.51
Algae	Algae (microscopic)	0.87
Characeae	Muskgrass	1.87
Miscellaneous		14.69
		74.36
Invertebrates		25.64

Source: From Crowder and Bristow (1988).

(1986) report that some half of the production in terrestrial ecosystems passes directly to decomposers. In a mixed deciduous forest, herbivores consume only 1%, and in grassland, herbivores consume about 8%. Similar low figures are found in wetlands. Herbivores consumed only some 10% of primary production in both peatlands (Miller and Watson 1983) and salt marshes (Wiegert *et al.* 1981), although Lodge (1991) reports higher values for effects of grazing on aquatic macrophytes. The preponderance of energy flow passes by grazers (Figure 1.20). The fate of this *ca.* 90% of the energy requires closer inspection of the activity of

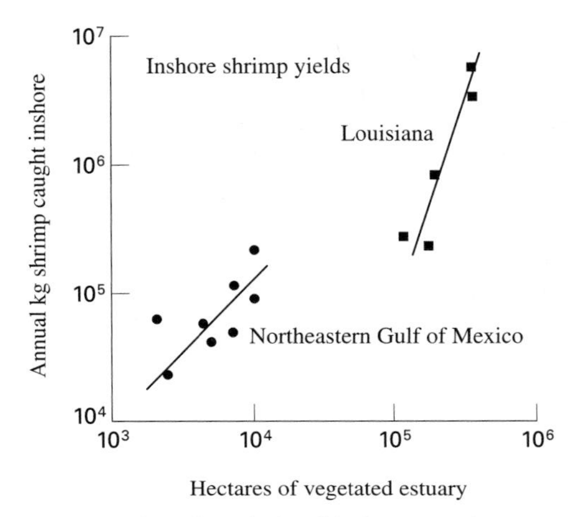

Figure 1.19 The relationship between the mean annual yield of shrimp caught inshore and the area of vegetated estuary (from Turner 1977).

decomposers. In salt marshes, decomposers are the base of an important food chain that supports estuarine and oceanic fisheries (Turner 1977), and a similar process appears to occur in rivers bordered by large floodplains (Welcomme 1976, 1986). In peatlands, the constant high water table and the acidic substrate reduce the activities of decomposers, so that a substantial proportion of the plant debris accumulates as peat (Gorham 1957; Miller and Watson 1983). In general, however, while the production of wildlife species is an important function of wetlands, the above examples show that the decomposers provide the critical link in converting plant production to wildlife biomass, as well as in controlling the accumulation of organic matter. Kurihara and Kikkawa (1986) conclude 'For most organisms, the concept of secondary production must incorporate the . . . role of decomposers in making the energy of primary production available to animals'. They emphasize, then, the bottom left arrow in Figure 1.20. The efficiency of decomposers in consuming primary production is illustrated by measurements showing that over 90% of the carbon fixed annually in peatlands is re-released as carbon dioxide (Silvola *et al.* 1996). Some other introductions to decomposer activity can be found in Polunin (1984), Heal *et al.* (1978), Good *et al.* (1978), Dickinson (1983) and Brinson *et al.* (1981).

Many animals only use wetlands for part of the year, but this role of wetlands as a seasonal refuge deserves wider appreciation. Consider, for example, the immense herds of grazing animals found on the East African

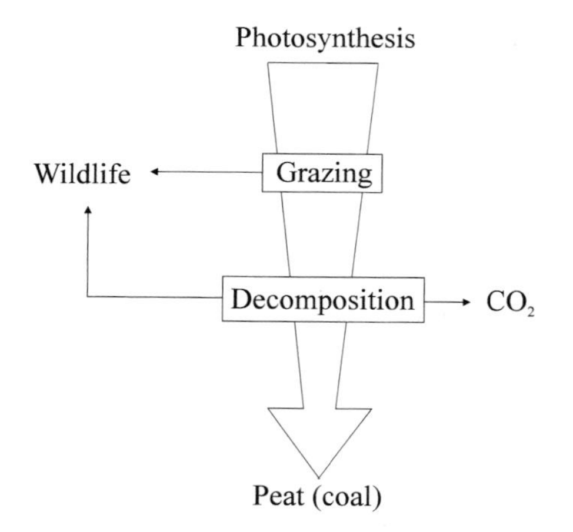

Figure 1.20 Some of the large primary production of wetlands is diverted to wildlife biomass by grazing or decomposer based food webs. The remainder is broken down by decomposers or else accumulates as peat.

plains. We can draw upon two complementary sources for the story; Denny (1993a) for the botanist's perspective, and Sinclair and Fryxell (1985) for the zoologist's. To appreciate the processes, we must understand that water availability in this region changes at two time scales; annual cycles driven by rainy seasons, and longer fluctuations driven by variation in mean annual rainfall (Sinclair and Fryxell 1985). In semiarid areas, the dry season forces grazing animals to converge on, and remain within, a 20 km radius of permanent water supplies such as rivers and swamps. In southern Sudan, for example, there are large areas of seasonally flooded and permanently flooded grasslands at the headwaters of the Nile (Denny 1993a). The deeper water areas may have the emergent *Cyperus papyrus*, but the shallower areas have grass genera such as *Vossia*, *Oryza*, *Echinochloa*, *Leersia* and *Panicum*, '. . . lush, nutritious grasses much favoured by herbivorous browsers'. Sinclair and Fryxell (1985) report that some 800 000 White-eared Kob occur here. Since the areas grazed during the wet season provide no permanent water, when the rains stop, the animals migrate from shorter grass areas into these ephemeral wetlands. This has important benefits for the food plants, since both the upland grasses and wetland grasses have a period each year without grazing when they can flower and set seed. In Southern Africa, similar wildlife migrations occur, but the dry season water supplies being pools that were filled by the previous year's rains, are less predictable. Sinclair and Fryxell conclude that migration between wet lands and dry lands rather than year-round residence patterns allows a landscape to support much larger mammal populations than would otherwise be possible.

The story has even broader implications than this, because the wetlands thereby maintain the integrity of the surrounding grasslands as well as pastoral human societies. Breman and deWit (1983) say '. . . the borders and flood plains of Lake Chad and the rivers Senegal, Niger and Chari, with vegetation dominated by perennial grasses, are indispensable for the pastoral people and their herds.' Sinclair and Fryxell (1985) go on to recount how failure to appreciate the natural dynamics of this system led to overgrazing and desertification. (We shall see in Chapters 4 and 6 how humans have failed to appreciate the dynamics of wetlands in particular, and ecosystems in general.) Noting that the pastoral peoples were continually moving to follow their herds, aid agencies developed projects to drill wells and provide water. This conveniently reduced unwanted movement of nomadic peoples across new political borders. It also allowed governments and aid agencies to provide fixed medical and veterinary facilities. A principal consequence of this was the replacement of traditional seasonal grazing by 'continuous intensive foraging'. All plant growth was completely eaten during the wet season, leaving little in reserve for the dry season. Further, the continual grazing forced the perennial grasses to draw upon energy reserves form their roots. As the roots were resorbed, they became too weak to reach water. As a consequence, deeply rooted perennial grasses were replaced by shallow rooted annual species that regenerated each year from seed, and were consequently much more sensitive to fluctuations in rainfall. Large areas of the Sahel then slowly became deserts. The traditional migratory patterns, in which wildlife congregated in wetlands each year, had not only allowed larger herds of animals to survive, but had actually ensured the integrity and persistence of the grasslands in the surrounding upland areas.

Climate regulation

The amount of carbon dioxide in the atmosphere controls the Earth's temperature through the greenhouse effect. As the data on primary production showed (Figure 1.18), swamps and marshes are ecosystems in which plants rapidly extract carbon dioxide from the atmosphere. In many cases, the secondary consumers and decomposers consume the plants and rapidly release carbon dioxide back into the atmosphere (Figure 1.20). Peatlands are an exception, being habitats in which consumers and decomposers are out of balance with the rate of primary production, with the consequence that carbon is stored in partially decayed plant material. The lack of decay is largely attributable to scarcity of oxygen. Some 500 million hectares (nearly 4% of the Earth's ice-free land area) now consist of peatlands (Gorham 1990). These peatlands store carbon that would otherwise be released to the atmosphere as carbon dioxide. One estimate suggests that 500 billion metric tons of carbon would be released into the atmosphere if all the peatlands on Earth were destroyed (Dugan 1993). If such peatlands are drained, or if warming climates shift soil processes towards aerobic decomposition, then we can expect significant climatic consequences (Gorham 1991; Woodwell *et al.* 1995), chiefly a further increase in mean global temperature. Further, Silvola *et al.* (1996) have shown that the CO_2 production of mires increases with higher temperature or with a lower water table. Warmer and drier summers will therefore speed up the rate of release from CO_2 from storage in peatlands, enhancing the greenhouse effect.

Methane is the most abundant organic chemical in the Earth's atmosphere, and because it absorbs infrared light, it is also an important greenhouse gas (Cicerone and Ormland 1988). Air samples extracted from dated ice cores suggest that methane concentrations have slowly increased from *ca.* 0.7 to 1.0 ppm over the last two millennia, with more rapid increases recently (1.5 to 1.7 ppm between 1978 and 1988). The major sources of methane are wetlands, mining/natural gas wells and ruminant animals such as cows and sheep. In wetlands, the methane source is methanogenic bacteria, an ancient group of micro-organisms in the Archaebacteria that are strict anaerobes and live in highly reduced conditions. Wetlands contribute from one-third to one-half of the methane released to the atmosphere each year (Cicerone and Ormland 1988; Whiting and Chanton 1993). This amounts to more than 100 Tg of methane (a Teragram = 10^{12} g); 25% of this comes from tropical and subtropical swamps and marshes, whereas 60% is released from high latitude peatlands. These emissions are roughly proportional to the area of these wetland types, closely corresponding with the two great global belts of wetlands, high latitude peatlands (from 50° to 70° N) and equatorial swamps and marshes (from 20° N to 30° S) (Matthews and Fung 1987). If rice paddies are included, mean methane emissions approach 200 Tg (Aselmann and Crutzen 1989). Rice paddies and natural wetlands produce roughly equivalent total emissions, although rice paddies have much higher emission rates on a m^2 basis, 300–1000 mg CH_4 m^{-2} day^{-1} as opposed to 1–50 for bogs or 137–399 for marshes (Table 1.12). In one peatland, Shannon *et al.* (1996) found that a majority (64–90%) of the

Table 1.12. *Global wetland methane emissions extrapolated from measured emission rates in field experiments*

Wetland category	Emission-rate (mg CH_4 m^{-2} day^{-1})	Area (10^{12} m^2)	Mean prod. period[a] (days)	Emission (Tg/a)
Bogs	15	1.87	178	5
Fens	80	1.48	169	20
Swamps	84	1.13	274	26
Marshes	253	0.27	249	17
Floodplains	100	0.82	122	10
Lakes	43	0.12	365	2
Natural wetlands		5.69		80
Rice fields				
1. Mean emission rate	310	1.31	130	53
2. Temperature-dependent emission rates	300–1000	1.31	130	92
Grand total		7.00		100–300

Notes:
[a] Mean CH_4 productive period comprising of both permanent and seasonal wetland areas in the respective wetland categories. The CH_4 productive period is determined either by monthly mean temperatures above 0 °C or by inundation. Swamps with unknown seasonality have been treated as permanent.
Source: After Aselman and Crutzen (1989).

methane produced in an ombotrophic peatland was emitted by one herbaceous plant *Scheuchzeria palustris*. The aerenchyma of the plant transported the methane produced by methanogenic bacteria from below the soil surface into the atmosphere. Other plants such as *Carex* spp., *Peltrandra virginica* and *Typha* are also known to emit methane.

On a larger time scale, consider the degree to which our civilization is based upon coal; in the 1980s we consumed in the order of 3 billion tons per annum (Manfred 1982). Even highly industrialized countries such as the United States still depend upon coal for roughly one-quarter of their energy consumption (Manfred 1982). Coal comes from swamps that existed long in the past (Figure 1.21). (Were it not for such coal swamps, it could be argued that the industrial revolution might never have occurred. It may therefore be that both agriculture (Figure 1.1) and the industrial revolution were dependent upon wetlands.) By burning the coal, humans are releasing carbon dioxide that was once extracted from

Figure 1.21 A carboniferous coal swamp (from Seward 1931).

the atmosphere by wetland plants. Humans are altering the linkages between carbon storage in wetlands and the CO_2 levels of our atmosphere.

Denny (1995) also suggests that wetlands can modify local climate, particularly in areas where the wetland is surrounded by arid lands, such as the Sudd swamp in the Sahara, or the Okavanga Delta in the Kalahari desert. In the Okavanga Delta, it is estimated that 95% of the water therein is lost through evaporation and evapotranspiration; this water is a presumed source for rainfall in the surrounding landscape. Similarly, in south east Uganda, where the valley hillsides are cultivated, the drainage of swamps in the valley appears to have lowered air humidity and early morning mists, leading to increased water stress in crops.

Environmental records

As a consequence of the low rates of decay in wetlands, plant and animal debris can accumulate, thereby recording the sequence of plant species that occupied a site over millennia. Exceptions include alluvial flood plains, where the sediments are constantly reworked by meandering rivers, so that the sedimentary record is lost (e.g. Nanson and Beach 1977; Salo *et al.* 1986). In peatlands, one frequently finds that the accumulations

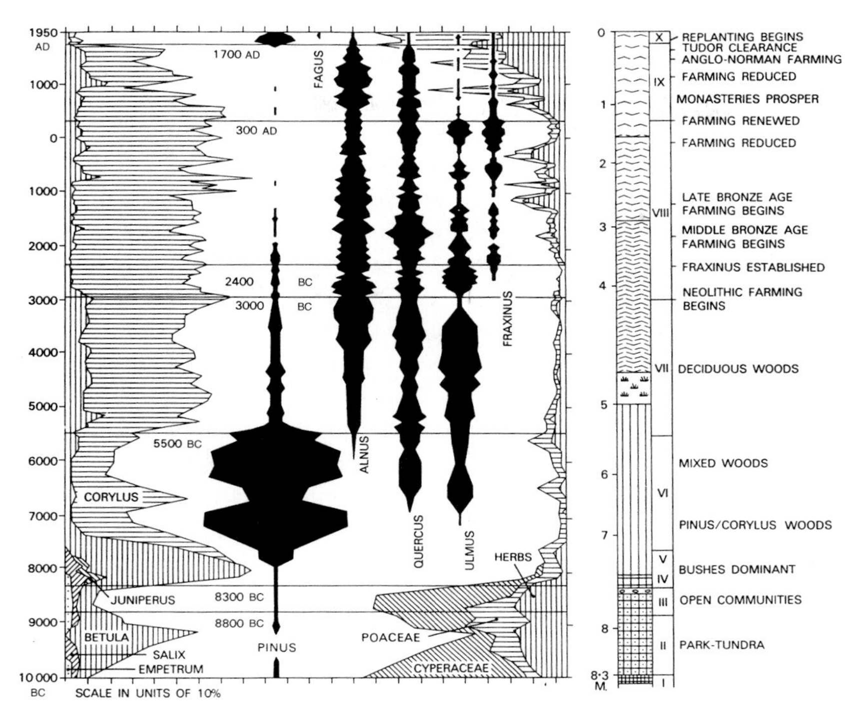

Figure 1.22 A 10 000 year profile through the Littleton Bog, England (after Mitchell 1965 from Taylor 1983).

of organic matter provide a nearly complete record of the plant associations that occurred on the site over thousands of years. This record most commonly takes the form of pollen and plant fragments, but can be supplemented by insect parts, charcoal fragments, archaeological artefacts and even rooted trees that have been buried over the years (e.g. Walker 1970; Moore 1973; Watts and Winter 1966; Godwin 1981; Delcourt and Delcourt 1988, 1991). These provide a record of the long-term relationships between vegetation and other factors such as climate, flooding and humans. For example, Mitchell (1965) studied the layers of pollen in a bog near Tipperary in south central Ireland (Figure 1.22). Now more than 8 metres of peat cover the original soil surface, which some 10 000 years ago was open tundra, as indicated by the abundant birch and sedge pollen. Pine woods developed some 8000 years ago, to be replaced by Elm–Oak woodland some 6000 years ago. This suggests a steady amelioration of climate. About 5000 years ago, the tree pollen declines and herb pollen increases; this appears to reflect woodland clearance by Neolithic farmers. About 3800 years ago the clearances become more extensive, apparently due to the arrival of Bronze Age farmers. At many sites, wooden trackways constructed from branches or split logs were apparently constructed to cross bogs and link farming communities (Godwin 1981). At about 300 AD, there was a reduction in intensity of farming, but since then there has been a steady increase in amounts of grass and herb pollen, indicating greater human impacts upon the Irish landscape.

Such records provide important opportunities to study long-term changes in vegetation and climate, the impacts of human cultures upon vegetation, and natural processes such as succession in wetlands. In many cases peatlands can be considered to be archives for adjoining regions of the Earth's surface (Godwin 1981). Changes in vegetation and land use are not the only records stored in bogs. A Danish almanac of 1837 records 'There is a strange power in bog water which prevents decay. Bodies have been found which must have lain in bogs for more than a thousand years, but which, though admittedly somewhat shrunken and brown, are in other respects unchanged.' Dieck (in Glob, 1969) has recorded 690 bodies recovered from peat bogs, the most famous of which are perhaps Lindow Man and Tollund Man; most of these which can be dated are from the period between 100 BC and 500 AD. They are distributed across Germany, Denmark, Holland, England, Scotland, Ireland, Norway and Sweden (Stead *et al.* 1986; Coles and Coles 1989). Men, women and children have been found, the outstanding feature being that they are so well preserved that they are sometimes first assumed to be the result of a recent murder. Some, such as the Tollund Man, were apparently strangled, with the plaited skin noose still attached to the neck; others appear to have been pegged down while still alive (Glob, 1969). The bodies give the appearance of having been tanned, a process now attributed to a polysaccharide (sphagnan) produced by *Sphagnum* (Painter 1991).

Nitrogen cycling in the biosphere

In his 1789 *Treatise on Chemistry*, published only a few years before he went to the guillotine, one of Lavoisier's first sections addressed the composition of the atmosphere:

> We have already seen that the atmospheric air is composed of two gases ... one of which is capable, by respiration of contributing to animal life ... the other, on the contrary, is endowed with directly opposite qualities; it cannot be breathed by animals, neither will it admit of the combustion of inflammable bodies, nor of the calcination of metals.

The former gas we call oxygen, the latter nitrogen (although Lavoisier preferred the term of azote). We now know some important further features of this azotic gas. First, the Earth's atmosphere differs from both neighbouring planets (Venus and Mars) by having this gas predominant in its atmosphere. Secondly, nitrogen is essential for the construction of amino acids, the building blocks of proteins and life. Thirdly, only a few organisms can remove nitrogen from the atmosphere, so that both plant growth and animal growth is limited by the availability of nitrogen (e.g. Raven *et al.* 1992; White 1993). Finally, the enzyme that catalyses the conversion of atmospheric nitrogen to biologically usable forms, nitrogenase, only functions under anoxic conditions, presumably because it originated early in the Earth's history when the atmosphere was still anoxic. Therefore, when cyanophytes (blue-green algae) reduce atmospheric nitrogen to a biologically usable form, they do so in special thick-walled cells called heterocysts in which the enzyme is protected from oxygen.

These peculiarities of the nitrogen cycle are an unusual feature of Earth: the limitation of organisms by a weakly reactive gas that is at once so common and yet so biologically precious. In Chapter 5 we shall further explore the effects of nitrogen availability on the distribution and abundance of plants and animals, but here let us keep to the global perspective. Wetlands fit into this story because soil chemical reactions are typical of anaerobic conditions, and wetlands are therefore the major reducing system in the landscape. Moreover, since the water level changes '. . . wetlands maintain the widest range of oxidation–reduction reactions of any ecosystem on the landscape. This allows them to function as effective transformers of nutrients and metals . . .' (Faulkner and Richardson 1989 p. 63). That is, wetlands are sites where elements are transformed among an array of chemical states (Rosswall 1983; Mitsch and Gosselink 1986; Patten 1990; Armentano and Verhoeven 1990). The complex biogeochemical cycle of nitrogen involves multiple biotic and abiotic transformations involving seven valency states ($+5$ to -3). In wetlands, most N is stored in organic sediments. There are two scales at which nitrogen movement and transformation can be studied. At the within wetland scale, the principal flows occur among three components: organic matter, the oxidized surface layer and deeper anoxic layers. At a landscape scale, there are flows among three other components: the surrounding terrestrial landscape, the wetland, and the atmosphere. Let us consider these two scales in turn.

Within a wetland (Figure 1.23), one of the principal steps controlling rates of nitrogen cycling is the rate at which organic nitrogen is mineralized to NH_4^+. This process of decomposition is controlled by factors

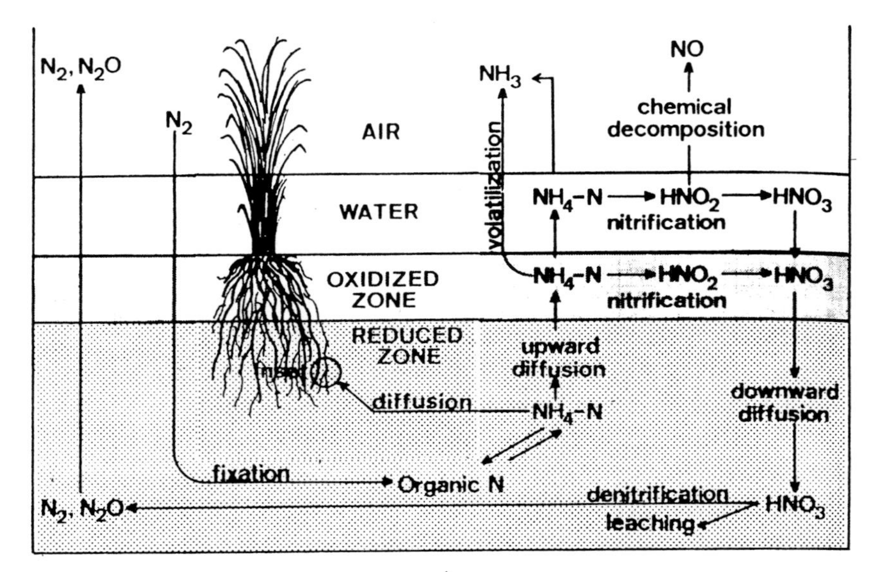

A

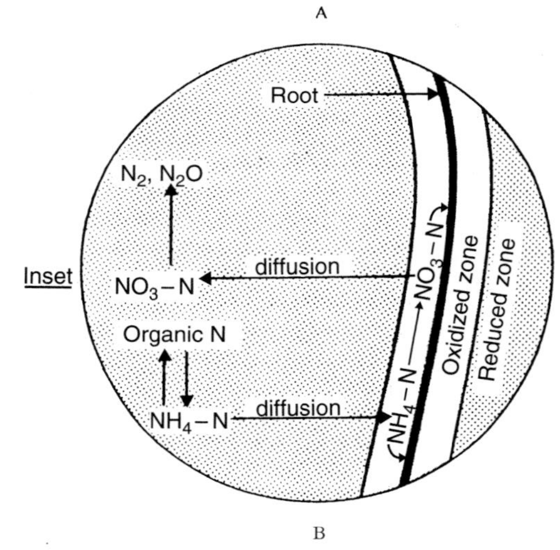

B

Figure 1.23 Key steps of the nitrogen cycle in flooded soils and sediments (from Reddy and Patrick 1984). The major stock of sediment nitrogen is the organic N component, and the rate of release of this nitrogen is largely dependent upon oxygen availability. (Reprinted with permission from Reddy & Patrick 1984.)

Table 1.13. *Nitrogen fixation and denitrification in wetlands*

Wetland type	N fixation		Denitrification	
	Mean rate ($g\ m^{-2}\ y^{-1}$)	Total ($Tg\ y^{-1}$)	Mean rate ($g\ m^{-2}\ y^{-1}$)	Total ($Tg\ y^{-1}$)
Temperate:				
Peat mires	1.0	3.0	0.4	1.2
Floodplains	2.0	6.0	1.0	3.0
Tropical:				
Peat mires	1.0	0.5	0.4	0.2
Swamp forest	3.5	7.8	1.0	2.2
Floodplains	3.5	5.2	1.0	1.5
Ricefields	3.5	5.0	7.5	10.8
Total		27.5		18.9
Total terrestrial		139		43–390

Note: $T = 10^{12}$.
Source: From Armentano and Verhoeven (1990).

including temperature, pH and the presence of oxygen. The availability of oxygen, as we have already seen (pp. 35–36), declines sharply with depth. The transformation of NH_4^+ to NO_3^- occurs in the shallow oxidized zone on the surface when chemoautotrophic bacteria oxidize the NH_4^+. The depletion of NH_4^+ sets up a concentration gradient which drives upward diffusion of nitrogen, NH_4^+ moving from deeper anoxic layers to the upper oxidized layer. At the same time, nitrogen in the form of NO_3^- flows in the reverse direction. In this case, the NO_3^- is depleted from the deeper anoxic layers by assimilation, denitrification or leaching. These two processes are largely controlled by the depth to water table and by the flow of oxygen through wetland plants into the rooting zone (Reddy and Patrick 1984; Faulkner and Richardson 1989).

At larger scales, inputs of nitrogen include fixation, runoff and precipitation. Outputs include runoff and gaseous nitrogen produced by denitrification. Let us consider the biological processes of fixation and denitrification in turn (Table 1.13). Recall from p. 36 that these processes are largely dependent upon the proximity of the surface oxidized layer to the anoxic regions deeper in the wetland (Faulkner and Richardson 1989). During nitrogen fixation, bacteria in the rhizosphere reduce atmospheric nitrogen (N_2) to ammonium (NH_4^+), providing a continual input of nitrogen to the soil. Since plants cannot use N_2, the rate at which micro-organisms convert N_2 to NH_3^+ generally has an important impact upon primary production. Rates of fixation in wetlands are, however, usually rather low (from 1.0 to 3.5 g $m^{-2}yr^{-1}$, Table 1.13). Exceptions may include rice fields, floodplains and wetlands such as the Everglades (Richardson pers. comm.) where cyanobacteria fix nitrogen. Some published estimates are considerably higher than those in Table 1.13; Whitney *et al.* (1981) estimated nearly 15 g^{-2} yr^{-1} for salt marshes in eastern North America. In general, however, it appears that the rates of denitrification exceed rates of fixation, so that wetlands can be thought of as sites where organic N arrives in run-off and detritus, being then returned to the atmosphere.

The principal organisms involved in nitrogen fixation are soil bacteria and cyanobacteria: they may either be free living, or in symbiotic associations with higher plants (Marschner 1995; Maseuth 1995). The free-living forms include cyanobacteria such as *Nostoc* and bacteria such as *Azotobacter* and *Clostridium*. Probably the best known are bacteria such as *Rhizobium* which form nodules on the roots of legumes; legumes, however, are rarely found in wetlands. In contrast, a group of filamentous bacteria known as actinomycetes forms nodules on the roots of some trees and shrubs associated with wetlands, notably the alders (*Alnus*) and wax myrtles (*Myrica*). *Rhizobium* is also associated with a family found in wetlands, the Ulmaceae. Finally, the cyanobacterium *Anabaena* often occurs in association with the floating water fern *Azolla*, and plays an important role in fixing nitrogen for rice paddies.

Denitrification, also carried out by micro-organisms living in anaerobic conditions, results in a loss of nitrogen from wetland soils, as NO^-_3 is converted back to N_2 or N_2O. These diffuse upward through the soil back into the atmosphere, with appreciable amounts also transported upwards by aerenchyma in rooted plants (Faulkner and Richardson 1989). Table 1.13 suggests that, in general, denitrification rates are slightly lower than fixation rates. In contrast, Bowden (1987) reports denitrification rates nearly an order of magnitude higher (30 gm^{-2} yr^{-1}), in which case the wetlands are efficiently transforming organic nitrogen to atmospheric nitrogen. Again, there are conspicuous exceptions, with Whitney *et al.* (1981) reporting 65 g $m^{-2}yr^{-1}$ in salt marshes of the eastern United States. More information on biogeochemical cycling of nitrogen and other elements can be found in Good *et al.* (1978), Rosswall (1983), Reddy and Patrick (1984), Mitsch and Gosselink (1986),

Faulkner and Richardson (1989), Armentano and Verhoeven (1990) and Gopal (1990).

Reduction of flood levels

Water levels in rivers change with time. In temperate zones, high water periods are associated with the melting of snow; in tropical areas, with rainy seasons. The properties of floodplains are a consequence of these seasonal floods, with floodplain species being dependent upon them (e.g. Junk *et al.* 1989). In Chapter 4 the many benefits of flooding will be examined. At the same time, the severity of flooding can be increased by human activities such as clearing of forests, building dykes to prevent water from escaping into low-lying areas, and draining wetlands associated with water courses (e.g. Barnard 1978; Klimas 1988; Miller and Nudds 1996). As a consequence, human development of watersheds often increases economic losses from floods. This is not a new problem, as illustrated by historical research into land drainage (Kelly 1975). When settlers moved into the deciduous forests of eastern North America, they first cleared forests in the soils most immediately useful for planting. Small wet patches could then be drained with ditches, and, as technology for drainage improved with the use of buried tiles, increasingly large areas of swamps could be undertaken. In southern Ontario, for example, large areas of swamps were under drained with tiles in the 1860s, thereby creating farmland described as 'first class lands . . . fit to produce any kind of crop'. But almost immediately these projects generated flooding in adjacent lower lands, and by 1873 a county council had petitioned the provincial legislature to set up a system of arbitration to settle disputes about flood damage (Kelly 1975).

Wetlands are now known to provide the function (or service) of floodwater retention: water may be stored within the substrate, as in peatlands, or if the wetland occupies a depression, then within the entire basin. Even floodplain wetlands (swamp, marsh in Figure 1.10), provide reduced flooding downstream by allowing flows to spread out over larger areas of landscape, thereby reducing both the velocity and the depth of discharge. When flood damages occur, the short-term measures taken to reduce flooding, such as constructing dikes, may exacerbate flooding in the long run by forcing floodwaters into increasingly narrow channels. Services such as floodwater retention are often ignored when decisions are made about land use within watersheds, and so Thibodeau and Ostro (1981) attempted to put an economic value upon development of 8500 acres of

Table 1.14. *Summary of the benefits of 1 acre of Charles River Wetland in New England*

Function	Low estimate of value	High estimate of value
Increases in land value		
Flood prevention	$33 370	$33 370
Local amenity	150	480
Pollution reduction		
Nutrients and BOD	$16 960	$16 960
Toxic substances	+	+
Water supply	$100 730	$100 730
Recreation and aesthetics		
Recreation	$2 145	$38 469
Subtotal	$153 000	$190 009
Preservation and research	+	+
Vicarious consumption and option demand	+	+
Undiscovered benefits	+	+
Total including visual–cultural benefits	$153 535 +	$190 009 +

Source: From Thibodeau and Ostro (1981).

marsh and wooded swamp in the Charles River basin in Massachusetts. The benefits from these wetlands were divided into categories including flood control, water supply, increases in nearby land value, pollution reduction, and recreation and aesthetics.

Flood control values were estimated by forecasting flood damage that would have occurred without wetlands. In one case, during a 1995 storm, the US Army corps of engineers estimated that the wetlands of the Charles River reduced peak river flows by 65% and delayed flooding over a period of 3 days after the actual storm. What property damage would have occurred if these wetlands had not been present? Thibodeau and Ostro estimate projected annual flood damage of nearly 18 million dollars, which translates into a value of about $2000 per acre of wetland (Table 1.14). An asset which yields $2000 in perpetuity has a present economic value of more than $33 000 per acre.

Many municipal wells in Massachusetts are located in, or adjacent to, wetlands. The value of the water supply function in wetlands can be calculated as the difference between the cost of wetland wells and the cost of providing water from the next best source. Continuing with American dollars and measuring system, an average acre of wetland can supply

100 000 gal per day at a cost of $7.44, which is $16.56 less than the cost of purchase from the regional agency which maintains reservoirs and sells water to cities. The difference of $16.56 per day works out to over $6000 per year, which, capitalized at 6%, exceeds $100 000 per acre.

Using flood control, water supply, and other values, Table 1.14 shows that the value of wetland exceeds some $US 150 000 per acre. 'Of course, a single private owner cannot capture most of these benefits. They are largely external benefits. It may well be to his economic advantage to fill the land, reaping its development value. When this happens, it is the town, the watershed, and the region which suffer the loss.' Thibodeau and Ostro are describing the Tragedy of the Commons (Hardin and Baden 1977), which, as Hardin (1968) first presented it for grazing communities, leads each citizen to make apparently rational decisions in their best short-term interest. Yet, when each individual in the community goes through the same decision-making process, and acts in this apparently rational manner, the result is destruction for the entire community. The property owner filling in the acre of wetland, the multinational logging executive felling the next tract of tropical forest, and the herdsman deciding to graze an additional animal upon the communal pasture, all are making a decision that produces short-term economic benefits to the individual or corporation, but which ultimately damages the larger community.

Each of the monetary estimates provided by Thibodeau and Ostro requires certain assumptions. In some circumstances these values may be lower, but in other cases, particularly in highly populated European landscapes, they may also be higher. The estimates do not, for example, include fisheries production, apart from its recreational value; yet some floodplains can produce 40–60 kg ha^{-1} yr^{-1} of fish (Welcomme 1976, 1986). Nor is carbon storage included as a service (Gorham 1991); on the global scale, carbon storage in the Charles River wetlands may be minor. Sustainable wood production from swamp forests is also omitted. The value of Thibodeau and Ostro's approach is that it attempts to provide a dollar value on many of the functions of wetlands that would otherwise be taken for granted, and they illustrate one approach to measuring the value of wetland functions (see also Larson 1990).

There is more that could be said about every one of these topics. But this is, after all, only the introduction. Chapters 2 and 3 will examine in more detail two fundamental properties of wetlands: the distribution of wetlands (zonation) and the relative abundance of species (diversity). Then we will explore the environmental factors that control such properties. These environmental factors are arranged in order of their

Figure 1.24 General aspect of Pantanal region. 1. Extensive area of seasonally inundated cerrado with forest islands interspersed. 2. Forested area showing gallery forest in foreground and semideciduous forest in background. 3. Area of flooded savanna and gallery forest. 4. A typical aquatic habitat showing a pond filled with *Eichhornia* and *Pontederia*. 5. Sandy margin of Rio Negro showing *Vochysia divergens* Pohl a tree characteristic of seasonally flooded areas. 6. A seasonally flooded savanna showing cattle which roam throughout the region in all upland and seasonally flooded areas (courtesy of Prance and Schaller 1982).

presumed decreasing importance in controlling wetlands: hydrology (Chapter 4), fertility (Chapter 5), competition (Chapter 6), disturbance (Chapter 7), herbivory (Chapter 8) and burial (Chapter 9). Chapter 10 introduces assembly rules as a general framework for exploring wetland pattern and process. Chapter 11 returns to wetland function and examines the possibility of simplifying research by assigning wetland species to functional groups. Finally, Chapter 12 discusses paths into the future: conservation, management and research strategies.

The Pantanal

To complete this brief introduction to wetlands, let us turn to one of the world's largest and least known wetlands, the Pantanal, an enormous floodplain (140 000 km^2 – roughly the size of England) near the headwaters of the Parana River basin in South America, a river which eventually enters the ocean at Buenos Aires. The word Pantanal is the Portuguese word for swampland, although much of it is actually marsh. Much of this vast flat area is a seasonally flooded grassland where water levels can rise and fall many metres over 1 year; rivers are lined with gallery forests and more elevated areas have semideciduous forests (Figure 1.24). As a consequence, aquatic and dryland vegetation (including cacti) are interspersed (Prance and Schaller 1982; Alho *et al.* 1988; Junk 1993). Seven types of wetland vegetation can be found, depending upon flooding regimes (Neiff 1986). Cattail swamps dominated by *Typha latifolia, T. dominigensis* and *Cyperus giganteus*, are inundated only by occasional floods (5–10 years), whereas annually flooded wetlands ('banados') have *Polygonum stelligerum, P. acuminatum* and *Ludwigia peploides*. Dominant genera overall include *Typha, Echinochloa, Panicum, Cyperus* and *Polygonum*. Further, a number of terrestrial vegetation types meet here: the cerrado of Central Brazil on the East, the semideciduous Amazonian forest to the north west and the chaco-like forest of Bolivia to the south-west. The wildlife is correspondingly rich, including jaguar, ocelot, giant anteater, giant otter, giant armadillo, crab-eating fox, pampas deer and swamp deer, all threatened or endangered (Alho *et al.* 1988). There are 13 species of herons and egrets, 5 species of kingfishers, and 19 species of parrots. More than 540 species of fish have been recorded for the river systems as a whole. The richest assemblages of fish occur in the Pantanal and the Upper Paraguay (Bonetto 1986) and their life cycles are closely tied to fluctuating water levels (Lowe-McConnell 1975).

The threats posed to this wetland, and the management problems that arise, typify the perils facing many of the world's wetlands. The list of threats to the Pantanal is ominously long. Ranching has traditionally been one of the major activities. While ranchers of the old *pantaneiros* moved their cattle in accord with rising and falling water levels, modern subdivision of farms and intensification of grazing are increasingly detrimental to the native plant communities. Prance and Schaller (1982) conclude that, while 'Little vegetation remains intact since virtually all has been modified by cattle, fire, man or a combination of all three' at the same time, '[t]he natural and interesting vegetation of the Pantanal can be maintained together with the ranches provided that the cattle population is not increased greatly, and that areas are set aside as reserves for conservation of the flora and fauna.' Poaching is another threat. Alho *et al.* (1988) report that one 2500 kg shipment of hides contained 70 000 individual skins including jaguars, maned wolves, caimans and snakes; the smugglers admitted that this shipment represented only 13% of all the skins (more than a half a million animals) which had been sent to Germany in the preceding 6 months. The added effects of removing large predators from this ecosystem are unknown. Other impacts include large-scale agriculture, alcohol distillation and deforestation (Alho *et al.* 1988; Junk 1993). Large-scale agriculture not only removes the vegetation, but also contaminates the river with herbicides and pesticides. Alcohol distilleries that produce biomass fuels add to the contaminant load in rivers. Deforestation continues, its rate accelerated by illegal sawmills and fires set by ranchers. On top of all of these growing problems, the five countries comprising the La Plata Basin have proposed the Hidrovia Project (Bucher *et al.* 1993), a scheme that will use dredging and canals to create a 3440 km long waterway starting at Puerto Cacéres in Brazil and ending at Nueva Palmira in Uruguay. The 1670 km of this project going straight through the heart of the Pantanal would be bound to cause major changes in hydrology.

Without a general appreciation of the factors that create and maintain wetlands, and lacking clear predictions for the effects of various human pressures, we end up only with this depressing list of threats and no obvious course of action. Creating reserves, for example, sounds promising, but if canals reduce annual flooding, then the wetlands in the reserve may dry out. This book therefore begins with inquiry into some basic features of wetlands such as the Pantanal, moves to an exploration of the handful of key environmental factors, and then returns to the challenging problem of how to manage wetlands wisely.

2 · *Zonation and succession: shorelines as a prism*

Given the remarkable diversity of wetland types, and the complexities of their processes and composition, how are we to begin scientific study? This is an important issue not only for those of us concerned with wetland ecology, but also for practitioners of ecology in general. Where and how do we start? One is reminded of the old Buddhist story about the blind scholars and the elephant. Asked to describe the elephant, the first scholar, touching the massive side, states 'It is like a wall'. The second scholar, holding the tail says 'No, it is like a piece of rope'. The third, holding the trunk, insists, 'You're both wrong. It is a kind of snake'. And so on. We constantly risk that scholarly understanding of the phenomenon will be distorted by our starting point, or by our own limited frames of reference. Yet, we must start somewhere. And the author must shoulder the responsibility for deciding whether it is best to start with the wall, the rope, or the snake. For this very reason, Chapter 1 tried to give a general sketch of the entire elephant.

Ecology is generally defined as 'the scientific study of the distribution and abundance of organisms'. Where better to start, then, with the two topics of distribution along gradients (zonation) and relative abundance of species (diversity)? This first focuses attention upon two key properties of wetlands. The next step is to dissect the environmental factors that affect those properties. That is, I begin with effects (patterns that can be observed) and only then move on to causes (which often require a mixture of inference and experimentation).

There is an added advantage to beginning with the topic of zonation. Gradients function like prisms. Prisms take ordinary light and spread it out into a spectrum for scientific study; a gradient does the same for a complicated ecological community. This spectrum provides us with a pattern we can study. Such patterns are necessary for initiation of scientific inquiry, and zonation provides ready-made patterns. There is a long-established habit of describing wetlands by sketching zonation patterns (Figure 2.1) and as the sketches remind us, zonation patterns summarize much of the spatial variation in wetlands. Further, many of our conceptual models in ecology are built around gradients and the distribution of species along them. The oft-repeated admonition by my PhD supervisor, Chris Pielou, was that ecologists should stop trying to find imaginary homogenous habitats and use the gradients that nature has provided. For all of these reasons, much of my own wetland work has involved the study of gradients, and one is also inclined to begin a difficult problem with familiar terrain.

A cost of beginning with zonation is that it may frustrate those who wish to leap immediately into topics like diversity or eutrophication. Others may feel, in spite of my argument, that shorelines seem to have little to do with wetlands; in my experience, a surprising number of people do not seem to connect shoreline zonation with wetlands at all. And then there is the risk that zonation may be seen as a plant phenomenon, thereby frustrating many zoologists. Finally, although we have many drawings of zonation, those more experimentally inclined may be frustrated at how little is actually known about mechanisms of zonation. At this point, an author is likely to fear alienating all but a handful of readers.

All the same, the prism analogy argues for beginning this enquiry with an examination of those situations in nature where wetlands are zoned. A shoreline can be thought of as a natural experiment (*sensu* Diamond 1983) where nature has set up a pattern of variation for us to investigate. Most zonation patterns summarize the consequences of differences in water level, from flood plains in Brazil (e.g. Junk 1986) to temperate zone peatlands in Asia (e.g. Yabe and Onimaru 1997) and from salt marshes in Spain (e.g. García *et al.* 1993) to lakeshores in Africa (Denny 1993b). A wetland spread out along a shoreline is not only like a spectrum, it may also be compared to a cadaver neatly spread out and already partly dissected in order to help a struggling medical student. Like medical students, we may find it useful to begin where nature has given a helping hand rather than leaping immediately into unattended surgery. For those eager to proceed to later topics, pp. 82–102 alone may provide an adequate coverage of zonation. Pages 102–123 go into further depth on mechanisms and patterns.

Shorelines as a model system for the study of wetlands

Vegetation on shorelines is closely connected with water levels (e.g. Pearsall 1920; Gorham 1957; Hutchinson 1975). The result is conspicuous

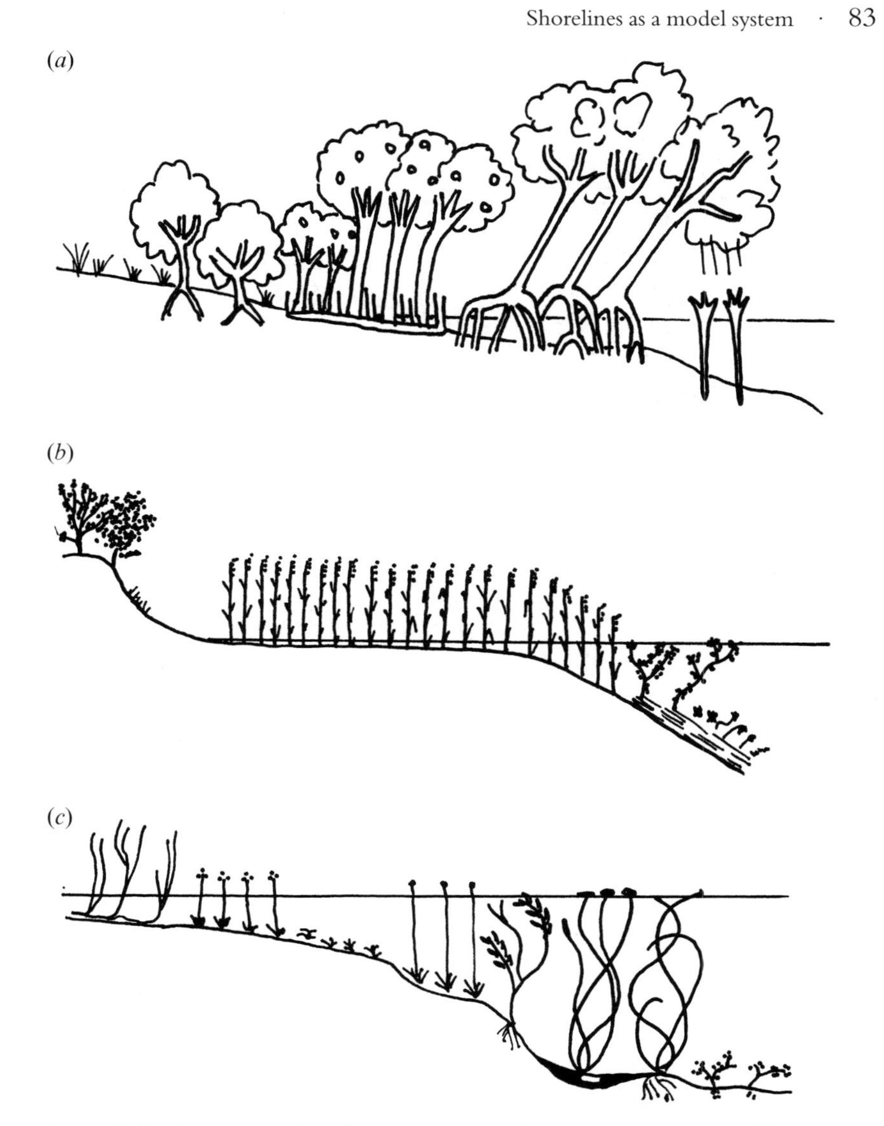

Figure 2.1 Some examples of plant zonation. (*a*) A mangrove swamp of the Caribbean (after Bacon 1978); (*b*) The eastern shore of Lake Kisajno, north-eastern Poland, a typical small-lake phytolittoral (after Bernatowicz and Zachwieja 1966); (*c*) A sandy shoreline (after Dansereau 1959).

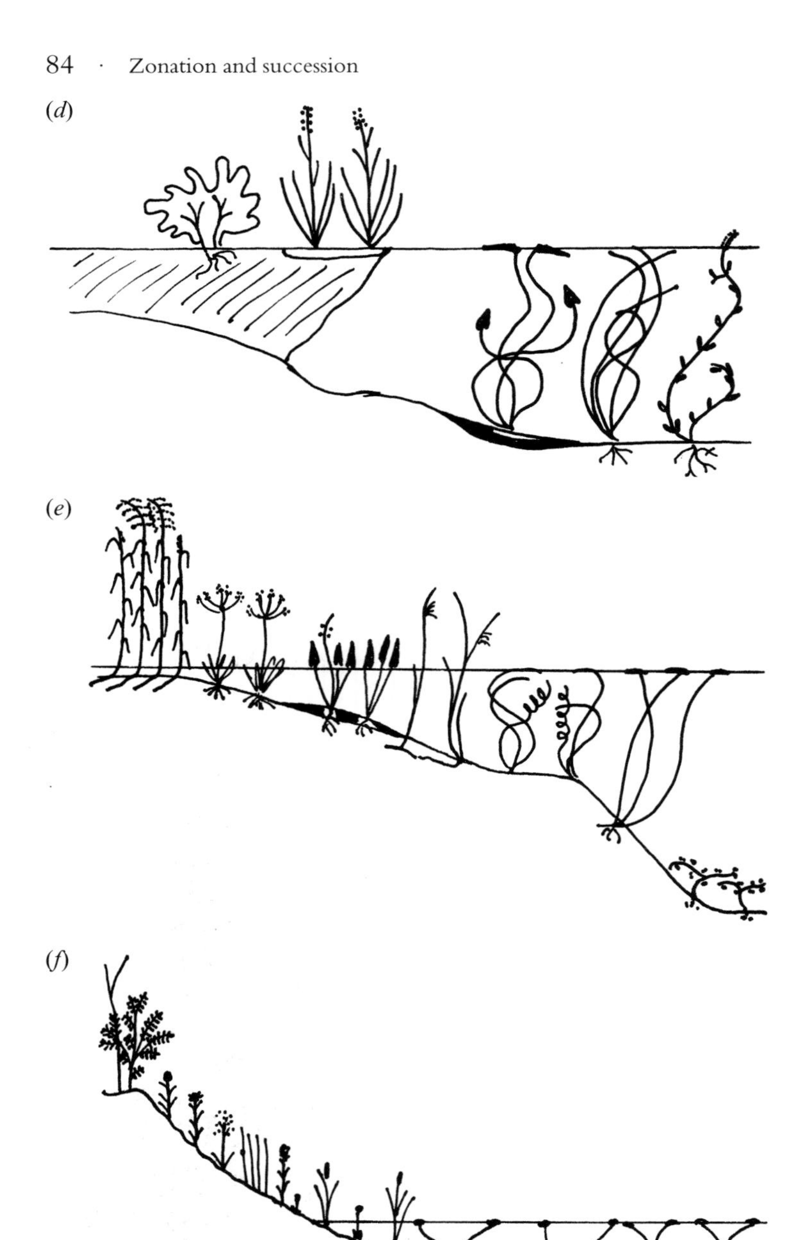

Figure 2.1 (cont.) (*d*) A bog (after Dansereau 1959); (*e*) The St Lawrence River (after Dansereau 1959); (*f*) Wilson's Lake, Nova Scotia (after Wisheu and Keddy 1989 b).

Table 2.1. *Large rivers and lakes provide extensive areas of shoreline with zoned plant and animal communities*

Ten largest rivers of the world	Country	Average annual discharge at mouth (m^3/s)
Amazon	Brazil	180000
Congo	Zaire/Angola	42000
Yangtze Kiang	China	35000
Orinoco	Venezuela	28000
Brahmaputra	Bangladesh	20000
Yenisei	USSR	19600
Rio de la Plata	Argentina/Uruguay	19500
Mississippi–Missouri	USA	17545
Lena	USSR	16400
Mekong	Vietnam	15900
Ten largest lakes in the world	**Country**	**Surface area (km^2)**
Caspian Sea	USSR	371000
Lake Superior	USA, Canada	83300
Lake Victoria	Uganda, Kenya, Tanzania	68800
Lake Aral	USSR	66458
Lake Huron	USA, Canada	59570
Lake Michigan	USA	57016
Lake Tanganyika	Zaire, Burundi, Zambia, Tanzania	34000
Great Bear Lake	Canada	31792
Lake Baikal	USSR	31500
Lake Nyasa	Malawi	30500

Source: After Czaya (1983).

zonation. The large rivers and lakes of the world provide extensive areas of such shoreline habitat (Table 2.1). It is natural that our first reaction to zonation is to pull out a field note book and make a sketch of it. Recall the six sketches of zonation patterns from studies in different parts of the world in Figure 2.1. One gains the impression that some wetland ecologists still think that, once a sketch of plants has been made, the scientific work is done. In truth, it has barely begun. Primary production, for example, varies among these zones reaching a maximum in shallow water emergent macrophytes (Figure 2.2). The distribution of animals is, in turn, related to the zonation of wetland plants (Figure 2.3).

Zonation of animals in wetlands has received less attention perhaps because animals are less visible and more mobile. But, we might expect

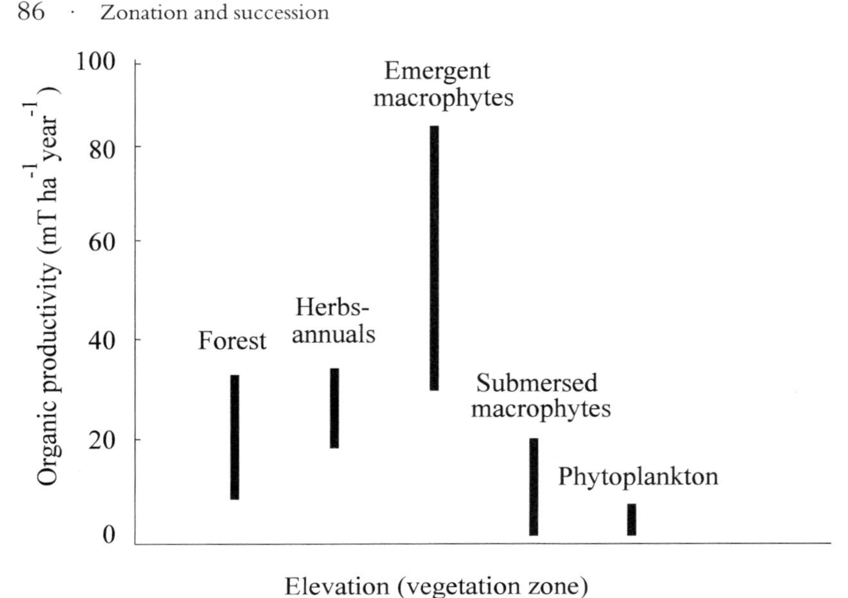

Figure 2.2 Changes in primary production with water level (after Wetzel 1989).

similar sorts of patterns elsewhere, if only because flooding can directly change food supplies, or indirectly change the habitat by changing the vegetation. For example, Price (1980) has documented zonation patterns of eleven species of foraminefera in salt marshes. Arnold and Frytzell (1990) found that flooding was an important factor in predicting the distribution of mink, with a strong tendency for them to select large semi-permanent and permanent wetlands with high water levels and irregular shorelines. The distribution of breeding birds also shows zonation, with species showing marked preference for certain vegetation zones in both freshwater marshes (Prince *et al.* 1992; Prince and Flegel 1995) and salt marshes (Weller 1994b).

To study zonation further, it may be helpful to recognize four vegetation types, and four plant growth forms, in relation to water level (Figure 2.4). Highest on the shore are wooded wetlands. These are only flooded for short periods of time each year and are dominated by trees and shrubs. At lower elevations with more flooding, woody plants give way to wet meadows. Although wet meadows are flooded for much of the year, they are uncovered for several months in each growing season, and so are occupied by plants that show only minimal modification to cope with flooding. As flooding increases further, wet meadows give way to emergent marsh, with plants that grow under flooded conditions; these sites may only be uncovered for a short time during drought periods, and as a result, show increasing morphological adaptation to flooding. Linear leaves and aerenchyma become conspicuous. Below this, plants occur that are truly aquatic, many with floating leaves.

Even peatlands, which sometimes seem to operate differently from marshes and swamps, are zoned in a similar manner. Both bryophytes and vascular plants change along elevation gradients (Vitt and Slack 1975, 1984), with the bryophytes being more sensitive to the water table than most vascular plants (Bubier 1995). Pools of water have herbaceous aquatic plants, and shallow depressions support emergent sedges. At higher elevations, shrubs become increasingly dominant, and if the peat is high enough above the water table, woody plants occur (Dansereau and Segadas-Vianna 1952; Gorham 1953; Glaser *et al.* 1990; Bubier 1995).

It will often be both convenient and possible to combine wetlands from northern fens to salt marshes to tropical floodplains to explore similarities in pattern and process. In other cases, it will be necessary to discriminate among kinds of wetlands. One of the most daunting aspects of ecology is the difficulty in discriminating between those situations where most is learned by the study of similarities, and those where most is learned by making distinctions. Since it is generally more difficult to take things apart than to put them back together (ask any amateur mechanic), I will deliberately err, if at all, on the side of emphasizing similarities.

Mechanisms of zonation

Ecological succession

Many interpretations of zonation emphasize that plant communities in Figure 2.1 typically appear to follow a temporal trend, the sequence of events that would occur as a wetland gradually filled in with detritus and turned into land. 'Zonation, therefore, is taken to be the spatial equivalent of succession in time, even in the absence of direct evidence of change' (Hutchinson 1975, p 497.) This view that zonation patterns are a profile through a successional sequence is widespread: it has been described for peatlands (e.g. Dansereau and Segadas-Vianna 1952) and small marshes along lakes (e.g. Pearsall 1920; Spence 1982). In all these circumstances, organic matter produced by the wetland, combined in some cases with sediment trapped by the vegetation, gradually increases the elevation of the substrate, turning shallow water into marsh, and marsh into land. Gorham (1953) traces this view back to at least the early 1800s. J. A. De

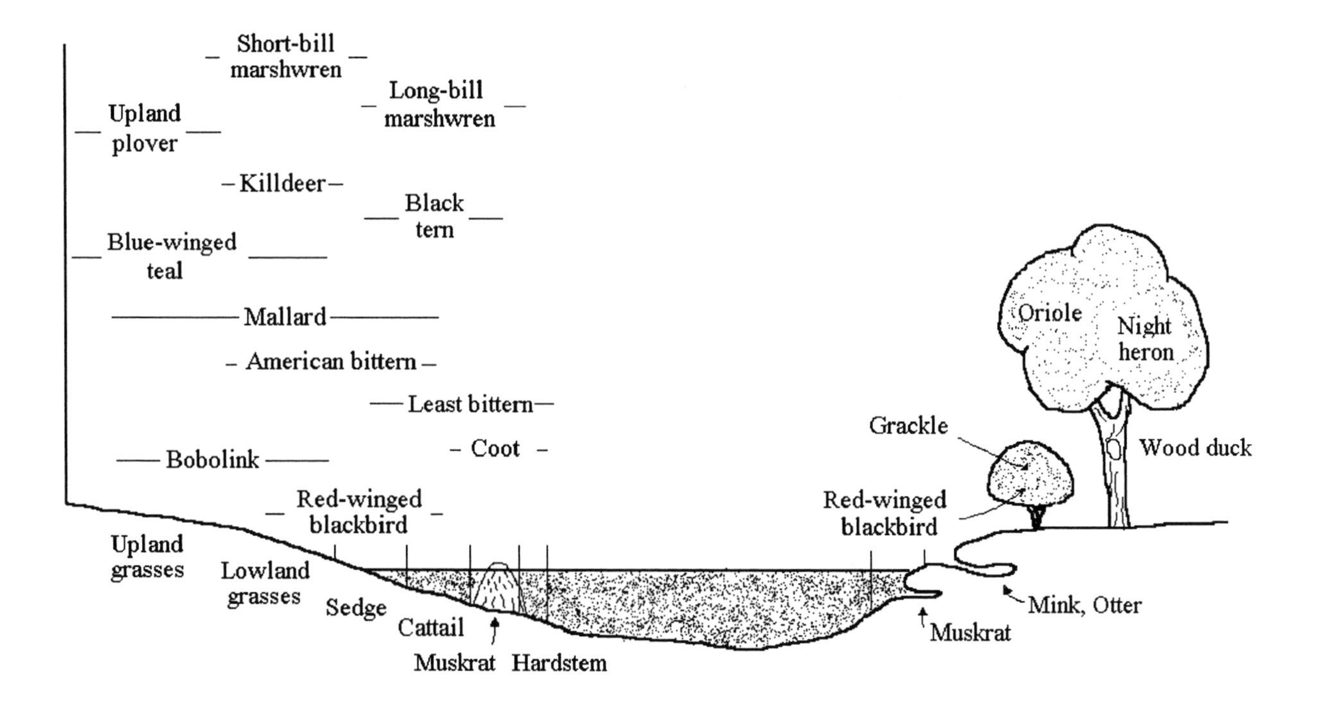
Short-bill marshwren
Upland plover
Long-bill marshwren
Killdeer
Black tern
Blue-winged teal
Mallard
American bittern
Least bittern
Coot
Bobolink
Red-winged blackbird
Upland grasses
Lowland grasses
Sedge
Cattail
Muskrat
Hardstem
Oriole
Night heron
Grackle
Wood duck
Red-winged blackbird
Mink, Otter
Muskrat

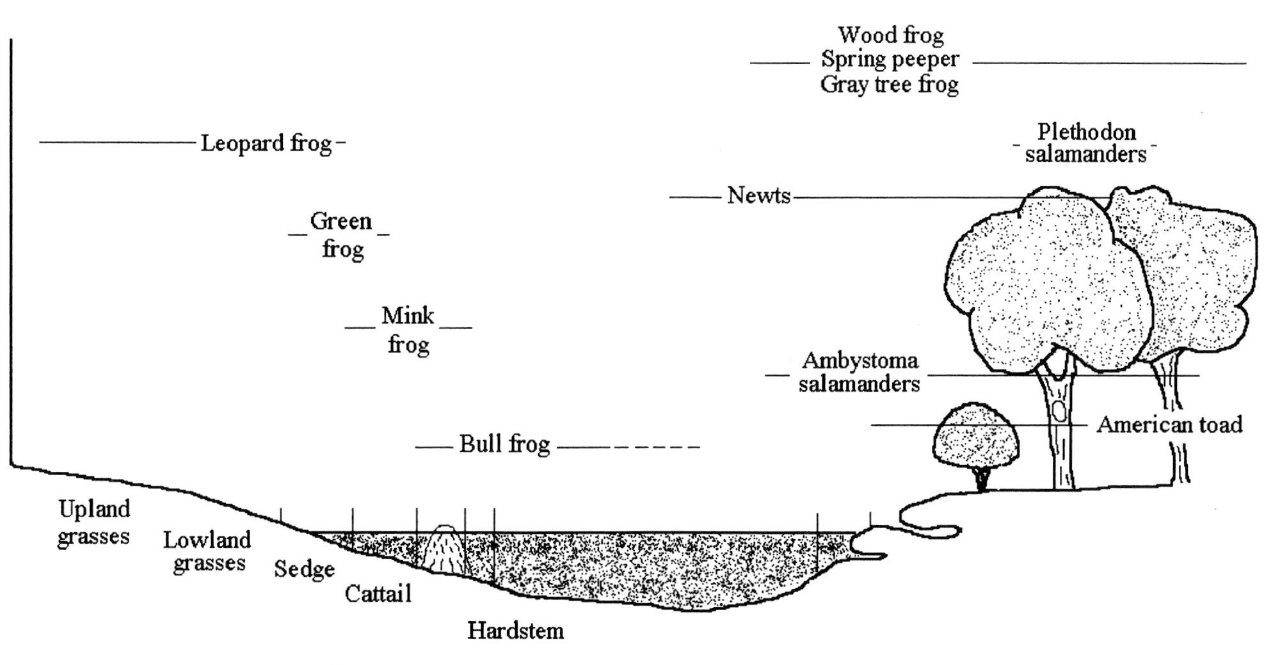

Figure 2.3 Zonation in some birds and mammals (*top*, after Weller 1994a) and amphibians (*bottom*) in relation to water level and vegetation.

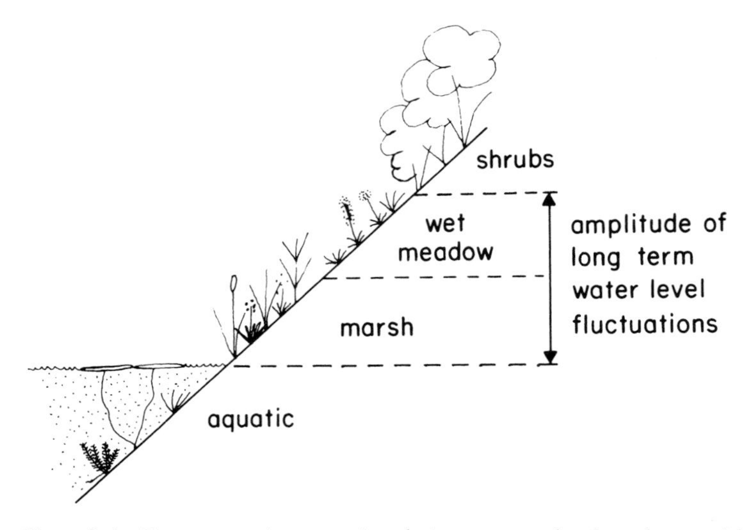

Figure 2.4 Four vegetation types in relation to water level on shores with fluctuating water levels (from Keddy 1991b).

Luc's book, *Geologic Travels*, published in 1810, recognized six discrete stages in the transformation of a lake into a peaty meadowland. Further, De Luc proposed that the rate of succession is greatest on shallow shores; on steep shores the vegetation zones are narrow and the process of change through time nearly non-existent. Walker (1970) also draws attention to Gough's account in 1793 of how lakes are converted to dry land by the accumulation of organic matter, so that 'the margin of the pond will be progressively advanced' and the land thereby produced 'will, in time, be covered with a bed of vegetable earth', the upper limit of which is set by dry periods because exposure to air will allow decomposition. Such observations were systematized as a successional sequence called a hydrosere by Tansley (1939). (Walker has thoroughly studied vegetation transitions in hydroseres using a series of 20 sediment cores, but we will defer examining his work on succession until Chapter 9.) The role of succession in all plant communities was emphasized and systematized by Clements (e.g. Clements 1916; Clements *et al.* 1929), and McIntosh (1985) has outlined both the development of his views and the development of opposing interpretations of ecological change. These are beyond the terrain of this book, except where wetland zonation is concerned.

The trend in wetland studies appears clear. As the concept of ecological succession was popularized with the growth of ecology in the mid-

1900s, the idea that ecological zonation was an example of succession became widely accepted and 'pond zonation' was frequently presented as 'pond succession' in introductory ecology texts. Zonation and succession may be closely linked in circumstances such as small ponds and peatlands, where organic matter accumulates, but even De Luc apparently understood that his generalization did not apply to the steep shores of lakes. With the explosion of ecological studies in the later 1900s, it became clearer that there are many natural forces that delay, or even restart such successional sequences. As the effects of fires, floods, storms, and droughts were better documented, many 'temporal' successional sequences could perhaps be better understood as dynamic balances between successional and disturbance (e.g. Pickett and White 1985). Simultaneously, population biologists were placing increasing emphasis on the mechanistic interactions among species, leading Horn (1976) to suggest that succession was best understood as a 'statistical result of a plant-by-plant replacement process'. The importance of disturbance, and the complexity of responses to it, has challenged many of the standard ideas about succession, stability and predictability in nature (Botkin 1990).

Along with the gradual shift towards the study of dynamics of populations in patches and landscapes, the ubiquity of buried reserves of seeds was becoming evident. Rather than buried seeds being mere detrital accumulations, it became apparent that, in many cases, disturbance was the trigger to re-emergence of species from pools of their buried seeds (e.g. Harper 1977). Charles Darwin himself had commented on the remarkable number of seedlings which emerged from a spoonful of mud, and increasingly, ponds and potholes were found to be vast repositories of buried seeds (e.g. Salisbury 1970; van der Valk and Davis 1976, 1978). This led van der Valk to propose that many zonation patterns were not successional sequences, but rather represented short-term responses of plant communities to local changes in the environment.

There has therefore been a progression of views on zonation, the past ones emphasizing long-term unidirectional succession, the later ones emphasizing the short-term responses of organisms to changing environmental conditions. Two specific examples nicely illustrate this shift in emphasis. In 1952, Dansereau and Segadas-Vianna could draw zonation profiles of peat bogs in eastern North America, and confidently relate them to succession diagrams ending in climax vegetation of *Picea mariana* or *Acer saccharinum* (these being named the *Pictum marianae* and *Aceretum saccharophori* associations). The many other vegetation types they named were considered to belong to one of three stages of bog succession:

pioneer, consolidation and subclimax, all leading, by means of peat accumulation, from open water to woodland. In 1996 Yu *et al.* described zonation through a single shoreline swamp at Rice Lake (just north of Lake Ontario), with a zonation sequence not unlike that discussed by Dansereau and Segadas-Vianna. Aided by sediment cores, and by studies of pollen and plant macrofossils, Yu *et al.* found two main stages in the vegetation history. An open marsh stage with sedge genera such as *Carex* and *Eleocharis* persisted for some 2700 years with no successional change, a situation they explain by fluctuating water levels. Then, about 8300 years BP, there was a transition to perennials associated with wet meadows (e.g. *Verbena hastata, Lycopus americanus* and *Carex* spp.) and by about 7500 years BP a transition to cedar (*Thuja occidentalis*) swamp. This change coincided with a regional period of warm and dry climate. The adjoining lake levels dropped and the swamp went dry about 1000 years later. During a cooler and wetter period, lake levels rose and the cedar swamp reappeared. Yu *et al.* conclude 'Paleoecological data for the past 11 000 years show that there were no significant successional changes of marsh communities for about 2700 years. . . . When change did occur, it was ultimately controlled by allogenic [external] factors such as climate and water level changes.' Further, when the climate changed 'The herbaceous marsh converted directly to cedar swamp without the shrub-marsh and(or) alder-thicket stages.'

These two studies illustrate the change in perspective that has occurred over the last 50 years. It would be far too easy, however, to merely conclude (as is often done) that a correct view (dynamics) has replace a wrong one (succession). Even Dansereau and Segadas-Vianna noted that fluctuating water levels could control vegetation succession, and that fire could cause vegetation to regress. And Yu *et al.* would have to concede that, at Rice Lake, trees are now growing in accumulations of 2 metres of peat and several more of organic silt, thereby elevating the swamp forest above what would otherwise be open water. Whether one therefore focuses upon succession or on short-term dynamics would appear to be somewhat a matter of emphasis and perspective: general patterns as opposed to site histories, large-scale processes as opposed to small-scale dynamics, and classification as opposed to process.

These developments leave us with two questions about zonation. What purpose or advantage is there in relating zonation to succession? If there is some value, then in what circumstances is this a correct parallel to draw? For the purposes of this book, I assume that viewing zonation as succession may be useful in habitats such as peatlands, where unidirectional change driven by the accumulation of peat is a powerful and useful generalization. In other cases, such as the shores of large lakes and rivers, the connection between zonation and succession is weak, and if anything, confuses rather than clarifies the causes of patterns seen in the vegetation.

Physical factors

The direct effects of physical factors have also been offered as explanation for zonation on shorelines and for the distribution patterns of most plants. Pearsall (1920) studied plant patterns in wetlands around English lakes, and concluded (p. 181) 'sediments become finer as water deepens; since sediments are zoned along lake shores and since they differ in chemical composition, we are justified in assuming that zonation of vegetation is a result of differences in soil conditions.' He placed particular emphasis upon soil organic matter, and silt and clay content. Spence's review (1982) added in some extra factors, such as lower light levels in deeper water, but still assumed as a first approximation, at least, that physical factors themselves produced the different distributions of species.

Further afield, Myers (1935) described the kinds of zonation he encountered along water courses in north eastern South America, and offered the explanation of physical factors as controls upon the different kinds of shoreline vegetation. Near the sea, he said, the zonation consisted of mangroves such as *Rhizophora mangle*, which mixed with and then, as the water freshens, gradually gave way to *Pterocarpus draco*. 'The distance to which the mangrove zone extends upstream is doubtless determined by the influence of brackish water, and this, in its turn, in these uniformly sluggish streams . . . depends chiefly on the size of the river.' The sequence from the ocean inland went as follows: (i) *Rhizophora*, (ii) *Pterocarpus* (often mixed with *Pachira aquatica*), (iii) mixed bank vegetation 'smothered by a dense curtain of creepers', (iv) swamp forest with no differentiated bank fringe, (v) tall rain forests with no differentiated bank fringe. Myers was of the opinion that the kind of zonation he saw could be explained by three main causes, the width of the stream, the character of the water and distance from the sea.

Since then, there has been increased sophistication in the study of plant response to flooding. As we have already seen (Chapter 1), flooding is associated with low soil oxygen levels. Below 1% oxygen, the aerobic metabolism of the plant is superseded by the glycolytic pathway, and the products of anaerobic metabolism accumulate (Crawford 1982). Plants that are intolerant of flooding then begin to accumulate ethanol in their

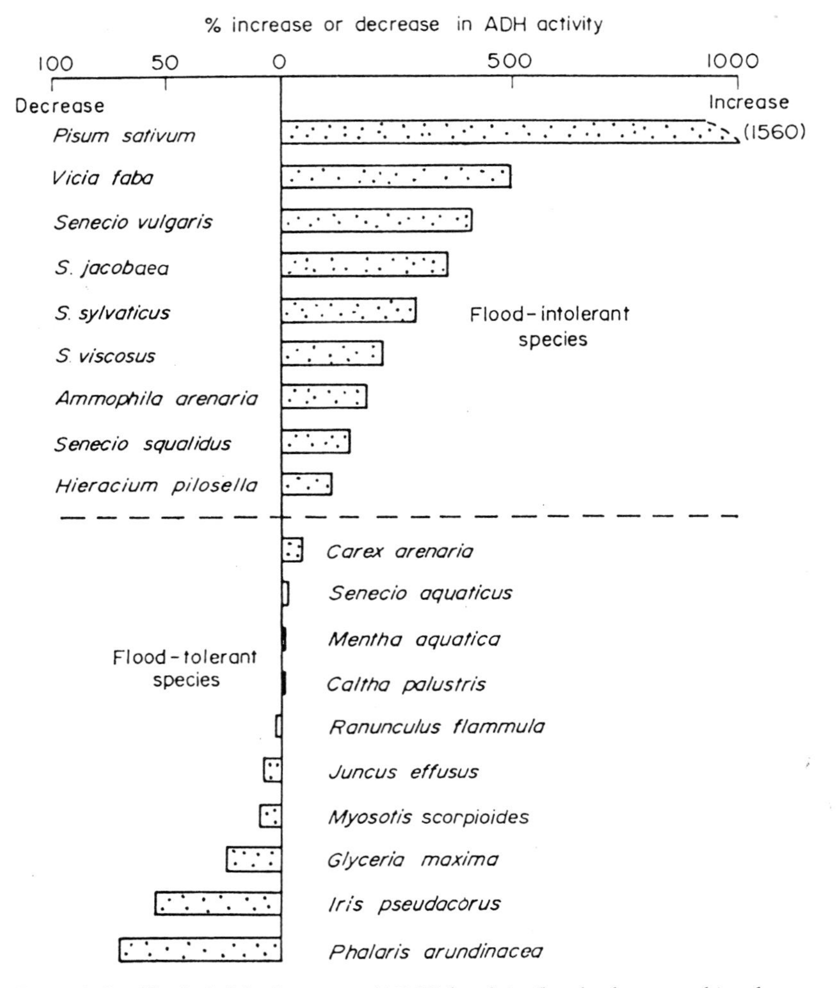

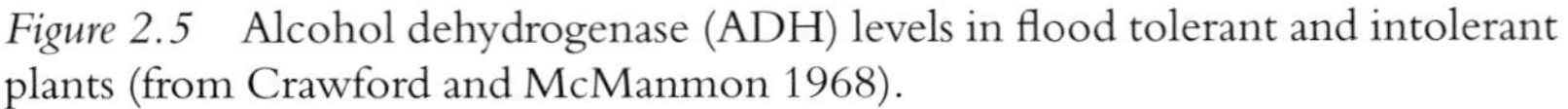

Figure 2.5 Alcohol dehydrogenase (ADH) levels in flood tolerant and intolerant plants (from Crawford and McMannon 1968).

roots. Associated with this is an increase in alcohol dehydrogenase activity (Figure 2.5). Physiological mechanisms to deal with anaerobic conditions appear to be widespread, as illustrated by some of the many molecules produced in flooded conditions by multicellular plants and animals (Table 2.2). The addition of unicellular life forms would probably lengthen this list considerably.

In salt marshes and mangal, the constraints of flooding are compounded by increased salinity (Figure 2.6). The growth of salt marshes through the

Table 2.2. *Substances reported to accumulate under anaerobic conditions in higher plants and animals*

Substance	Animal occurrence	Plant occurrence
Lactic acid	Vertebrate skeletal muscle	Germinating seeds, tubers
Pyruvic acid	Vertebrate skeletal muscle	Willow roots
Formic acid	Parasitic helminths	–
Acetic acid	Bivalve molluscs, cestodes	–
Acetoin	Nematodes	–
Propionic acid	Molluscs, cestodes	–
Butyric acid	Parasitic protozoa	–
Succinic acid	Bivalve molluscs	Seeds
Malic acid	–	Roots of marsh plants
Shikinic acid	–	Iris and water lily roots
Glycolic acid	–	Willow roots
Ethanol	Parasitic protozoa, helminths	Flood-intolerant roots and seeds
Sorbitol	insects	–
Glycerol	insect	Alder roots
Alanine	Sea turtles, molluscs	Flood-tolerant roots
Aspartic acid	Marine annelids	Flood-tolerant roots
Glutamic acid	Marine annelids	Flood-tolerant roots
Serine	–	Flood-tolerant roots
Proline	–	Flood-tolerant roots
Octopine	Cephalopods	Crown-gall tissues
γ-aminobutyric acid	–	Tomato roots and radish leaves
Methyl butyrate	Parasitic nematodes	–
Methyl valerate	Parasitic nematodes	–
Glycerophosphate	insects	–
Hydrogen	Parasitic protozoa	–
Ethylene	–	Roots and fruits

Source: After Crawford (1982).

accumulation of peat and silt has been described by Frey and Basan (1978) and Niering and Warren (1980). Figure 2.7 illustrates some possible interrelationships of environmental factors controlling zonation. The salinity of the soil water has received particular attention from ecologists. A detailed consideration of the effects of salinity upon plant distribution and zonation is beyond the topics considered here, but can be found in standard reference works (e.g. Chapman 1974; Poljakoff-Mayber and Gale 1975; Tomlinson 1986; Adam 1990). One might naively begin an enquiry into salt marsh plant physiology by assuming that roots somehow pump

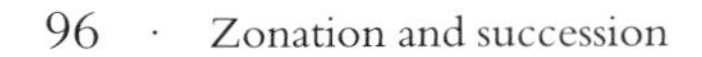

Figure 2.6 Inundation with saline water produces distinctive wetland communities (*left*, brackish marsh in St Lawrence River; *right*, salt marsh Petpeswick Inlet, Nova Scotia). The zonation of species in these habitats is a response to both flooding and salinity

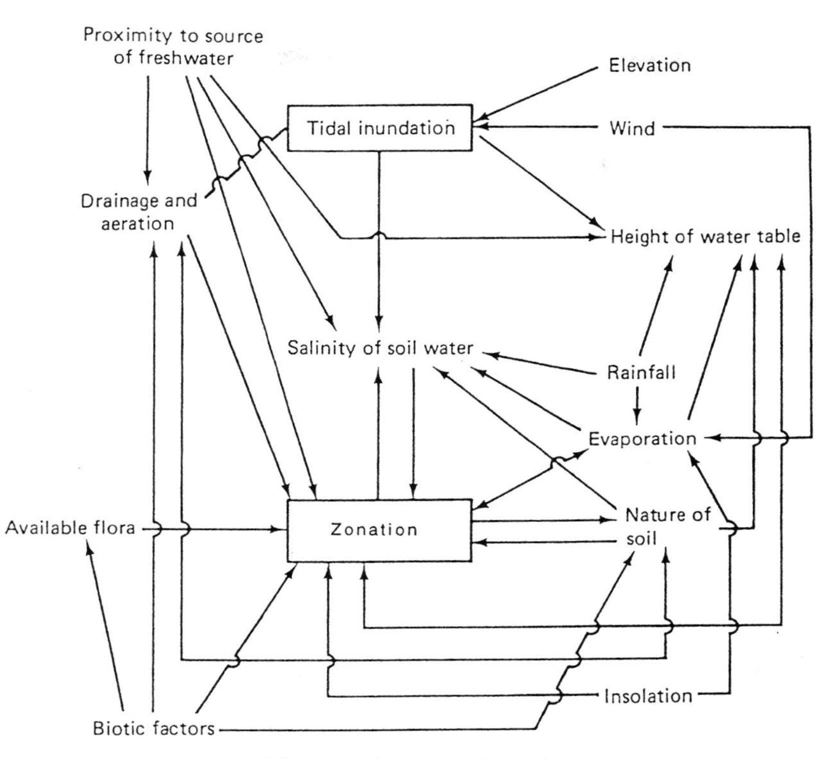

Figure 2.7 Environmental factors influencing saltmarsh zonation (modified from Adam, 1990 and Clarke and Hannon, 1969).

water out of the soil, but in fact, nearly the reverse is happening (Canny 1998). Evapotranspiration creates osmotic gradients within plant tissues. The water deficit is transmitted down the plant through the xylem, thereby causing water to diffuse into roots (Salisbury and Ross 1988). The less water in the soil, or the greater the salinity, the stronger this osmotic gradient must be to extract water from the soil. One can measure these water deficits in photosynthetic tissues using a pressure bomb (Scholander *et al.* 1965), which yields readings of xylem tension in megapascals. Plants growing in salt water have much more negative tension in their xylem (Figure 2.8); presumably this directly reflects the difficulty in withdrawing water from saline solutions. Root pressure also occurs, as does tissue pressure from the adjacent phloem, but their primary role appears to be maintaining and repairing water transport caused by evapotranspiration (Canny 1998). It is tempting to assume therefore, that the distributions of species in zoned wetlands (whether fresh water or salt water) are directly a

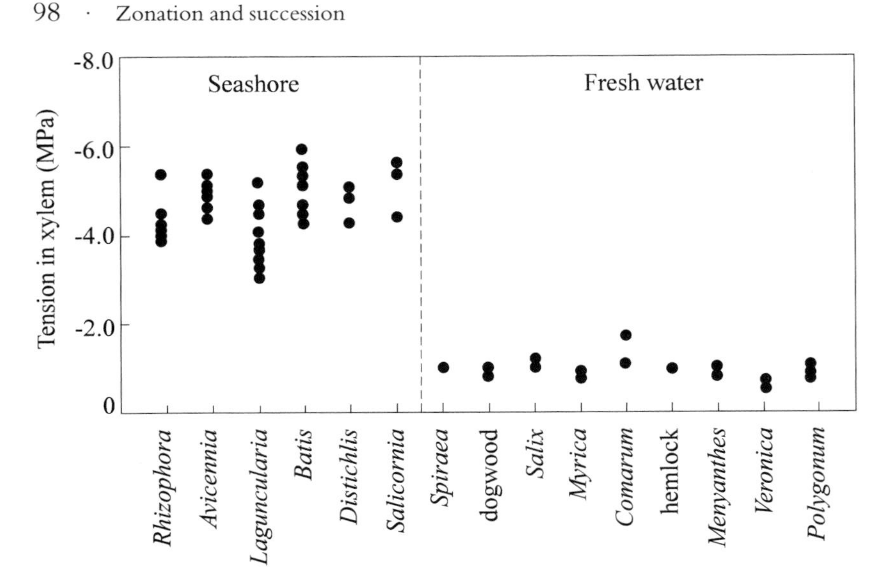

Figure 2.8 Xylem tension in plants of two contrasting wetland habitats (after Scholander *et al.* 1965).

consequence of their abilities to cope with physical constraints imposed by flooding and salinity.

Biological interactions

There is an important and frequently overlooked problem with the foregoing physiological explanations based upon plant responses to physical factors: while they demonstrate how organisms can respond to flooding, or salinity, they rarely take the next step and demonstrate a correlation between physiological properties and zonation in the field. For example, the often encountered Figure 2.5 requires us to assume that there is a relationship between ADH and flooding, but does not test for it. Even if this next step were taken, it still does not demonstrate that such processes cause zonation. One could, for example, entirely reverse the suggested cause and effect, and argue that some completely independent factor such as grazing produces the patterns, and that the observed physiological properties are simply responses to the conditions that are present. That is, while few would deny that physiological properties of organisms represent evolutionary responses to habitat, they still do not show us whether the ecological processes cause zonation or merely are responses to it. Field experiments are required for this.

More recently therefore, attention has shifted to experiments which have investigated how biological interactions can produce zonation. One obvious interaction is competition. If one observes the zonation that occurs on disturbed shorelines, say where a boat launch or trail enters the water, one can often find marsh and wet meadow species growing further landward than elsewhere on the shoreline. This suggests that terrestrial plants may normally restrict the landward occurrence of aquatic plants. There are only two basic ways to explore this: pot experiments and field experiments. In either case, plants are grown with, and without, the presence of neighbours, thereby showing whether, at least in experimental conditions, one species can displace another.

Let us begin with an example from salt marshes. In Alaskan salt marshes, the zonation of vegetation is closely connected to flooding (Jefferies 1977; Vince and Snow 1984). Four zones can be delineated with increasing elevation: outer mudflat (*Puccinellia nutkaensis*), inner mudflat (*Triglochin maritimum*), outer sedge marsh (*Carex ramenski*), and inner sedge marsh (*Carex lyngbyaei*). At one extreme, the outer mudflat with *Puccinellia nutkaensis* is flooded some 15 times per summer for periods of 2–5 days each, leading to a soil water salinity of some 15–35%. The inner sedge marsh with *Carex lyngbyaei* is flooded only twice per summer, when a new or full moon coincides with the perigee (although this single flood may last more than 5 days) and soil salinity is only 6–11%, slightly below the 12% for flooding seawater.

Reciprocal transplant experiments across these four zones (plus a fifth, *Poa eminens* that grows on riverbank levees), showed that all species could grow in all zones when neighbouring plants were removed (Snow and Vince 1984). Further, the *Puccinellia nutkaensis* from the outer mudflats actually grew nearly four times larger when transplanted upslope to the inner mudflat than when transplanted back to its own zone. The two species from the highest elevations (*Carex lyngbyaei* and *Poa eminens*), did, however, show reduced growth when transplanted downslope to the outer mudflats. When the same five species were grown at different salinities in pots, all grew best in waterlogged but low salinity conditions. Thus, in spite of conspicuous zonation, the limited distributions of these species cannot be accounted for simply by narrow tolerances to salinity or flooding. Zonation therefore, must be partly produced by biological interactions; in general, Snow and Vince (1984) suggest '. . . species occurring in zones along a physical gradient are often limited by physiological tolerance toward one end of the gradient, and by competitive ability towards the other . . .'. Bertness and Ellison (1987) similarly used transplants to study zonation in New England salt marshes; although the species were different (*Spartina alterniflora, S. patens, Distichilis spicata, Salicornia europea*), the conclusions were similar: 'the performance of each species was lowest in the low marsh and highest on the terrestrial border of the marsh' (p. 142).

Other studies in salt marshes report similar results. For example, on the east coast of North America, the woody species *Iva frutescens* (marsh elder) occurs at higher elevations in marshes. Bertness *et al.* (1992) found that, when these shrubs were transplanted to elevations lower than those they normally occupied, all died within one year. Since death occurred in both cleared and uncleared plots, the lower limit of *I. frutescens* is apparently set by physiological constraints rather than competition. We shall examine some other experimental studies of competition and salt marsh zonation in Chapters 6 and 7.

Similar results have been found in freshwater marshes. Grace and Wetzel (1981) studied two common and widespread species of cattails (*Typha latifolia* and *T. angustifolia*); together these species comprised 95% of the biomass at their study site, a small experimental pond, in the American midwest. While both species are relatively tall and spread from extensive mats of rhizomes, the taller of the two, *T. angustifolia* usually occupies the lower areas of shoreline in the deeper water. Is this just because each species grows exactly where it is best suited by physiology? Both species were transplanted to pots and grown without neighbours at a range of elevations from 15 cm above the water line to 100 cm below it. The transplants grew successfully over a much wider range of elevations than the natural population did, and the interaction was strongly one-sided (Figure 2.9). The *T. latifolia* was only marginally affected by *T. angustifolia*, whereas *T. angustifolia*, which naturally occurred in the deeper water, was apparently excluded from higher elevations by *T. latifolia*. Here is evidence that zonation can be attributed to competition, and that a dominant competitor can exclude a weaker neighbour forcing it into a zone that is physiologically suboptimal.

Another widespread zonation pattern in wetlands is the transition from woody plants at higher elevations to herbaceous plants at lower elevations. This produces the characteristic zonation pattern encountered throughout, for example, the northern temperate zone including North America (Keddy 1983), northern Europe (Spence 1964; Bernatowicz and Zachwieja 1966), and Asia (Yabe and Onimaru 1997). Is each species simply responding to the flooding regime, or are biological factors such as competition contributing to these patterns? Suspecting a situation similar

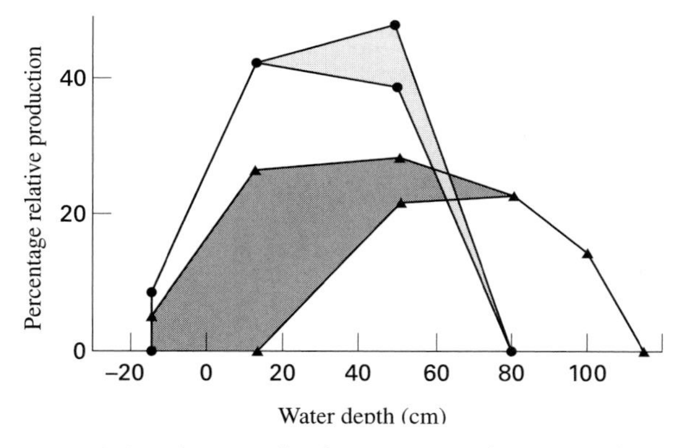

Figure 2.9 The growth of two species of *Typha* as a function of water depth; the shaded region shows reductions caused by the presence of the other species (after Grace and Wetzel 1981).

to that of cattails, Keddy (1989b) experimentally removed these shrubs from many areas of shoreline to test whether herbaceous plants would move landward. Figure 2.10 shows that, after several years, the number of shoreline plants in cleared plots was significantly higher than in uncleared plots. If one creates artificial clearings, and zonation does not change, it may be only that the species had low rates of spread. This could be tested by not only removing woody plants, but also by transplanting wetland plants into these clearings and by comparing their performance to those of wetland plants grown in association with woody plants. Figure 2.11 summarizes the results of such a study.

It is therefore clear that at least some of the herbaceous plants found in flooded conditions can actually grow under considerably drier conditions. This has two possible explanations. It may be that the species that spread up the shore are merely spilling over from their preferred habitat into marginal conditions created by removing shrubs. It is also possible that these plants are moving to higher elevations which are not marginal, but which actually are better for growth than the lower elevations are. Returning to Figure 2.9, *Typha angustifolia* actually can grow better at the water line (where it is naturally absent),than it can in 100 cm of water (where it is naturally present). In nature it apparently occupies a habitat which is less suitable than adjoining drier habitats occupied by neighbouring species.

In conclusion, early studies on zonation made the simple assumption that each species grows to the limits permitted by its own physiology. The importance of competition from neighbours in controlling species

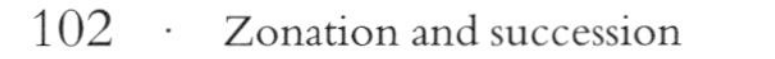

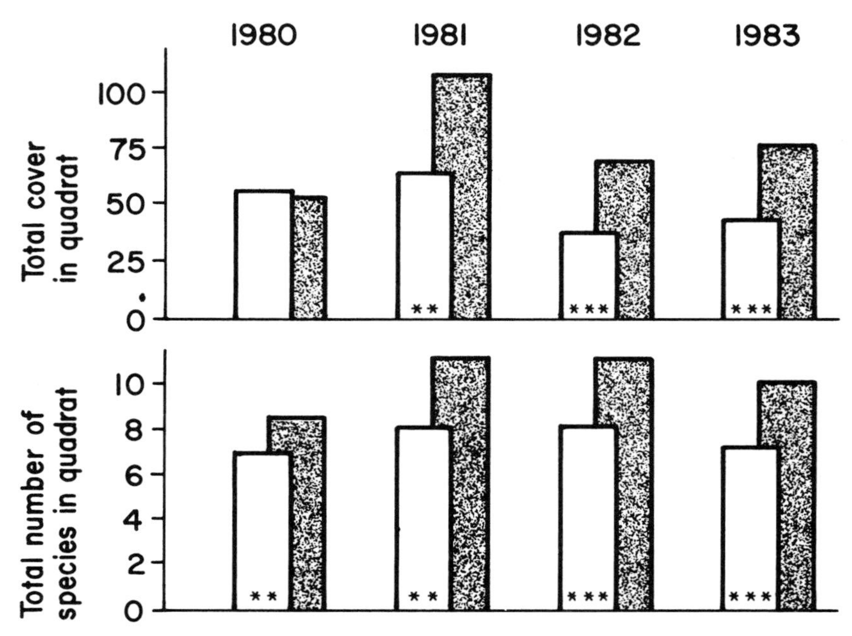

Figure 2.10 Experimentally removing shrubs increases the cover and number of species of herbaceous wetland plants (from Keddy 1989b).

distributions has since been implied in studies from a wide range of habitats and species (Miller 1967; Muller-Dombois and Ellenberg 1974; Colwell and Fuentes 1975; Keddy 1989a). Recent experiments demonstrate that neighbours can exert a significant influence upon distributions. That is to say, zonation is an ecological, not just a physiological phenomenon. Further exploration of the mechanisms requires the introduction of some new terminology.

Some relevant theory: ecological and physiological response curves

One-sided competition on shorelines

The terminology of ecological and physiological response curves (Muller-Dombois and Ellenberg 1974) (or, equally, the terms realized and fundamental niches (Pianka 1981)) clarify the biological and physiological interactions that produce zonation. The *ecological response curve* (realized niche) is the distribution pattern of a species in the field with neighbours present. Zonation patterns such as those in Figure 2.1 report

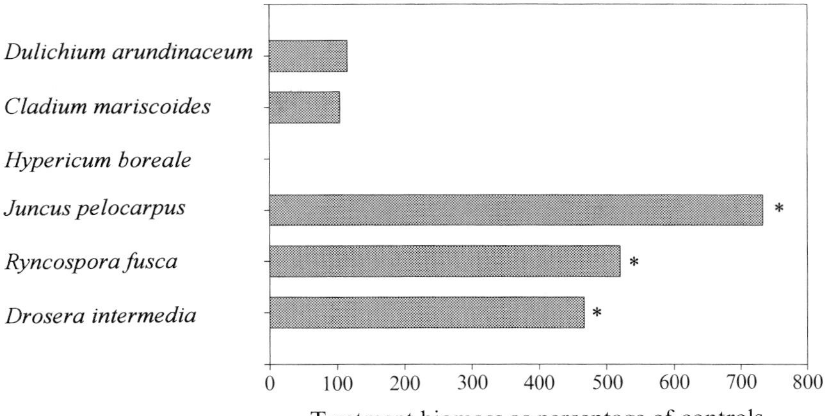

Figure 2.11 The growth of six herbaceous wetland plants in their natural habitats (controls) compared with their growth when transplanted into cleared plots at approximately 20 cm higher elevation on a shoreline (treatments). The experiment used 48 ramets of each species, collected in June 1984 and replanted in a randomized block design. At each of eleven sites around Axe Lake (Keddy 1983), two ramets of each species were replanted into their normal depth (submersed by *ca.* 25 cm of water in June), and two more were transplanted into clearings at 10 to 20 cm higher on the shoreline (*ca.* 5 cm below water levels in June). All plants were exposed by falling water levels over the summer; they were harvested 2 months later, dried and weighed. Significant P values ($* = < 0.05$) are from a paired t-test.

only on ecological response curves. In contrast, the *physiological response curve* (fundamental niche) is the distribution of a species when neighbours are removed, in which case the distribution is presumably explained by the direct effect of physical factors. In the majority of cases studied, the distribution of organisms expands when neighbours are removed – the physiological response curve is usually greater than the ecological response curve. The greater the difference between the two, the greater the effects of competition from neighbours in producing zonation. (Note that if there are commensal or mutualistic relationships, the removal of neighbours may lead to narrower physiological response curves, e.g. Bertness and Leonard 1997.)

In the zoological literature, zonation and ecological response curves are often referred to as 'resource partitioning' (Schoener 1974), and it is generally assumed that ecological and physiological response curves (or, equally, realized and fundamental niches) are very similar. However, in the absence of field experiments, this is pure speculation (Keddy 1989a,

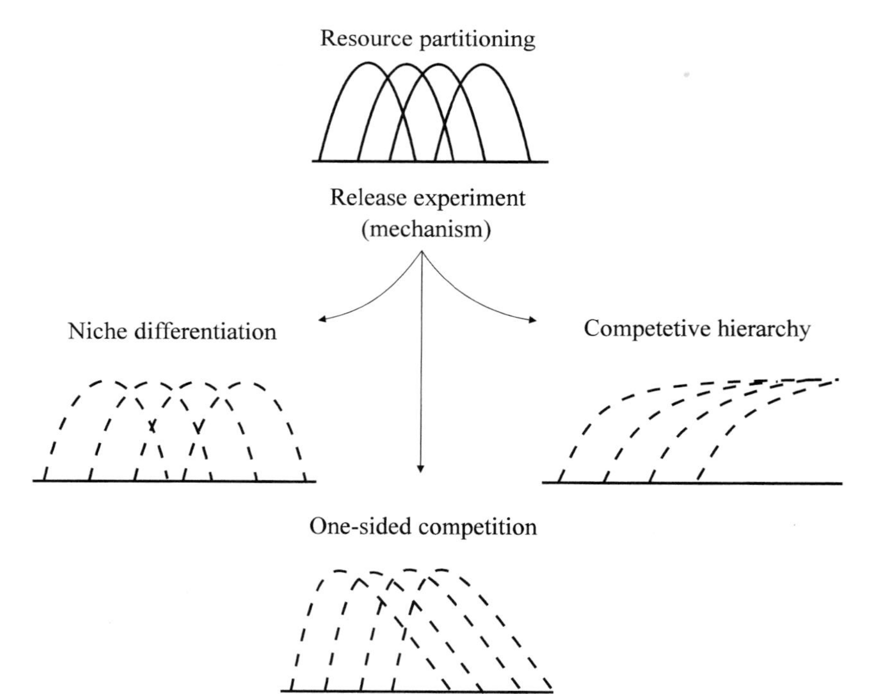

Figure 2.12 A zonation pattern (*top*) can be caused by at least three different competitive mechanisms (*bottom*). Only experiments can separate among the various mechanisms.

Wisheu 1998). When experiments on partitioning are carried out, two extreme situations are possible. In one case (Figure 2.12 *left*) the physiological curves remain nearly identical to the ecological. In such circumstances, competition plays a minor role in producing zonation patterns. In the other case (Figure 2.12 *right*), the physiological response curves shift and nest one within another, a situation termed 'inclusive niches' or 'shared preference'. In such circumstances, competition plays a major role in producing field distributions. Here is a situation where we need fewer ecologists drawing zonation patterns or writing about 'ecotones', and more conducting field experiments.

It may be possible to connect these ideas to the strategy concepts developed by Grime (1977, 1979) and Southwood (1977, 1988). Using their terminology, one group of species can be classified as stress tolerators. Stress tolerators occupy habitats that are chronically unproductive, and they occupy those sites not because they are better competitors, but simply because they can tolerate the extremely undesirable conditions better than other species. There appear to be many examples of wetland species occupying marginal habitats that are beyond the physiological tolerance limits of most other species, including *Zostera* (eelgrass) in salt marshes, and umbrids (mudminnows) in low pH bog ponds.

Perhaps, then, many emergent and submersed plants are stress tolerators. That is to say, they are excluded by competition to a marginal habitat that is unsuitable for other species. Although they tolerate flooded habitats, they may actually grow better under less stressful conditions. In order to tolerate the extreme conditions, they must reduce rates of photosynthesis or divert photosynthate away from foraging, growth and reproduction into those adaptations crucial to tolerating the stress. Deeply buried rhizomes, aerenchyma and reduction in leaf surface area could all be interpreted as costs imposed by flooded conditions. A further cost of such traits could be inherently low growth rates – recall that aquatic plants have inordinately low assimilation rates, usually less than 10 mol CO_2 m^{-2} s^{-1} (Sand-Jensen and Krause-Jensen 1997).

It may seem to be outrageous to suggest that wetland plants would generally grow better in drier areas. But, there is a long list of costs associated with flooding. The biomass invested in petioles increases steadily with water depth in all floating leaved plants that are rooted; this biomass could instead be allocated to leaves or seeds if the plants were growing in shallower water or on mud flats. There are lesser, but surely measurable costs associated with the production of aerenchyma; even if aerenchyma could be produced nearly without costs, the presence of aerenchyma confirms the metabolic constraints imposed by flooded soils. One could therefore postulate a scenario in which wetland plants tolerate extremes of flooding but do not physiologically require them. In this scenario, flooding has the primary role of killing the terrestrial plants that would otherwise invade the site and exclude the wetland species. That is, the requirement for flooding may be as much ecological as physiological.

The assumption that organisms are best adapted to the sites they occupy still is often automatic, particularly in physiological studies, but there is growing evidence that many species occupy habitats that are physiologically suboptimal in order to escape the higher costs of occupying habitats where better competitors are already established. We have already seen data suggesting that many wetland plant species would grow better under less flooded conditions than those in which they are naturally found (e.g. Figures 2.9–2.11). And, while competition is an important and pervasive force in wetlands (e.g. Keddy 1990a; Gopal and Goel

1993) a number of recent experiments suggest that competition is relatively less important in areas that are permanently flooded (e.g. McCreary *et al.* 1983; Wilson and Keddy 1991), leading Grace (1990) to conclude independently that deep water offers a refuge for weaker competitors. Postulating that wetland plants would all have shared preference for mud flats or wet meadows may be an unrealistic extreme, but a third alternative is possible, one-sided competition (Figure 2.12 bottom). Here each species' physiological response curve is shifted in the same direction relative to its ecological response curve. In the case of wetland zonation, one may postulate that, in the absence of neighbours, each species would extend landward of its field distribution. Were this to be the case, wetland plants might be sorted into competitive hierarchies where the best competitors have excluded other species into progressively deeper water, forcing such species to adopt increasingly costly adaptations to tolerate flooding (recall Figure 2.9).

An apparent exception from arid salt marshes

The controls on salt marsh zonation may be quite different in more arid climates, where high rates of evapotranspiration produce a salinity gradient that increases with elevation; hence flooding and salinity are uncoupled as controlling factors. Thus, in Mediterranean-type salt marshes, it is less likely that there is an elevation gradient with one that is benign and one that is stressful. In southern California, for example, three zones can be recognized: a low zone with *Salicornia virginica*, a middle zone with *Arthrocnemum subterminale*, and an upper hypersaline salt flat. Transplant experiments (Pennings and Callaway 1992) showed that the intermediate elevations (upper *Salicornia* and lower *Arthrocnemum* zones) were the most suitable for growth of both species, presumably because of lower flooding combined with lower salinity. The two species were therefore crowded into one region favourable to both: 'since the border between *Salicornia* and *Arthrocnemum* occurs in prime habitat for both, the competitive interactions were not one sided but rather represented a stand-off: each species excluded the other from the portion of the superior habitat in which it was the dominant competitor. Results of the competition experiments are mirrored by the patterns of standing biomass across the marsh. For each species, standing biomass was higher near their border than farther away, corroborating our conclusion that the area near the border represents the best habitat for both species.'

Positive interactions in salt marshes

Positive interactions may also influence zonation. We will examine several examples of positive interactions in Chapter 7, but three can be briefly mentioned here. The anoxic soil conditions at low elevations may be alleviated by oxygen transported by aerenchyma, stimulating the growth of neighbours (Bertness and Ellison 1987; Bertness and Shumway 1993). High soil salinities may be ameliorated by neighbours, allowing seedlings of *Iva fructescens* to establish (Bertness and Hacker 1994). *Juncus gerardi* both shades and aerates intertidal soils, stimulating the growth of neighbouring species (Hacker and Bertness 1999). Overall, then, the zonation found in salt marshes appears to result from both positive and negative interactions (Figure 2.13).

Zonation, competition and fertility

The zonation patterns that competition produces may be modified by fertility. The striking changes in plant zonation with changing soil fertility have been best documented on freshwater shorelines (e.g. Pearsall 1920; Keddy 1983), but the best experimental test comes from salt marshes. Levine *et al.* (1998) fertilized a series of plant competition experiments involving typical salt marsh plants: *Spartina alterniflora, S. patens, Juncus gerardi* and *Distichilis spicata*. After two years, the competitive interactions in fertilized treatments were the reverse of those in the controls. It appeared that *Spartina alterniflora*, which normally occupied the low elevation zone closest to the ocean, was able, when fertilized, to invade higher areas of the shoreline and to exclude both *S. patens* and *J. gerardi*.

Empirical studies of zonation

Although there are vast numbers of papers with sketches of species distributions along gradients, virtually none has taken the next logical step: the quantitative study of zonation and its control by other environmental factors. Such studies are even rarer than experiments. There are at least four reasons for the statistical properties of zonation. (i) Although there are many pictures and a vast literature describing zonation, there is no way to compare such studies without measurable properties. (ii) The zoological literature abounds with theories of resource use (e.g. Miller 1967; MacArthur 1972; Pianka 1981), and zoned communities provide

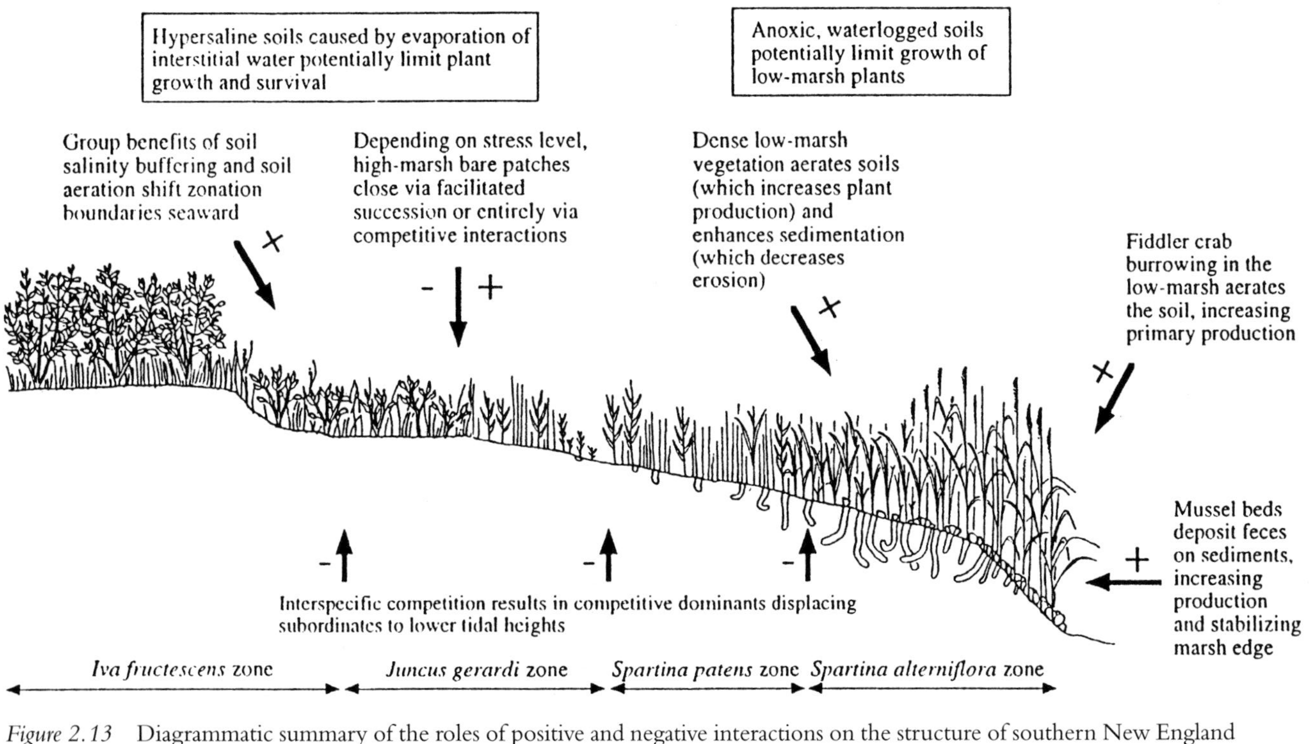

Figure 2.13 Diagrammatic summary of the roles of positive and negative interactions on the structure of southern New England saltmarsh plant communities (from Bertness and Leonard 1997).

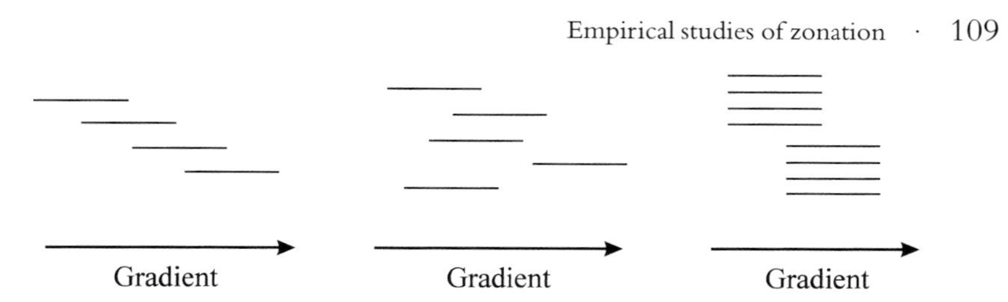

Figure 2.14 Species may be distributed along gradients in a manner that is overdispersed (*left*) like shingles on a roof, random (*middle*), or underdispersed (*right*) like pages of a book. Underdispersed boundaries are usually called clustered boundaries. Statistical tests can distinguish among these possibilities.

an exquisite opportunity to test hypotheses about such phenomena. (iii) The issue of whether ecological communities are continuous or discrete has raged on for decades without resolution, and there is no way to slay this dragon except to actually measure the manner in which communities change along gradients. (iv) Science requires measurable properties. Without actually measuring species distributions along gradients, we can only tell entertaining tales about them.

What properties might we measure on zonation? Here are four, with a brief rationale for each.

(i) The degree to which species distributional limits are clustered ('boundary clustering'). At one extreme (Figure 2.14 left) they may be over-dispersed, like the shingles on a roof; at the other extreme (Figure 2.14 right) they may be clustered (Pielou 1975; Underwood 1978). The middle case is a random distribution. Since one can analyse landward and waterward (or upper and lower) distributional limits independently, there are actually two properties here.
(ii) The range of elevation that each species occupies in a transect. This is a crude measure of realized niche width for a species; averaged over all species, one obtains a measure of mean niche width on that gradient.
(iii) Species richness. Some shorelines have many species on them, others have very few. By counting the number of species in transects of standard width, one can relate richness to other properties.
(iv) Exposure. It has been widely observed that zonation patterns change with exposure to waves (e.g. Pearsall 1920; Bernatowicz and Zachwieja 1966; Hutchinson 1975). By measuring the position of transects along gradients of exposure to waves, one can determine how properties (i) to (iii) are affected by waves and ice scouring.

Given the great theoretical interest in the effects of disturbance upon ecological communities (e.g. Connell 1978; Huston 1979; Grime 1979), it seems remarkable that more studies have not made use of these circumstances.

We now have at least four quantitaive properties that we can measure along gradients: boundary clustering, niche width, species richness and exposure. The next step is to explore some relationships among them.

The first example, Pielou and Routledge (1976), examined data on species distributions in five sets of salt marshes at different latitudes in eastern North America. In many of the transects, species boundaries were significantly clustered, that is, there were zones composed of sets of species with similar distributional limits. Salt marsh zonation therefore, looks similar to the right side of Figure 2.14. Moreover, the upper limits were more clustered than the lower limits, irrespective of latitude (Figure 2.15). This pioneering study showed that with proper sampling methods and appropriate null models, it was possible to find measurable patterns in zoned communities.

The causes of such patterns cannot be deduced solely from statistical analyses. None the less, Pielou and Routledge did find evidence that biological interactions were responsible for some of the species distributions. Their logic was as follows. If these patterns were solely the result of physiological responses to salinity and inundation, then distributional limits of species would be independent. If, however, one species set the limits of another through competition, then there would be a tendency for distributional limits to coincide. That is, species distributional limits would tend to abut one another. In terms of physiological and ecological response curves (pp. 102–106); Pielou and Routledge suggest that physiological and ecological factors will create different kinds of zonation patterns. Using a set of 40 transects near Halifax, Nova Scotia, they found that distributional limits tended to coincide ($P < 0.001$). Therefore, they concluded that competition produces some of the observed clustering of zonation in salt marshes. Regrettably, the test was too crude to test for changes in the intensity of competition among latitudes.

A subsequent attempt to explore patterns among these properties used data on zonation from a small sandy lake typical of many near the Great Lakes. This lake had an array of zonation patterns including those associated with open sand beaches, sheltered fertile bays, and floating bog margins (Keddy 1981, 1983). The flora of this lake, and the array of

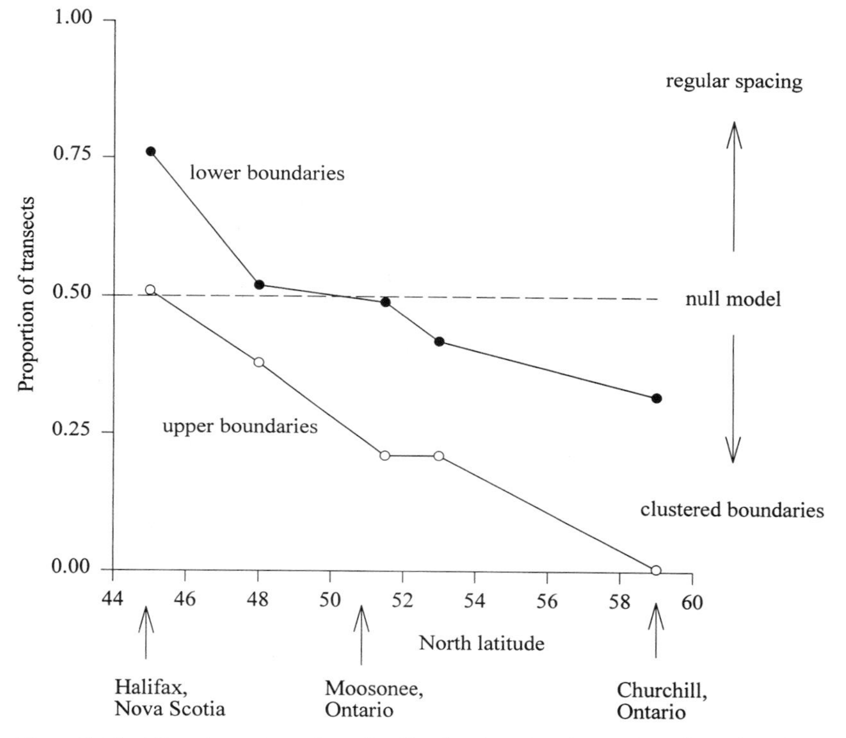

Figure 2.15 The clustering of species distributions in salt marshes plotted against latitude. Notice that the lower the measure of clustering, the more species distributional limits coincide (after Pielou and Routledge, 1976).

vegetation types, appears in many ways typical of the northern temperate zone. The following patterns were found:

(i) Both the upper and lower boundaries of species were clustered. That is, just as Pielou and Routledge (1976) showed, there were certain elevations where more species reached their distributional limits than would be expected by chance alone. This is shown in Figure 2.16 where the measures of boundary of clustering for each of 25 transects fall below zero.

(ii) The degree to which species distributions were clustered (that is, the intensity of the zonation on a shoreline) increased with exposure to waves. This occurred because exposure to waves increased the clustering of upper boundaries (Figure 2.16 top); lower boundaries were unaffected (Figure 2.16 bottom).

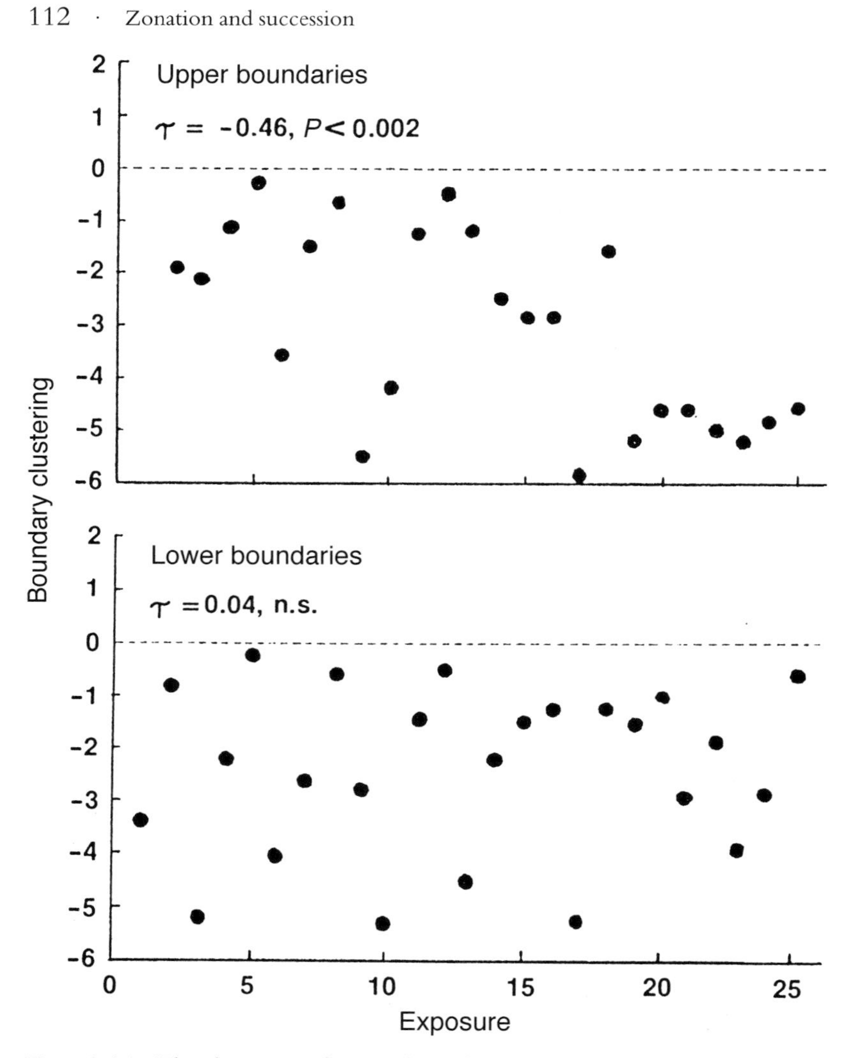

Figure 2.16 The clustering of species boundaries plotted against exposure to waves in a small lake. The dashed line presents the null model (from Keddy 1983).

(iii) Species' distributions were pushed up the shoreline as exposure increased. Figure 2.17 shows how aquatics such as *Lobelia dortmanna* moved up the shoreline as exposure to waves increased. This pattern also showed up in the joint distribution of species as a landward shift in distributional limits (Figure 2.18).

(iv) Mean niche width did not significantly increase as the number of species in a transect increased. That is to say, more species were not

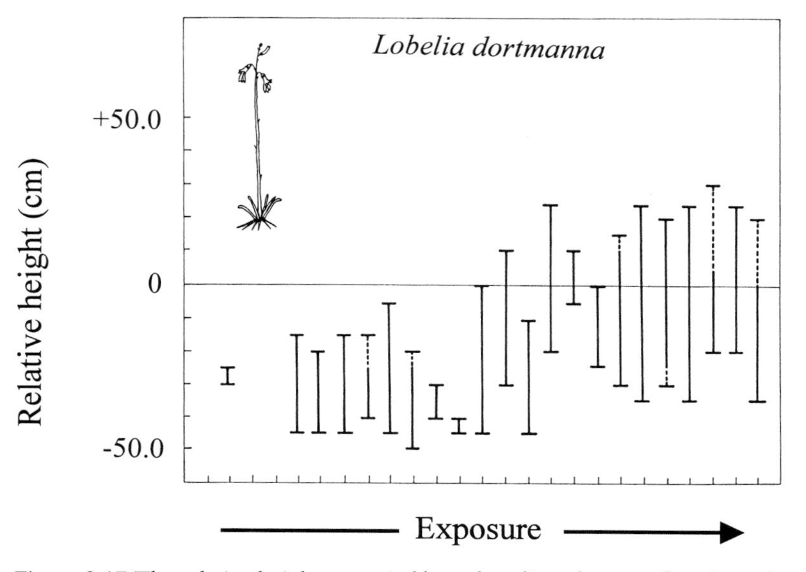

Figure 2.17 The relative height occupied by a shoreline plant as a function of exposure to waves; zero marks the August water line (after Keddy 1983).

packed in by increased specialization of each species in the community. However, while the mean may be the same, exposed shores have significantly greater variation in niche widths. That is, some species have very narrow distributions, and others have much broader distributions, than in sheltered bays.

The above patterns occurred in a lake in Ontario. How general are they? As a first test for biogeographic generality, the same questions were posed for a lake in Nova Scotia, a lake in a different biographic region, with a substantially different flora and a different type of bedrock (Keddy 1984). Similar patterns were found (Table 2.3), except that the intensity of clustering did not increase with exposure. These patterns, and their relative consistency across eastern North America, suggest that it may be possible to group zoned wetlands into categories having specified patterns. Further, some of these properties may be related to broader debates over the kinds of communities that occur in nature, and the manner in which species are packed into them. Several more recent studies have added to both the empirical data base and its conceptual interpretation, so let us continue with the theme of statistical investigations of zonation patterns.

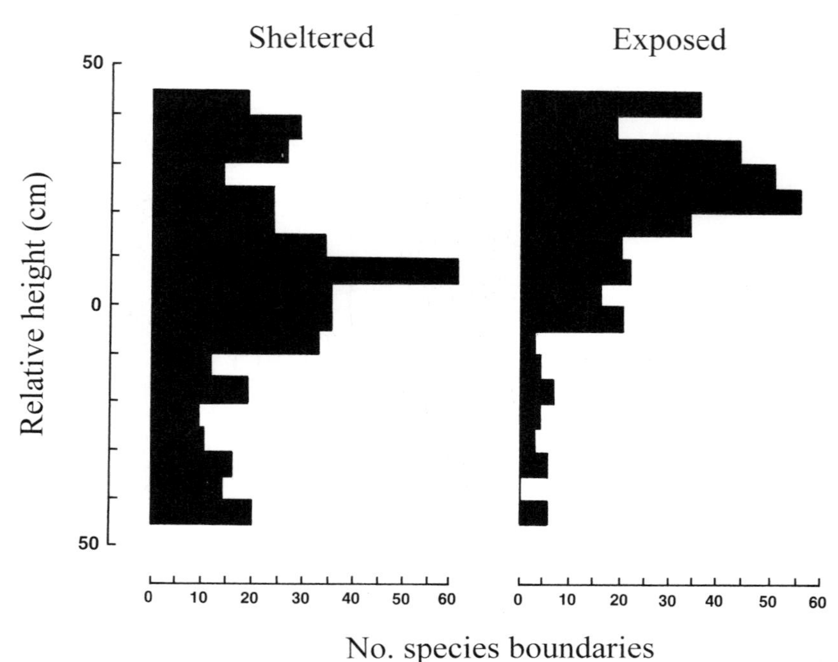

Figure 2.18 The relative height of species distributional limits (upper and lower boundaries combined) for ten sheltered transects (left) and ten exposed transects (right) (from Keddy 1983).

On the nature and existence of communities

We have just seen that it is possible to measure different properties of zonation, and to test whether these measured values are different from those which would arise by chance. Measurement is an important first step in science. Measurement discloses that there are non-random patterns in zonation. Further, we have seen that the degree of non-randomness (or, if you prefer, the intensity of the patterns) sometimes changes along environmental gradients. These empirical relationships provide tools for the quantitative exploration of zonation patterns.

Such measurements and relationships are of far greater significance if they can be related to broader theoretical issues, or to general models for how ecological communities are assembled. One of the most persistent themes of enquiry in community ecology over the last century has revolved around the existence of communities. The first question has been rather general: (i) do ecological communities exist? The second has

Table 2.3. *Clustering of species boundaries on a lakeshore in eastern North America. Data consisted of 30 transects with 117 species on the shoreline of Gillfillan Lake (Lat. 43°57′, Long. 65°48′) in the Tusket River Valley of Nova Scotia*

Boundary	Are boundaries clustered?	Does intensity of clustering change with exposure?	Does location shift with exposure?
Upper	Yes ($t = -9.12, P < 0.001$)	No ($t = 0.0, P = 1.00$)	Landward 40 cm
Lower	Yes ($t = -3.16, P < 0.01$)	No ($t = 0.06, P = 0.64$)	Landward 20 cm

Source: Data collection and analysis as in Keddy (1983).

been more empirical, (ii) what non-random patterns occur in assemblages of species? In practice, these two questions are often mixed together, the assumption being that non-random patterns prove the existence of communities. The general question has therefore been: are living organisms organized into discrete communities as opposed to random assemblages (e.g. Whittaker 1967; Connor and Simberloff 1979; McIntosh 1985)?

There have been two basic approaches to search for evidence of communities. The first has used rigorously defined null models for species composition and compared observed composition to that which would occur randomly (e.g. Connor and Simberloff 1979). In some cases, non-random composition of communities has indeed been detected (Harvey *et al.* 1983; Weiher and Keddy 1995). The second approach, and this is the one more relevant to zonation, has created null models for species distributions along gradients and compared real communities against these null models (Pielou 1975). A vast majority of the published studies on the existence of communities examines island data (Harvey *et al.* 1983), and very few have followed Pielou's suggestion to exploit the power of null models for zonation patterns, although, conveniently for us, most of the latter have been done in wetlands. Let us therefore consider the use of zonation patterns in more depth.

The debate concerning the nature of community organization has continued sporadically for more than 70 years. Colinvaux (1978) provides an entertaining introduction to this controversy, and Whittaker (1962) a more technical view. A convenient starting point is Clements (1916), who proposed that there are relatively discrete ecological units, called communities, that tend to repeat across landscapes. His view, also called the 'community-unit concept' (Whittaker 1975) was accepted by the majority of ecologists during the first part of this century. Gleason (1926, 1939) argued, instead, that each organism is distributed individually, and that communities are not discrete, but rather intergrade. His view came to predominate, in part, because the description of zonation patterns appeared to show patterns of species replacement that were inconsistent with the patterns predicted by Clements (McIntosh 1967; Whittaker 1967). However, all such studies suffered from the problem of using subjective methods of analysing the observed patterns; they failed to employ inferential statistics to compare empirical data to the proposed models. The relative merits of the individualistic as opposed to community unit views therefore remained to be statistically evaluated.

It is possible to test between these alternatives by casting them into testable form using species' distributions patterns (Shipley and Keddy 1987). The 'community-unit' concept proposes that, when species distributions are plotted along some gradient or gradient-complex whose rate of change is constant, there exist groups of species, 'communities', which replace themselves along the chosen gradient (Whittaker 1975). Within each grouping, most species have similar distributions, and the end of one group coincides with the beginning of another. The individualistic concept, in contrast, proposes that 'centres and boundaries of species' distributions are scattered along the environmental gradient' (Whittaker 1975). No distinct groups of species are predicted to exist. These alternatives are illustrated in the top part of Figure 2.19.

Following Pielou (1975, 1977), explicit hypotheses of these two concepts can be formulated using upper and lower boundaries of species along gradients.

The *community-unit hypothesis* states that:

(i) there should be significantly more boundaries (both upper and lower) in some intervals of the gradient than in others, i.e. boundaries are clustered;
(ii) the number of upper and lower boundaries per interval should increase and decrease together along the gradient.

The *individualistic hypothesis* states that:

(i) the average number of boundaries (both upper and lower) in each interval of the gradient should be equal except for random variation about the mean;
(ii) the number of upper boundaries per interval of the gradient should be independent of the number of lower boundaries.

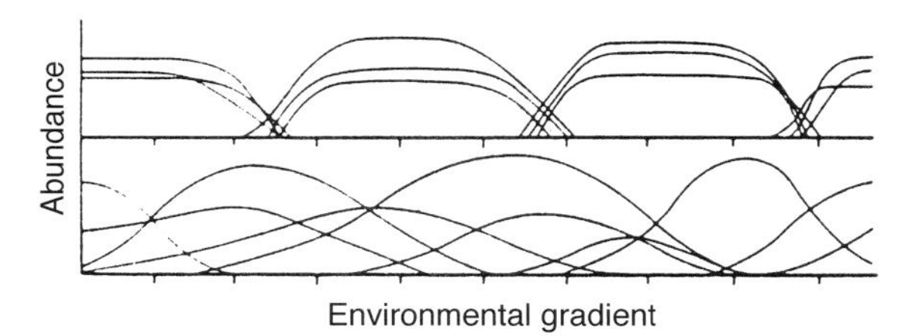

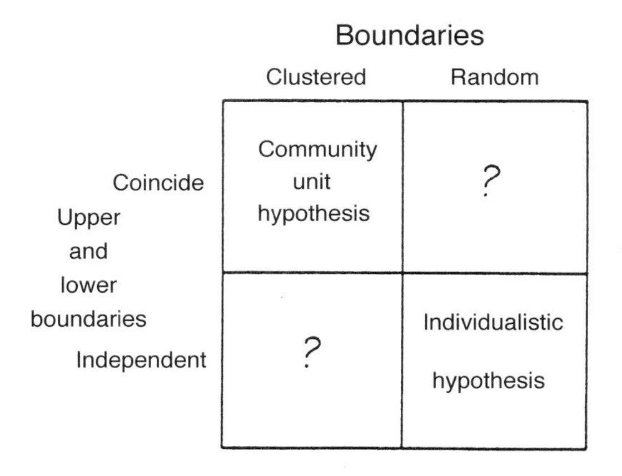

Figure 2.19 The individualistic and community unit hypotheses recast into a testable form (after Shipley and Keddy 1987).

The patterns generated by these hypotheses are shown in Figure 2.19. Note that the 2×2 table suggests at least two other possibilities for patterns along a gradient, as pointed out by Whittaker (1975). Shipley and Keddy (1987) collected data on species boundaries from 13 transects located in a freshwater riverine marsh. As with the example from Axe Lake, the distribution of species boundaries was tabulated for 5 cm increments of elevation. Along this gradient the dominant species changed from *Carex crinita* to *Acorus calamus* to *Typha angustifolia*. These data were analysed using analysis of deviance, which is analogous to analysis of variance, but does not assume normality in the error structure of the model. They found that both upper and lower boundaries were clustered (Figure 2.20). This was clearly contrary to the individualistic concept, but they also found that the pattern of clustering was different between upper and lower boundaries, a result inconsistent with their formulation of the community unit concept. They therefore concluded that, rather than a simple dichotomy between two models, the data suggested the need to

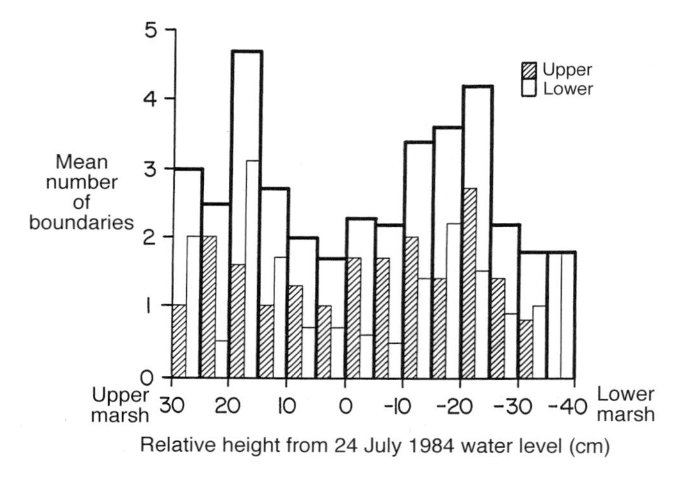

Figure 2.20 Zonation in a riverine marsh. The mean number of species boundaries in each 5 cm height interval is plotted against relative height. Within each rectangle the mean number of upper boundaries (hatched) and lower boundaries (clear) are shown (from Shipley and Keddy 1987).

erect multiple models for the kinds of communities that exist in nature. In other words, more than 50 years of debate about pattern had dragged on, in part, because the patterns were not expressed in clear testable form. This example illustrates the power of zonation patterns as a research tool in ecology.

This study, however, also had two significant weaknesses. First, it tested a broad general model with data from a single wetland. Secondly, it used only data on the distributional limits of species. Hoagland and Collins (1997a) have tried to rectify these deficiencies. First, they collected data from 42 wetland sites. Secondly, they measured three attributes of zonation patterns: (i) boundaries of species distributions, (ii) modes of species response curves and (iii) nested structure. The use of the three properties not only provides a more powerful way to test among competing models, but it also allows the creation of new kinds of community models. Hoagland and Collins trace the origins of four contrasting models of zoned communities:

(i) The highly deterministic community unit model of Clements (1936) could be interpreted to imply that plant communities are comprised of distinguishable associations of species with little overlap in species distributions among associations. This model can be portrayed as a series of species response curves in which the starting and ending points of species distributions are clustered (Figure 2.21(*a*)).

(ii) Other interpretations of this community unit model are possible. Clements (1936) described the occurrence of 'predominants', species that were dominant and spanned one or more association. Figure 2.21(*b*) shows a model in which boundaries and modes of response curves are clustered yet some species response curves are nested within the curves of other, more dominant species.

(iii) The individualistic distribution of species (Gleason 1926) and the continuum concept of vegetation (Whittaker 1967) are represented in Figure 2.21(*c*) as a series of broadly overlapping species response curves with randomly distributed starting and stopping boundaries, and modes, along an environmental gradient.

(iv) Dominant species may be regularly spaced and encompass several curves of subordinant species; the hierarchical continuum model predicts that modes and boundaries of species response curves are random, but because distributions are hierarchical, this model predicts that species distributions are nested (Figure 2.21(d)).

Three test statistics were used to discriminate among these models in the 42 wetland sites. The three test statistics were as follows: Morisita's index (Hurlbert 1990) was used to determine whether or not species boundaries were clustered:

$$I = Q\sum_{i=1}^{Q}\left(\frac{n_i}{N}\right)\left(\frac{n_i-1}{N-1}\right)$$

where Q is the number of quadrats, n_i is the number of starting and stopping boundaries in the *i*th quadrat, and *n* is the total number of boundaries. The degree of aggregation (*P*) of species modes was determined using the sample variance of distance between modes (Poole and Rathcke 1979).

$$P = \frac{1}{k+1}\cdot\sum_{i=0}^{k}\{y_i = 1 - y_i - [1/(k+1)]\}^2$$

where *k* is the number of species, y_{i+1}-y_i is the distance between modes, and $1/(k+1)$ is the mean of y_{i+1}-y_i. If $P=1$, modes are randomly distributed, if $P<1$, modes are regularly distributed, and if $P>1$, modes are aggregated.

Nestedness was determined by using the index of Wright and Reeves (1992):

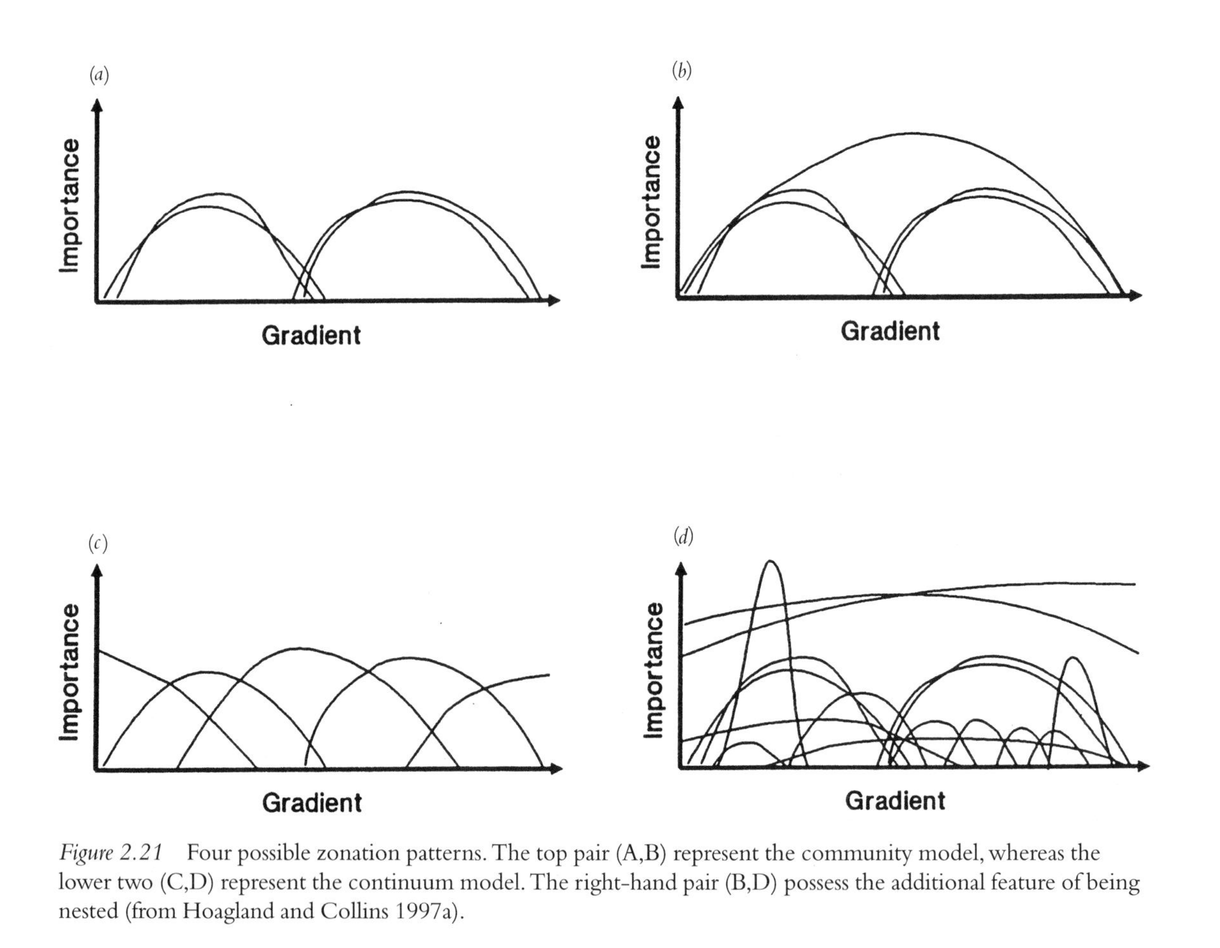

Figure 2.21 Four possible zonation patterns. The top pair (A,B) represent the community model, whereas the lower two (C,D) represent the continuum model. The right-hand pair (B,D) possess the additional feature of being nested (from Hoagland and Collins 1997a).

$$N_c = \sum_{i=1}^{K-1} \sum_{m=i=1}^{K} \sum_{j=1}^{S} X_{ij} X_{mj}$$

where S is the total number of species, K is the number of quadrats, and $X_{ij} = 1$ if species j is present at quadrat 1 and 0 if it is absent. This index counts the number of times that a species' presence in a quadrat correctly predicts that species' presence in quadrats that are more species rich.

The value of N_c was then used to calculate a relative nestedness index:

$$C = \frac{N_c - E\{N_c\}}{\max\{N_c\} - E\{N_c\}}$$

where $E\{N_c\}$ is the expected value and max $\{N_c\}$ is the value of N_c for a perfectly nested matrix. C ranges from zero (complete independence) to 1 (perfect nestedness). Cochran's Q, was used to test for significance of nested species distributions.

All 42 transects were nested. This is an important generalization; Hoagland and Collins interpret this as evidence for 'hierarchical' community structure. Given the many uses of the word hierarchy, it may be more useful to simply use the descriptive result: nested patterns are the rule in zoned vegetation.

Clustering of boundaries occurred in only 10/42 transects; thus the continuum model is more prevalent than indicated by Pielou and Routledge (1976), Keddy (1981) or Shipley and Keddy (1987). Unfortunately, the use of Morisita's Index rather than previously used indices raises the possibility that the prevalence of the continuum model in Hoagland and Collins (1997a) data may be an artefact of the test used. Such problems emphasize the need for methodological consistency.

More than half of the transects did not fit into any of these four categories (Table 2.4). Seven had clustered boundaries but unclustered modes, whereas 16 had clustered modes but unclustered boundaries. This work shows the merit of applying a battery of tests to zonation patterns. The differences among the transects, and among published studies, suggest that ecologists need to use a number of different models to describe the kinds of zonation patterns in nature.

General lessons from analysis of zonation

We may now be digressing well beyond the realm of wetland ecology, but there is perhaps a general lesson here. The continuum hypothesis of Gleason is now generally regarded as correct; Colinvaux's entertaining

Table 2.4. *Summary of models of distribution along gradients (based on distribution of boundaries of species response curves, modes of species response curves, and degree of nestedness of species distributions) and the prevalances of these models in a set of 42 transects from Minnesota and Oklahoma*

	Boundaries clustered	Modes clustered	Distributions nested	Examples found
Community-unit	Yes	Yes	No	0
Nested community-unit	Yes	Yes	Yes	3
Alternative model	Yes	No	Yes	7
Alternative model	No	Yes	Yes	16
Continuum	No	No	No	0
Nested continuum	No	No	Yes	16

Source: After Hoagland and Collins (1997a).

review (1978) states that Clements was wrong because Whittaker's data show that 'zone boundaries could not be found. Instead the data showed clearly that individual species of plants came and went with gentle gradualness as one ascended a mountain, that there was that endless blending of species that should result if each kind of plant did its own free thing, without benefit of social organization.' (p. 71) Remarkably, this important conclusion was reached based on two sources of evidence. The first was ordination of plant communities (e.g. McIntosh 1967, Whittaker 1967), (sometimes called 'indirect gradient analysis') a technique that by its very nature assumes the continuum hypothesis. The second source was the visual inspection of species' distributions along real environmental gradients (e.g. Whittaker 1956, 1967), an approach that was carried without creating null models, and without using techniques developed by Pielou (1975, 1977). An important unifying theme in ecology was therefore evaluated with inadequate (or even inappropriate) techniques, leading to a conclusion that now appears to be just wrong. For whatever the general zonation patterns in forests may be, the few herbaceous zonation patterns that have been carefully analyzed and tested against null models (e.g. Pielou and Routledge 1976; Keddy 1983; Shipley and Keddy 1987; Hoagland and Collins 1997a, this volume, Table 2.3) do seem to show that species boundaries occur in clusters. The interpretation of these clustered boundaries is still unclear. It may be that there is some discontinuity in the underlying gradient, such as a sudden transition from aerobic to anaerobic conditions, or the upper limit of ice scour on a shoreline.

It may also be that a competitive dominant sets the distributional limits for a group of weaker competitors. Or perhaps a few dominant species are distributed as Gleason postulated, each dominant having a group of subordinates and commensals associated with it. The cluster of upper boundaries observed at Axe Lake (Figure 2.18) occurred where shrubs began to occur, suggesting that one of the latter two explanations may account for the patterns there. But, even if the last explanation – a dominant with subordinates and commensals – were the mechanistic explanation for clustered boundaries, is this not more in accord with Clements than Gleason (Figure 2.19 top)? Zoned wetland communities may indeed have important lessons for the entire discipline of community ecology.

Conclusions

We began with the need to find patterns in nature. Zonation makes pattern obvious, and therefore provides a powerful tool for the investigation of wetland communities. Zonation patterns are superficially simple and obvious. Closer inspection, however (using quantitative analyses) has yielded both significant new views on the nature of communities as well as interesting patterns that tease us with potential generalities. Experiments have revealed that early views on physiological causes of zonation were simplistic, and that competition is an important factor in causing the observed patterns. Moreover, there is growing evidence from experiments in fresh and salt marshes that this competition is strongly one-sided. Competition, however, interacts with positive effects arising from neighbours, so that the relative importance of both positive and negative interactions varies with elevation. As Pielou (1975) reminds us, instead of seeking for mythological uniform habitats, we would be better off to look for and study gradients.

3 · *Diversity*

A fundamental property of any ecosystem or habitat is the number of species it contains. Diversity has therefore long been of keen interest to ecologists (e.g. Pielou 1975; May 1986; Huston 1994) and to conservation biologists (Ehrlich and Ehrlich 1981; Groombridge 1992). Let us therefore turn to an examination of diversity in wetlands.

Figure 3.1 shows the number of bird species in selected North American wetlands. Higher numbers of species occur in riparian forests, swamps, and shrub-dominated wetlands. Presumably this reflects a relationship between the number of bird species and habitat structure. As is so often the case, we know rather less about invertebrates; Botch and Masing (1983) list more than 58 species of invertebrates known from bog habitats; Table 3.1 provides a breakdown of the main groups of invertebrates in fens.

Many more such tables and figures could be presented, but there are limits to the value of species lists for wetlands. Such information will, of course, be useful for managing particular sites. Indeed, a complete inventory is often one of the first steps in collecting baseline data for protected areas. We could, however tabulate such data endlessly. What more can be done?

We must first clarify some terms. The number of species in a sample is called *alpha diversity*, *species density,* or *species richness*. In contrast, the number of species in an entire community or larger geographical area is referred to either as *biodiversity* or the *species pool* for that area. The term *diversity* is often used synonymously with all of these, but, precisely speaking, it includes the relative abundance of species (Peet 1974; Pielou 1975, 1977). If all species are equally abundant, then diversity and richness are the same. But this is rarely observed in nature; usually a few species dominate any site or sample.

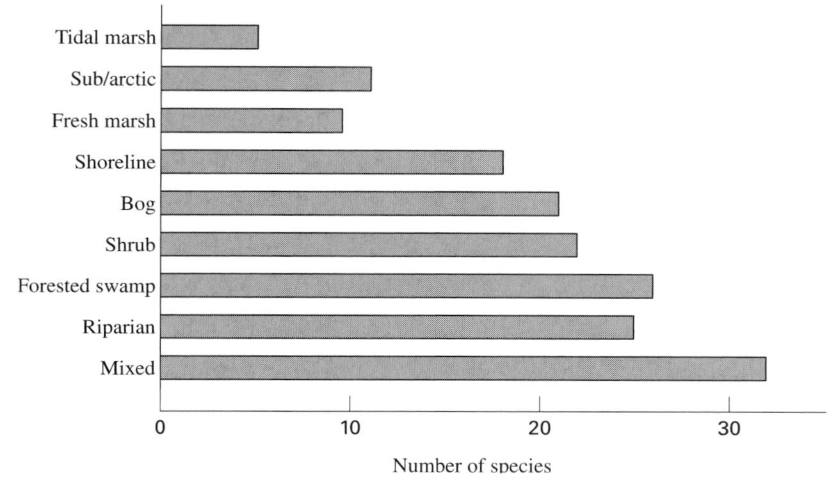

Figure 3.1 Median number of species of birds occupying different wetland vegetation types in North America (after Adamus 1992).

Dominance and diversity

A list of species from a sample or habitat will usually create a quite misleading impression of biological composition; only a few species on the list are usually abundant, and many of the others are often rare. To express these differences in relative abundance of species in a site or sample, data are often presented in a ranked abundance list, also called a dominance-diversity curve (Peet 1974). Figure 3.2(*a*) shows that increased dominance is associated with increased flooding. Figure 3.2(*b*) shows a similar trend to increased dominance, except in this case, it is associated with higher salinity. (Dominance is not only a pattern of established plant communities; similar patterns occur in seed banks: Keddy and Reznicek (1986).) In wetlands, as in other vegetation types, however, dominance is the rule. That is, plant communities are not socialist, where resources are shared more or less equally; they are, in fact, so skewed that they could be called monarchical.

The causes of such patterns are still obscure. Some postulated mechanisms and their associated statistical models are beyond our area of concern here. The famous (or notorious) broken stick model is one of the better known, least realistic models, which provides a way to examine relative abundance lists such as these. The necessary readings can be found in

Table 3.1. *Numbers of individuals of the main groups of invertebrates in fens as % of total*

Group of invertebrates	Moss layer	Field layer		Shrub layer
		rich fen	poor fen	
Oligochaeta	7.1	–	–	–
Mollusca	18.3	2.2	–	2.1
Araneae	19.9	14.9	8.1	6.8
Homoptera, Cicadina	4.7	37.1	16.6	17.6
Psyllina	–	–	–	10.8
Hemiptera	3.7	5.1	9.4	3.4
Thysanoptera	–	7.5	2.1	–
Coleoptera (larvae + adults)	19.5	4.3	7.3	20.7
Lepidoptera (larvae)	–	0.9	1.3	2.3
Hymenoptera	–	3.9	7.7	6.2
Terebrantia	11.2	1.2	1.6	2.1
Formicidae				
Diptera, Brachycera	+[a]	10.7	18.1	17.1
Nematocera	3.8	5.2	24.0	8.3
Mean number of individuals	95.5	222	315	240

Notes:
[a] Small larvae only present, not counted.
Source: From Botch and Masing (1983).

Preston (1962 a, b), May (1981, 1986) and Pielou (1975, 1977). When all is said and done, however, such statistical models have led to few advances in understanding beyond the obvious one in Figure 3.16 that a few species monopolize resources wherever living organisms are found. The exceptions, such as high diversity lakeshores, coastal plain ponds, or tropical floodplains are of great interest precisely because they are exceptions.

Factors controlling the number of species in wetlands

Having counted the number of species in a series of sites, it is natural to enquire how the number of species is determined by environmental factors. To answer this question, we could use the number of species in a sample as the dependent variable, and then look for environmental factors that can predict it. Although there are a good many studies that have used this approach (Huston 1994), rather few have been done in wetlands.

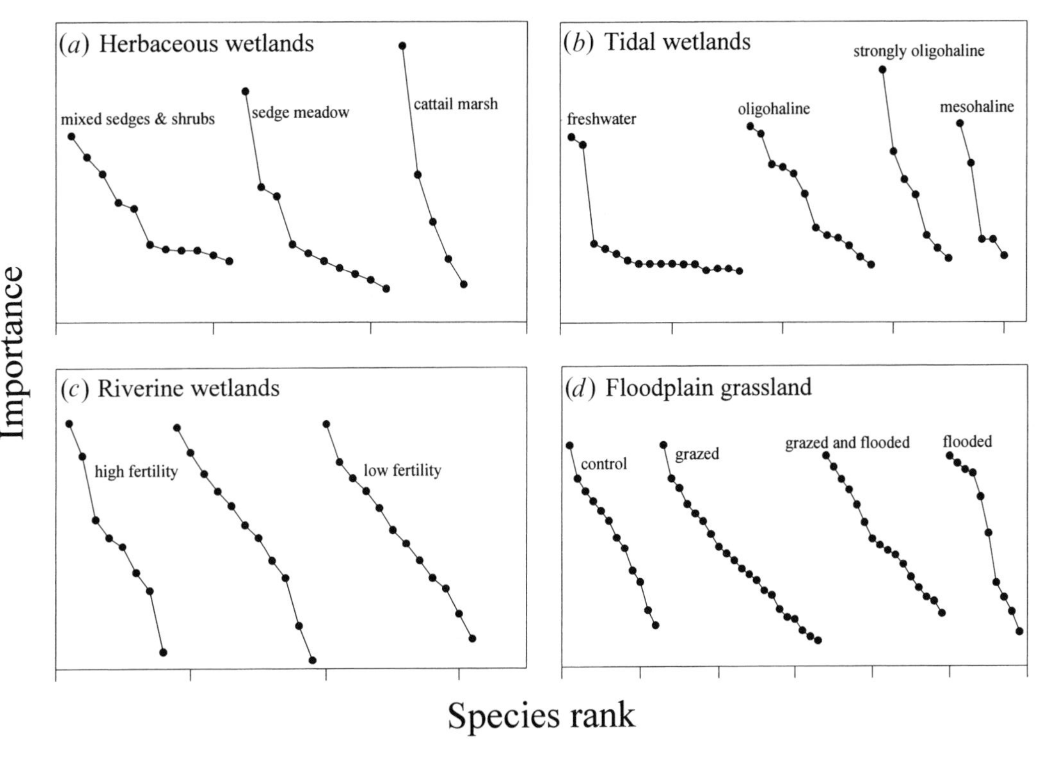

Figure 3.2 Ranked abundance lists (dominance diversity curves) for four different wetlands; (*a*) herbaceous (after Gosselink and Turner 1978), (*b*) tidal (after Latham *et al.* 1994), (*c*) riverine (data courtesy of E. Weiher), and (*d*) floodplain grassland (after Chaneton and Facelli 1991). (Figure courtesy of E. Weiher.)

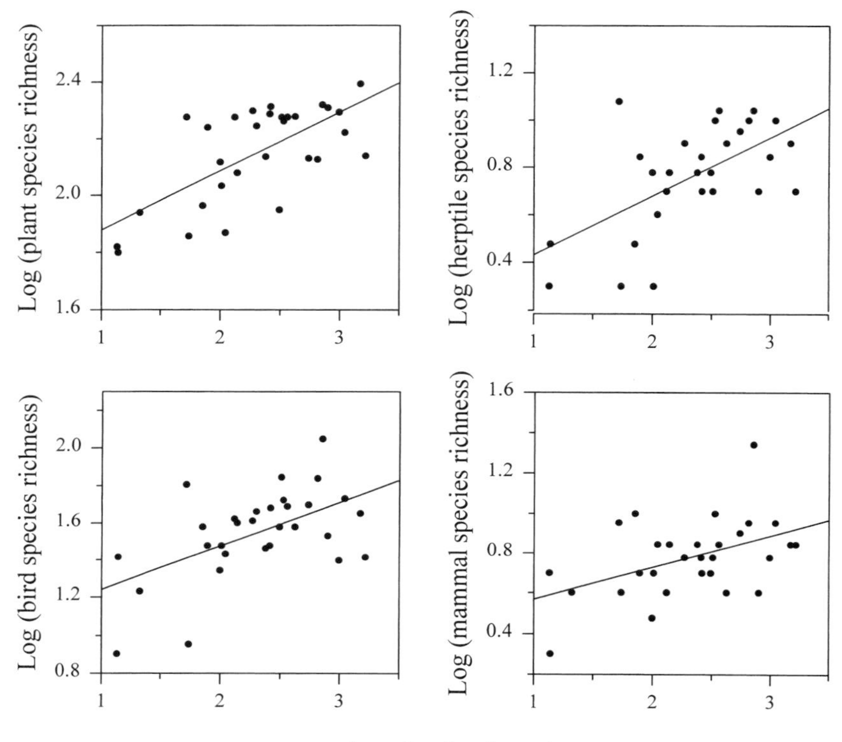

Figure 3.3 Species richness against area for birds, mammals, herptiles and plants in south eastern Ontario wetlands (Findlay and Houlahan 1997).

The starting point for any such study is species–area relationships. It is well established by now that the number of species in any habitat increases with area (e.g. Arrhenius 1921; Connor and McCoy 1979), and decreases with isolation (Darlington 1957; MacArthur and Wilson 1967). This is true for wetlands as any other ecosystem, the pattern applies to all major taxa including mammals, herptiles, and plants (Figure 3.3). Weiher and Boylen (1994) compared an array of published studies from aquatic habitats. Using the standard regression model $S = cA^z$ (where S is the number of species, c is constant, A is area and z is the slope of a log/log regression) they were able to compare the slopes of the species–area relationships for fish, molluscs and plants (Table 3.2). The slopes all fell between 0.20 (fish in Ontario lakes) and 0.29 (aquatic plants in Danish ponds). These species/area relationships generally had lower slopes than

Table 3.2. *The slope (z) of the species-area relationship* $S = cA^z$ *or* $\log(A) + \log(c)$ *for various biota*

Biota	z
Aquatic plants in Adirondack lakes	0.225
Aquatic plants in Danish ponds (two districts)	0.289
	0.266
Fish in Ontario lakes	0.20
Fish in Wisconsin lakes	0.29
Fish in New York lakes	0.24
Molluscs	0.23
Various studies ($n = 90$)	0.31

Source: After Weiher and Boylen (1994).

those found in 90 other studies reviewed by Connor and McCoy (1979). Weiher and Boylen suggest that the lower slopes in lakes may be a result of smaller species pools for freshwater organisms. Having addressed area effects, one can then proceed to explore the predictive value of other environmental factors. Let us consider some examples from different groups of wetland organisms.

Fish

At large geographic scales, climate, salinity and area are the predominant factors. 'Of the 20 000 recent fish species known to science, over 40 per cent live in freshwaters, and the majority of these live . . . within the tropics . . .' (Lowe-McConnell 1975, p. 4). In rivers, the number of fish species is closely related to the area of the river basin, with the Amazon (>2000 species) vastly ahead of other world rivers (World Conservation Monitoring Centre, 1992). A similar but weaker effect of area pattern is found in lakes (Barbour and Brown 1974). In African rivers, river basin area predicts the number of fish species according to the following formula (Welcomme 1979, 1986)

$$S = 0.449A^{0.434}$$

The Niger River has, for example, 160 species (the above equation predicts 190). This relationship does not apply to other continents, since similar sized rivers in South America (e.g. Parana, Orinoco rivers) have about 370 species and in Asia, the Mekong has some 600.

Returning to the Amazon, something more needs to be said about this fauna, the largest fish fauna in the world. Lowe-McConnell (1986) notes that many of its thousand tributaries have yet to be surveyed. The majority of the described species, some 85%, belongs to the superorder Ostariophysi with characoids and silurnoids represented about equally (Figure 3.4). This fauna appears to be derived from a few ancestral forms descended from immigrants originating in Africa (Darlington 1957). The characoids are mostly laterally compressed, silvery, open water fish that are active by day. They have undergone spectacular adaptive radiation, include both the fruit-eating *Colossoma* and carnivorous piranha and are probably 'one of the most diverse groups of living vertebrates' (Lowe-McConnell 1975, p. 38). The siluroids ('catfish'), by contrast, are mostly bottom living and nocturnal. They include piscivores, planktivores, and even parasites. Apart from the characoids and siluroids, the other notable group is the gymnotoids, the nocturnal electrogenic fish. They use electric signals to sense the environment and communicate with each other, and the 'electric eel', in addition, uses electricity to defend itself and capture prey.

Perhaps the most significant group is the fruit-eating characoids, which live in flooded forests and feed upon fruits and seeds. Goulding (1980) describes their abundance, and their important role in seed dispersal of floodplain plants. Many of these species, such as *Colossoma*, are commercially important; this species comprised nearly half the 31 000 t landed at Manaus in 1976. It is captured throughout the year with gill nets in flooded forest.

At more local scales, but still within the Amazon, up to 50 species can occur in one water body, although only a subset of these will be common, for as with most other groups, a few species dominate each sample (Lowe-McConnell 1986). There is also turnover along habitat gradients; the fast-flowing Andean streams have a specialized fauna of algal-grazing catfish, whereas the estuarine reaches have largely marine species. Lowe-McConnell also suggests that an added component of diversity can be attributed to the use of forest foods, since this increases both the volume and array of food relative to rivers in which plankton are the main source of primary production.

Leaving the Amazon, and the tropics, for temperate zone lakes, Tonn and Magnuson (1982) examined 18 small lakes in northern Wisconsin. The top row of Table 3.3 shows that, even at this scale, richness increased with lake area. Beyond this, Table 3.3 shows that fish species richness in the summer was significantly related to lake pH ($r = 0.70$) and vegetation structure ($r = 0.69$). Within smaller bog pond lakes, vegetation diversity accounted for more than half of the variation in the summer ($r = 0.84$), but these patterns disappeared during the winter. In larger lakes with Cetrachids, winter oxygen and depth were the best predictors ($r = 0.59$), and in the small bog ponds with Cyprinids, substrate type and vegetation together were the only significant predictors of the number of fish species. Production can be examined using similar tools. Whillans (1989) used a larger pool of lakes to test for relationships among fish production and wetland area for inland lakes in Ontario. Not surprisingly, wetland area can be best predicted from lake properties such as depth and area. Table 3.4 shows that yields of sport fish are positively related to total wetland area, as well as to the area of each main wetland type. Whillans reports that, in conjunction with other variables (such as fishing effort), it is possible to predict up to 59% of the variation in fish yield.

Birds

Birds are the best-known animals, with some 8600 species, and flight reduces the importance of barriers to their dispersal, there being 'no final limit to the width of barriers that some birds cross' (Darlington 1957, p. 240). As with fish, amphibians and reptiles, the tropical fauna is richer, with some 85% of all species or subspecies being tropical (Darlington 1957). Passerines far outnumber all other bird groups on every continent, although few occur on freshwater and none is marine. Major wetland groups include the Anatidae (ducks), Ardeidae (herons), Rallidae (rails), Charadrininae (Plovers) and Scolopacinae (sandpipers).

At the smaller scale, Figure 3.1 showed that the number of birds changed with the type of wetland. What might be the cause of this? Let us begin by considering the effects of salinity and vegetation type in herbaceous wetlands. Many birds use coastal marshes during migration, and these coastal marshes are strongly patterned by salinity gradients. Weller (1994b) reports that the number of bird species changes among these vegetation types, with *Scirpus olneyi* marshes having more species and individuals than other vegetation zones (Figure 3.5). Largely unvegetated mud flats, however, have both the highest number of individuals and species of birds; this habitat is particularly favoured by shorebirds foraging for marine invertebrates.

Such examples might therefore lead us to expect that the number of bird species in a wetland will correspond rather well with the number of vegetation types. In general, this appears approximately true, but the

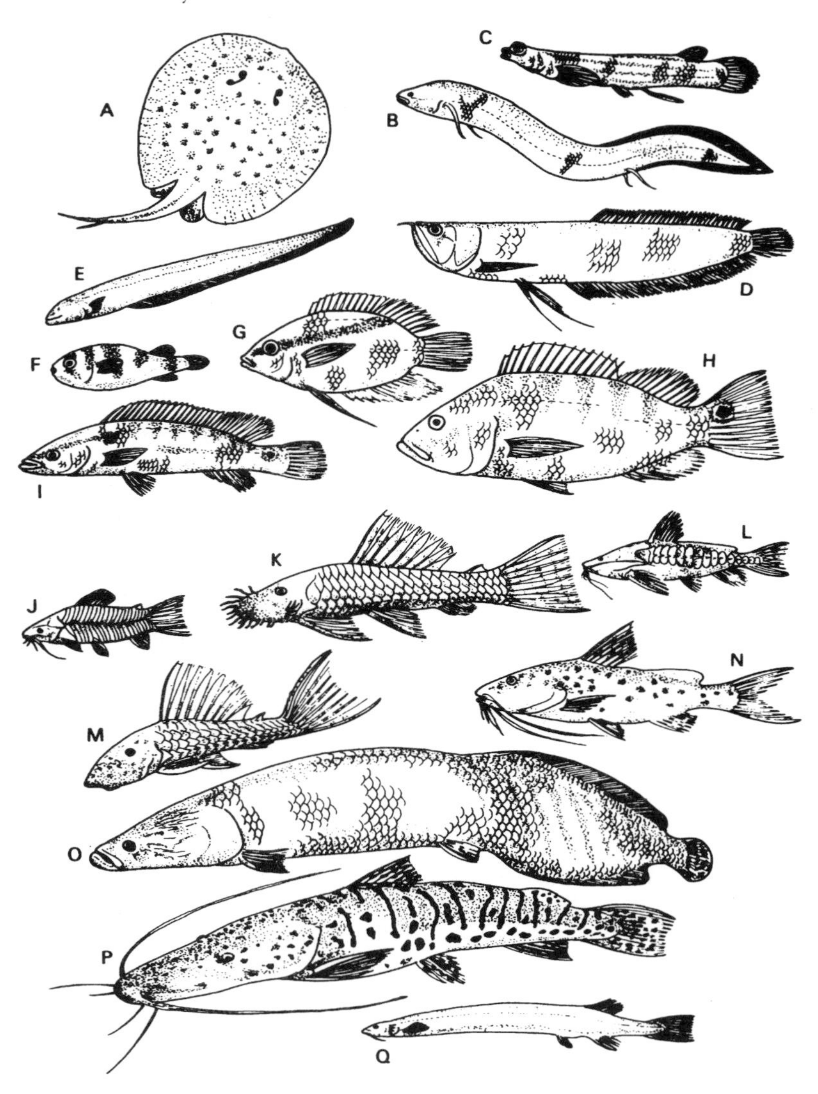

Figure 3.4 (*a*) South American freshwater fishes I. (*a*) *Potamotrygon* stingray (30 cm, Elasmobranch); (*b*) *Lepidosiren* lungfish (50 cm, Dipnoi); (*c*) *Anableps* foureyefish (15 cm, Anablepidae); (*d*) *Osteoglossum* (50 cm, Osteoglossidae); (*e*) *Electrophorus* electric eel (60 cm, Electrophoridae); (*f*) *Colomesus* pufferfish (Tetraodontidae); (*g*) *Cichlasoma* (10 cm, Cichlidae); (*h*) *Cichla* (40 cm, Cichlidae); (*i*) *Crenicichla* (25 cm, Cichlidae); (*j*) *Hoplosternum* (15 cm, Callichthyidae); (*k*) *Ancistrus* (15 cm, Loricariidae); (*l*) *Megalodaros* (70 cm, Doradidae); (*m*) *Hypostamus* (15 cm, Loricariidae); (*n*) *Pimelodua* (30 cm, Pimelodidae); (*o*) *Arapaima* (150 cm, Osteoglossidae); (*p*) *Pseudoplatystoma* (120 cm, Pimelodidae); (*q*) *Vandellia* (4 cm, Trichomycteridae). (Reproduced from Lowe-McConnell, 1987).

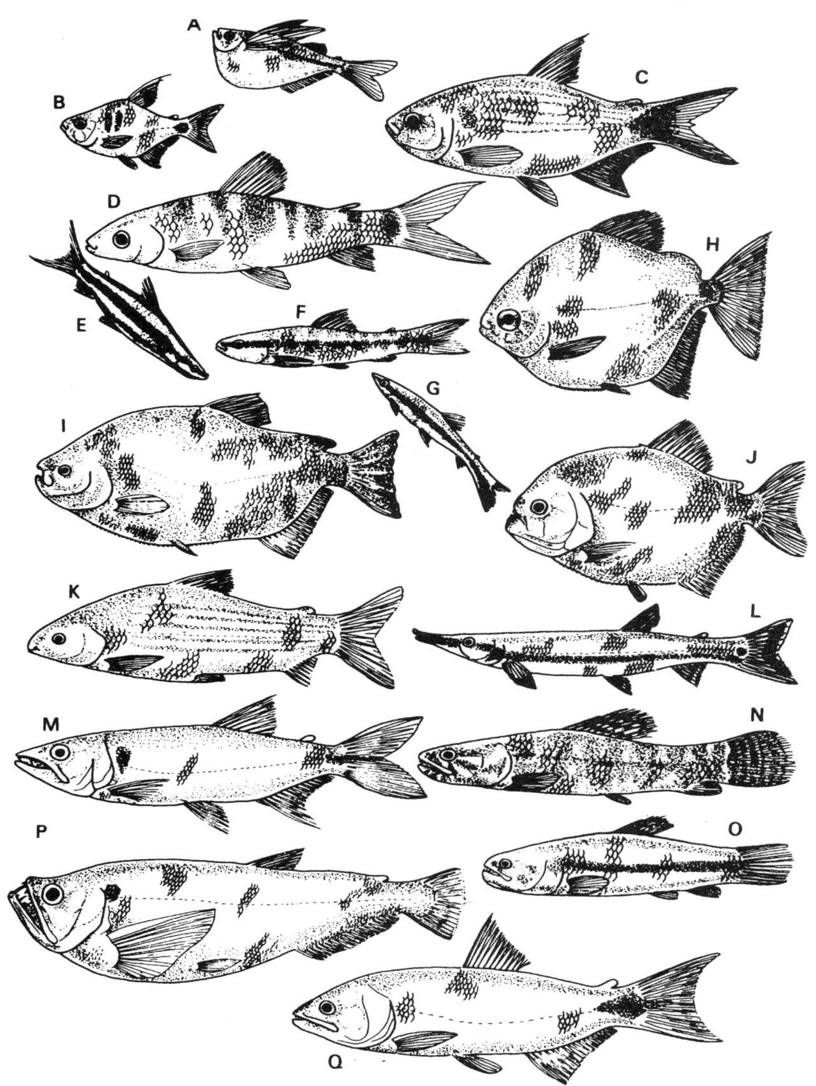

Figure 3.4 (*cont.*) (*b*) South American freshwater fishes II, illustrating the adaptive radiations in characoid fishes. (*a*) *Gasteropelecus* (6 cm); (*b*) *Tetragonopterus* (12 cm); (*c*) *Brycon* (50 cm); (*d*) *Leporinus* (30 cm); (*e*) *Anostomus* (12 cm); (*f*) *Characidium* (4 cm); (*g*) *Poecilobrycon* (4 cm); (*h*) *Metynnis* (12 cm); (*i*) *Colossoma* (50 cm); (*j*) *Serrasalmus* (30 cm); (*k*) *Prochilodus* (40 cm); (*l*) *Boulengerella* (45 cm); (*m*) *Acestrorhynchus* (20 cm); (*n*) *Hoplias* (30 cm); (*o*) *Hoplerythrinus* (25 cm); (*p*) *Hydrolycus* (60 cm); (*q*) *Salminus* (50 cm). (Reproduced from Lowe-McConnell, 1987.)

Table 3.3. *Factors predicting species richness of fish in Wisconsin Lakes. Correlation coefficients* (r) *and linear regressions for summer and winter species richness* (y) *vs. each of 15 environmental factors* (x)

	Summer				Winter			
			$y = a + bx$				$y = a + bx$	
Independent variable	r	($P \leq 0.05$)	a	b	r	($P \leq 0.05$)	a	b
All lakes ($N = 18$)								
1. Log (lake area)	0.69	–	1.86	3.50	−0.08	NS	3.14	−0.26
2. Log (maximum depth)	−0.47	–	8.25	−5.50	0.04	NS	2.69	0.34
3. Log (connectedness + 1)	0.60	–	3.58	1.96	−0.30	NS	3.64	−0.67
4. Log (alkalinity)	0.66	–	1.58	3.87	−0.02	NS	2.94	−0.09
5. Log (conductivity)	0.60	–	−7.70	7.20	−0.06	NS	3.82	−0.52
6. pH	0.70	–	−9.98	2.39	0.14	NS	0.58	0.34
7. Log (total dissolved solids)	0.42	NS	−0.58	3.95	−0.07	NS	3.59	−0.45
8. Log (winter oxygen + 1)	−0.42	NS	7.48	−2.83	0.02	NS	2.78	0.11
9. Substrate diversity	−0.08	NS	6.48	−0.66	−0.27	NS	4.04	−1.48
10. Vegetation diversity	0.69	–	2.66	3.93	0.00	NS	2.83	0.00
11. Depth diversity	−0.12	NS	7.33	−1.61	0.19	NS	1.36	1.72
12. Depth and substrate	0.08	NS	4.92	0.72	−0.02	NS	2.98	−0.10
13. Depth and vegetation.	0.58	–	−0.47	4.22	0.07	NS	2.30	0.35
14. Substrate and vegetation	0.50	–	0.73	3.67	−0.16	NS	3.98	−0.81
15. Depth, substrate, and vegetation	0.57	–	−2.16	4.25	−0.08	NS	3.58	−0.39

Notes:
Multiple regressions are as follows: Summer richness = 3.75 + 4.56 log area – 3.84 substrate diversity ($r^2 = 0.67$, $P \leq 0.05$); Winter richness = −3.15 + 1.14 pH – 1.30 log (watershed + 1) ($r^2 = 0.24$, $P > 0.05$)
Source: From Tonn and Magnuson (1982).

Table 3.4. *Significant relationships between wetlands and fish in 66 inland lakes in north eastern North America*

Dependent variable	Independent variable	Lake trout lakes		Non-lake trout lakes		All lakes	
		N	*r*	*N*	*r*	*N*	*r*
Total wetland area	Lake area	34	0.34	32	0.87	66	0.80
	Log_{10} morphoedaphic index	–	–	–	–	63	0.39
	Mean depth	–	–	32	−0.39	66	−0.36
Emergent vegetation	Lake area	–	–	32	0.90	66	0.80
	Log_{10} morphoedaphic index	–	–	–	–	63	0.37
	Mean depth	–	–	32	−0.35	66	−0.35
Submersed and floating vegetation	Lake area	34	0.41	32	0.77	66	0.72
	Log_{10} morphoedaphic index	–	–	32	−0.42	66	−0.34
	Mean depth	–	–	–	–	63	0.39
Treed wetland (swamp)	Lake area	–	–	–	–	63	−0.36
	Log_{10} morphoedaphic index	–	–	–	–	63	0.27
	Mean depth	–	–	–	–	66	0.33
Sport fish yield	Lake area	28	0.77	32	0.70	58	0.70
	Log_{10} morphoedaphic index	27	0.42	–	–	55	0.34
	Mean depth	27	−0.38	–	–	58	−0.31
	Wetland area	–	–	32	0.51	58	0.44
	Emergent vegetation	–	–	32	0.58	58	0.50
	Submerged and floating leaf vegetation	–	–	–	–	58	0.29
	Log_{10} treed wetland	–	–	–	–	55	0.28

Source: After Whillans (1989).

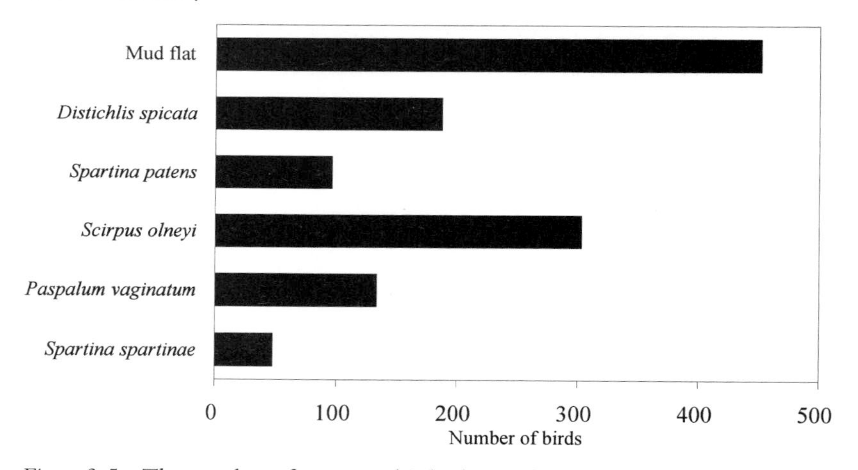

Figure 3.5 The number of common birds observed in six vegetation zones along the Texas coast (from data in Weller 1994b).

overriding factor is, more precisely, the diversity of physical structure in a habitat. This physical structure appears to have both a vertical and horizontal component. With respect to the vertical component, bird diversity often increases with the complexity of plant structure (e.g. MacArthur and MacArthur 1961; Huston 1994) and this means that, in general, forested wetlands will have higher bird diversity than herbaceous wetlands. Adamus (1992) has summarized data from breeding bird census reports; forested riparian wetlands have a median of 25 bird species whereas freshwater marshes had only 9.5. The horizontal component of structural diversity refers to the patchiness of habitat. Wetlands with patches of vegetation interspersed with patches of open water are considered most desirable for waterfowl (e.g. Weller 1978; Kaminski and Prince 1981; Ball and Nudds 1989). The census data from Adamus also illustrate this horizontal component; vegetation types reported as 'mixed' had a median of 32 bird species, the highest number of species reported (Figure 3.1).

In addition to spatial variability, there is temporal variability to consider. Bethke and Nudds (1993) explored habitat variability as a possible predictor of duck richness in prairie and boreal wetlands. Over half the variation in duck richness could be predicted from percentage of area covered by wetlands, annual run-off, and the means and variances of precipitation and temperature. The relationship was, however, non-linear; maximum richness of 15 species occurred in sites judged to have intermediate variability. The assumed mechanism was displacement of ducks from rigorous and variable habitats into relatively more benign ones.

Amphibians

There is much less published work on amphibian diversity than upon bird diversity (see Beebee 1996; Pearman 1997). Since amphibians have moist, permeable skin, and eggs and larvae that are sensitive to desiccation, environmental factors related to moisture are likely to be important predictors of abundance (Darlington 1957) and, at least at large geographical scales, the number of amphibians is strongly related to potential evapotranspiration (Arnold 1972; Currie 1991). Other studies have found that species composition and richness change with precipitation, soil moisture, altitude, and forest structure (Guyer and Bailey 1993; Pearman 1997). Longer periods of standing water will often increase diversity (Pechmann *et al.* 1989), so long as predation by fish does not become a factor (Wilbur 1984). A number of studies have used multiple regression to seek quantitative relationships between environmental factors and the richness of amphibians in wetlands. In eastern North America, wetland area is an important factor contributing to herptile diversity, as is the proportion of the landscape in forest cover (Findlay and Houlahan 1997). The length of roads near the wetland has a negative effect upon diversity (Findlay and Houlahan 1997). A study of 19 wetlands in western North America (Richter and Azous 1995) examined factors including area, vegetation classes, presence of predators and hydrology. *Ambystoma gracile* and *Rana aurora* were the most widespread species. Overall, wetland area had no effects, but wetland area in this study only varied between 0.5 and 13 ha, whereas Findlay and Houlahan examined a range of areas across three orders of magnitude (Figure 3.2). The percentage of the watershed urbanized was negatively correlated with the number of amphibian species, and sites with higher water level fluctuations had fewer species.

Amphibian species richness in temperate zone ponds is probably closely related to the array of water depths and vegetation types (Figure 2.3 bottom). At one end of the water depth gradient, small *Hyla* species call from open grassy areas with temporary shallow water; at the other end of the gradient, large *Rana* species call from floating aquatics in deep water. Other *Hyla* species call from the woody plants which emerge from the water. As with birds, the diversity of frogs in a wetland therefore appears to be the result of an array of structural vegetation types rather than the presence or absence of particular plant species.

In wet tropical forests the amphibian fauna is much larger; one reserve in Ecuador combining floodplain and forest with winding streams had 75 species of amphibians (Pearman 1997). In this reserve, both the richness of amphibians, and the proportion of *Eleutherodactylus* frogs, (frogs which produce young adults from eggs) declined near pastures. Hylid frogs were relatively richer in disturbed forests (although canopy dwelling hylids were not sampled in this study), whereas the number of species of *Eleutherodactylus* frogs was higher in pristine areas.

Plants in freshwater wetlands

Since it is well documented that the number of species in most groups of organisms including fish, amphibians, birds and mammals increases with decreasing latitude (e.g. MacArthur 1972; Pielou 1975), we would expect the same pattern in plants. In terrestrial plants, this is indeed what we find. For example the average 0.1 hectare plot of lowland neotropical forest has from 53 to 265 species, whereas equivalent areas of temperate forest have some 20 to 26 plant species (Grubb 1987; Gentry 1988). Similarly, while Costa Rica has a flora of some 8000 to 10 000 species, the Carolinas, which are four times as large, have a vascular flora of only 3360 (Radford *et al.* 1968). Against this background, Crow (1993) reports a remarkable observation: such patterns do not occur in aquatic plants, and if anything, the flora is richer in the temperate zone. Since this result is quite remarkable, a few more details are necessary. Crow compared the floras of a number of different types of wetlands using published and original data for tropical and temperate regions. New England had 89 species of aquatic plants, and the Carolinas 65, whereas Costa Rica had only 38 and Panama just 35. In the sedge family (Cyperaceae), the same patterns occur: north eastern North America 217 species, the Carolinas 231 and Central America a mere 94. The middle Amazon, by comparison, has only 37 species of Cyperaceae (Junk and Piedade 1994).

Crow then provides a breakdown on a habitat by habitat basis. Consider marshes first. Palo Verde National Park in Costa Rica, a large marsh situated along Rio Tempisque, had a flora of 66 species; in contrast, a single two acre marsh on the shores of the Great Lakes had a wetland flora of 128 species (Stuckey 1975) and a mere 1800 m^2 of Longleaf Pine savannah in the southeastern United States can have 140 species (Peet and Allard 1993). Temporary aquatic habitats in the Guanacaste Province of Costa Rica yielded a flora of 32 species, compared to 42 for vernal pools in southern California. In the recently glaciated peatlands of North

America, studies of individual sites yield floras of some 100 species, whereas in the Cordillera de Talamanca only 20 species were found. The same trend occurs in coastal habitats: New Hampshire salt marshes have a vascular flora of 81 species, whereas mangrove habits in the Caribbean have less than three species each. Admittedly, the knowledge of tropical floras is insufficient, and not all the data are strictly comparable, but Crow's compilation suggests that something is fundamentally different about global patterns of richness in wetland plants. I will resist the obvious temptation to speculate what this might be.

Now let us consider patterns at the more local scale. We will spend some time on temperate areas, but eventually return to the tropical. Grubb (1987) states that, at the meso-scale (0.1–10 ha) and in the temperate zone, 'there can be no doubt that the greatest number of tree species is to be found in the broad flood plain forests of the great rivers such as the Mississippi, Rhine and Danube.' In central Europe, for example, Grubb reports that 27 of the 45 angiospermous trees north of the Alps occur in remnants along the Rhine floodplain. However, his data showed that, at least on a stand-by-stand basis, floodplains have a lower richness than upland forests (Table 3.5), presumably reflecting the intolerance of trees to waterlogging of soils. Grubb notes that we still do not understand why trees have been unable to evolve better means to tolerate flooding.

Elevation is also very important on freshwater shorelines. Figure 3.6(*a*) shows species richness plotted against depth; the number of plant species is highest in the seasonally flooded region. This appears to be a general trend in shoreline wetlands. Figure 3.6(*a*) shows that exposure to waves also influences the number of species, and interacts with the water depth gradient. It appears that richness reaches a maximum at intermediate levels of disturbance from waves (Figure 3.6(*b*)). We can explain this by suggesting that in bays, a few species manage to dominate the shoreline and exclude other species. On very disturbed shores, damage from waves eliminates many species. In the intermediate region, neither of these forces predominates. There is a growing literature on this phenomenon, as we shall see in Chapter 6.

Multivariate techniques have also been used to explore diversity patterns in wetlands. The data in Table 3.5 and Figure 3.6 are considered a form of direct gradient analysis (*sensu* Whittaker 1967) in that the dependent variable, in this case the number of plant species, is plotted directly against environmental gradients. Instead of selecting one gradient in the field, Auclair *et al.* (1976a) collected measures of richness and diversity

Table 3.5. *Mean numbers of species (mostly 100–400 m^{-2}) of ten major forest-types in southern Germany and nearby northern Switzerland*

	Mean number of species				Number of stands sampled
	Herbs	Shrubs	Trees	Climbers	
Summer-dry forests rich in half-shade species					
Pubescent oak (*Lithospermo-Quercetum*)	32	9.6	8.5	1.3	12
Oak-hornbeam (*Galio-Carpinetum*)	27	3.1	6.5	1.0	8
Scots Pine (*Cytiso-Pinetum*)	27	4.9	3.6	0.1	16
Ridge-top and slope forests of beech not so subject to summer drought					
On calcareous soils (*Carici-Fagetum*)	22	3.7	5.6	1.1	18
On moderately acidic soils (*Asperulo-Fagetum*)	13	0.3	4.4	0.8	19
On strongly acidic soils (*Luzulo-Fagetum*)	8	0.1	3.5	0.5	11
Floodplain forest					
Elm–ash–oak (*Querco-Ulmetum*)	20	5.9	8.0	2.1	27
Gulley forests with moist, heavy-textured soils not subject to year-round waterlogging					
Ash–maple (*Aceri-Fraxinetum*)	26	3.8	5.9	0.7	18
Streamside forests with soils waterlogged much of the year					
Ash–alder (*Carici-remotae-Fraxinetum*)	16	1.3	2.8	0.1	14
Swamp forest					
Alder (*Carici-elongate-Alnetum*)	12	1.4	2.2	0.0	5

Source: Modified from Grubb (1987); original sources therein.

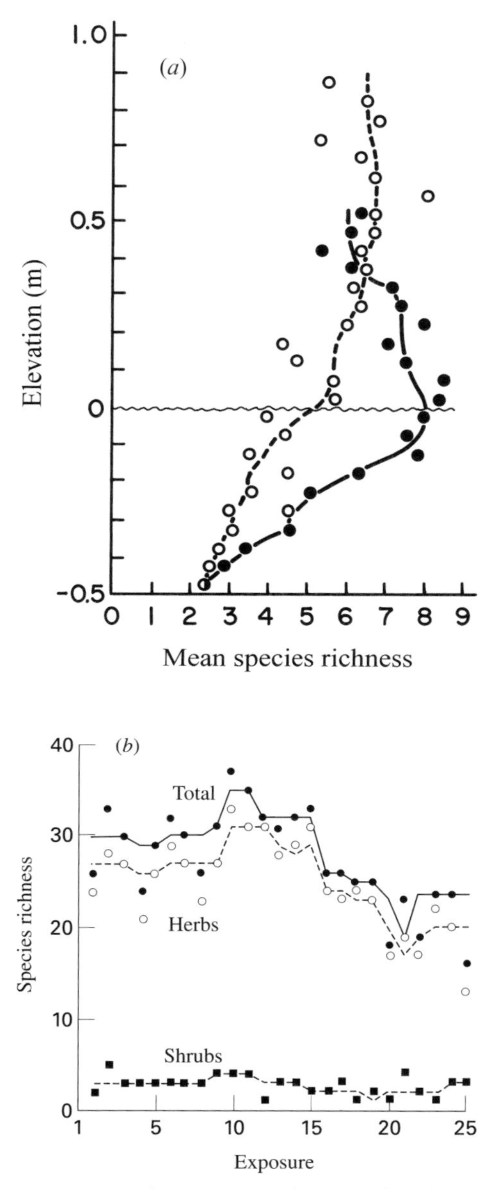

Figure 3.6 Plant species richness on lakeshores (*a*) richness against elevation on exposed shores (open circles) and sheltered shores (solid circles) (from Keddy 1984), (*b*) richness plotted against exposure to waves (from Keddy 1983).

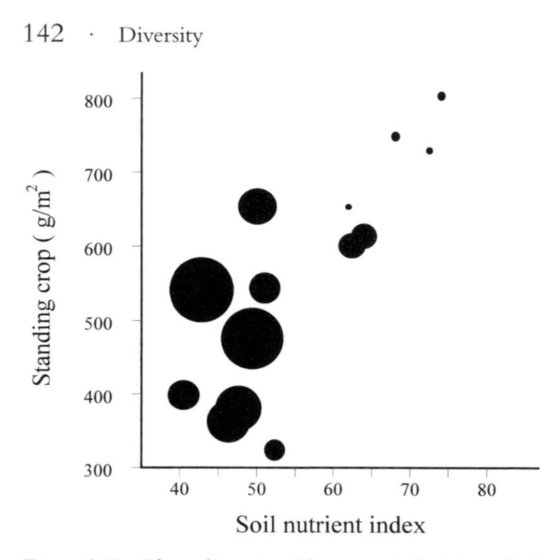

Figure 3.7 Plant diversity (Shannon index) in a freshwater marsh related to standing crop and soil nutrients; circle diameter proportional to diversity values (after Auclair *et al.* 1976a).

from a riverine marsh and analysed them along with many environmental factors. Diversity was greatest where soil nutrients and standing crop were low (Figure 3.7). (As we shall see in Chapter 7, they also found that diversity increased with disturbance by fire, and deceased with amounts of litter.)

Let us consider one final example, this one from a less-well studied part of the world, a large wetland complex (>2500 km^2) on the edge of the Tibetan plateau. In Asia, wetlands dominated by herbaceous vegetation range from temperate to tropical (Hou 1983). In the temperate zone, common graminoid genera include *Phragmites* and *Calamagrostis*, whereas in subtropical and tropical areas *Eulalia*, *Miscanthus*, *Rhynchospora*, *Fimbristylis* and *Sporobolus* occur. Hou is of the opinion that many of these wet meadows are not virginal, nor a climatic climax, but rather a serial stage of succession formed after the felling of forests by humans. Admittedly, China has a long history of human habitation, but given the importance of flooding in producing herbaceous meadows elsewhere, one suspects that flooding is also implicated, and, as in so many other landscapes, may have been overlooked. The largest wetland area in China is a mixture of peatlands and marshes in the Ruoergai wetland in northern Sichuan Province at the headwaters of the Yellow River (Tsuyuzaki *et al.* 1990). A series of 122 1×1 m quadrats arrayed along an elevation gradient tallied 135 plant species comprising eight major vegetation types. At the lower elevations, frequently flooded, one principal community was dominated by *Carex enervis* and *Equisetum limosum*, whereas the other had *Carex meyeriana*, *Equisetum limosum* and *Potentilla anserina*. Drier areas had communities dominated either by *Ranunculus pedicularis* or *Polygonum sphaerostachyum* and *Trollium rananculoides*. The mean number of species per quadrat increased along this gradient from 3.5 to 10, illustrating the impact of flooding in restricting particularly the number of forbs. Further west, near occupied Tibet, grazing by yaks, sheep, goats and horses is a common use of wetlands. In a marshland having similar vegetation to Ruoergai, but intermediate amounts of flooding, *Blysmus sinocompressus*, *Poa chalarantha* and *Kobresia tibetica* dominated. The mean number of species m^{-2} declined from 8 to 4 with increased grazing intensity (Tsuyuzaki and Tsujii 1990). In these wetlands, peat accumulation reached 4 m depths on the bottom of slopes; it is therefore appropriate to now turn to diversity patterns in peatlands.

Plants in peatlands

Glaser (1992) thoroughly studied patterns in the species richness of raised bogs, and provides a model of the sort of analyses of richness needed for other habitat types. His sample of 65 raised bogs from across eastern North America yielded a flora of 81 species, with values ranging from 13 to 50 for individual bogs. There were four major floristic regions; raised bogs in the southern continental region (north of the Great Lakes) contained fewer than 20 species and had the most impoverished floras of eastern North America, whereas bogs in the maritime region had the richest flora of from 32 to 50 species. Multiple linear regression showed that the most important factors controlling richness were precipitation (mean annual precipitation, $r^2 = 0.605$) and growing season (annual freezing degree-days $r^2 = 0.570$), each of which accounted singly for more than 50% of the variation in species richness among bogs. Other important factors were longitude, latitude, Na, Mg and Na/K ratios. In contrast with many other studies (e.g. Connor and McCoy 1979; Barbour and Brown 1974; Welcomme 1979), there was no discernible effect of bog area. Nor was there any apparent effect of disturbance by fire, since bogs with a fire history had no obvious differences from those without evidence of fires.

Further analysis showed that there were important climatic thresholds of 1000 mm of annual precipitation, and 1000 freezing degree–days separating species rich maritime flora from those of other regions. Many of the impoverished continental bogs were less than 2500 years old, whereas the maritime bogs ranged in age from 600 to 9000 years. The higher richness of maritime bogs seems to result in part from the presence of additional species that otherwise occur in fens (e.g. *Drosera intermedia*) or in uplands (e.g. *Cornus canadensis*) in more continental regions. Glaser suggests that, as bogs age, the topography may become more variable, thereby providing more different microhabitats for bog species. It may also be that freezing acts as an additional filter which removes part of the bog flora, just as salinity removes species from other wetlands. Glaser concludes by invoking the hypothesis that evolution determines the size of the pool (e.g. Ricklefs 1987; Eriksson 1993), in this case by determining the number of species able to tolerate the harsh environment of the raised bog.

At more local scales, multivariate studies of peatlands have found nearly the reverse pattern of the freshwater marshes shown in Figure 3.7. Plant diversity increased significantly with pH, calcium and nitrogen in a set of vegetation samples representing a series of bogs, fens and conifer swamps in boreal Canada (Jeglum and He 1995). One explanation for this apparent contradiction might be the size of the samples; these peatlands had woody plants, necessitating the use of larger 10×10 plots. The most likely explanation for this difference, however, lies in the non-linear relationship of plant diversity with soil fertility. As we shall see in the next section, intermediate levels of fertility generally allow the most plant species to co-occur. In very infertile peatlands, then, the number of species would be expected to increase with fertility, while in eutrophic riverine wetlands, the number of species may decrease with fertility. Over the range of fertility levels within peatlands alone, the pattern found by Jeglum and He seems typical – similar results are reported from Alberta (Vitt and Chee 1990) and Minnesota (Figure 3.8).

Since one of the defining characteristics of peatlands is the richness of the bryophyte flora, this component of the vegetation can be examined independently of the others. A study of 96 peatlands of continental western Canada found 110 bryophyte species representing three groups: 64 mosses, 26 hepatics and 20 Sphagna. The number of bryophytes increased with latitude ($r^2 = 0.088$) and decreased with mean annual temperature ($r^2 = 0.145$) (Vitt *et al.* 1995). None of the surface water variables was correlated with richness, and by far the most important factor was the number of microhabitats ($r^2 = 0.455$) as measured by hummocks, pools, tree bases, shaded tree bases, etc. To explore the effects of proximity to the coast, Gignac and Vitt (1990) studied 27 peatlands, from the coastal

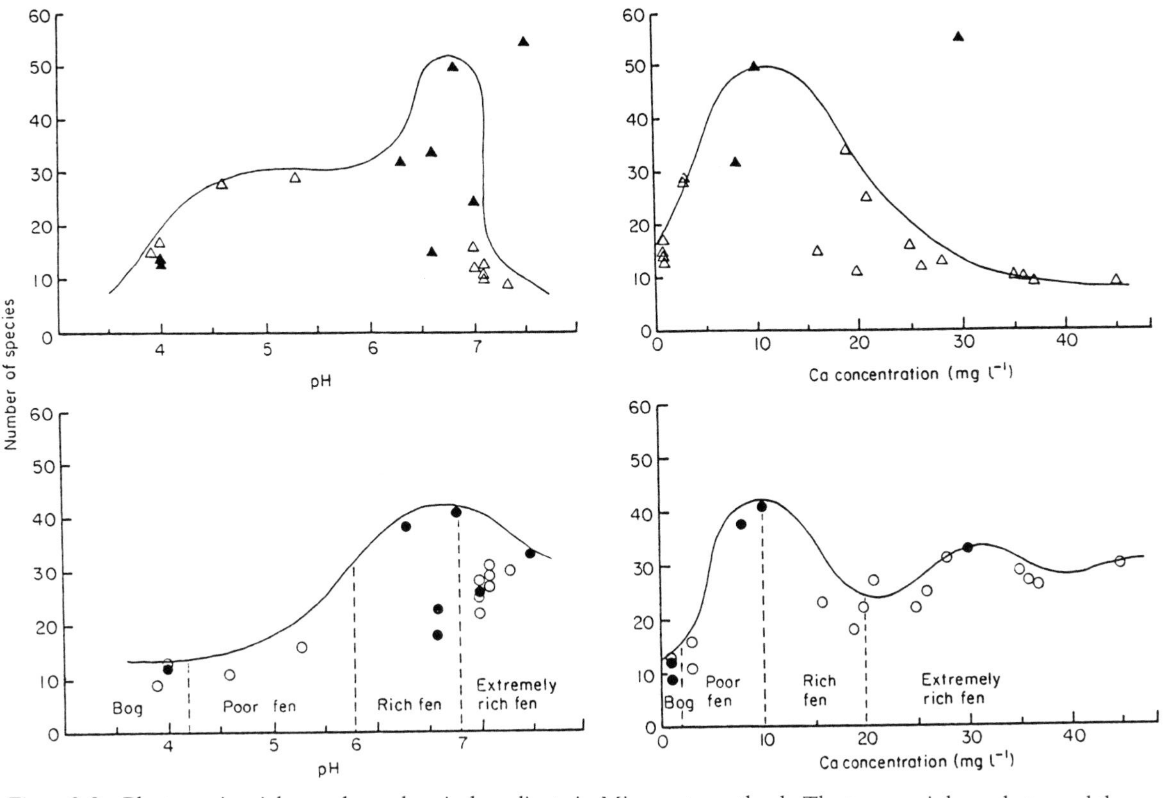

Figure 3.8 Plant species richness along chemical gradients in Minnesota peatlands. The top row is bryophytes and the bottom is vascular plants (from Glaser *et al.* 1990).

islands of British Columbia to the interior of Alberta. *Sphagnum fuscum* was the most widespread species, and appeared to be little affected by climate and surface water chemistry. Of the 18 bryophyte species found, seven were limited to oceanic areas (e.g. *S. pacificum*). Some such as *S. lindbergii* occurred largely in oceanic sites, but extended inland in poor fens. Within individual sites, *Sphagnum* species are also restricted to specific zones along topographic gradients (Vitt and Slack 1975, 1984; Vitt 1994), but this is unlikely to affect their geographic distributions. Overall, fens at high latitudes with low temperatures, high microhabitat diversity and pH between 5 and 7 have the highest bryophyte diversity (Vitt *et al.* 1995).

Plants in intertidal environments

Saline environments appear to pose a nearly insurmountable obstacle to plant adaptation, and one of the primary consequences of high salinity is reduced plant diversity. For illustration, consider the woody plants found in tropical intertidal forest (Tomlinson 1986); the usual nomenclature here is that the species are referred to as *mangroves*, whereas the vegetation is called *mangal*. Tomlinson reports that there are only 9 major genera of mangroves and some 34 species in the world, with 11 minor genera contributing a further 20 species. Including the associates of mangroves, this list can be extended a further 60 species. (To put these figures into perspective, some 90 woody plant species have been found in a single 500 m^2 transect in a tropical riparian floodplain of central America (Meave *et al.* 1991); that is, one riparian transect yielded the same number of species as the world mangal flora). The constraints imposed by salinity are further illustrated by comparisons within plant families; there are some 200 genera and 2600 species in the palm family, but only four are commonly found in mangal (p. 30, 295). The family Myrsinaceae has over 1000 species in about 30 genera distributed throughout the tropics and subtropics, but only four species occur in mangal (p. 284). Similar sorts of calculations could be done with herbaceous families. The question as to why so few plants have evolved salt tolerance remains an open one, particularly since it is evident that some species have found physiological solutions to the stresses imposed by the intertidal habitats. It may be that high rates of decomposition produce extremes of hypoxia, while high rates of evaporization produce extremes of salinity, the combination of these two extremes being nearly insurmountable (Bertness pers. comm.). We may nevertheless conclude that, in saline habitats, the predominant limitation upon diversity is the restricted number of species available that can tolerate intertidal environments.

There are two more detailed studies of diversity in salt marshes. García *et al.* (1993) studied the Guadalquivir River delta, some 1500 km^2 in the south west of Spain on the shore of the Mediterranean. The total flora for the site comprised just 87 plant species, including widespread perennial herbaceous genera such as *Scirpus, Juncus, Phragmites, Cynodon, Polygonum* and *Senecio*, some of these genera suggesting near freashwater conditions; at higher elevations, there were many more annuals with scattered stands of chenopoid shrubs. Species richness in 1/4 m^2 quadrats ranged from 2 to 26. All of the high richness quadrats occurred at low salinity. Half of the variation in richness could be accounted for by salinity alone; in contrast, biomass of the quadrats could explain only one-quarter of the variation.

Plant diversity was explored in 36 marsh communities along the coast of the Gulf of Mexico (Gough *et al.*1994). Most of these plant species were perennial, including widespread genera such as *Aster, Eleocharis, Scirpus* and *Spartina*. The best predictor of plant richness was elevation, which would be a surrogate for flooding and hypoxia; it alone accounted for 52% of the variation in the number of plant species (Figure 3.9(*a*)). At the lowest elevations, there was only slightly above one species per square metre, while at high elevations, the number approached nine. Species richness also decreased with salinity (Figure 3.9(*b*)) and biomass (Figure 3.9(*c*)), but increased with soil organic matter (Figure 3.9(*d*)). Overall, a multiple regression analysis incorporating elevation, salinity, soil organic matter and biomass accounted for 82% of the variation in species density.

Woody plants and environmental constraints

In terrestrial ecosystems, tree diversity decreases with latitude and aridity, but increases with topographic and geological variation (Gentry 1988; Specht and Specht 1993; Latham and Ricklefs 1993; Austin *et al.* 1996). As we have already seen in Chapters 1 and 3, flooding and salinity strictly constrain the growth of woody plants, in which case, these large-scale patterns reported from terrestrial ecosystems may provide little useful information on wetlands. Keogh *et al.* (1998) therefore collected a data set consisting of 245 plots from forested wetlands described in 77 studies, along with 12 new stands from northern temperate floodplains, and found that latitude was a comparatively weak factor, accounting for less than 10% of the variation in tree species richness. In tropical areas, terrestrial forests had >120 species per plot, whereas freshwater wetlands averaged only 31 species per plot. Figure 3.10 shows that temperate climates, peat substrates, and salinity all further reduced the number of species of trees found. Some

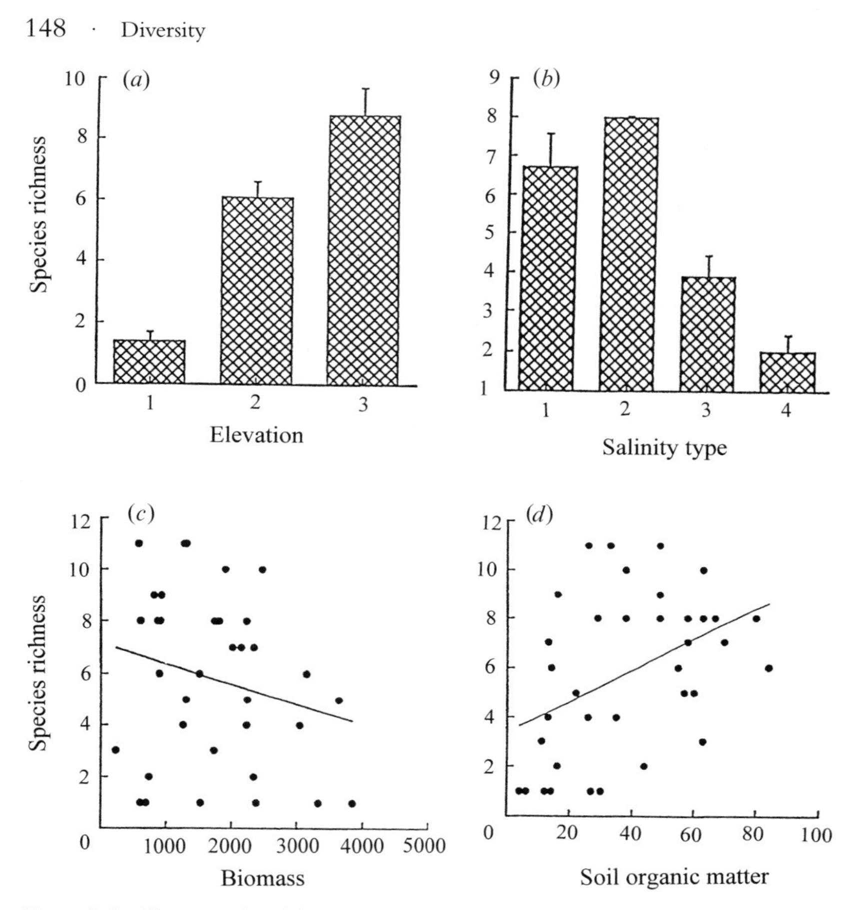

Figure 3.9 Plant species richness patterns in Gulf coast wetlands; effects of (*a*) elevation, (*b*) salinity, (*c*) biomass and (*d*) soil organic matter (from Gough *et al.* 1994).

combinations of constraints, such as northern temperate climate and salinity (salt marsh), were incapable of supporting any trees at all. Keogh therefore suggests that the additive effects of local environmental constraints are probably the most important factors controlling tree richness in wetlands. Each constraint appears to reduce the number of species by some 2/3, with salinity exerting a much stronger effect than peat, cold or flooding.

Plant species richness and resource specialization

It has long been believed that, all other things being equal (which, of course, they rarely are), higher numbers of species can co-exist in a given range of habitats if they each use a narrower range of resources (e.g.

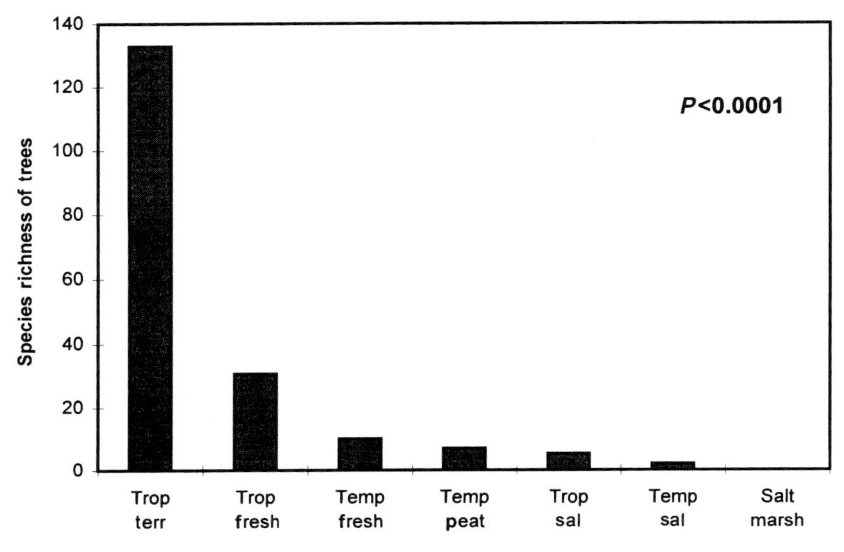

Figure 3.10 The species richness of trees in five types of forested wetlands compared to tropical terrestrial forests (*left*) and saltmarshes (*right*). $N=257$, Kruskall–Wallis ANOVA, five categories, $P<0.0001$ (from Keogh *et al.* 1998).

MacArthur 1972; Schoener 1974; Pianka 1981). Taking advantage of zonation patterns, one can measure the range of elevations occupied and species richness for many shoreline transects and test for the predicted relationship. We expect there to be more specialists where more species co-exist, but there is no evidence to support this hypothesis (Figure 3.11). If anything, the opposite is true. Similar results were found in Keddy (1983). Such results suggest that co-existence in plants is likely to be explained by factors other than co-evolution along gradients.

Biomass[1] and diversity: a general model for herbaceous plant communities

The above studies show that a relatively small subset of environmental factors can often successfully predict the numbers of species occupying a wetland. For fish the most important factors are pH, oxygen levels and vegetation structure. For plants, the factors are elevation, salinity and fertility.

[1] There is an unfortunate synonymy in terminology; When they measure the dry weight of plants in a sample, Europeans tend to call it standing crop, or standing crop and litter, whereas North Americans tend to call it biomass. I have decided here not to force all terminology of text and figure captions into consistency.

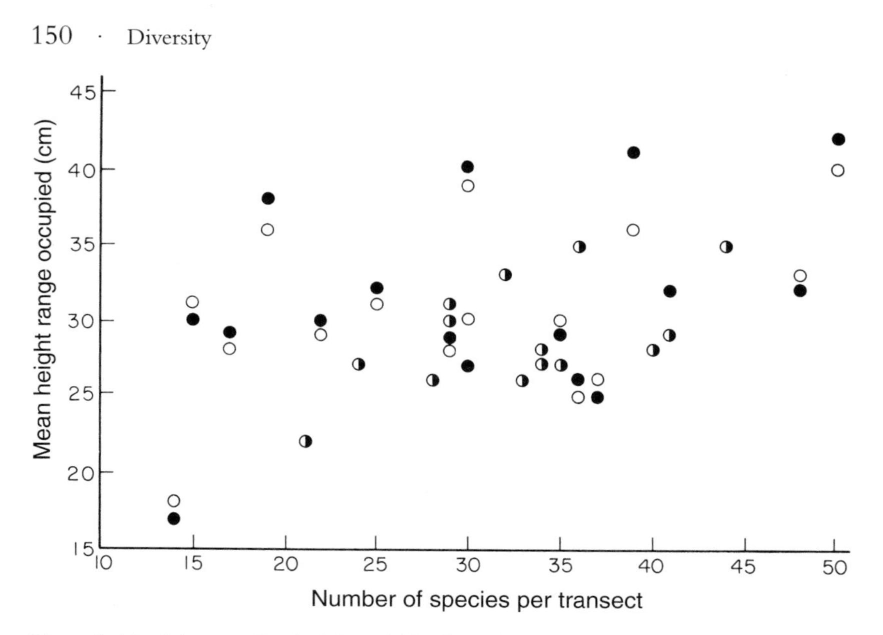

Figure 3.11 Mean realized niche width plotted against species richness in 30 lakeshore transects. Open circles include all lakeshore species; solid dots are shoreline species alone (from Keddy 1984).

The primary goals in studies of biological diversity are to first uncover these factors controlling richness, and then to rank them in order of importance. Once such lists are available for a wide array of habitat types and organisms, the task of comparison can begin.

Another approach is to ask whether some common underlying process would allow us to combine studies from different habitats. Is there, for example, some way in which pH and oxygen for fish is comparable with salinity and elevation for plants? There is a growing body of evidence to suggest that many of these factors can be related back to productivity and/or the biomass of the organisms involved. Since biomass is relatively easy to measure in the field, let us focus on a model that predicts species richness from biomass. Owing to the general importance of this model, and its potential application to many taxa, I will trace the historical evolution of this work in added detail. Models seeking the predictors of diversity are now an active area of research (for overviews, see Huston 1994 and Grace 1999), but only a small subset of the published studies use wetlands and are therefore considered here.

Let us begin the story with Grime's (1973) observation in British grasslands that habitats with intermediate levels of biomass appeared to

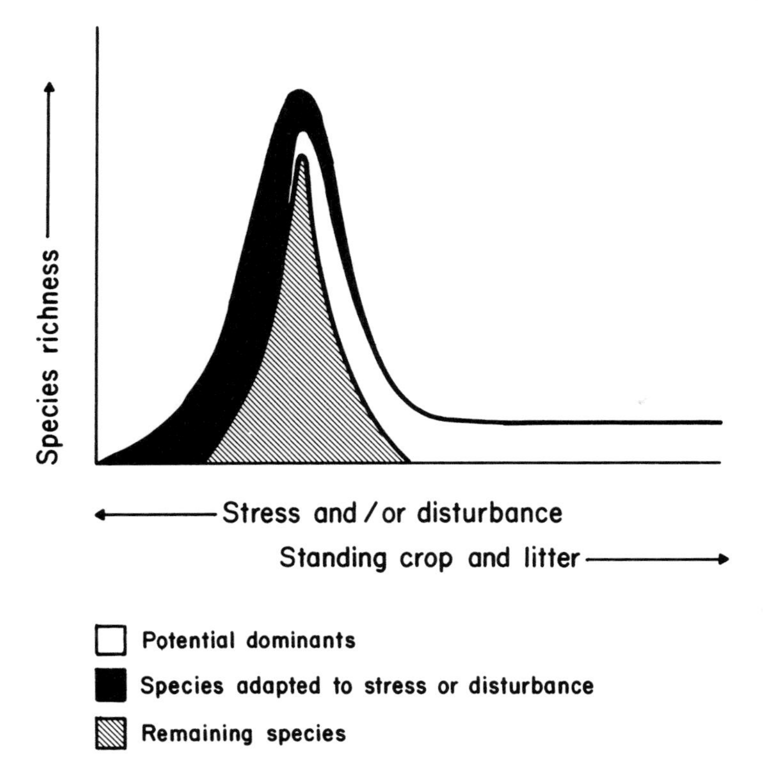

Figure 3.12 Species richness along a gradient of standing crop and litter: the concept (from Wisheu *et al.* 1990, adapted from Grime 1979).

have the largest numbers of plant species. He postulated that there is a general relationship in vegetation between species richness and standing crop of the form shown in Figure 3.12. Moreover, he postulated that at one end of the gradient, species richness is low because of high levels of stress or disturbance, whereas at the other end, species richness is low because of dominance from a few strong competitors.

Wheeler and Giller (1982) first showed that a similar relationship might occur in wetlands by examining plant diversity in an array of British fens subjected to different types of management (Figure 3.13(*a*)). These fens were located in the Norfolk Broadland, some 3300 ha of fen along water courses draining into the sea at Great Yarmouth. Their work raised the promising possibility that one could develop a general predictive model that would relate richness to biomass. If so, we could have more than a postulate about pattern, we would have a model relating an

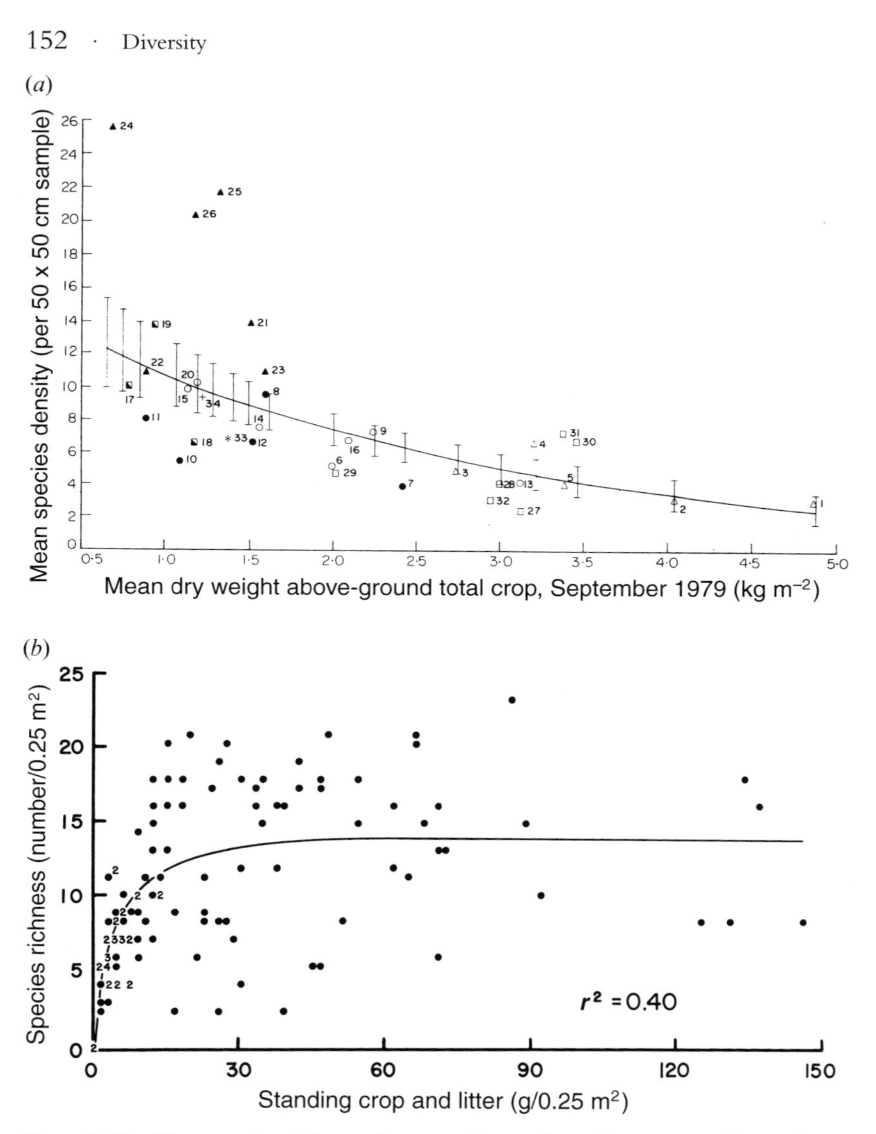

Figure 3.13 Plant species richness along gradients of standing crop and litter: the tests (*a*) fens (from Wheeler and Giller 1982), (*b*) lakeshores in Nova Scotia (from Wisheu and Keddy 1989a)

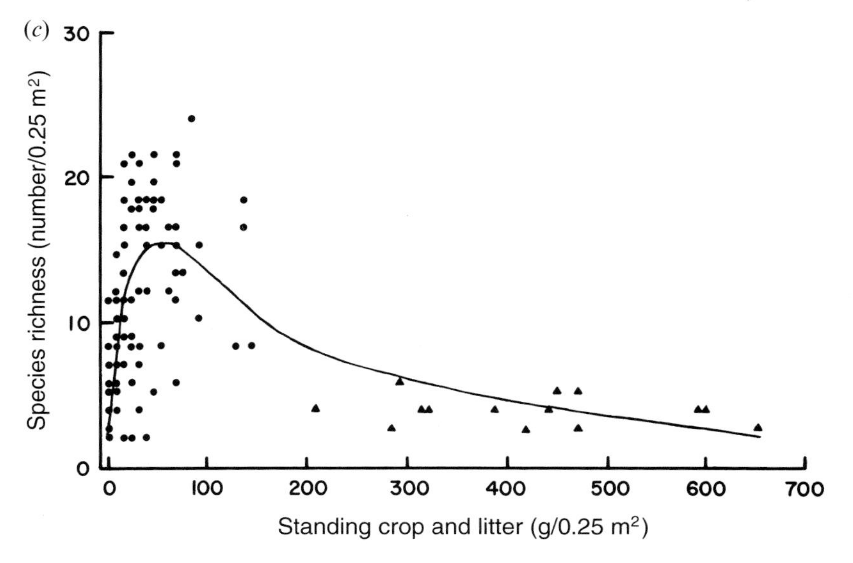

Figure 3.13 (*cont.*) (*c*) lakeshores in Nova Scotia (dots) and high biomass *Typha* marshes in Ontario (triangles) (from Wisheu and Keddy 1989a).

easily measured and fundamental independent variable (standing crop) to an important dependent variable (species richness). In the next extension, Wisheu and Keddy (1989a) tested whether this 'intermediate diversity' model applied to wetlands on another continent, and further tested whether the pattern was the same in four different vegetation types. A somewhat similar pattern emerged (Figure 3.13(*b*)), but the coefficients of the regression lines also varied among the four habitats. Thus, while the overall shape of the relationship had broad generality, the specific shape was apparently less consistent. The absence of high biomass sites was thought to be responsible for failure to detect a decrease in richness at high biomass; Figure 3.13(*c*) shows that when high biomass sites were added, the typical Grime-type curve emerged.

The level of biomass associated with high species richness is of particular interest. Does it vary among vegetation types? That is, even if the shape of the curves vary somewhat, does the region of high diversity always occur in similar biomass levels? Table 3.6 compares a series of published studies; the results are similar enough to hold promise, but not as consistent or tidy as one might hope. Perhaps a much larger data set with many more quadrats is necessary. Data from more than 400 quadrats in Eastern North America (Figure 3.14(*a*)) show that wetlands with biomass in the

Table 3.6. *Standing crop and litter values associated with maximum species richness in an array of wetlands*

Site	Habitat type	Standing crop and litter (g/m^2) at maximum species richness
Wilsons Lake, NS, Canada	Occasionally flooded gravel lakeshore	200
	Occasionally flooded boulder lakeshore	140
	Frequently flooded gravel lakeshore	260
	Frequently flooded boulder lakeshore	80
Eastern Canada	Marshes and wet meadows	60–400
Axe Lake, Ont., Canada	Lakeshore	50–300
Ottawa River, Ont., Canada	Riverine wetland	300
Green Swamp, NC, USA	Pine–wiregrass savannas	280
Gelderse Vallei, Netherlands	Reed swamp, grassland, road verges	400–500
Westbroekse Zodden, Netherlands	Fens	400–500
	Wet grassland	425
	Herbaceous fen	1500
Norfolk Broadland, Britain	Chalk grasslands	150–350

Notes:
Note that standing crop and litter are expressed as g/m^2 and not $g/0.25\ m^2$. Maximum richness occurs at approximately 500 g/m^2 in terrestrial vegetation.
Source: From Wisheu and Keddy (1991).

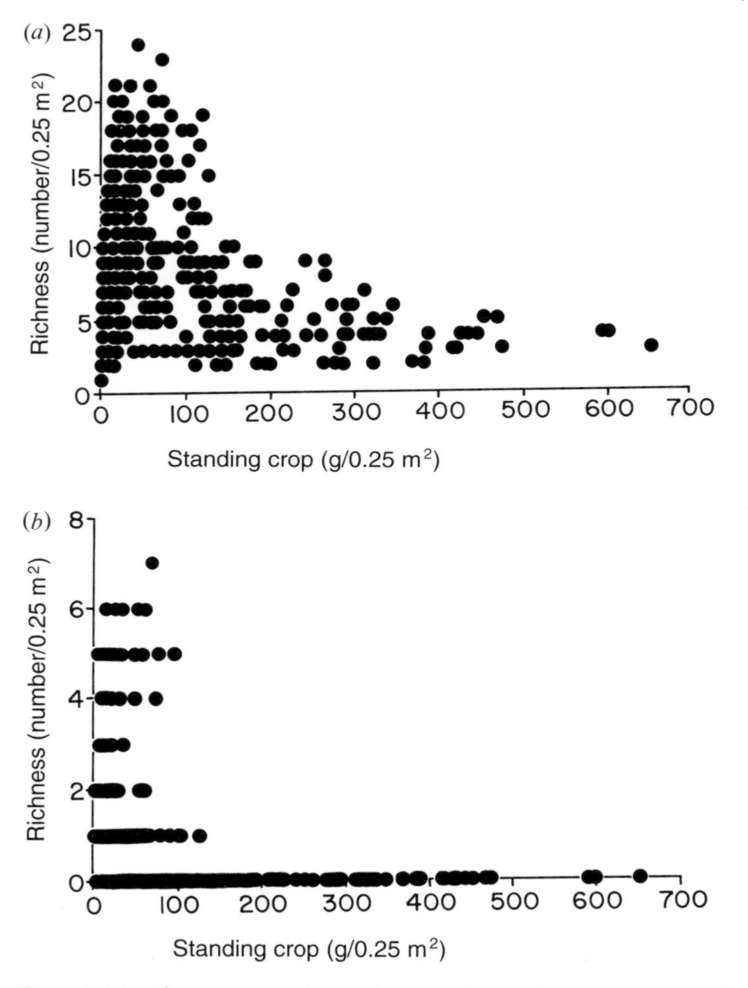

Figure 3.14 Plant species richness along gradients of standing crop (*a*) for 401 0.25 m^2 quadrats in eastern North America (*b*) same quadrats but nationally rare species only (reprinted from Moore *et al.* 1989).

range of 50 g/0.25 m^2 had the highest richness across a wide range of wetland sites. Figure 3.14(*b*) shows, moreover, that if rare species alone are considered, the vast majority of nationally rare wetland species occur in the very low standing crop habitats, less than 100 g/0.25 m^2. Similarly, Wheeler and Shaw (1991) collected data from 86 herbaceous fens, finding that the number of species ranged from 2 to 50 in 0.25 m^2 quadrats, and standing crop and litter ranged from 80 to $>$2900 g m^{-2}. The plot of

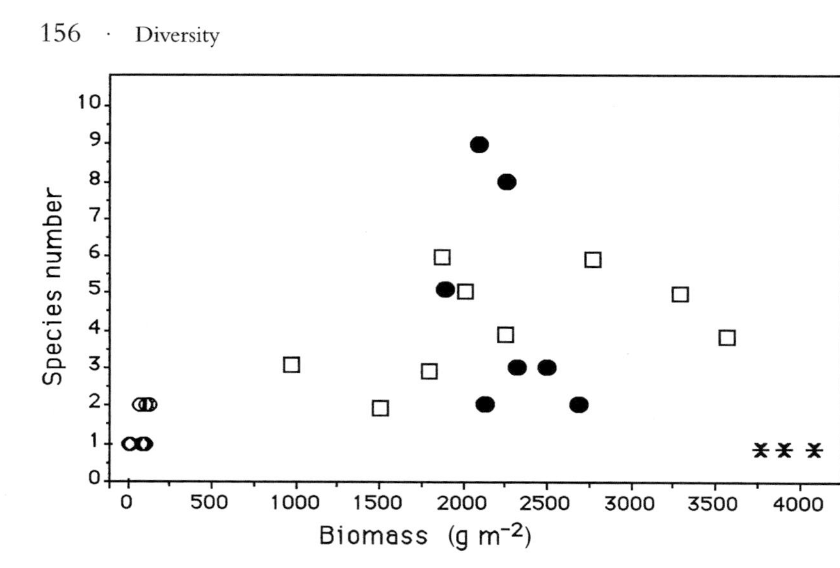

Figure 3.15 Plant species richness plotted against biomass for marshes in Belize. Circles are *Eleocharis cellulosa*, empty squares are *Cladium jamaicense,* solid circles are *Typha domingensis*, asterisks are *T. domingensis* stands in the Everglades. Plots are 5×5 m^2 (from Rejmankova *et al.* 1995).

richness against biomass was very similar to that in Figure 3.14, heavily skewed to the left with maximum richness around 1000 g m^{-2}. This region had vegetation types dominated by *Cladium mariscus* fens from the Norfolk Broadland. The highest biomass and lowest richness occurred in sites dominated by *Phragmites australis*, and sites with high biomass had few, if any, rare species. They conclude 'Maintenance of low-growing swards of herbaceous fen vegetation is particularly important for the conservation of many rare fen species.' At this large scale (wetlands representing a broad range of habitats), the pattern is clear and consistent.

Rejmankova *et al.* (1995) then added to the story by looking for similar patterns in tropical herbaceous wetlands. They sampled marshes from floodplains and depressions in northern Belize; these marshes are usually mostly flooded all year long, but are occasionally disturbed by fire and desiccation. In general, plant diversity is remarkably low, with a mean of some five plant species in 5×5 m quadrats! Plant richness still reaches a maximum at intermediate levels of biomass in *Typha domingensis* marshes (Figure 3.15). Further, if one adds in high biomass reference sites (as in Figure 3.13(*c*)), a decline in richness at high biomass levels becomes evident. Thus, a similar pattern to temperate zone marshes emerges, with two key exceptions: overall richness is strikingly lower, and the region of highest richness is shifted to the right. Similarly, Crow (1993) describes how large clonal perennials such as *Eleocharis mutata* and *Typha domingensis* form extensive nearly pure stands in floodplain marshes along the Rio Tempisque in Costa Rica. This tendency for dominance by a few such species is, suggests Crow, one possible explanation for the low diversity of herbaceous wetland plants in the tropics.

There remains a great deal of scatter around the lines in Figures 3.9 and 3.13. There are at least three ways to try to reduce variance and increase precision. First, one could try to reduce the residual variance by adding in a second predictor variable besides biomass. If biomass is a crude measure of productivity, then disturbance might be an appropriate second factor to use. There is no easy way to measure disturbance, but Shipley *et al.* (1991a) used the proportion of species that were functional annuals as a surrogate, in which case the proportion of flora being perennials becomes an inverse surrogate of disturbance. Using both biomass and the proportion of plants that are perennial it was possible to predict plant richness with greater success (Figure 3.16). Seventy-five per cent of the variance could be accounted for in a set of 48 quadrats from Quebec, and 45% of the variation in a larger data set of 224 quadrats from southern Ontario.

A second approach to greater precison would be to try reducing unexplained variation by dividing the data into smaller units. Instead of comparing across many vegetation types, one might look for better relationships within single vegetation types. In many cases this would also shorten the biomass gradient because long biomass gradients are usually the result of comparison across many vegetation types. If the data are divided this way, the Grime model cannot be detected at all (Figure 3.17). Apparently, the factors and processes that produce the intermediate diversity pattern, occur only in large scale comparisons across vegetation types.

A third approach to greater precision requires reconsidering the statistical methods. In his original work, Grime proposed that the relationship applies to *potential* species diversity. A significant problem with this verbal formulation is that 'potential diversity' cannot be measured. Most workers have used curve fitting techniques which assume that all observed data points have equal weighting. However, taking the original wording precisely, Grime proposed that there is an outer envelope or upper limit to species richness. It was only the outer limit of the relationship he drew in Figure 3.11. Regrettably, there are few if any techniques for fitting upper limits or envelopes (but see Scharf *et al.* 1998), and most of us are now so driven by statistical tools (or at least intimidated by others who profess to know more) that we would fear to hand draw a line

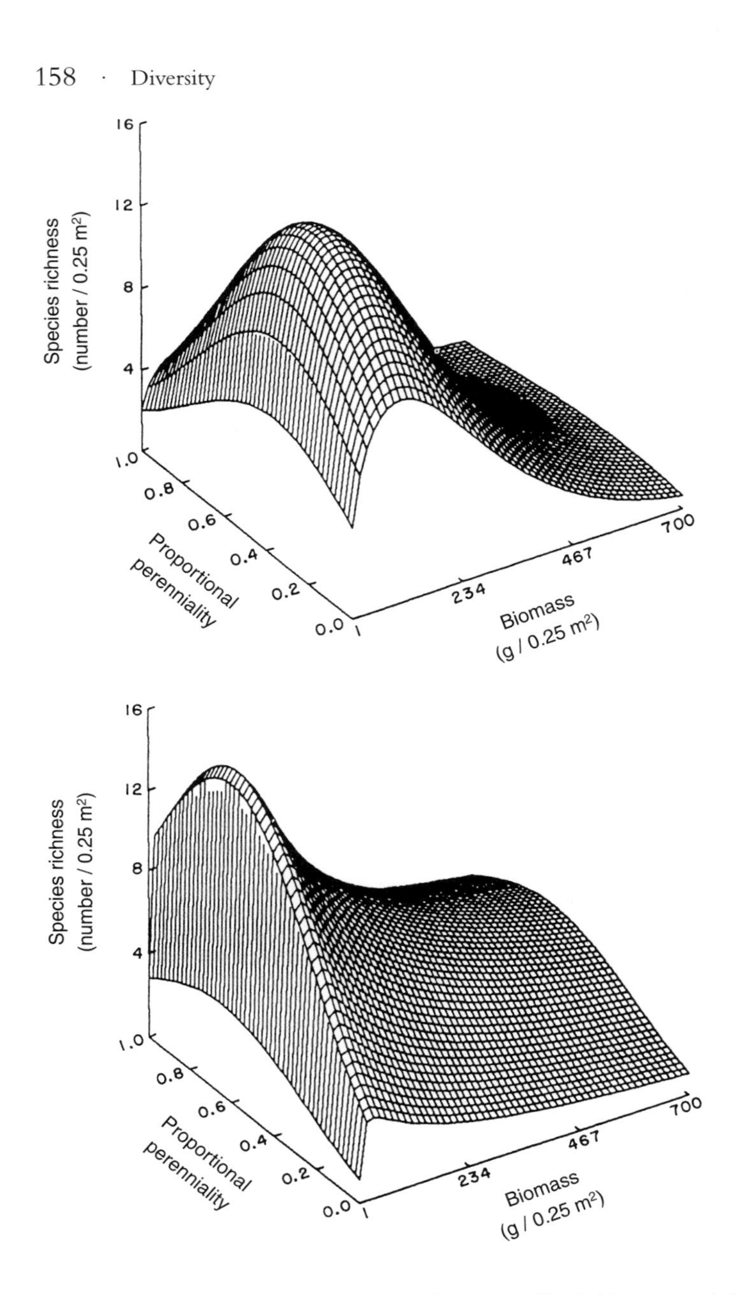

Figure 3.16 Plant species richness as a function of both biomass and the proportion of species that are perennial for two geographic locations, Quebec (*top*) and Ontario (*bottom*) (from Shipley *et al.* 1991a).

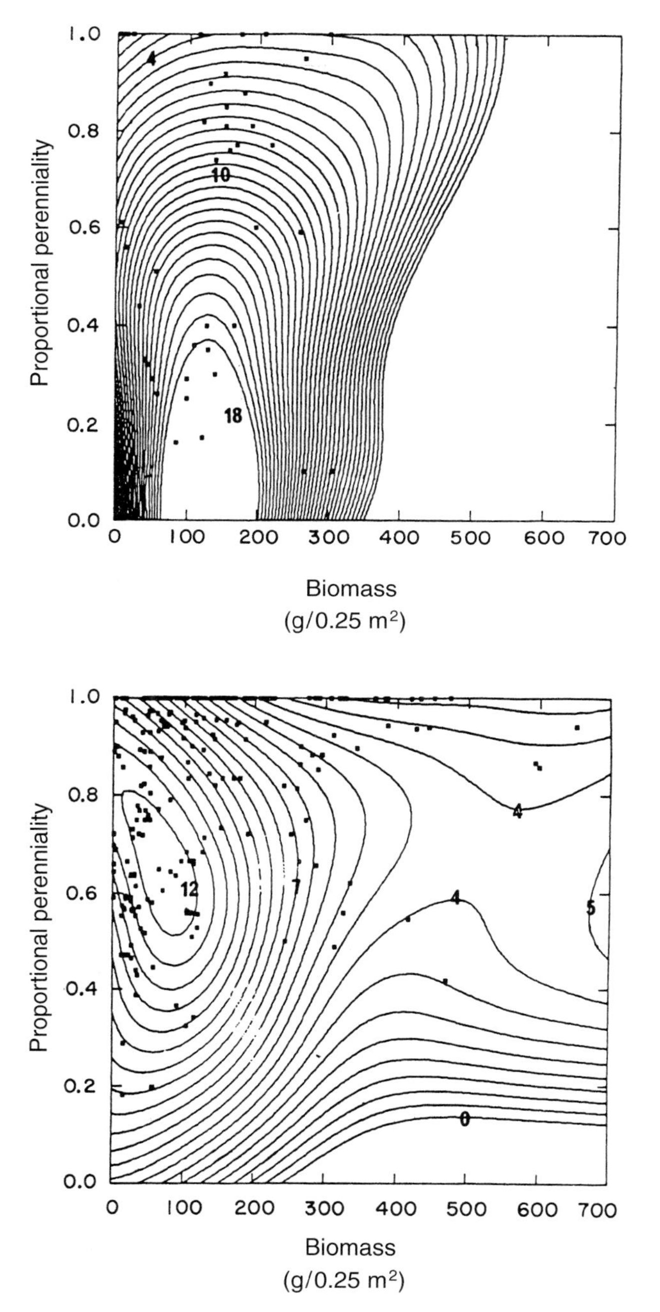
1.0
0.8
0.6
0.4
0.2
0.0
Proportional perenniality
0
100
200
300
400
500
600
700
Biomass
(g/0.25 m2)
4
10
18
1.0
0.8
0.6
0.4
0.2
0.0
Proportional perenniality
0
100
200
300
400
500
600
700
Biomass
(g/0.25 m2)
12
7
4
4
5
0

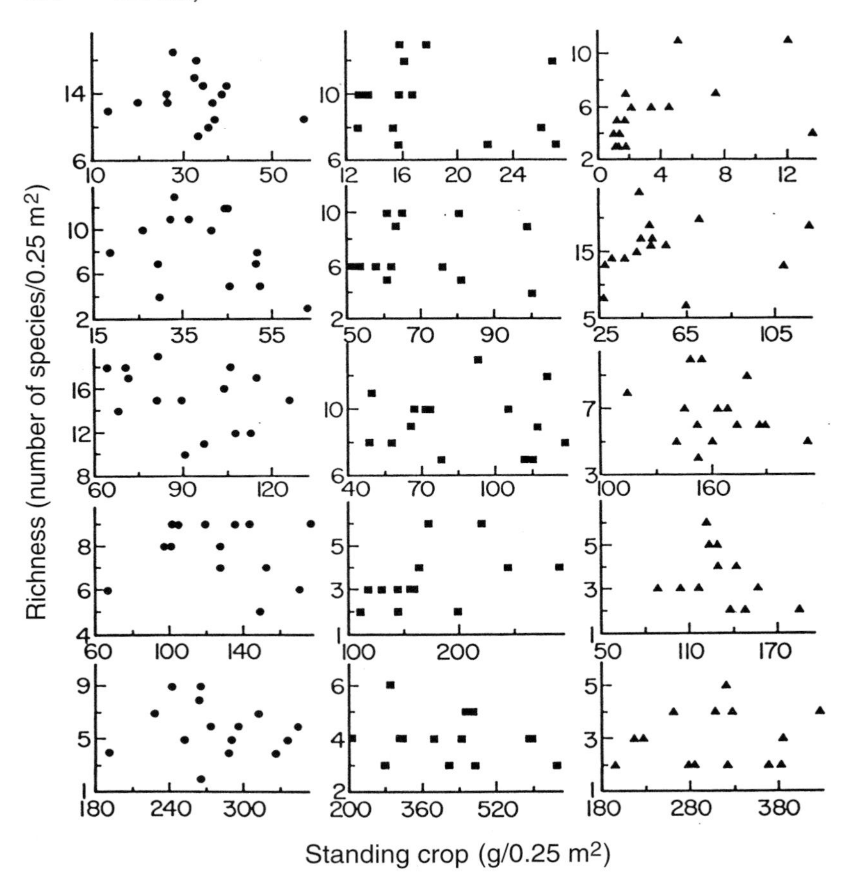

Figure 3.17 Plant species richness as a function of standing crop within 15 vegetation types (from Moore and Keddy 1989).

around our data. Perhaps one can get away with this in a book. If so, one generates the line in Figure 3.18. Moving the line in this way changes the issue of precision. Instead of trying to account for scatter around the line, as Shipley *et al.* (1991a) attempted to do, we are, instead, challenged to explain why many sample units fall below the line. Perhaps stochastic events of dispersal and extinction must be invoked to explain why all sample units are not simultaneously at the maximum.

The consideration of potential richness requires us to reconsider the salt marsh work by Gough *et al.* (1994) shown in Figure 3.9. The quite remarkable result of this study was the importance of physical factors as opposed to those of biomass. Does this indicate that the biomass-based

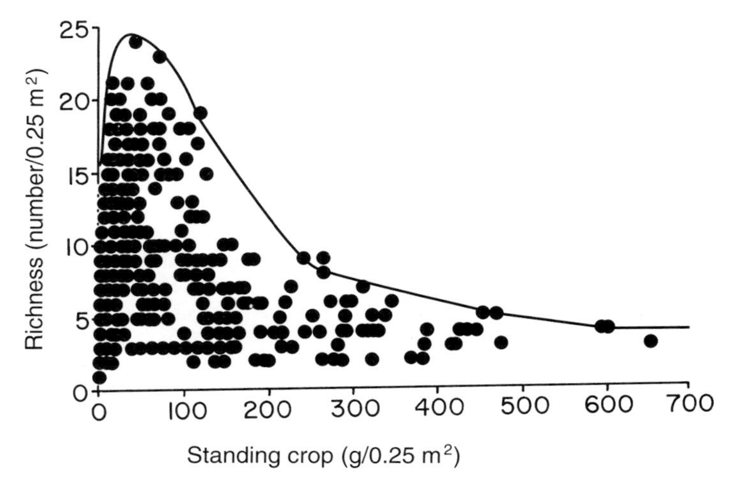

Figure 3.18 The hand-drawn line marks the upper limit of richness in herbaceous wetlands (after Moore *et al.* 1989 in Wisheu *et al.* 1990).

models are untenable? In part, such results may arise from the inclusion of elevation, since most of the studies preceding Gough *et al.* did not combine both elevation and biomass as predictors. Wisheu and Keddy (1989a), for example, explicitly sampled only two elevations in their study. As well, Gough *et al.* used wetlands with strong gradients in salinity, whereas the other studies used freshwater wetlands. Thus Gough *et al.* had two major physical factors varying in their data. If one excludes certain extreme environments – lowest elevations, saline and brackish marshes, a much stronger relationship between richness and biomass emerges, one accounting for 34% of the variance. Even so, the pattern is simply one of decreasing richness with increasing biomass, with no evidence of a curvilinear component. Gough *et al.* conclude their interpretation by resurrecting the concept of potential richness. In their scenario, physical factors control potential richness by controlling the pool of species available (that is, the set having the appropriate physiological tolerance limits, Figure 3.19). If salinity and elevation control the pool of species potentially able to occupy the site, biomass maybe a surrogate for the intensity of competitive exclusion. Higher biomass produces higher intensity of competition, killing many neighbours and leaving high biomass sites with few species. Finally, the size of the pool (potential richness) is controlled not at local scales, but by larger and longer-term processes such as history, biogeography and elevation (Ricklefs 1987).

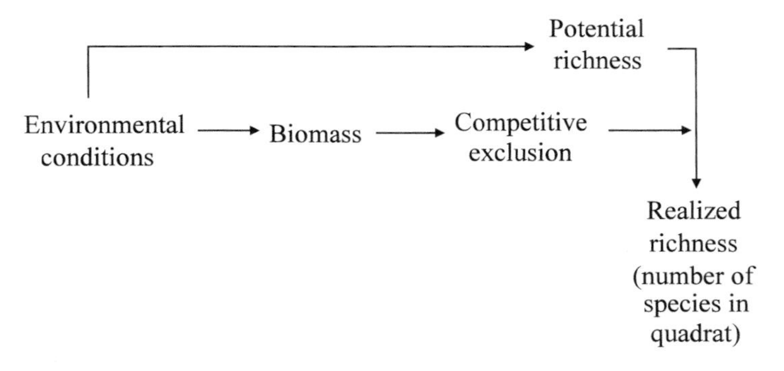

Figure 3.19 Two pathways may control the species richness of a quadrat. Environmental conditions at a site may regulate either the potential richness (that is, the pool of species potentially able to tolerate a site) or the biomass (that is, the intensity of competitive exclusion) (after Gough *et al.* 1994).

The variation encountered in the previous studies might be a source of despair – how can we tolerate this amount of variation in models of nature? Solace may be drawn from the variation within the Hertzsprung–Russell diagram (Figure 3.20) which is a fundamental relationship in astrophysics. I have discussed the merits of this diagram, and the lessons for ecologists, elsewhere (Keddy 1994), so I will not repeat the material here. (I am not, like Dickens, being paid by the word.)

Experiments may also help to reduce uncertainty by clarifying causal relationships. Hacker and Bertness (1999) extended these descriptive studies of salt marshes by experiments carried out across intertidal zones in New England salt marsh. In this system, plant species number is low in the high and low intertidal and high in the middle intertidal. They established plots with and without plant neighbours (pp. 330–333 as in Wilson and Keddy (1986b)), finding that competition was more important in the high intertidal habitat whereas physical factors were critical in the low intertidal. In the middle intertidal, direct positive interactions apparently influenced richness – three of the four species died (*Atriplex patula, Iva futescens* and *Solidago sempervirens* but not *Limonium nashii*) without neighbours but had minimal mortality with neighbours. These positive effects were due to one particular facilitator species, *Juncus gerardi*, which ameliorated the soil conditions that develop in its absence. It shades the soil, decreasing evaporation and the resulting salt accumulation and oxygenates the soil with its aerenchyma. They concluded that the higher species richness in the middle intertidal is dependent on three co-occurring conditions; the absence of a competitive dominant, less harsh physical

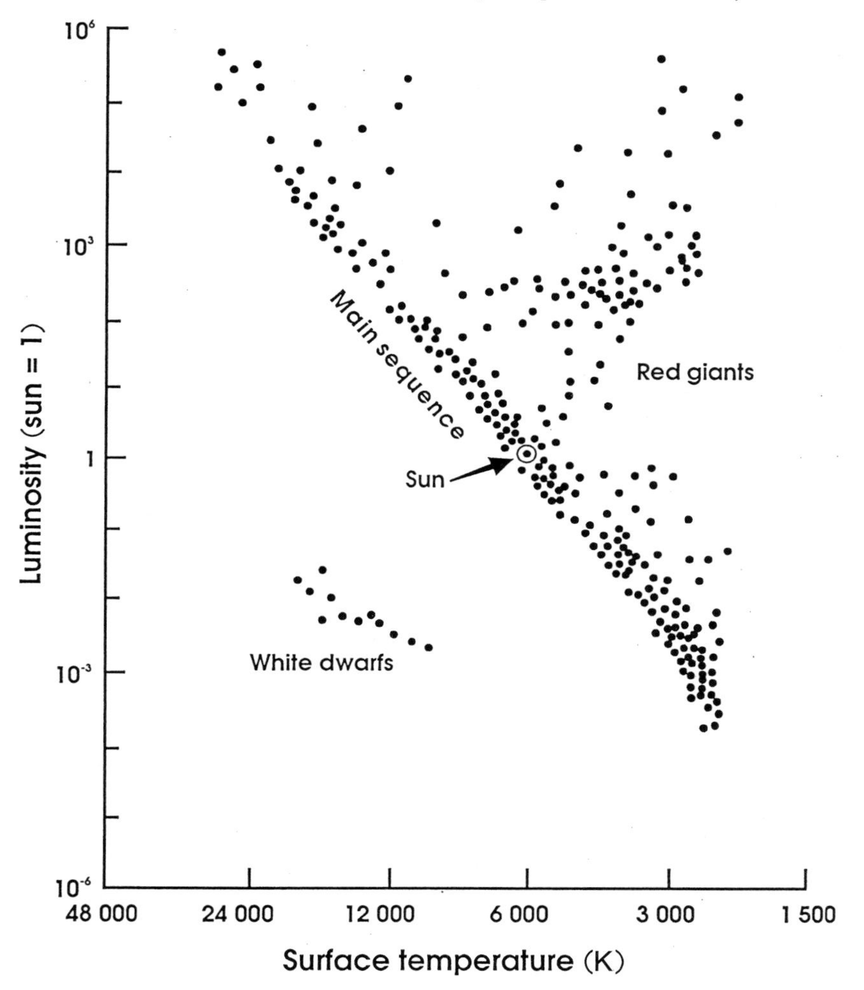

Figure 3.20 The Hertzsprung-Russell diagram summarizes fundamental relationships among stars (from Keddy 1994).

conditions than the lower intertidal, and the presence of a facilitator species.

Species pools and biodiversity

The foregoing studies concentrated largely upon predicting the number of species in small samples or small pieces of habitat. When planning parks and nature reserves, there is less interest in the number of species per sample unit, the focus being the total number of species protected in a reserve system. Quadrat samples will reveal something about local diversity patterns. Species-area relationships may tell us how many species to expect in any given reserve. The number of protected species in a set of reserves can only be obtained by pooling the data from many individual parks. Hence the origin of the term 'species pool', an approximate synonym for the more popular term 'biodiversity'. There are few studies of species pools that relate specifically to wetlands.

This distinction among the number of species in a sample unit, the number in one park, and the number in a pool, is important for both theoretical and applied purposes. It is self-evident that as we increase the size of each sample unit (e.g. quadrat), or as we increase the number of such units, more species will be encountered. The relationship has been well documented, and is generally asymptotic – that is to say, the number of new species encountered declines with sampling effort until, in theory, all of the species have been encountered and tallied (e.g. Pielou 1975). Figure 3.21 presents a wetland example. The asymptote is a good estimate of the number of species in the habitat type. In practice, if sample units become large enough, or numerous enough, new habitats are encountered, in which case the number of species tends to continue increasing with the sampling effort. Further, as the scale of sampling is expanded, the controlling processes are likely to change, with local processes such as competition being eclipsed by geographic processes such as history or speciation (Ricklefs 1987). It is therefore important to specify in advance exactly what habitat type or geographic locale is being inventoried. The word *pool* then describes the complete list of species for that habitat, park, or geographic locale.

Eriksson (1993) has provided a simple model to help explore the relationships between community richness (the number of species in a local community), and the number in the pool. Let the number of species in a particular community S be a function of local colonization and extinction, just as the number of species in the pool N is a function of speciation and extinction. The local colonization rate is then proportional to the number of species in the species pool N minus the number of species that is already present in the community S – that is, the number of potential colonizers left in the pool. The local extinction rate, however, is a function only of the number of species already present. Adding in two proportionality constants for the rates of colonization c and extinction e,

$$\frac{dS}{dt} = c(N - S) - eS$$

and at equilibrium the value of S is

$$S^\star = N(c/c + e)$$

This simple expression yields some predictions. First, if the local extinction rate e is very low, then $S^\star$ will be close to N; that is, the local community will contain most of the species in the pool. Secondly, if the rates of colonization and extinction are equal, that is $c = e$, then $S^\star = \frac{1}{2}N$; the local community will have one-half of the species in the pool. Thirdly, if extinction rates are much higher than colonization rates, the community will be species poor and yet the number of invading species per unit time will be relatively high. One simple explanation, then, for communities having many species is the existence of a large species pool; that is to say, S is large because N is large. Eriksson calls this the species pool hypothesis. Depending upon the values of e and c, however, a variety of other scenarios can be postulated.

In many cases, ecologists have data not on the entire community, but only on the number of species in a series of samples from that community, such as the number of species in a set of quadrats, traps, nets or transects. The total number of species encountered in the sampling therefore becomes a working estimate of the pool; the greater the sampling effort, the more reasonable the assumption becomes. Dissecting all these possible sample types a 'sample unit' for the sake of linguistic convenience, we can ask what relationships might exist between the number and compostion of sample units and the species pool. In the extremely unlikely case where all sample units are identical in composition, then the species richness of any one sample unit will be the same as the pool. The greater the compositional difference between each pair of sample units, the greater pool size will be relative to the number of sample units. We have already seen (pp. 150–162) how species richness (or number of species per sample unit) varies along biomass gradients. Suppose we now take the same gradient, but ask instead how pool size varies along it.

Several relationships between pool size and environmental conditions might be postulated. From one point of view (e.g. Connell and Orias 1964), pool size may be highest in high biomass wetlands because these have the most energy to allocate among species (Figure 3.22*a*). Alternatively, pool size may simply mirror alpha diversity (Figure 3.22*b*),

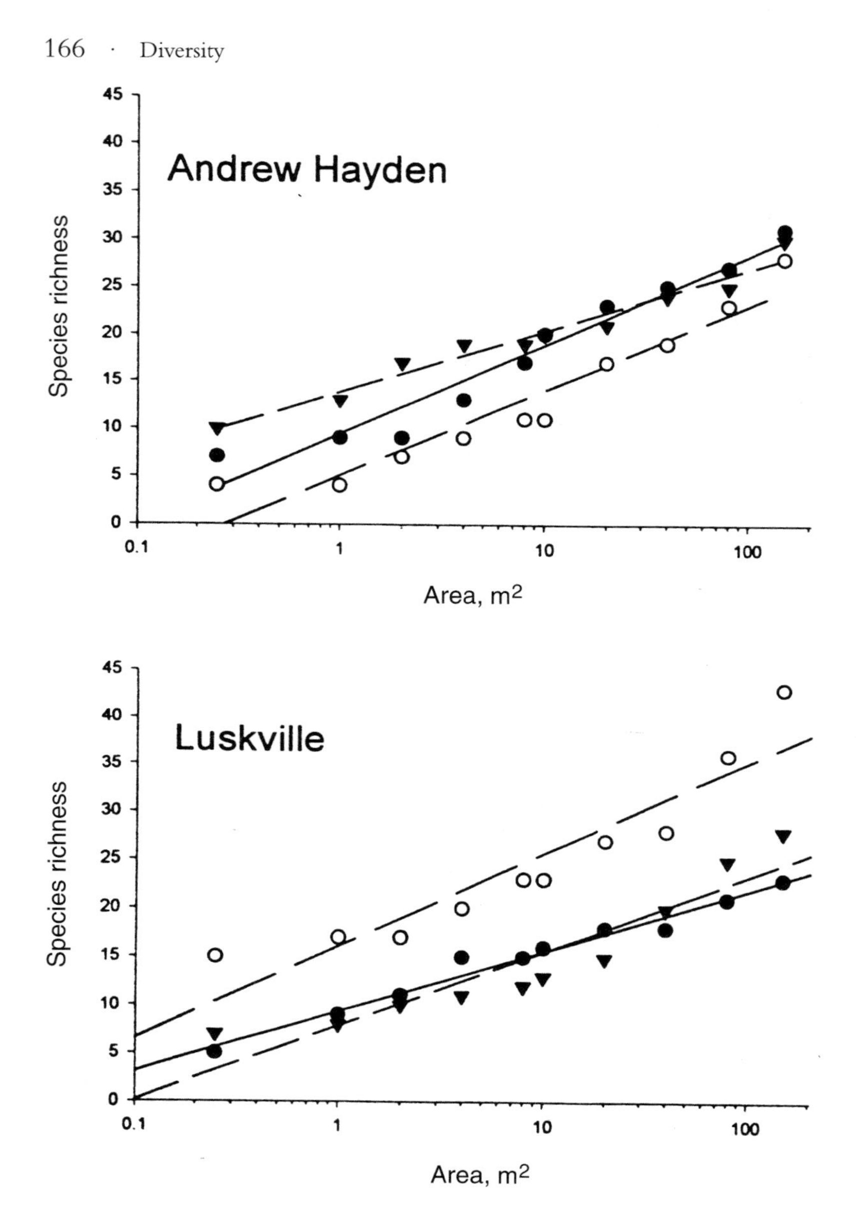

Figure 3.21 Species–area relationships in herbaceous marshes along the Ottawa River, Canada. (Common species include *Lythrum salicaria*, *Scirpus americanus*, and *Eleocharis erythropoda*, see Day *et al.* 1988). Filled circles – high biomass; open circles – intermediate biomass; triangles – low biomass. All slopes are significantly ($P < 0.0001$) different from zero. Slope increased significantly with biomass and soil organic content, whereas the intercepts decreased significantly with biomass and organic content (from Weiher, 1999).

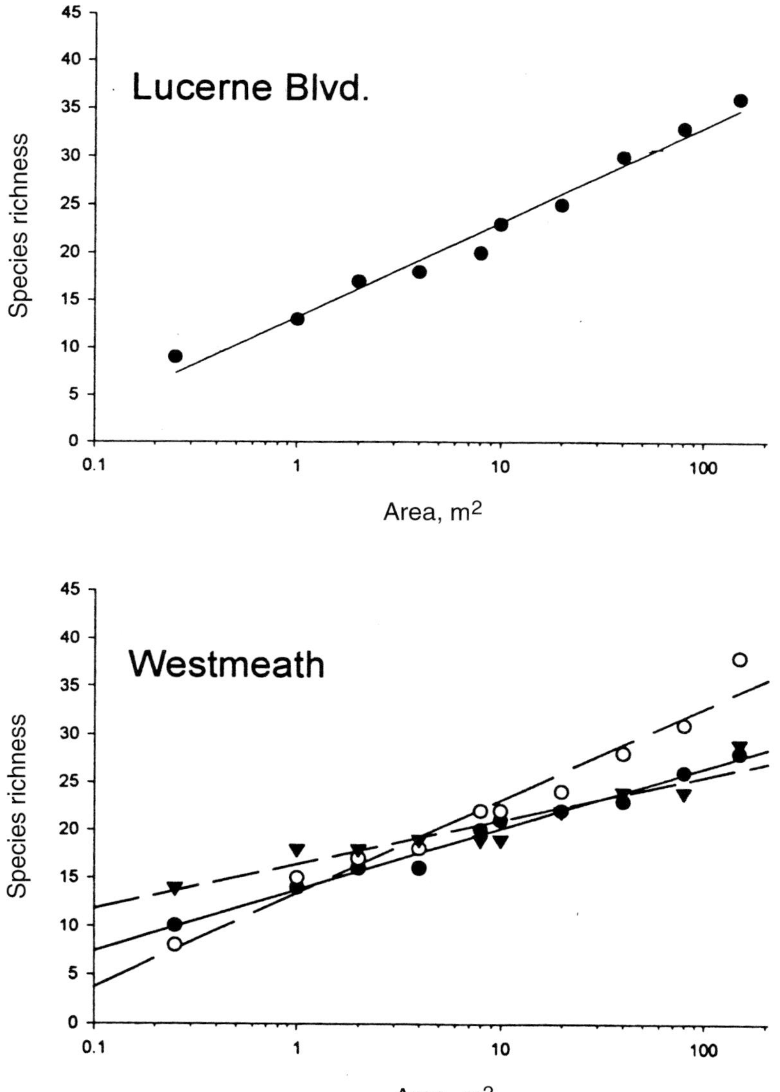
Lucerne Blvd.
Species richness
Area, m2
0
5
10
15
20
25
30
35
40
45
0.1
1
10
100
Westmeath
Species richness
Area, m2
0
5
10
15
20
25
30
35
40
45
0.1
1
10
100

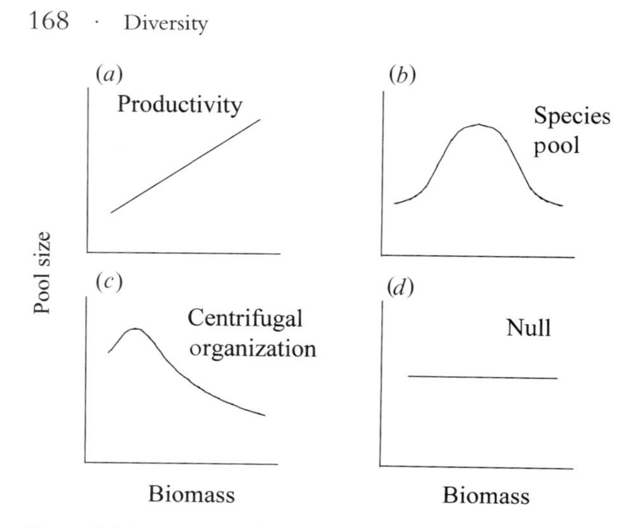

Figure 3.22 Four possible patterns in pool size along a biomass gradient (after Wisheu 1996).

high pool diversity may be occurring where alpha diversity is also high (e.g. Preston 1962a,b; Taylor *et al.* 1990; Eriksson 1993). Although this latter hypothesis may perhaps seem the most likely, particularly based upon what we know about global diversity patterns, it is founded upon the assumption that the degree of similarity among quadrats does not differ along gradients. If the average difference between quadrats is greater in low biomass conditions, then the pool size curve will be shifted to lower biomass habitats (Figure 3.22(*c*)). This would likely be the case with centrifugal organization (Keddy 1990a, b). Wisheu and Keddy (1996) tabulated data for 640 quadrats from shoreline marshes across eastern North America, plotting both species richness and pool size against biomass (Figure 3.23). Both reached maxima in similar habitats with approximately 50 g/0.25 m^2. Alpha diversity was a nearly constant percentage of pool diversity, irrespective of the biomass of the sites examined.

The ecological processes that cause this pattern in pool size remain unclear. We cannot even say with certainty which mode of causation operates. Is the pool pattern (produced by large scale evolutionary processes) causing the well-known intermediate diversity model? Or, is the alpha diversity pattern (produced by local ecological processes such as stress, disturbance and competition) causing the pool pattern? These sorts of basic questions have important consequences for the way in which we both design and manage nature reserve systems.

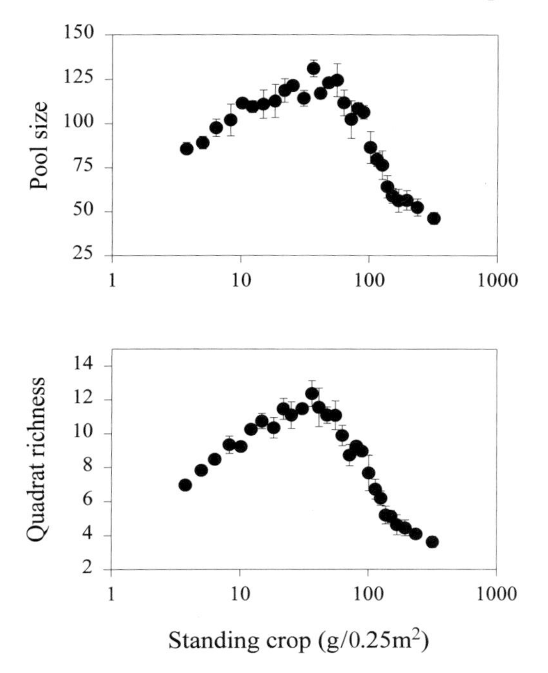

Figure 3.23 Patterns in species richness at two scales: pool size (*top*) and quadrat richness (*bottom*) (after Wisheu 1996).

Comparable data for other wetlands are not available, so we do not know how general the patterns in Figure 3.23 are. Pool sizes in tropical bottomland forests must be remarkable, since Meave *et al.* (1991) found that the number of plant species continued to climb linearly as quadrat size approached 500 m^2 (Figure 3.24). Such tropical riparian floodplain forests may be extremely important centres for the protection of biological diversity (Salo *et al.* 1986; Meave and Kellman 1994). This stands in sharp contrast with Crow's (1993) observations on the restricted size of species pools in aquatic vascular plants. Riparian corridors in general, appear to have high numbers of plant species and may support a significant proportion of the entire flora of an area; Nilsson and Jansson (1995) found that just four free-flowing rivers supported 366 plant species, or 18% of the entire Swedish flora of vascular plants. Meave and Kellman (1994) report 292 plant species from a mere 1.6 ha, and suggest that riparian corridors may have provided reservoirs of diversity for rain forest plants during periods of drier climate.

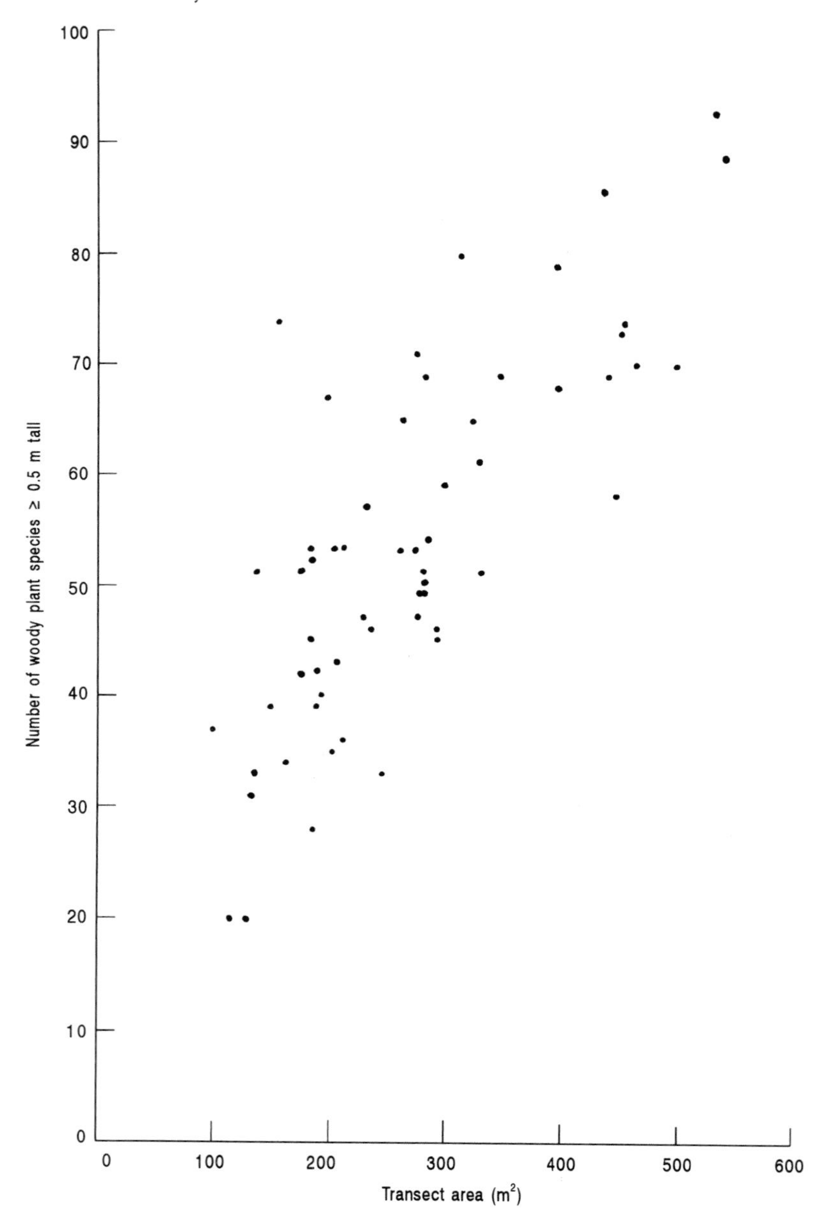

Figure 3.24 Patterns of species richness of woody plants occurring in tropical riparian forests (from Meave *et al.* 1991).

Table 3.7. *Species occurring at pristine and enriched coastal plain sites in the New Jersey Pine Barrens*

	Number of species	% carnivorous	% non-native
Pristine sites	26	12	12
Enriched sites	72[a]	0	96

Notes:
[a] Actual count was 73. One species was unidentifiable as to being native or non-native.
Source: From Wisheu and Keddy (1992) after Ehrenfeld (1983).

Conservation of biological diversity

There is a very important, and still frequently misunderstood, connection between local species diversity and pool size. Conserving 'biodiversity', which is an important conservation goal (e.g. World Conservation Monitoring Centre 1992; Reid *et al.* 1993; Noss 1995), requires us to think about the full species *pool* for an area. At the largest scale, our challenge is to maintain the species pool of the entire planet or at least of a region, a topic to be more fully addressed in Chapter 12. But most managers must focus on maintaining or increasing diversity at only one location. It is entirely possible that our attempts to increase local diversity can actually reduce biodiversity, that is, reduce the species pool. To understand this apparent paradox, consider the following two examples.

We shall see in the next chapter that altering the hydrology of salt marshes with dykes can increase the number of bird species found there. This might at first appear to be a worthy contribution to conservation of biological diversity. The added birds in impoundments, however, are generally birds that occur in other freshwater wetlands, whereas the birds in natural salt marshes are more specialized species such as clapper rails and sharp tailed sparrows. Impounding a salt marsh may simply increase numbers of common birds and decrease uncommon ones. From the global perspective, diversity is decreased.

A similar management outcome can be seen in the New Jersey Pine Barrens where there are many rare plants, in part because the habitats are very infertile. This infertility means that carnivorous plants are particularly well represented in the flora. Human activities often increase the fertility of sites. Sites that are thus enriched, support nearly three times the number of species as pristine sites (Table 3.7). Enrichment therefore, increases local biodiversity, but the increase is almost entirely accounted for by the invasion of exotic species better adapted to higher nutrient levels. Infertile sites, in contrast, have flora that is 88% composed of native species, 12% of which are carnivorous. Again, the local increase in diversity occurs because common species are increasing at the expense of uncommon ones.

The general rule then, is that, if increasing local diversity is achieved by increasing populations of globally common species, or decreasing globally rare ones, the local management is being counterproductive. Measuring species richness or diversity is no substitute for thinking. Management for maintaining 'biodiversity' only makes sense if one takes a global perspective as the context for evaluating local actions.

Conclusions

This chapter began with simple lists of species for a few wetland habitats. In spite of the fact that species richness is an important and frequently measured property of wetland communities, there has been only limited success in predicting it from environmental factors. Moreover, a vast majority of the examples used in this chapter used descriptive approaches, experimental studies being much less common. I have reserved a few experimental studies for later chapters (Willis 1963; Riley and Bookhout 1990; Bertness 1991; Weiher and Keddy 1995), and they illustrate the potential of properly designed field experiments to inform us of factors controlling richness, but they also illustrate the limits of experimentation. Most existing experimental studies suffer from two limitations. The first is limited scale, usually small quadrats or microcosms, although Carpenter *et al.* (1995) remind us that large-scale experiments are possible, albeit at the considerable cost of limited replication. The second is purpose, the dependent variable often being productivity (of a few selected species), in which case consequences for diversity often cannot be ascertained. In other cases, one can infer changes in diversity, but ecologists carrying out studies of production often overlook minor species which would greatly increase sampling effort and be truly insignificant in terms of production. It is precisely these minor species which play an important role in enhancing richness and diversity (e.g. Figure 3.2). Some of these difficulties can be overcome by altering design or choice of dependent variables, but others may be inherent in the practical constraints imposed by experiments (Diamond 1983). If there is a lesson here, it is likely the need for a wise and discriminating mixture of descriptive and experimental studies. Meanwhile, our inability to predict such simple characteristics of wetlands has two lessons. First, we are remarkably ignorant of factors controlling wetland properties. Secondly, this requires an attitude of humility; if we cannot yet predict basic properties, humans need to approach wetland management with caution given the real risk of making things worse rather than better. The common assumption that increased local diversity must be beneficial (even when it decreases global diversity) is a perfect example of widespread misapplication of theory in management. Given the rising number of endangered species in the world, there is a great deal remaining to be done.

Part II

Factors controlling properties of wetlands

4 · *Hydrology*

Gilgamesh

Flooding has been a part of human experience ever since the first settlements were built on floodplains. Stories of floods go back at least to the book of Genesis, which recounts how it rained 40 days and nights, how Noah built an ark which floated on the floodwaters, how '. . . the waters prevailed exceedingly upon the earth; and all the high hills, that were under the whole heaven were covered' (7:19).

Recent archaeological work has provided some interesting context for this early flood story of both Jewish and Christian traditions. Sandars (1972) recounts how, in 1839, a young Englishman, Austen Henry Layard spent some years excavating archaeological sites in Mesopotamia (recall Figure 1.1). One of the most significant discoveries were thousands of broken tablets from the palace of Ninevah. Ninevah was an Assyrian city that fell in 612 BC to a combined army of Medes and Babylonians. The destruction was so complete that the city never rose again. Included in the ruins was the entire library of Assurbanipal 'King of the World, King of Assyria'.

Over 25 000 broken tablets from this library were taken to the British Museum, and when deciphered revealed an epic immeasurably older than Homer's Iliad, the fragments of which are now known as 'The Epic of Gilgamesh'. One section of this epic narrates how there was a flood. 'The rider of the storm sent down the rain . . . a black cloud came from the horizon; it thundered within where Adad, lord of the storm, was riding . . . For six days and nights the winds blew, torrent and tempest and flood overwhelmed the world . . .' (pp. 110–111). This epic also has a boat full of survivors who come to rest on a mountain and who release a dove to search for land.

Then, as now, floods are depicted as tragedies. Says the epic 'Would that famine had wasted the world rather than the flood. Would that pestilence had wasted mankind rather than the flood.' (p. 112). This attitude is not dissimilar from the accounts of floods presented by reporters using satellite communications systems millennia later. One of the objectives of this chapter is to tell the rest of the story and put floods in their proper ecological context.

Changing water levels

Water levels change over many time scales. Large rivers around the world show marked seasonality in discharge rates (Figure 4.1). Such seasonal changes in water level are characteristic of temperate zone rivers, where floods are produced each spring by the rapid melting of the accumulated precipitation of an entire winter. Equally, Figure 4.1 shows that seasonal changes in rainfall produce flooding in tropical and subtropical rivers. Seasonal variation is only part of the picture; there are also pronounced differences in water levels among years: Figure 4.2 shows changes in annual mean water level in four of the Great Lakes; over a little more than a century, the yearly mean level has ranged across nearly 4 m. At longer time scales yet, we know that lakes and wetlands have come and gone; the size and shape of the Great Lakes has changed dramatically over the last 10 000 years (Figure 4.3).

In this chapter, it will become apparent that the amplitude and frequency of water level fluctuations control the characteristics of wetlands, just as fire intensity and frequency control characteristics of forests. High water periods create wetlands by destroying any existing terrestrial plants and allowing wetland species to become established. Water level fluctuations may destroy existing vegetation, terrestrial or wetland, and permit re-establishment from reserves of buried seeds (e.g. Salisbury 1970; Harris and Marshall 1963; van der Valk 1981). Because woody plants are particularly sensitive to flooding (Junk 1983; Kozlowski 1984a), high water periods tend to eliminate this growth form and replace it with herbaceous species (Keddy and Reznicek 1986). Although this chapter concentrates upon water level fluctuation over relatively short time scales (that is, roughly less than a century), we should not forget that water levels fluctuate on much longer times scales as well (Figure 4.3, see also Street and Grove 1979; Finney and Johnson 1991), a topic to which we shall return in the last chapter.

Let us begin the study of fluctuating water levels with one of the world's most important wetlands, the Amazon River Basin. The Amazon has a catchment basin of 7 000 000 km^2 and it carries approximately one-fifth of

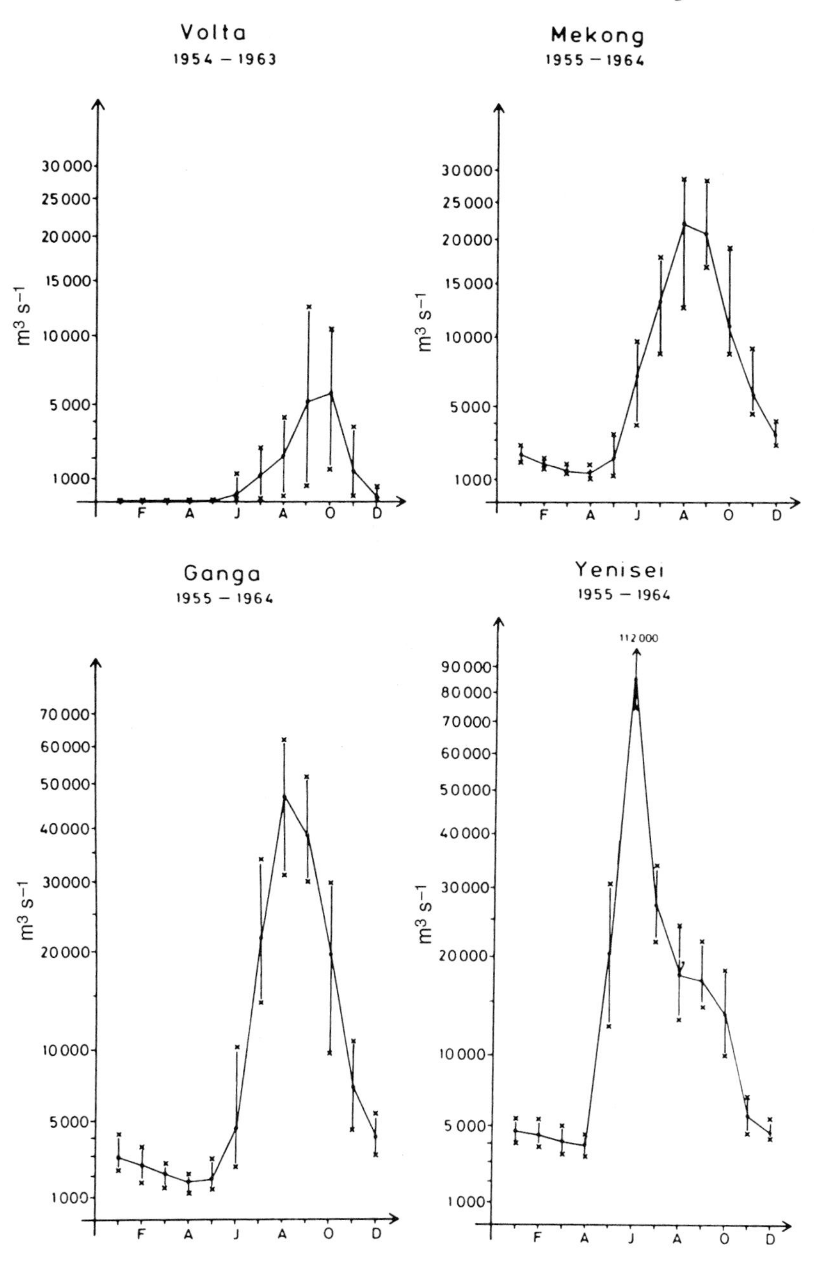

Figure 4.1 Patterns of seasonality in the flow of four large rivers (from Junk and Welcomme 1990).

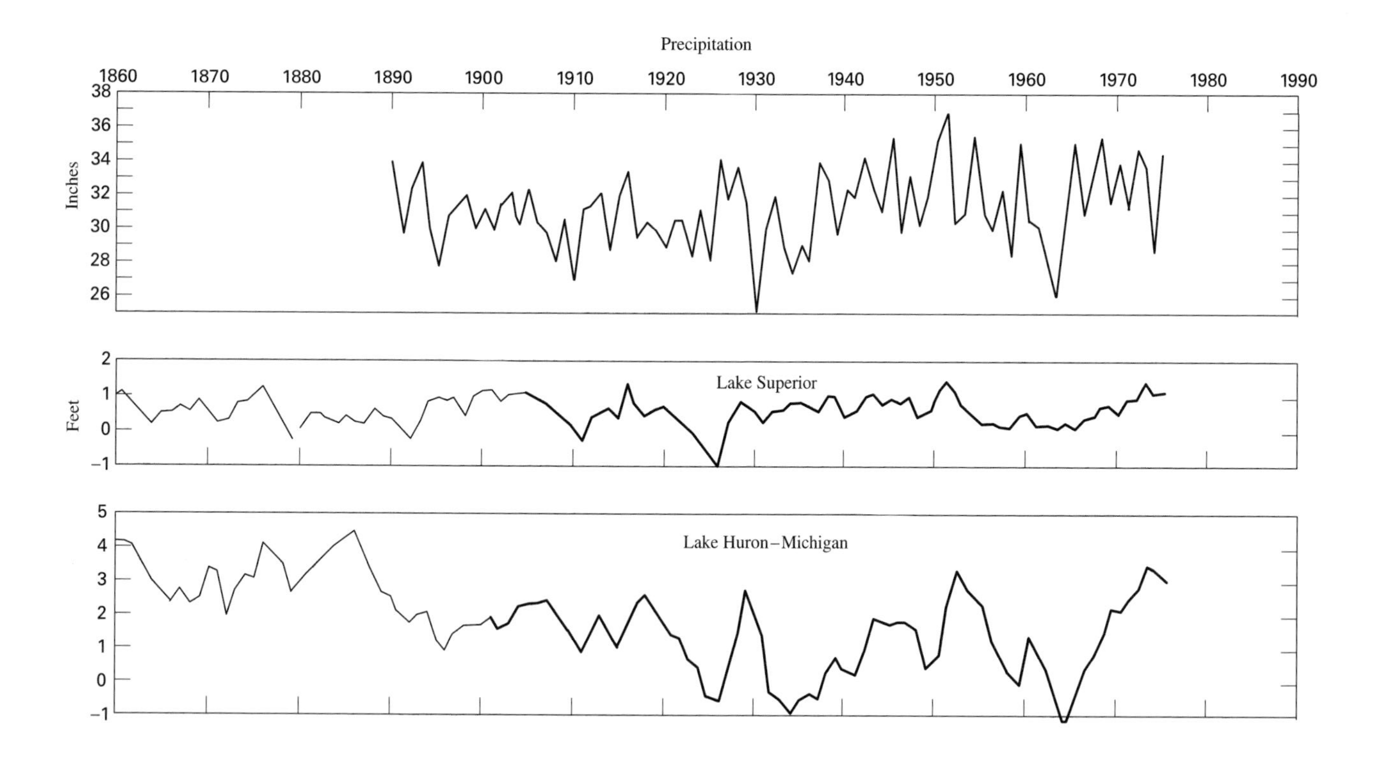
Precipitation
Inches
38
36
34
32
30
28
26
1860
1870
1880
1890
1900
1910
1920
1930
1940
1950
1960
1970
1980
1990
Feet
2
1
0
−1
Lake Superior
5
4
3
2
1
0
−1
Lake Huron−Michigan

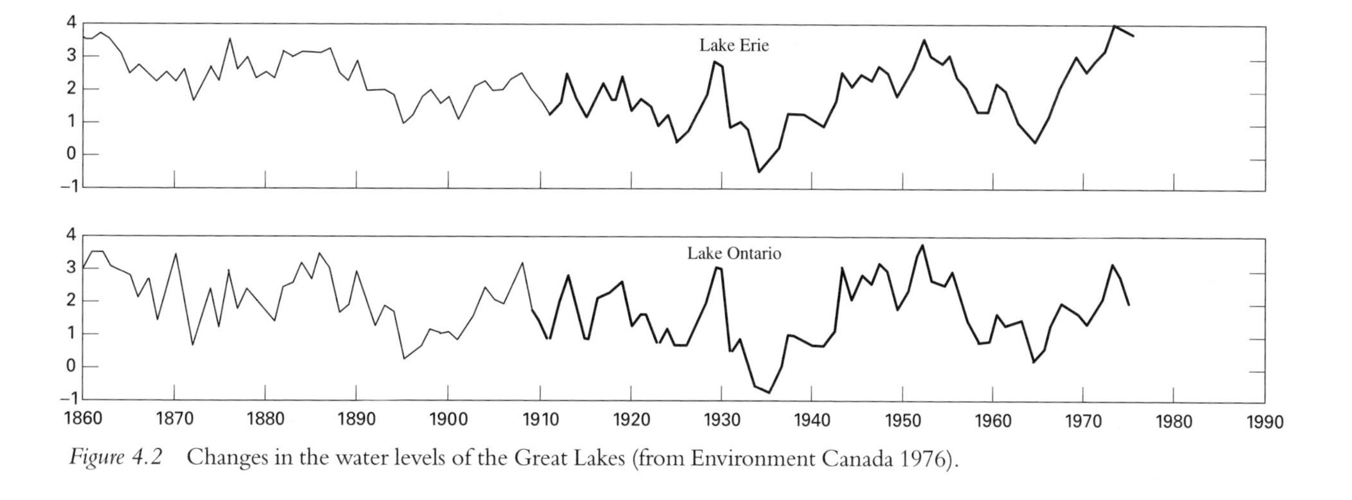

Figure 4.2 Changes in the water levels of the Great Lakes (from Environment Canada 1976).

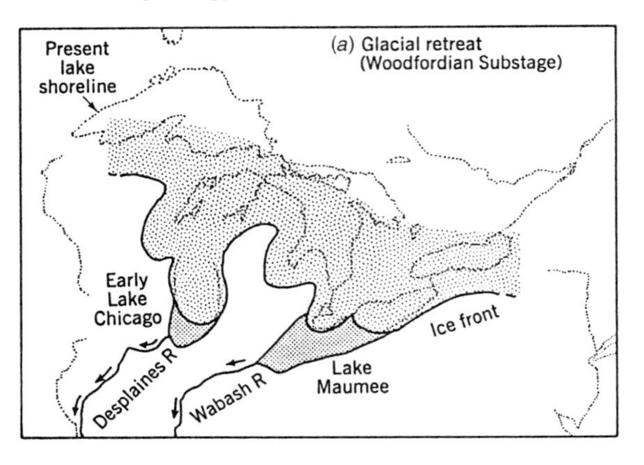

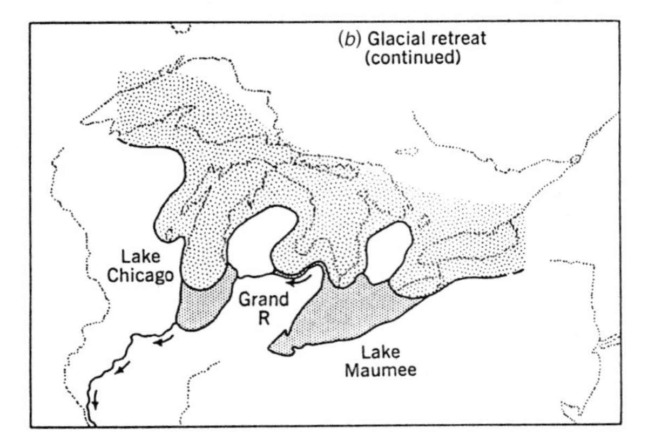

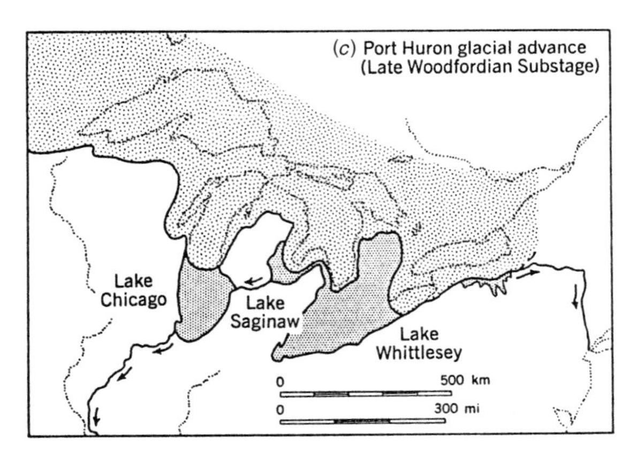

Figure 4.3 Changes in the water levels of the Great Lakes over millennia (from Strahler 1971).

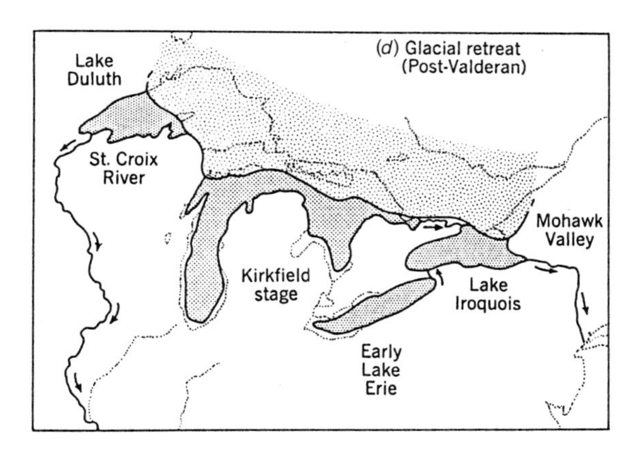
(d) Glacial retreat
(Post-Valderan)
Lake Duluth
St. Croix River
Kirkfield stage
Mohawk Valley
Lake Iroquois
Early Lake Erie

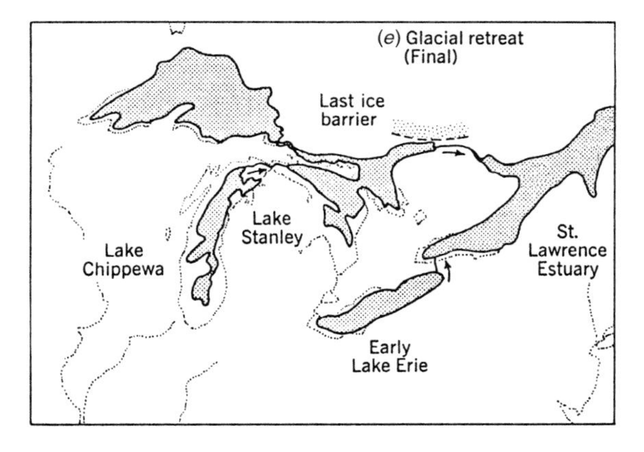
(e) Glacial retreat
(Final)
Last ice barrier
Lake Stanley
Lake Chippewa
St. Lawrence Estuary
Early Lake Erie

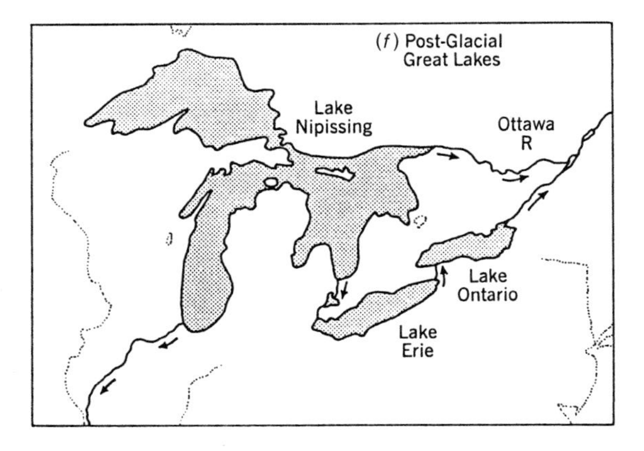
(f) Post-Glacial
Great Lakes
Lake Nipissing
Ottawa R
Lake Ontario
Lake Erie

the Earth's total freshwater run-off. In his review of the ecology of swamps in the Amazon basin, Junk (1983) tells us that, within 1 year, water levels may fluctuate more than 10 m! 'The large water-level fluctuations' he says, 'are a very important factor in the development of the flora and fauna colonizing the floodplains.' Owing to the dynamics of erosion and sedimentation, lakes and cut off channels are formed. Some of these have enormous water level fluctuations, but others, being partially cut off by sediment banks, remain filled and do not follow the same downward trend as the river. Junk argues that most of the variation in plant and animal communities can be explained by such differences in hydrology interacting with erosion and sedimentation (Figure 4.4).

In Chapter 1, we saw that the three main sources of water to wetlands were precipitation, ground water and surface flow, and that these produced bogs, fens and riverine wetlands respectively. Since bogs and fens tend to occur where there is rather little variation in water levels, at least when compared to wetlands fed by surface flow (Figure 1.10), this chapter will focus upon riverine and lacustrine wetlands. These include some of the world's largest wetlands from the Peace River/Athabasca Delta in north western North America to the Amazon River Basin and the Pantanal in South America through to the Sudd and the Congo River basin in Africa.

Some biological consequences of changing water levels

The major effect of water level fluctuations upon wildlife is the creation and maintenance of different habitats, and this linkage between water levels and habitats will be the primary focus of this chapter. But, let us first briefly consider three examples (insects, birds and fish) of the direct effects of water levels upon wildlife. Then, we will return to examine broader linkages between water levels and habitat in different types of wetlands. Finally, we will explore how humans alter these processes by building dams on rivers and reservoirs.

Water levels and invertebrates

Aquatic invertebrates are important both as decomposers and as food for other aquatic life. Table 4.1 compares invertebrate communities in different managed wetlands that have been converted to annual plants (*Polygonum lapathifolium* and *Echinochloa walteri*) for migrating waterfowl. The table shows two habitat types: the marsh bottom (benthic habitat)

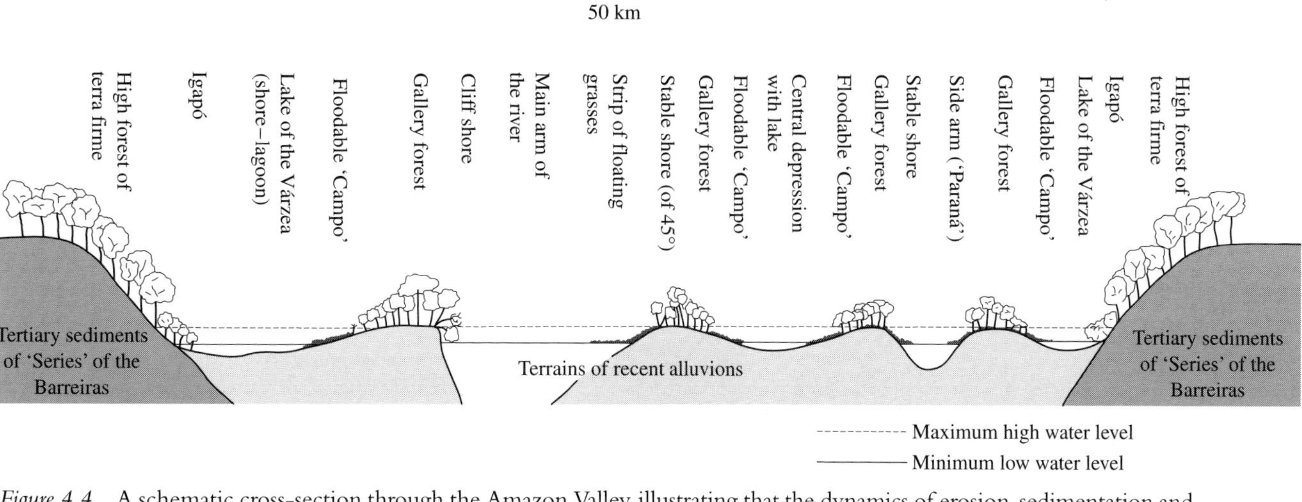

Figure 4.4 A schematic cross-section through the Amazon Valley, illustrating that the dynamics of erosion, sedimentation and subsidence that provide the template for biological diversity (from Sioli 1964).

Table 4.1. *Invertebrate responses to water depth (shallow vs. deep) in freshwater marshes*

Invertebrate taxa		Density (number/m^2)			
		Benthos samples		Water column samples	
		Shallow	Deep	Shallow	Deep
Class Crustacea	Amphipoda	760	531	8⋆	1
	Cladocera	3581	4775	893⋆	116
	Eucopepoda	–	–	438⋆	11
	Copepoda	1955	1520	–	
Class Gastropoda	Basommatophora (Physidae)	1061	1061	5	3
Class Insecta	Hemiptera (Corixidae)	1061	1061	22⋆	4
	Coleoptera (Dystiscidae)	1061	1061	7	5
	Diptera (Chirronomidae)	3682	796⋆	17⋆	<1
Class Oligochaeta		3797	5128	2	<1
Total		25610	7138	2921⋆	142

Note:
⋆ means significantly different between the years.
Source: After Riley and Bookhout (1990).

and the water column. The density of macroinvertebrates on the bottom of the marsh, roughly 25 000 per m^2, was nearly ten times that of those active in the water column. In both habitats, however, marshes with shallow rather than deep water in the spring had significantly more invertebrates of groups including Cladocerans, Eucopepods, Corixids and Chironomids. No invertebrate groups were significantly more abundant in deep water marshes. Those invertebrates attached to plants were uncommon (densities around 50 to 200 m^2) and were not affected by water levels. The abundance of invertebrates in the water column, and possibly in the benthos, increases during periods of shallow water. This should both increase food supplies for predators such as fish and waterfowl, and increase rates of decomposition and nutrient cycling.

Water levels and birds

Hydrology also has a major impact upon birds. Water levels naturally rise and fall in coastal salt marshes exposed to tidal influences. It is possible to both raise water levels, and reduce salinity, with water control structures such as stop ditches and impoundments. Burger *et al.* (1982) studied six salt marshes on the east coast of North America. There were two control sites of typical salt marsh, two impoundments and two areas that were partially flooded by plugging drainage ditches. The impoundments were shallow ponds 15 to 30 cm deep, with fresh rather than saline water, and typical salt marsh plants such as *Spartina* had therefore been replaced by scattered emergent plants. The impoundments had more than five times as many birds as the natural tidal marshes. Figure 4.5 shows that red-winged blackbirds, gulls and terns, shorebirds and waterfowl were more common in impoundments, whereas salt marsh species such as clapper rails, seaside sparrows and sharp-tailed sparrows were absent. Thus changes in the water level regime not only changed the abundance of birds, they also caused dramatic shifts in composition.

Fish in floodplains

Flooding has an important impact upon fish ecology; recall Welcomme's work (Chapter 1) documenting the quantitative relationship between floodplain area and fish production in tropical rivers. In her monograph, Lowe-McConnell (1975) says that the annual cycle of tropical fish is closely tied to periods of inundation. 'In both Africa and South America where much of the land is very flat peneplain, the rivers inundate

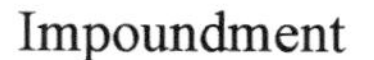

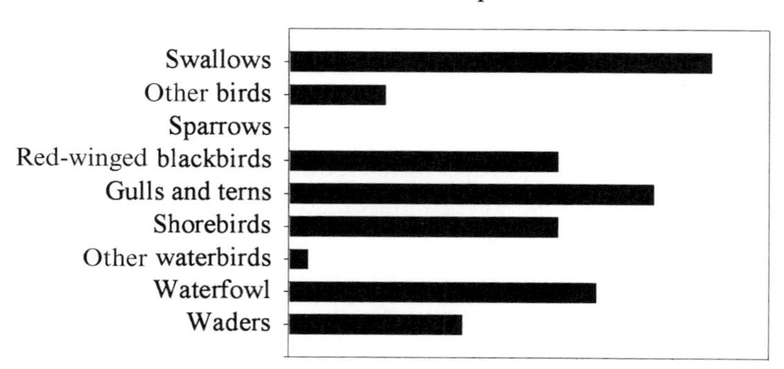

Stopped ditch

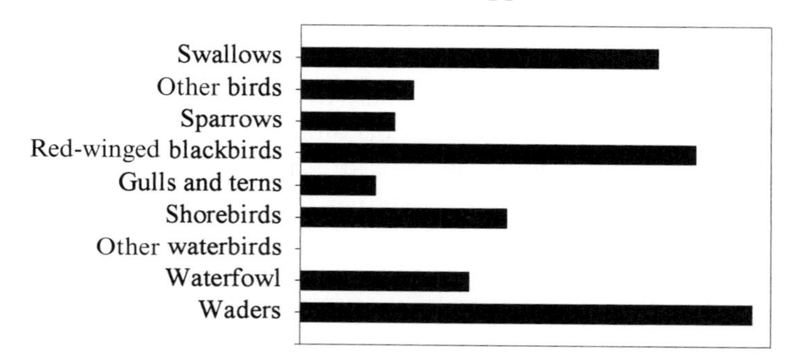

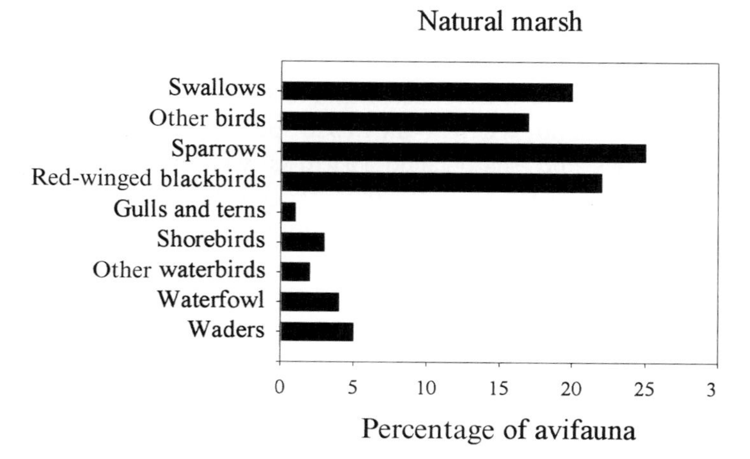

Figure 4.5 The composition of birds in three different kinds of coastal marsh: impounded marsh, marsh with stopped ditches, and natural salt marsh (from data in Burger *et al.* 1982).

immense areas, on a scale unknown in temperate regions. Submerged seasonally and drying out for part of each year, these floodplains are interspersed with creeks, pools and swamps, some of which retain water throughout the year' (p. 90). Although rains occur in the summer, flood peaks occur well after the rains have started; the delay depends upon the origin of the main floodwater and the time taken to travel downstream. As the rising water floods up channels and creeks, it releases fish imprisoned within ponds and swampy areas. Still higher levels then create a vast sheet of water. This water is enriched in nutrients from decaying organic matter, including the droppings of grazing animals, perhaps first baked by sun or fire. 'This leads to an explosive growth of bacteria, algae and zooplankton, which in turn supports a rich fauna of aquatic insects and other invertebrates. The aquatic vegetation, both rooted and floating, grows very rapidly.' (p. 92). Many fish then migrate upstream and move laterally onto the floodplain to spawn. The eggs hatch within a few days, so the young appear when food is plentiful. 'The highwater time is the main feeding, growing and fattening season for nearly all species' (p. 93) but as nutrients are depleted and water levels fall, the fish move back into the main river. Some fish are killed by being stranded in drying pools, and predators often hunt the mouths of channels leading back to the main stream. Even ungulates such as peccaries (*Tayassu pecari*) have been observed visiting floodplains to feed upon eels and often fish trapped by falling water (Fragoso 1998). Fish may mature within one or two years and life cycles are generally short, so that catches can be closely connected to single good or bad spawning years. Some fish make long migrations between feeding and spawning areas; movements of 600–700 km in each direction have been recorded form the Parana and 125–400 km in the Niger. The same general sequence of events occurs in rivers throughout the tropics including Africa, South America and Asia (Figure 4.6).

The general relationship between wetlands and water level fluctuations

Chapter 1 introduced the six principal types of wetlands: bog, fen, swamp, wet meadow, marsh and aquatic ecosystems. The last four of these (swamps, wet meadows, marshes, aquatic) represent a sequence of vegetation types associated with increasing duration of flooding. From the perspective of wetland classification, these are distinctive types of wetlands, but from the perspective of wetland dynamics and water level fluctuations, these are merely four regions in a continuum of communities that are

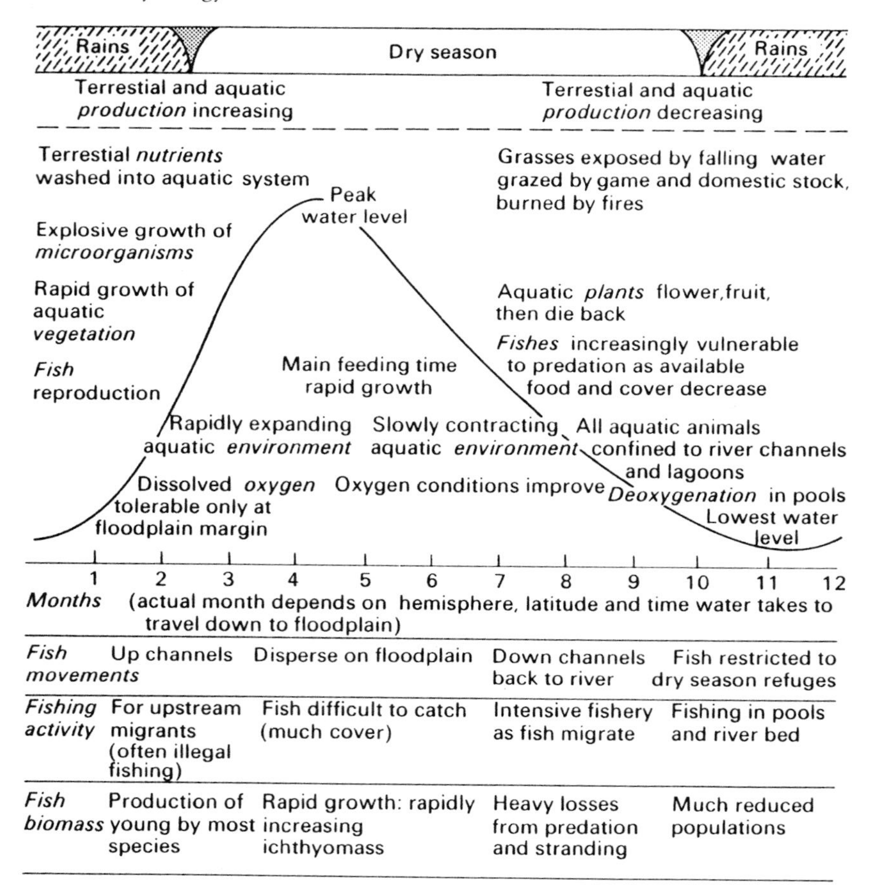

Figure 4.6 The effects of flooding upon fish using the floodplains of tropical rivers (from Lowe-McConnell 1975).

ever-changing, short-lived responses to recent water levels. Let us reintroduce these four wetland types as they relate to changing water levels; Figure 4.7 places these four zones in the context of yearly mean water levels in Lake Erie, but they could apply equally to any lake, river or wetland.

Swamps

The highest elevations are only periodically flooded; this is the zone of woody plants. Such habitats are referred to as swamps, bottomland forests riparian forests or floodplain forests. At higher elevations (landward), they grade into upland species; at the other end (waterward), they are killed by

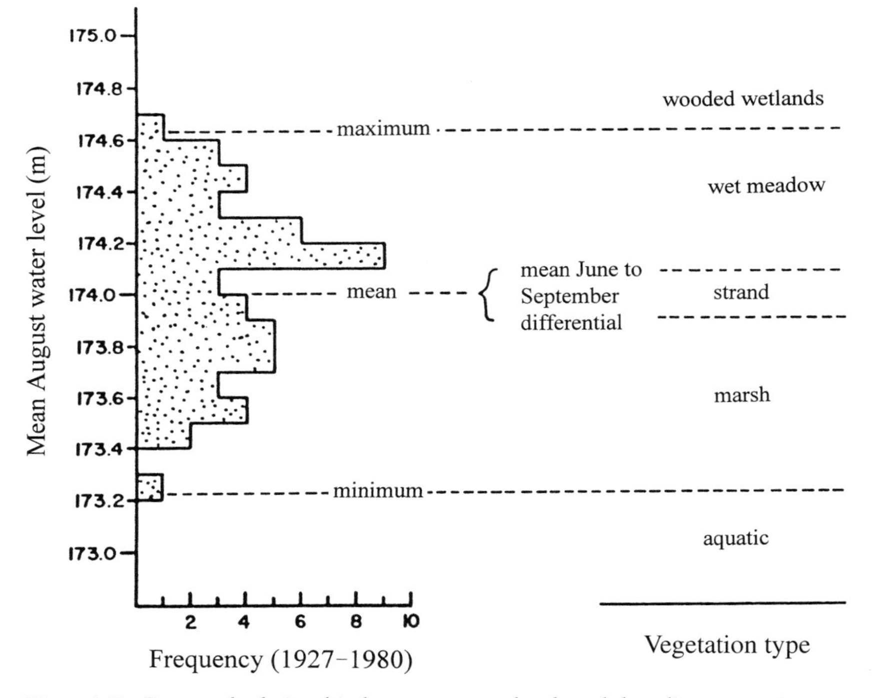

Figure 4.7 Proposed relationship between water levels and shoreline vegetation type, illustrated with long-term water level data from Lake Erie (from Keddy and Reznicek 1986).

prolonged flooding and replaced by more flood-tolerant herbaceous plants. Large areas of the world fall between these two extremes: flooded enough to exclude terrestrial plants, but not enough to kill trees (Lugo *et al.* 1990). Table 4.2 shows some survival times for woody plants in flooded soils. Extensive areas of floodplains along rivers are dominated by riparian forests (e.g. Junk 1983; Denny 1985; Sharitz and Mitsch 1993; Messina and Connor 1998). The boundaries of swamps are controlled by high water levels, but the exact duration and depth of flooding needed to cause the transition from terrestrial to swamp, or swamp to marsh, is an interesting unresolved issue (e.g. Robertson *et al.* 1978; Kozlowski 1984b; Toner and Keddy 1997).

Wet meadows

At the lower elevations, where the duration of flooding is sufficient to kill woody plants, swamps are replaced by wet meadows. Wet meadows usually support more plant species than any other vegetation type. In the

Table 4.2. *Survival time under inundation of some flood-tolerant trees*

Species	Survival time (yrs)
Quercus lyrata	3
Q. nuttalii	3
Q. phellos	2
Q. nigra	2
Q. palustris	2
Q. macrocarpa	2
Acer saccharinum	2
A. rubrum	2
Diospyros virginiana	2
Fraxinus pennsylvanica	2
Gleditsia triacanthos	2
Populus deltoides	2
Carya aquatica	2
Salix interior	2
Cephalanthus occidentalis	2
Nyssa aquatica	2
Taxodium distichum	2
Celtis laevigata	2
Quercus falcata	1
Acer negundo	0.5
Craetagus mollis	0.5
Platanus occidentalis	0.5
Pinus contorta	0.3

Source: From Crawford (1982).

absence of periodic flooding, this zone is invaded and dominated by woody plants. Occasional flooding, however, kills the woody plants and allows regeneration of wet meadow plants from buried seeds (Figure 4.8). The abundance of these species in the buried seed bank (Keddy and Reznicek 1986), and their dependence upon disturbance for regeneration, are two primary properties of wet meadow vegetation. Two other processes are associated with the formation of wet meadows. Scouring by ice and waves can also retard re-invasion by shrubs (e.g. Raup 1975; Keddy 1989b); this is a theme we will return to in the chapter on disturbance. Infertile substrates also may extend the area of wet meadows, since the lower the soil fertility, the longer it may take for herbaceous competitive dominants and woody plants to re-establish and exclude the rich herbaceous flora. This may be why infertile wet meadows tend to have the particularly rich floras (Moore *et al.* 1989).

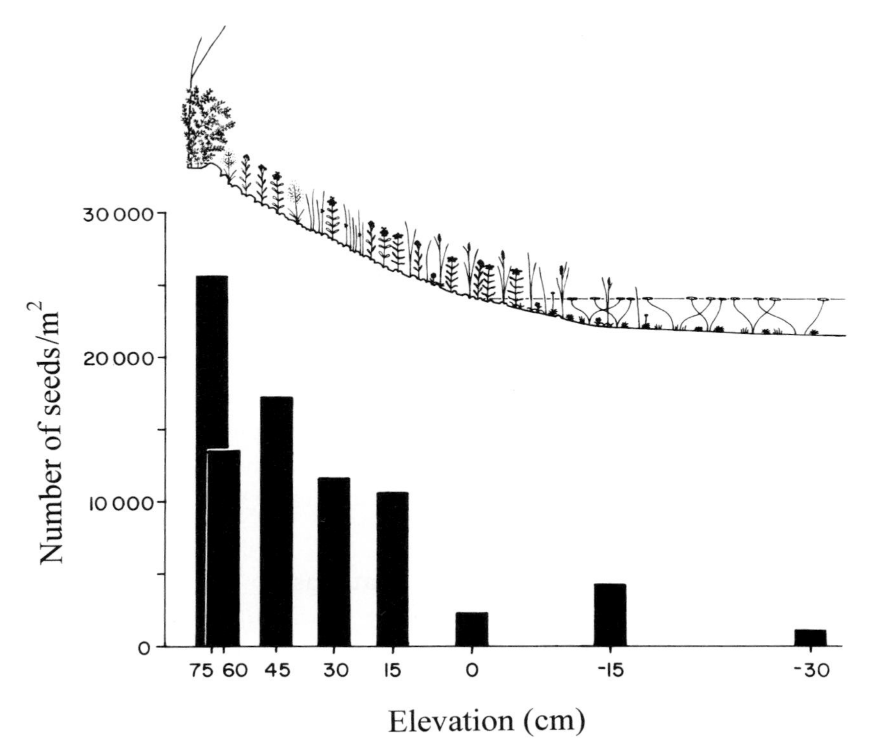

Figure 4.8 Wet meadow vegetation on a lakeshore and the number of buried seeds as a function of elevation (from Wisheu and Keddy 1991).

Marsh

Marshes are flooded for longer periods of time than wet meadows; wet meadows may be inundated only during flood peaks, whereas marsh vegetation is flooded for most of the growing season. As a consequence, marsh plants have traits for flood tolerance such as aerenchyma (recall Figure 1.15). While marsh species can tolerate flooding, most still require occasional dry periods to regenerate from seeds (van der Valk and Davis 1976, 1978; Smith and Kadlec 1983). Thus, while both marshes and wet meadows are produced by flooding, the duration of flooding and its timing differs markedly between them.

Aquatic

At the lowest elevations flooding is more or less continuous; the species here are not dependent upon water level fluctuations for their regeneration. Chapter 1 listed a number of features that allow aquatic plants to tolerate prolonged flooding. These included (Figure 1.15) the presence of aerenchyma to transmit oxygen to the roots, reinforced floating leaves, dissected leaves, and greatly modified flowers. Many of the studies on this zone are found in the limnology literature (e.g. Wetzel 1975; Hutchinson 1975; Sculthorpe 1967).

This brief introduction already introduces two important principles. (i) The greater the long term amplitude of water level fluctuations in a landscape, the more extensive the area of wetland will be. (ii) The relative abundance of these four wetland types in a landscape will depend upon the frequency and duration of flooding. These important themes will be illustrated in many ways in the following examples. What duration and frequency of flooding cause the transition from one type to another? In the absence of such knowledge we are limited to coarse scale predictions: wet meadow and marsh will increase with the amplitude of long-term water level fluctuations, and, if water level fluctuations are eliminated, the wet meadow and marsh zones will disappear (Figure 4.9).

The ubiquity of water level fluctuations

Rivers and watersheds

Large fluctuations in water levels are certainly the rule in natural watersheds. We have already encountered Junk's observations of 10 m changes in the water levels of the Amazon, and seen the seasonal variation in flow of major rivers (Figure 4.1). These seasonal floods produce the extensive bottomland forests that occur along rivers in Europe (Grubb 1987; Palczynski 1984), central North America (Robertson *et al.* 1978), the Amazon River basin (Duncan 1993) and Africa (Denny 1985; Petr 1986). Figure 4.10 illustrates the original extent of such floodplain forests in south eastern North America. Floodplain forests often intergrade with other wetland types. For example, in eastern Africa, there are extensive wetlands (The Sudd or the Upper Nile swamps) at the headwaters of the Nile, composed of some 16 000 km^2 of permanent swamp, some 15 000 km^2 of seasonal swamp and a further 70 000 km^2 of seasonal floodplain (Denny 1993b). Seven different vegetation zones can be related to the different flooding regimes. The driest are rain-flooded grasslands dominated by *Hyparrhenia rufa* and the wettest are permanently flooded areas of open water with aquatic plants such as *Najas pectinata* (Figure 4.11). Thompson and Hamilton (1983) say that Africa has several of the world's largest swamps, listing not only the Upper Nile swamps, but also the 90 000 km^2 Chari-Logone seasonal floodplains that drain into Lake Chad and the 80 000 km^2 central basin of the Zaire River; their seven sites

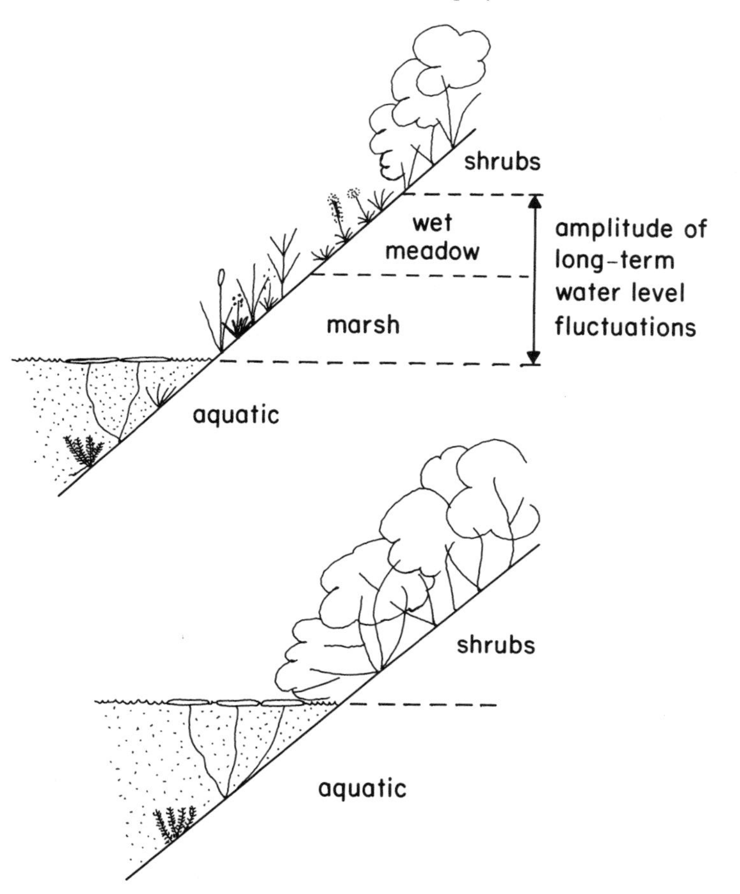

Figure 4.9 Stabilizing water levels compresses wetlands from four zones (*top*) to two zones (*bottom*) (from Keddy 1991a).

alone account for 60 000 km^2 of permanent and 400 000 km^2 of seasonally inundated wetland. In contrast, peatlands are rare in Africa, except for unusual cases such as high altitudes in Central Africa and at sea level in South Africa.

Given the ubiquity of dams and reservoirs today, we may fail to appreciate the enormous expanses of swamp forests and herbaceous wet meadows that once accompanied free-flowing rivers. One of the few free-flowing rivers remaining in Europe is the Torne River in Sweden. Figure 4.12 shows the wide expanses of herbaceous wetland that occur between the forested floodplain and normal summer water levels. In

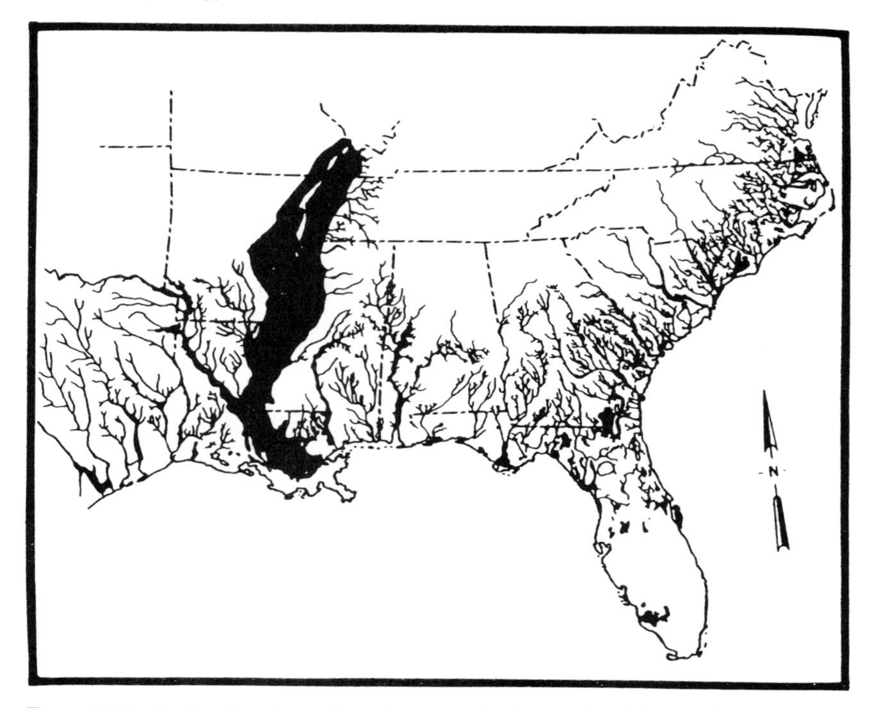

Figure 4.10 Spring floods produce the extensive bottomland forests that accompany many large rivers, such as those of the south eastern United States of America (from Mitsch and Gosselink 1986).

North America, the endangered Furbish Lousewort (Figure 4.13) occupies similar habitats in the St John River Valley that flows along the Canada–United States border. The Lousewort appears to require further disturbance in the form of ice scour and slumping of soil along river banks, and is restricted to these unstable sites (Menges and Gawler 1986). The population dynamics appear to combine fairly rapid increase in appropriately disturbed sites, with catastrophic losses from ice scour. Changes in hydrology associated with land development, forestry practices or dams, may reduce the distribution of such species or eliminate them entirely.

What is the duration of flooding that sets the lower limits of forested wetlands and thereby sets the upper elevational limits of these wet meadows? We could try to answer this in two steps: first we explore relationships among different components of hydrology, and secondly, we could try to relate them to vegetation. Such a study was carried out in

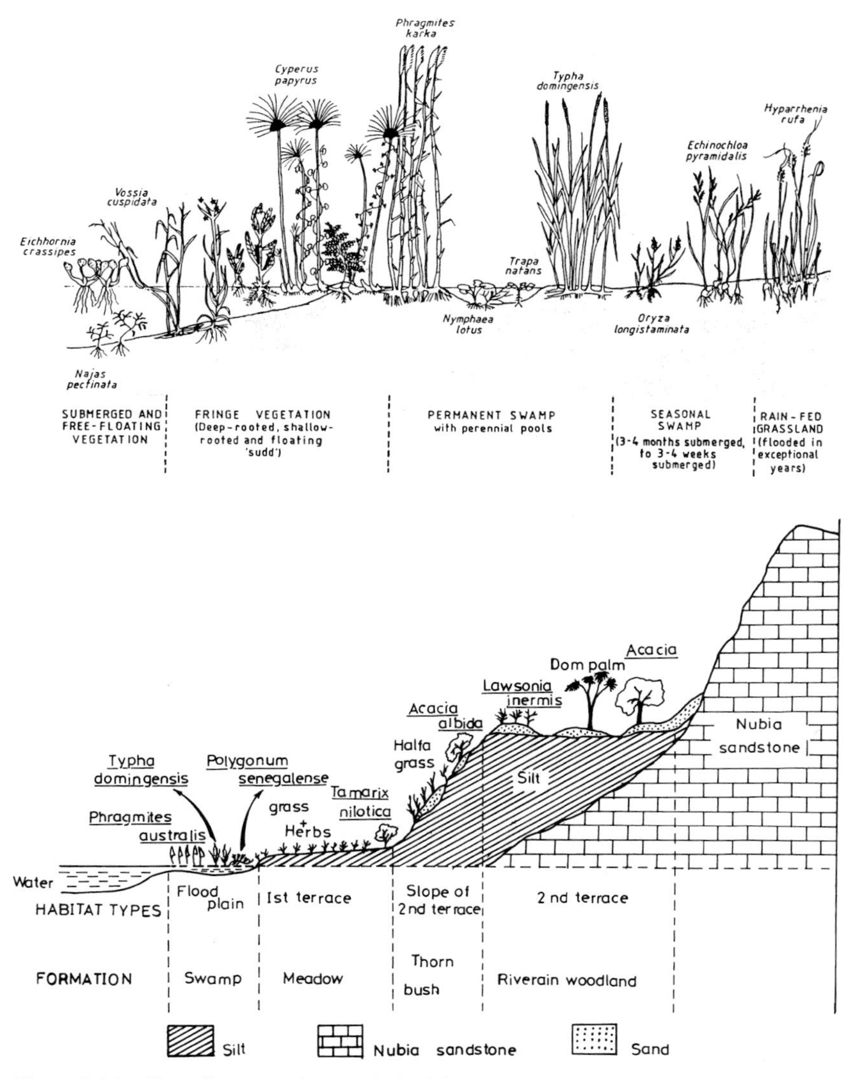

Figure 4.11 Shoreline zonation typical of the extensive Upper Nile Swamps of the Sudd (from Thompson 1985); also for comparison, the lower Nile floodplain (from Springuel 1990).

Figure 4.12 Annual water level fluctuations of 3.5 m create a broad wet meadow along the unregulated Torne River in Sweden (courtesy of C. Nilsson).

swamp forests and marshes along the Ottawa River in Canada (Toner and Keddy 1997), where, as in most temperate zones, there is a pronounced spring flood (April to June) when the winter accumulations of snow melt. Table 4.3 shows that most hydrological variables are strongly correlated, particularly those which are related to duration of flooding. In contrast, the depth of flooding was largely independent of duration. Logistic regression analysis showed that the best predictor of the occurrence of swamp as opposed to marsh combined two hydrological variables: the end of the first flood, and the beginning of the second flood. Figure 4.14 shows that the probability of finding a wooded, as opposed to herbaceous wetland, is a function of these two factors.

The duration of the first flood is probably critical. The period of 70 days flooding corresponds to roughly one third of the average growing season at this latitude. Further south, woody species such as *Acer saccharum* and *Fraxinus pennsylvanica* can tolerate 100 and 160 days of flooding (Robertson *et al.* 1978). Timing of the second flood is probably important for the following reason: if a second flood follows closely on the first, it represents simply a prolongation of the unfavourable conditions. If, on the other hand, there is a significant gap between floods, there is a period

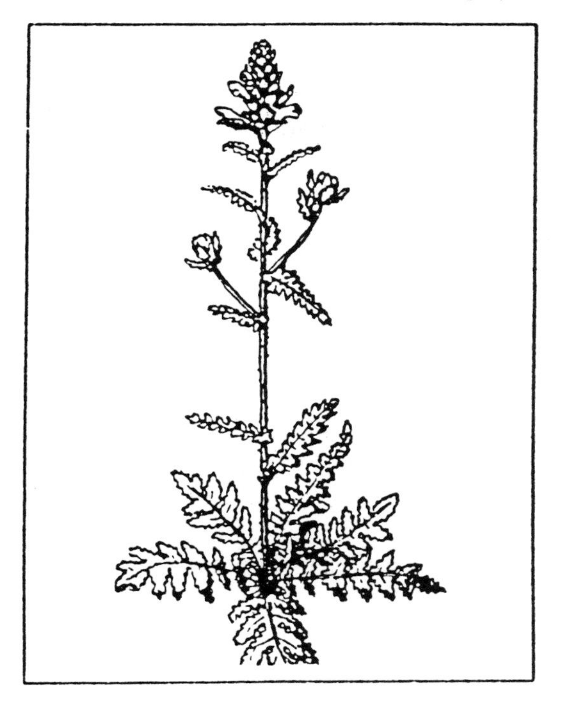

Figure 4.13 Endangered plants such as the Furbish Lousewort are restricted to eroded areas in wet meadows along fluctuating rivers (from Menges and Gawlor 1986).

of more favourable conditions, and the longer this favourable period, the greater the opportunity for plants to recover enough to withstand a second adverse period. Thus, the greater the period between the first flood and the second flood, the higher the probability of the survival of trees and the occurrence of wooded wetlands (Toner and Keddy 1997). These sort of quantitative studies provide the type of information that can be used to predict the consequences of dam construction on wetlands of major watersheds. At the same time, different models may be needed for different parts of the world, particularly where differences in climate may interact with tolerances to flooding (e.g. Poiana and Johnson 1993; Johnson 1994).

The wet meadows maintained by flooding may support entire associations of rare plants. The Tusket River, on the east side of the continent of North America, is the only one north of Cape Cod with rich wet meadows supporting significant Atlantic coastal plain species such as

Table 4.3. *Correlation of hydrological variables within the Ottawa River*

Variable	Fraction of growing seasons over which flooding occurred	Last day of the first flood	Length of the second flood	Mean depth of flooding	Number of floods per growing season	Number of days without flooding proceeding	First day of the second flood
Fraction of growing seasons over which flooding occurred	1.00						
Last day of the first flood	0.83^{***}	1.00					
Length of the second flood	-0.35^{***}	-0.52^{***}	1.00				
Mean depth of flooding	0.13	0.30^{***}	-0.60	1.00			
Number of floods per growing season	0.64^{***}	0.52^{***}	-0.33^{***}	$0.19^{0.78}$	1.00		
Number of days without flooding proceeding	-0.60^{***}	-0.67^{***}	0.34^{***}	-0.21^{**}	-0.57^{***}	1.00	
First day of the second flood	0.62^{***}	0.73^{***}	-0.54^{***}	$0.75^{0.27}$	0.22^{***}	-0.48^{***}	1.00

Notes:
$N=22$ $P<0.01$ (★★★) or $P<0.01$ (★★).
Source: From Toner and Keddy (1997).

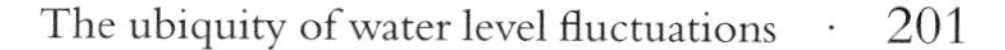

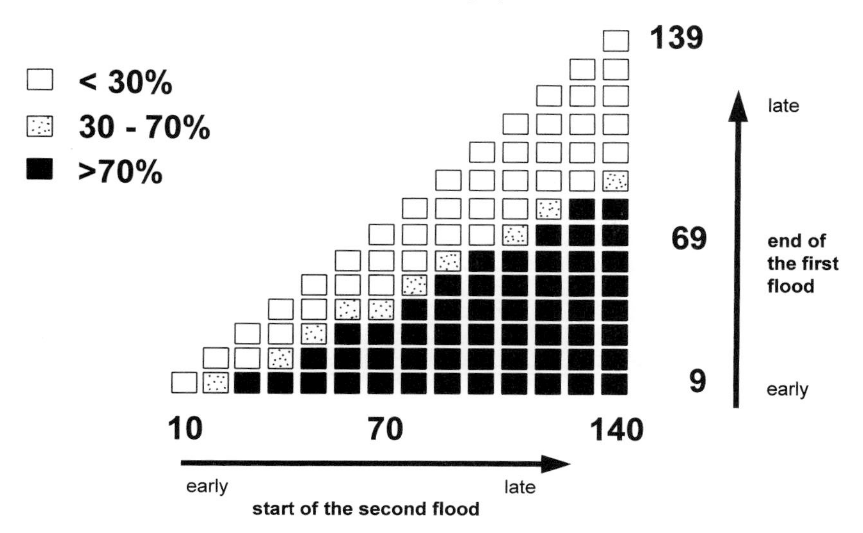

Figure 4.14 The probability of finding woody as opposed to herbaceous plants in a wetland can be predicted by knowing when the first flood ends and when the second flood (if any) begins (from Toner and Keddy 1997).

Plymouth Gentian and Pink Coreopsis (Figure 4.15). These species co-occur with many other Atlantic coastal plain species that are considered rare, threatened or endangered in Canada (Keddy and Wisheu, 1989). The wet meadows in which they occur are created by fluctuating water levels in the Tusket River; some of the high water periods are attributable to melting snow in the spring, but others are caused by the occasional severe storm. One severe thunderstorm in the summer of 1983 raised water levels some 75 cm over 2 days, completely submerging the wet meadows. In other years periods of drought can expose extensive areas of river and lake bottom. Both the Plymouth Gentian and Pink Coreopsis once occurred in the lower lakes of the Tusket River watershed, as shown by the pre-1925 records of herbarium specimens from Vaughn, Gavels, and Kings Lake. However, they are no longer present there. The explanation for this decline appears to be that all three of these lakes were converted to reservoirs for the Tusket Falls generating station in 1929, and the rare plant species now survive only in those lakes where water level fluctuations are unaffected by reservoirs. Although the maps show only two indicator species, most other typical wet meadow species including other rare species (e.g. *Xyris difformis, Rhexia virginica, Panicum longifolium*) are also absent from the reservoirs.

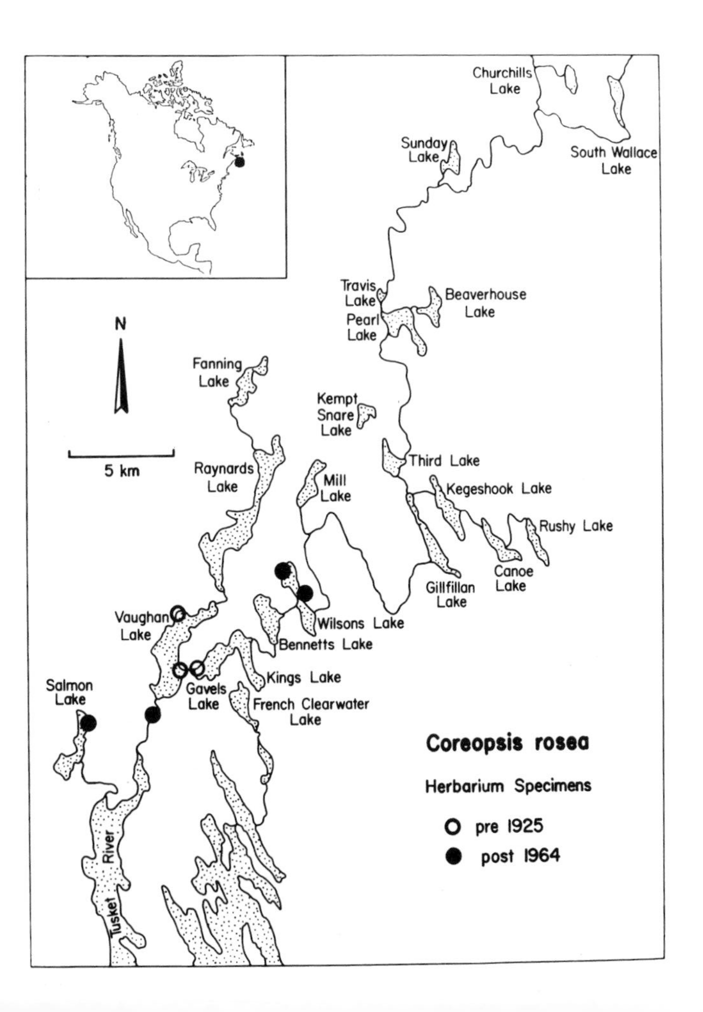
Churchills Lake
South Wallace Lake
Sunday Lake
Travis Lake
Beaverhouse Lake
Pearl Lake
N
Fanning Lake
Kempt Snare Lake
5 km
Raynards Lake
Third Lake
Mill Lake
Kegeshook Lake
Rushy Lake
Canoe Lake
Gillfillan Lake
Vaughan Lake
Wilsons Lake
Bennetts Lake
Salmon Lake
Kings Lake
Gavels Lake
French Clearwater Lake
Coreopsis rosea
Herbarium Specimens
pre 1925
post 1964
Tusket River

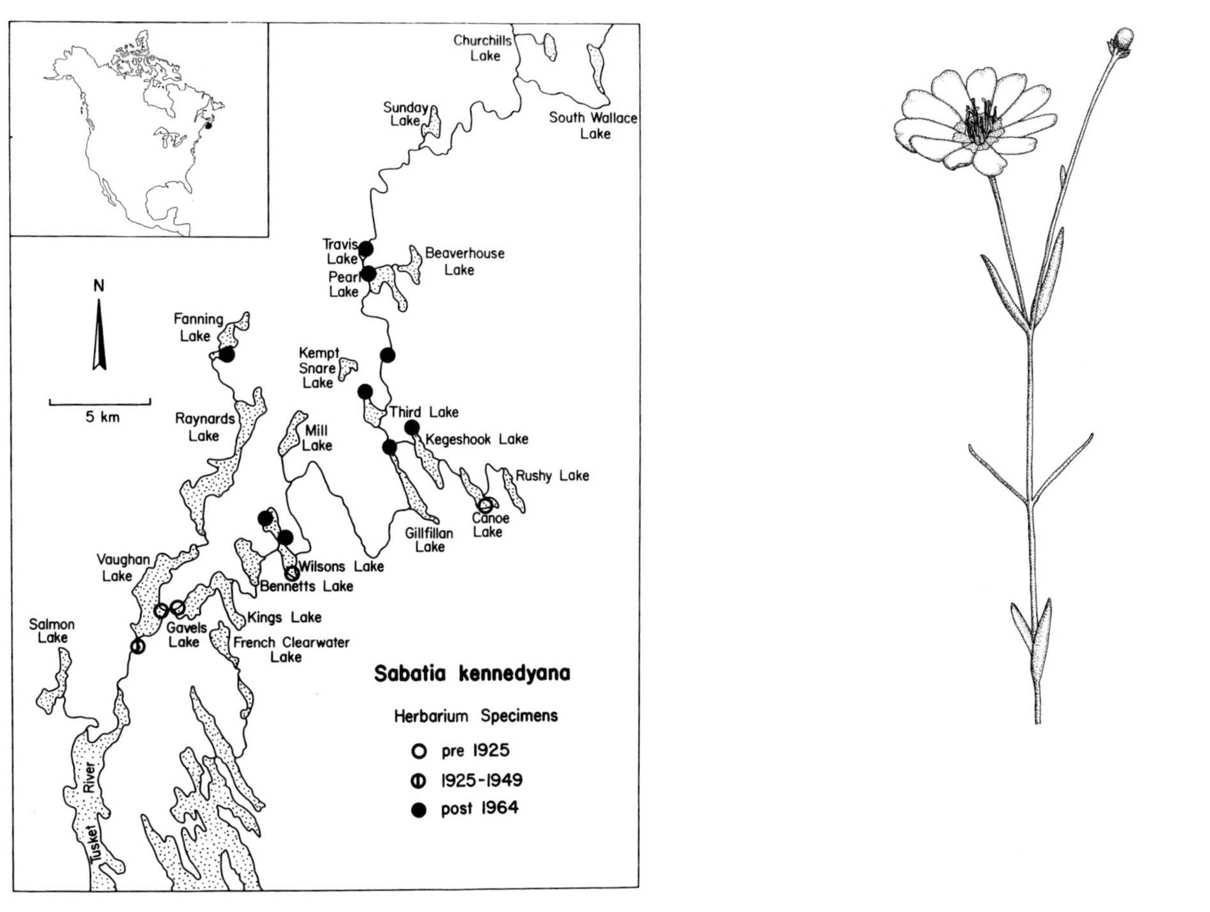

Figure 4.15 Two threatened plants that grow in wet meadows along the Tusket River, Nova Scotia (Pink Coreopsis, *top* and Plymouth Gentian, *bottom*). The open circles show sites from which the species disappeared after the construction of a hydro-electic generating station in 1929 (after Keddy 1985a).

The same wet meadow floras were examined more systematically in a survey of 37 coastal plain lakes in Nova Scotia (Hill and Keddy 1992). The number of plant species in these herbaceous wet meadows is strongly correlated with watershed area (Figure 4.16(*a*)). If we examine only the subset of rare coastal plain species in the shoreline flora, watershed area alone accounted for more than 64% of the variation in their frequency of occurrence. In contrast, woody plants were negatively correlated with watershed area ($r = -20.44$). The mechanism underlying these patterns appears to be the link between watershed area and the amplitude of water level fluctuations (Figure 4.16(*b*)). In this sample of lakes, the larger the watershed area, the larger the amplitude of water level fluctuations, the broader the examples of wet meadows and the higher the plant diversity.

Lakes

Flooding and drying cycles seem typical of many water bodies, ranging from the smallest beaver ponds to some of the largest lakes. Changes in vegetation with falling water level have been frequently observed. The American naturalist and philosopher Thoreau was a keen observer of nature. In 1854 he wrote:

> This rise and fall of Walden [Pond] at long intervals serves this use at least; the water standing at this great height for a year or more though it makes it difficult to walk around it, kills the shrubs and trees which have sprung up about its edge since the last rise . . . and, falling again, leaves an unobstructed shore; for, unlike many ponds and all water which are subject to a daily tide, its shore is cleanest when the water is lowest . . . By this fluctuation the pond asserts its title to a shore, and thus the shore is shorn, and the trees cannot hold it by right of possession.
>
> (H. D. Thoreau 1854)

Beginning our survey with large lakes, Raup (1975) studied the vegetation of the Athabasca – Great Slave Lake region of north western Canada during the period 1926 to 1935. He too emphasized the importance of water level fluctuations in producing wet meadow communities, and made observations on shoreline succession. However, high water levels in 1936 drowned all the shore 'successions' that were described in his earlier work. Thus, in his words, 'successions in this vegetation are reduced to fragments which, if they exist at all, have indeterminate beginnings and ends'. Had he not been so constrained by the framework of succession, he might have been able to articulate more clearly the dynamic nature of

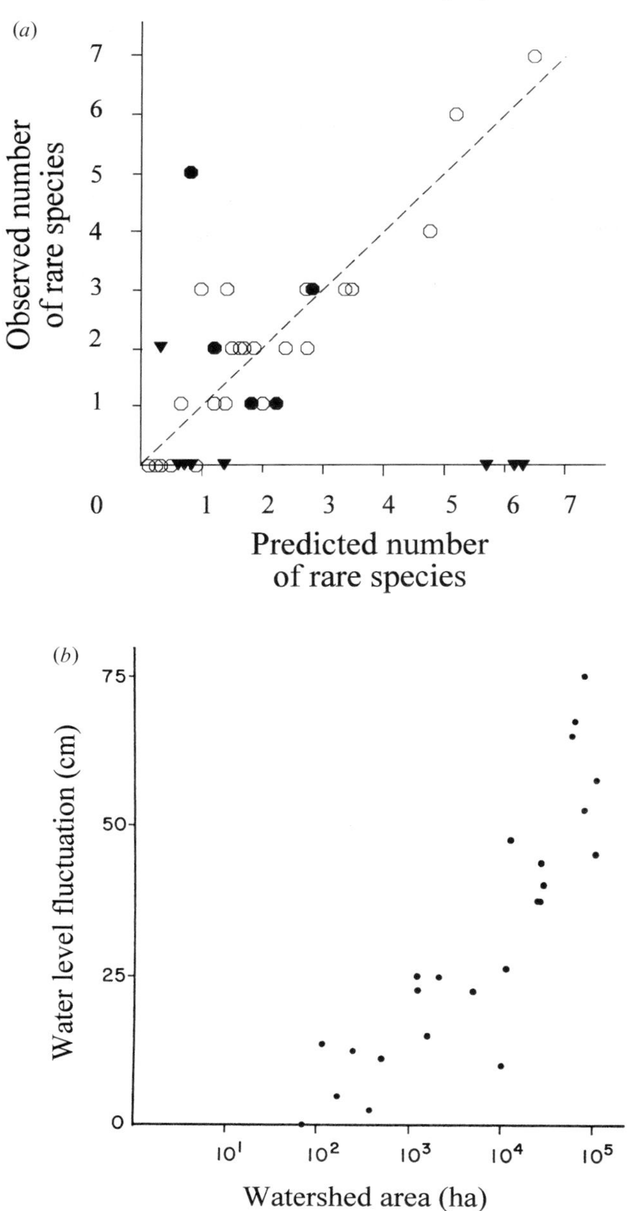

Figure 4.16 Effects of water level fluctuations on rare plant species (*a*) observed values vs. predictions from knowledge of watershed area (solid dots and triangles are lakes with dams; from Hill *et al.* 1998) and (*b*) amplitude of water level fluctuations as a function of watershed area (from Hill and Keddy 1992).

Figure 4.17 During a low water year in Lake Erie there was dense regeneration of *Scirpus* and *Sagittaria* plants in Metzger Marsh (courtesy of Doug Wilcox).

wetlands, and the linkages between water level fluctuations and the area of wetlands.

The Great Lakes have received much more attention, probably as a consequence of their size and their nearness to the populated regions of eastern North America. The water level fluctuations shown in Figure 4.2 cause dramatic changes in shoreline vegetation (Keddy and Reznicek 1986; Reznicek and Catling 1989). Rich wet meadows and prairies are maintained by periodic flooding, which kills woody plants. Many of the common genera in these habitats (e.g. *Carex, Cyperus, Juncus, Polygonum*) are known to produce persistent seed banks. In similar marsh vegetation types, seed densities are typically 3000 seeds m^{-2}, and in one of the few wet meadows studied, exceeded 38 000 m^{-2}. Low water periods allow these seeds to germinate (Figure 4.17). These dynamic changes in water level and vegetation maintain a rich flora of some 450 plant species. The short-term effects of high water levels can be deleterious; Farney and Bookhout (1982) observed declines in waterfowl nesting, as well as drastic declines in muskrats. However, these long-term fluctuations are necessary to maintain the full array of wetland communities and their associated wildlife (Prince and D'Itri 1985; Reznicek and Catling 1989; Smith *et al.* 1991).

Over the last 15 000 years and the retreat of the Wisconsinan ice sheet, the Great Lakes have changed dramatically in area, distribution and drainage (Figure 4.3). Lake Agassiz, Lake Algonquin, Lake Minesing, Lake Chicago, Lake Tonawanda, Lake Warren and the Champlain Sea have come and gone in this process (Karrow and Calkin 1985). Figure 4.3 does not show Lake Agassiz, a 350 000 km^2 body of meltwater that covered much of southern Manitoba and western Ontario about 10 000 BP. This lake was partly created by an ice dam on its eastern outlet. When this ice dam melted, Lake Agassiz drained along the Laurentide ice margin, entered Lake Superior (arrow Figure 4.3(*e*)) and then discharged through the North Bay outlet of the Great Lakes. Calculations of discharge rates yield values of 200 000 m^3 s^{-1}, which, to put it in context, is 13 times the mean discharge of the Mississippi River at New Orleans today, and some 1000 times the current outflow from today's Lake Huron (Teller 1988). At this rate it could have taken only 2 years for Lake Agassiz to fall 12 m in level.

On the other side of the world, Talling (1992) reports on changes in the water levels of African lakes; these changes are caused by the seasons as well as long-term changes in rainfall. He noted that the rise in water levels is often rapid from the influx of floodwaters, whereas the subsidence is slower, resulting from evaporation and reduced run off. In Lake Chilwa, changes in elevation of 2.5 m were recorded over one decade. Owing to the high rates of evaporation, salinity had important impacts on aquatic life. The high production of shoreline plants such as *Cyperus papyrus* and *Phragmites australis* is illustrated by biomass levels exceeding 5–7 kg m^{-2} dry weight. Similarly Denny (1993b) describes how Lake Victoria has an amplitude of fluctuations of about 1.5 m each year. *Cyperus papyrus* often develops a floating margin that can rise and fall with changes in water depth of about a metre; if greater than this, he reports that the floating section will break away to form a floating island. Floating islands also occur along other great *Papyrus* swamps (e.g. Figure 11.29 in Thompson and Hamilton 1983). As in the Great Lakes of North America, water levels have changed with climate (Kendall 1969). Before 12 500 BP Lake Victoria had low levels and lacked an outlet, probably being surrounded by swamps and savannah. Yet fossil beaches encircle the lake at 3, 12 and 18 m above current levels, indicating that there were much higher water levels, probably within the past 10 000 years.

Equally dramatic changes occur in small ponds and lakes. During low water periods, rich shoreline floras may emerge from reserves of buried seeds (Salisbury 1970; Mandossian and McIntosh 1960; Keddy and

Figure 4.18 During low water periods a rich wet meadow flora emerges from the seed bank of Matchedash Lake, Ontario. Typical high water levels are indicated by the tall grasses at left; the canoe marks the summer water level in a low year (from Keddy and Reznicek 1982).

Reznicek 1982). Intervening high water periods kill woody plants. Indeed, there are striking similarities between the herbaceous meadows along the Torne River in Sweden (recall Figure 4.12), those of the Great Lakes, and those of the much smaller Matchedash Lake in Canada (Figure 4.18). In New Mexico, Wright and Bent (1968) were able to describe 16 'discrete' vegetation types around a small lake in the Chuska Mountains; these were related to differing periods of inundation as well as erosion and deposition of shoreline sediments. As with Raup, they observed that a high water period reversed successional sequences. Other examples are found in Lieffers (1984) and Wilcox and Meeker (1991).

Beaver Ponds

Beavers obstruct water flow in streams, thereby flooding the forest and creating small ponds (Figure 4.19). Occasionally, under the right physical conditions, they can also regulate the water levels of lakes. Before the arrival of Europeans, the beaver population of North America was estimated to be 60 to 400 million individuals, with a range stretching from

Figure 4.19 Beavers can produce water level fluctuations by building dams that periodically break or are abandoned (courtesy of Friends of Algonquin Park).

arctic tundra to the deserts of northern Mexico. The ponds that beavers create cause changes in forest structure, nutrient cycling, decomposition rates and properties of water downstream (Naiman *et al.* 1988). C. G. Jones *et al.* (1994) point out that beavers are only one of many organisms that creates, modifies or maintains natural habitats. They propose that we recognize a class of organism called ecosystem engineers. In wetlands, these would include beavers building ponds, alligators excavating wallows

Table 4.4. *The ten most common species that germinated from the mud in a series of beaver ponds in Canada*

Species	Number of seedlings
Juncus effusus	388
Leersia oryzoides	355
Scirpus cyperinus	224
Juncus brevicaudatus	155
Ludwiga palustris	89
Hypericum boreale	87
Unknown dicot	66
Elocharis obtusa	57
Galium palustre	56
Hypericum majus	49

Source: Le Page and Keddy (1998).

(p. 290), plants aerating anoxic soil (pp. 351–353) and *Sphagnum* mosses building peat bogs (pp. 396–401). Other examples would include coral reefs, termite mounds and prairie dog colonies.

Beavers create cyclical disturbance and succession in the landscape with two different frequencies. The short-term cycle is one of dam destruction and repair. Dams may be washed out during floods, or holes may be punched by mammals such as otters, or predators may kill the beavers maintaining the dam. In such cases, water levels suddenly fall and many plant species regenerate from buried seeds. Although nearly 40 species of plants are known from beaver pond seed banks (Le Page and Keddy 1998), Table 4.4 shows that a few genera of moncotyledons (e.g. *Juncus*, *Leersia*, *Scirpus*) are most common. There will be a surge of regeneration in these marsh and wet meadow species unless the dam is repaired. Breaking and repair cycles produce short-lived periods of low water; this probably explains the abundance of annuals and facultative annuals in such conditions.

If food supplies collapse, the beaver pond may be abandoned. Abandonment of the dam will result in a short-lived mud flat, a longer period of marsh formation, and then, as the beaver meadow gradually dries, woody plants will re-invade. It may take many decades for the trees upon which beavers feed to re-establish. This longer cycle of beaver ponds alternating with swamp forest probably has a frequency of centuries rather than decades (Figure 4.20).

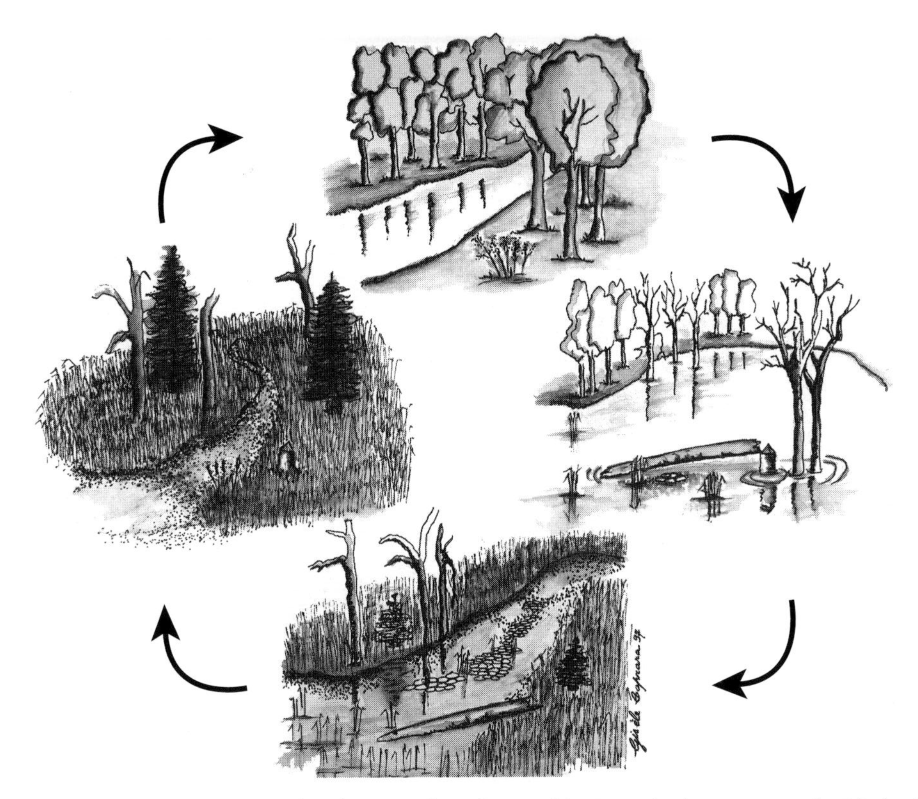

Figure 4.20 The beaver pond cycle going from forest with stream (*top*), to new pond with dead trees (*right*) to established pond with aquatic plants (*bottom*). When the food supply diminishes, indicated by the presence of conifers, the dam bursts and a beaver meadow forms (*left*). Eventually, the forest re-invades (*top*). Temporary events can cause a short circuit of the cycle.

Aerial photographs might provide one method to quantify rates of pond formation and abandonment by beavers. In much of eastern North America, beaver populations were very low at the turn of century, and then expanded rapidly after 1940. Johnston and Naiman (1990) examined six sets of aerial photographs (covering 1940 to 1986) from the Kabetogama Peninsula of Voyageurs National Park in northern Michigan, and found that the number of pond sites increased from 71 to 835. The mean area was 4 ha (range from <1 to 45 ha), and decreased over time, suggesting that later ponds were established in less suitable areas. During the first half of the period (1940 to 1961) ponds were created at the rate of 25 yr^{-1}, but later (from 1961 to 1986) the rate declined to 10 yr^{-1}. These figures, however, include all areas with vegetation altered by past or present ponds, both active ponds and beaver meadows covered in grasses and sedges. The sequence of events they describe is thus a progressive increase in area of patches influenced by beavers, to some 3000 ha by 1986. Since over the 40-year period, no patches were lost, it is not possible to calculate rates of pond abandonment that might result from food shortages, nor what the equilibrium patch densities might be. The rapid spread of beavers in this region is, in part, explained by their biology: the average litter size is 3–4 kits yr^{-1} and at the age of 2 years, kits generally disperse within a 16 km radius, so that over a 46-year period, it would be possible for them to colonize areas more than 700 km from an initial nucleus.

Beaver ponds provide important habitat for amphibians, mammals and birds. In a sample of 70 wetlands representing three classes of beaver activity Grover and Baldassarre (1995) found a total of 106 species of spring birds, with between 9 to 39 species per wetland. Larger wetlands had more species, as did wetlands with active beaver colonies. Active ponds had more open water, more dead standing trees, more flooded emergents and a higher habitat diversity index. Nineteen obligate wetland species were found in active ponds compared with 12 in the inactive ones, and there were 18 facultative species that used cavities in the standing dead trees. They conclude that beaver ponds provide habitat for more than half of the regional avifauna.

Prairie potholes

So far this chapter has focused on wetlands that are heavily influenced by annual flooding from moving water. Some wetlands that are exposed to seasonal flooding are not connected to large river systems, being dependent instead upon local sources of water. Prairie potholes were formed by the glaciers that covered and then retreated from the continent of North America, leaving behind millions of depressions across over a half million km^2 of prairie stretching from Alberta south eastward to Iowa. These depressions are filled by melting snow and, depending upon depth, water table and summer rainfall. The depressions range from ephemeral to permanent wetlands. Froebel, an early explorer on an expedition under General Sully against the 'hostile Sioux' wrote in 1870 'The entire face of the country is covered with these shallow lakes, ponds and puddles, many of which are, however, dry or undergoing a process of gradual drying out'. These wetlands contained a 'central portion' surrounded by 'concentric circles of different species of plants' (in Kantrud *et al.* 1989).

More than 20 million ducks representing 12 common species breed in this region, and it is considered the most important production habitat in North America for most species. The most common species are Mallard, Blue-winged teal, Northern pintail, Northern shoveler and American widgeon (Batt *et al.* 1989). The dominant herbivores are muskrats and waterfowl (Murkin 1989); we will return to them in Chapter 8. The rich wetland flora ranges from wet prairie to submersed aquatics, and is controlled by three main environmental gradients: water regime, salinity and disturbance (Walker and Wehrhahn 1971; Shay and Shay 1986; Adams 1988). Buried seeds allow plant species to survive inhospitable periods and then re-emerge after prolonged flooding or drought (van der Valk and Davis 1978; van der Valk 1981). Dominant plant species include *Phragmites communis, Typha latifolia, Scirpus validus, S. maritimus* and *S. acutus* (Shay and Shay 1986). The vegetation cycles make classification systems difficult to apply, but workers generally recognize at least four zones familiar from other wetlands: wet meadow, shallow marsh, deep marsh (emergent marsh) and shallow water. The first are flooded for a few weeks each spring, whereas the last are permanently flooded except during severe drought (Kantrud *et al.* 1989). The major water loss is evaporation, which can reduce water levels nearly a metre in a dry summer; fluctuations in the water table are as high as 2 to 3 metres. Since below ground flow is relatively slow (0.025–2.5 m yr^{-1}), the ponds are hydrologically isolated from one another in the short term (Winter and Rosenberry 1995). Superimposed upon the vegetation types produced by hydrology are those produced by salinity; based upon specific conductance, water can range from fresh ($<$500 μS m^{-1}) through brackish (5000 to 1500 μS m^{-1}) to saline ($>$45 000 μS m^{-1}); across the region conductance values exceeding 10 000 μS cm^{-1} are common (LaBaugh 1989).

One classification scheme uses the above four vegetation zones combined with three other features: a low prairie zone for the prairie immediately adjacent to the wetland, fen zones where ground water seeps into the pothole, and an 'alkali zone' corresponding to saline water (Figure 4.21). Kantrud *et al.* (1989) review other classification systems that have also been applied to these habitats. Woo *et al.* (1993) have used the duration of inundation to recognize four main hydrological groupings: ephemeral, intermittent, semi-permanent and permanent types of potholes; the close relationship between snowfall and water levels in these sloughs suggest that future conditions for breeding waterfowl can be estimated from the snowfall available to fill them.

Some half of these potholes have already been drained and ploughed, but apart from this obvious loss of habitat, there is the risk of a gradual decline in the water table. In the Nebraska sand hills, for example, irrigation from wells is increasing in extent; by 1985 there were over 70000 irrigation wells registered, with an estimated volume of water removal approaching 10^8 m^3 yr^{-1} (Novacek 1989). The water table not only falls many metres during the pumping season, but it also appears that longer-term declines are in progress. Wet meadows, says Novacek, would be most at risk.

Further to the south in North America, in the arid high plains, large circular depressions that retain moisture from the spring rainy season are called playas (Bolen *et al.* 1989; Hoagland and Collins 1997b). Some 22000 playas occur in Texas and New Mexico. The dominant species are perennial grasses, largely *Agropyron smithii, Buchloë dactyloides* and *Panicum obtusum*, with *A. smithii* most common in the low sites. The dominant environmental factors are elevation and unpredictable wet–dry cycles, with fire and grazing thought to be of secondary importance. Playas were once used as wallows by herds of bison and antelope in a manner reminiscent of the African megafauna. Seed banks again play an important role in allowing plants to tolerate fluctuating water levels (Haukos and Smith 1993, 1994).

In the south eastern United States, the flat coastal plain areas are predominantly Longleaf Pine (*Pinus palustris*) woodlands, apparently as a result of coarse soils, summer drought and fire. In the infertile sandy depressions scattered through these woodlands, one can find an array of unusual wetland types, from savannah to seepage bog, with extremely rich floras (Peet and Allard 1993). Frequent herbaceous plants include *Sarracenia* spp., *Drosera* spp., *Rhexia* spp., *Rhynchospora* spp., *Lachnanthes caroliniana, Lophiola aurea, Xyris* spp., *Eriocaulon* spp., and shrubs include

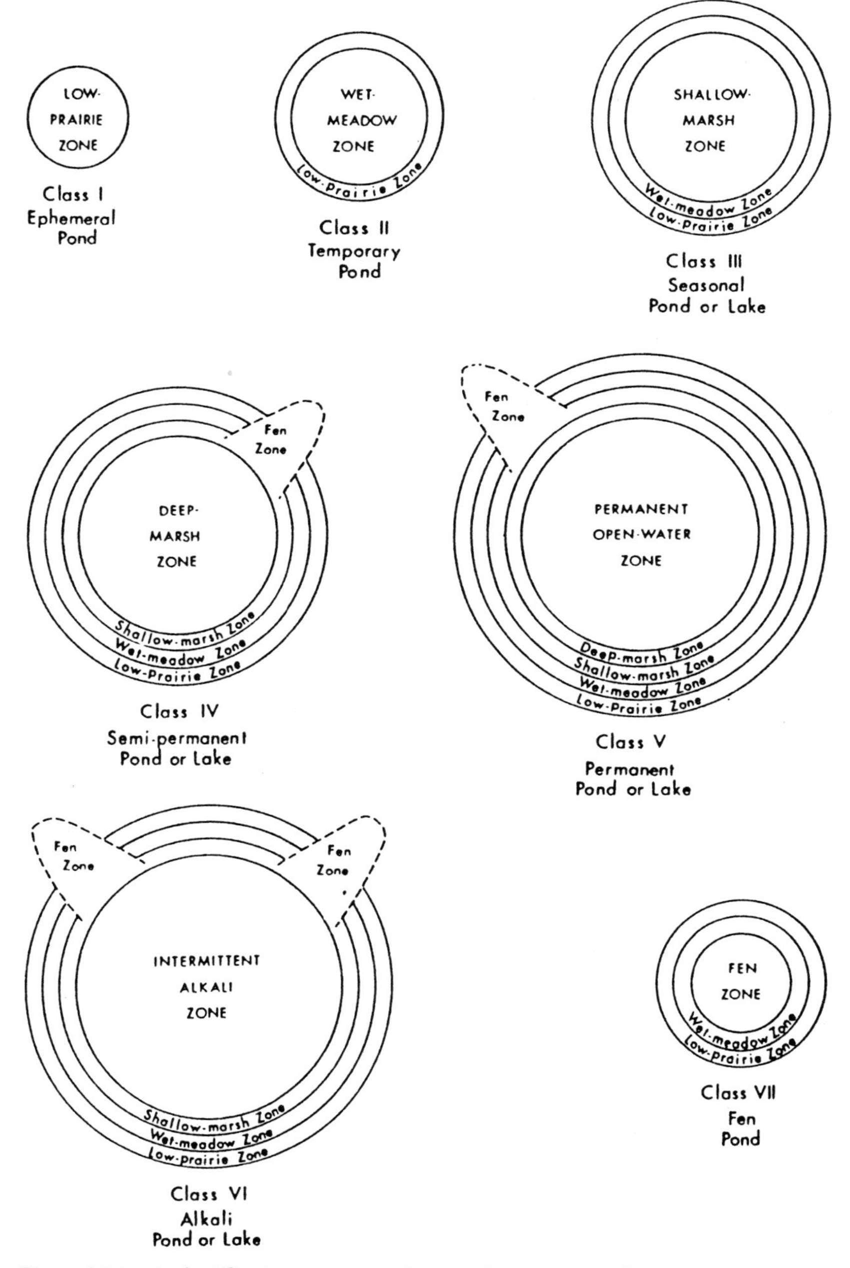

Figure 4.21 A classification system and vegetation zonation for seven types of prairie potholes (from Stewart and Kantrud 1971 in van der Valk 1989).

Clethra alnifolia, Cyrilla racemiflora, Aronia arbutifolia and *Smilax* spp. More continually flooded sites are dominated by *Nyssa sylvatica, Acer rubrum* and *Taxodium distichum*. In the savannas and seepage bogs, one can find as many as 40 species in a single square metre, and up to 140 in 1000 m^2. Peet and Allard report that water level fluctuations, infertile soils and fire are three of the main natural factors responsible for this high diversity.

Under other arid regions, small wetlands may arise in vernal pools. Winter rains can fill small depressions for periods of 3 to 5 months, producing pools ranging from 20–250 m^2 in area and up to some 30 cm deep. In western North America, for example, vernal pools can be found in an area extending from southern Oregon though central California into northern Baja Mexico (Bauder 1989). These pools have a unique flora termed vernal pool ephemerals, many of which are annuals endemic to the California floristic province. This flora is absent when water stands for more than 6 months. Using experiments with the endangered annual species of mint, *Pogogyne abramsii*, Bauder (1989) was able to show that competition appeared to have a negligible effect on restricting these species to the vernal pools; watering, however, significantly increased survivorship. If we can extrapolate from *P. abramsii*, the vegetation of these pools thus seems to consist of a group of species dependent upon high levels of soil moisture in an otherwise arid area.

The arid Cape Peninsula in south west Africa has some of the highest levels of plant diversity and endemism recorded in the world (Cowling *et al.* 1996a, b). There is high topographical heterogeneity, long and steep rainfall gradients, a variety of infertile soils and frequent fire. The predominant vegetation is a fire-prone sclerophyllous shrubland called fynbos, with Restoid and Proteoid floras, but where rainfall and topography permit, wetlands occur. Seasonably waterlogged sites have distinct floras (Table 4.5), but little is known of depressions with standing water called vleis.

Much less is known about these types of wetlands in Eurasia. In some semidesert regions of Russia shallow undrained basins are called firths, padinas or saladas depending upon the depth and duration of flooding (Zhulidov *et al.* 1997). Many of these are found in the loess deposits to the north east of the Black Sea. Firths are shallow (2–4 m) undrained basins which maintain lush meadow vegetation during summer droughts. Padinas, shallower than firths, but 0.2–5 km in diameter, are filled by melting snow, whereas saladas are associated with salt domes and contain saline water. Genera such as *Phragmites, Scirpus, Eleocharis, Carex,* and *Calamagrostis* occur in the firths and padinas, whereas saline firths can

Table 4.5. *The major wet vegetation types of the Cape Peninsula*

Vegetation type	Structural characteristics[a]	Common species	Environment
Wet restoid fynbos	Low restoid herbland	*Ischyrolepis cincinnata, Tertaria cuspidata, Elegia filacea, Thamnochortus lucens, Cliffortia subsetacea, Erica imbricata, Leucadendron laureolam, Pentaschistis curvifolia, Restio quinquefarius, R. bifurcus*	Shallow seasonally waterlogged sands on sandstone at low altitudes: MAR = 634 mm Mean slope = 13° Mean aspect = WSW
Upland restoid fynbos	Low restoid herbland	*Thamnocortus nutans, Chrondropetalum ebracteatum, Ursinia nudicaulis, Restio bifidus, Ehrharta setacea, Watsonia barbonica* subsp. *borbonica, Penae mucronata, Cliffortia ruscifloliа, Erica hispidula, Chondropetalum mucronatum*	Shallow seasonally waterlogged sands on sandstone MAR = 1404 mm Mean slope = 32° Mean aspect = SW
Wetlands	Medium-height ericoid shrubland with an ericoid and restoid understory	*Penaea mucronata, Berzelia abrotanoides, Platycaulos compressus, Leucadendron laureolum, B. lanuginosa, Pentaschistis curvifolia, Osmitopsis astericoides, Watsonia tabularis, Psoralea pinnata, Restio quinquefarius*	Seepage sites with shallow-medium depth sandy soils with high organic matter over sandstone bedrock; MAR = 779 mm Mean slope = 12° Mean aspect = SSW
Vleis	No data	No data	Freshwater lake and vleis (small, shallow water bodies) MAR = 844 mm Mean slope = 5° Mean aspect = SE

Notes:
[a] Low = < 1 m, medium = 1.3 m, tall = > 3 m.
Ericoid = shrubs with leptophyllous smaller leaves with revolute margins.
Restioid = wiry, aphyllous and evergreen graminoids belonging largely to the Restionaceae.
Source: Modified from Cowling *et al.* (1996b).

include *Puccinellia, Artemisia, Juncus, Suaeda* and *Salicornia*. Zhulidov *et al.* (1997) do not describe these as occurring in the adjoining steppe zone, emphasising instead the reed thickets (plavni) that form in continuously flooded lowlands, particularly those associated with river deltas.

Water levels in peatlands: a study in contrasts

If a site is continually wet, decomposition may lag behind primary production, allowing peat to accumulate. This is the essential process, or, more precisely, the difference between rates of two processes, that sets peatlands apart from other wetlands. Periodic droughts would allow decomposition to increase, or fires to burn the peat; hence the frequency of peatlands in cool and wet climates. In order for peat to accumulate, water levels must be relatively stable; otherwise it would be removed by erosion during high water periods and decomposition or fire during low water periods. Based upon his studies of Asian wetlands, Yabe (1993) observes

> There is a strong relationship between the stability of the water level . . . and the development of . . . bogs. The reason why the stable water level is necessary for the growth of sphagnums is that there must be enough moisture for preventing desiccation, and . . . the water must not flood the plants small ones of only a few centimeters tall (p. 46).

Further, he says (Yabe and Numata 1984), the two most important factors determining the type of wetland are water level fluctuations and mean water level, since fluctuating water levels allow the decomposition of organic matter, reducing rates of organic matter accumulation. Similarly, Tallis (1983) observes that the range of water table fluctuations is an important factor determining the type of wetland that forms and reports, following Spence (1964), that the tolerance of water level fluctuations is lower for fens than for marshes.

As peat accumulates, the water table can become more stable, as illustrated by the presence of water at or near the surface of most mires. In a series of wetland types in Japan, valley fens had some of the lowest summer changes in water level (SD $<$ 10 cm) whereas marshes had some of the highest (SD $>$ 20 cm). (Floodplains, however, also had large fluctuations in water level (SD $>$ 30 cm), but in this case the peat mats often float during high water (Yabe and Onimaru 1997)). In ombrotrophic bogs, the accumulation of peat is determined by the balance between inputs from rain and losses through evaporation and seepage (Ingram 1982, 1983). The ability of the peat to retain rainfall therefore appears to determine the shape of the bog surface (Ingram 1982). Further, the critical water balance for these calculations is the driest period through which the bog survives without irreversible desiccation. In a dry summer at Wicken Fen in England, the water table dropped as much as 48 cm below the mire surface, although declines of −20 to −4 cm were more typical. A dry summer in Finland caused the water table in a peatland to drop some 25 cm (Kurimo 1984). In Labrador, Canada, pools on the surface of raised bogs may entirely dry out, forming a mud bottom that cracks from desiccation (Foster and Glaser 1986). Further south, along the coast of Nova Scotia, mid-summer water levels fell 8–16 cm on the top of a plateau bog, and 20–30 cm on the edge of the bog and in the lagg. These changes in water table can be clearly connected with the type of vegetation present (Damman and Dowhan 1981; Foster and Glaser 1986).

Low water periods can also have major effects on the local and geographical distribution of plants. Santelmann (1991) studied the factors controlling the distribution of *Carex exilis*, a widespread sedge that, in coastal areas, occupies ombrotrophic bogs (Newfoundland, Nova Scotia, New Brunswick and Maine) whereas inland it is restricted to minerotrophic fens. Reciprocal transplants across this large scale, ranging from the Forchu Bog in Nova Scotia to Alborn Fen in Minnesota, showed that all populations were able to survive elsewhere, excluding ecotypic differentiation as an explanation for this distribution. Further, in hydroponic experiments, neither salinity nor pH had significant main effects upon growth; all treatments receiving N and P were highly productive, regardless of salinity or pH. Water table was then examined; irrespective of seed source or peat type, germination strongly decreased as the water table was reduced from 0 to 20 cm below the peat surface. Further, seedlings in the high water treatment had several tillers, whereas those in the low water treatment had single tillers with one to two leaves. *Sphagnum* moss also grew best in the high water treatments. Santelmann concludes that the water table could account for the presence of *Carex exilis* in those peatlands in which the water table remains relatively close to the surface, and its absence from mid continental bogs, where the water table normally drops 20 cm or more below the moss surface. The proximal cause, however, results because *Sphagnum capillifolium* overgrows *Carex exilis* when the water table height is 20 cm or more below the moss surface. 'Water table height thus interacts with competition for light to exclude *Carex exilis* from midcontinental bogs' (p. 2036).

If the water regime of a site is changed, say by removing the tree cover and reducing evapotranspiration, peat accumulation may be triggered.

Bakker *et al.* (1997) studied peat accumulation in different successional stages, reporting that accumulation rates were highest in emergent marshes ('reed swamps', 1.12 kg m^{-2} yr^{-1}) and lowest in later phases of quaking fen and carr (0.49 and 0.58 kg m^{-2} yr^{-1}). Over time, the influence of mineral rich water declines so that, as peat accumulates, the fen becomes a bog. In peat profiles, this transition can be recognized by changes in the nature of the peat, particularly the appearance of *Sphagnum*. Stratigraphic sequences document that accumulation of sedge peat, and then transition to forest peat or sphagnum peat (Tallis 1983; Kuhry *et al.* 1993). The water table in peatlands can slowly increase with time, as the water table rises with the accumulating organic remains of plants. Mires can then form over previously forested land, grassland or even bare rock, a process known as *paludification* (van Breman 1995). Frenzel (1983) proposes that stable water flows in rivers are an essential first step for this process to occur, '. . . paludification begins when the river flow becomes stable and when the amount of water transported [is] sufficient to influence the accompanying lowlands by seepage.' In some cases paludification may be attributed to changes in the climate, but in other cases it is driven by local circumstances related to hydrology (Walker 1970; Frenzel 1983), as we saw earlier in the discussion on succession.

As peat accumulates and a raised bog forms, the underlying topography has increasingly less impact upon vegetation; there is therefore a gradual shift away from control by local site factors towards control by climatic factors (Foster and Glaser 1986). Spring snow melts are one climatic factor which is closely related to hydrology. In regions where snow accumulates, and then melts rapidly in the spring, the flush of oxygenated water can increase rates of oxidation and produce mineral rich water flows above the anaerobic zone; this reduces raised bog development and instead maintains minerotrophic fens. In such cases, raised bogs become restricted to areas with minimal water flow such as water divides.

When water tables fall, the rate of decomposition and rate of CO_2 production increases (Silvola *et al.* 1996). In contrast, a large data set from northern peatlands (Bubier 1995) shows that methane production declines as the water table falls below the peat surface; shrubs, he concludes, are good predictors of low methane production because shrubs occur on drier sites. Warmer and drier conditions, it appears, will lower the water table and simultaneously reduce methane emissions. The influence of changing water tables upon the balance between CO_2 and CH_4 fluxes is an important issue for scenarios of global warming (Gorham 1991; Hogg *et al.* 1992). Fire frequency may also increase (Hogg *et al.* 1992). While the details of water flow within individual peatlands still pose many questions, the gross hydrology of peatlands is, in most ways fundamentally different from that of other wetlands, both with respect to the amplitude of fluctuations, frequency of fluctuations, source of water, and mineral content of the water. The ombrotrophic bog, with its relatively stable water table dependent upon local precipitation, may therefore be regarded as being nearly the exact opposite of the many examples of forested floodplain where distant rainfall causes the water levels to fluctuate dramatically.

Reservoirs, dams and floodplains: the consequences of altered hydrology

Humans are far more effective at constructing dams than are beavers. The world is now covered with reservoirs created by *Homo sapiens*. Final world maps are not yet available, but Figure 4.22 shows a representative continent. The construction of dams on rivers is now a global phenomenon (Figure 4.23); as a result, natural water level fluctuations are being disrupted around the world. Alteration of hydrology is believed to be one of the three major causes of damage to aquatic animals; in the United States of America alone, there are now more than 75 000 large dams (higher than 8 m) and 2.5 million small ones (Richter *et al.* 1997). The nature of the alteration to natural patterns will vary depending upon the purpose of the dam. The effects will also differ between the reservoir upstream from the dam and the river downstream. The list of negative impacts from the construction of dams includes destruction by flooding, mercury contamination, release of greenhouse gases (CO_2, CH_4), and damage to migratory fish species (Rosenburg *et al.* 1995). A nearly universal effect of dams is the reduction in spring flooding, and a consequent reduction in wetland area in the remaining floodplain, along with conversion of wet meadows to swamps.

The effects of dams upon hydrology can be summarized in four categories (Klimas 1988).

Water levels stabilized

The near permanent inundation or saturation of substrates that were formerly periodically exposed.

Shifted flood timing

Reservoirs cause lags in the release of water, so that flooding can be delayed by months, sometimes until well into the growing season of vegetation that normally would develop after flood peaks had passed.

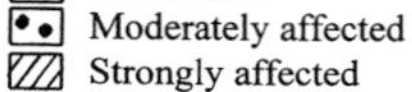

Figure 4.22 River regulation in North America (after Dynesius and Nilsson 1994).

Increased flooding

Levees can be considered as linear dams; they tend to increase flood peaks by constraining the flow of water onto adjacent floodplains.

Decreased flooding

This is a major consequence of many projects. The intent may be to deliberately reduce flooding downstream, or to maintain a steady flow of

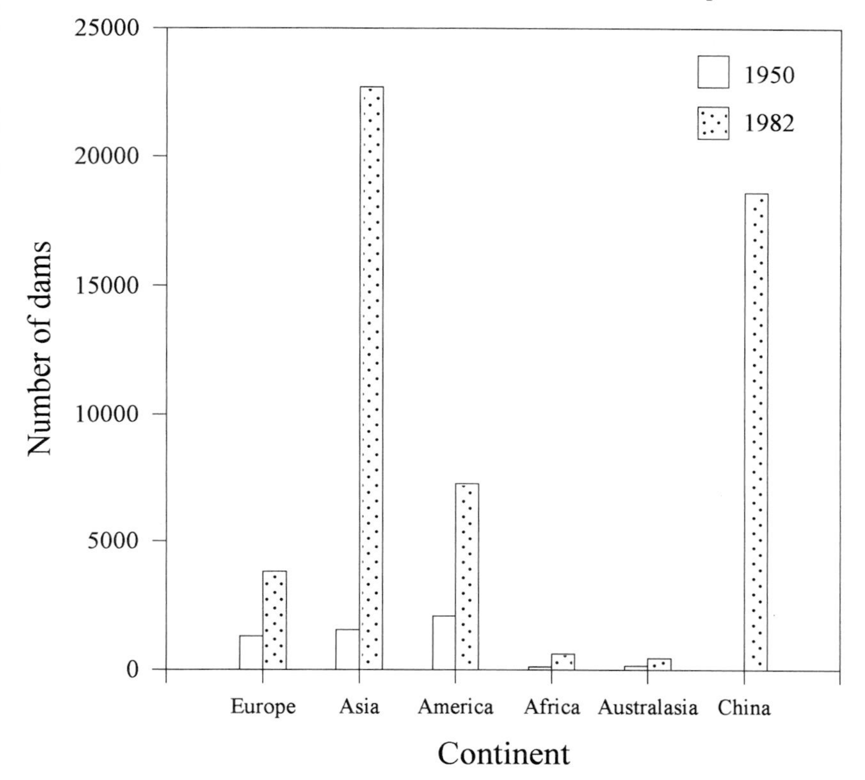

Figure 4.23 The construction of dams is altering hydrology around the world (data from Van der Leeden 1990).

water for hydroelectric generation or irrigation. By holding back run-off during normal flood periods, the duration, frequency and area of flooding are all reduced.

Upstream effects: the reservoir

The frequency of water level fluctuations may be decreased within the reservoir if the reservoir is a lake that has been dammed to stabilize water levels for recreation or shipping, or it may increase dramatically if the reservoir is being used to provide pulses of water for daily peaks in power demand. Similarly, the amplitude may be reduced if the dam reduces summer low water periods, or it may be increased if the dam is used to regulate water levels in headstock ponds for power generation. Every reservoir may be considered to have its own water level personality, determined in part by the purpose of its construction, and in part by the personalities of the people that maintain it.

Consider an extreme example: the Gardiken Reservoir in northern Sweden covers an area of 84 km^2 on the Umea river in northern Sweden (Nilsson 1981). Water levels may fall as much as 20 m, exposing approximately 56 km^2 of shoreline. This reservoir is of particular interest because it represents some sort of biological limit to how much one can alter water levels in wetlands; the water levels are almost exactly the reverse of natural cycles. It is lowest in spring (in order to begin storing melt water) and it is highest in the autumn. It is then progressively lowered during the winter in order to generate hydroelectric power, which, in Sweden, is needed most during the winter. One might expect that reversing the entire water levels cycle would have dramatic effects upon vegetation. One would be right. Most of the reservoir shore is barren, except for a 1–2 m wide strip of sparse vegetation close to the high water level. That is, this combination of frequency and intensity is so extreme that no plants can tolerate the conditions over most of the reservoir.

Within such circumstances, there are two basic questions we need to answer: (i) what are the main patterns in the controlling factors, and (ii) what is the relationship between the controlling factors and the vegetation? One of the nice features of reservoirs is that they generally have good water level records. Historical records on key factors are of great value to ecologists. Table 4.6 takes the water level variation and breaks it into two components: the magnitude of effects, and the variation in them. Within the magnitude category, duration of flooding, maximum depth of flooding and longest flooding period are all significantly correlated. Within the variation category, date of last flood, date of first flood and number of floods are also significantly correlated, but these account for a very small proportion of the total variance in the data set.

How does the vegetation respond to these factors? There are two basic properties we could try to predict: the abundance of plants and the richness of plants. The abundance of plants decreases with greater duration of flooding, but increases with the number of floods. The richness of plants shows similar patterns, but also declines with the date of last flood. A multivariate regression model showed that the two best predictors of the vegetation were (i) the composition of the vegetation the preceding year and (ii) the duration of the flooding the preceding year. If the duration of flooding was around 60 days, the vegetation remained more or less unchanged. If duration of flooding was less than this, abundance of plants increased, and if it was greater than this, the abundance of plants decreased (Figure 4.24).

Table 4.6. *Patterns in the Gardiken Reservoir in northern Europe*

	Magnitude			Variation		
	Duration of flooding	Maximum depth of flooding	Longest flooded period	Longest dry period	Date of first flood	Date of last flood
Magnitude						
Duration of flooding	1.00					
Maximum depth of flooding	0.71	1.00				
Longest flooded period	0.97	0.71	1.00			
Longest dry period	−0.68	−0.42	−0.61	1.00		
Variation						
Date of first flood	−0.66	−0.40	−0.60	0.99	1.00	
Date of last flood	0.51	0.50	0.44	0.01	0.05	1.00
Number of floods	−0.20	−0.26	−0.37	−0.07	−0.07	−0.07
Vascular plants						
Abundance	−0.14	–[a]	–	–	−0.04	−0.05
Species richness	−0.26	–	–	–	0.03	−0.16
Bryophytes and lichens						
Abundance	−0.28	–	–	–	0.07	−0.10
Species richness	−0.30	–	–	–	0.11	−0.12

Notes:
[a] Not used further because of strong correlation with first column, duration of flooding.
Correlations among flooding factors are shown at the top half. Correlations between vegetation and flooding are shown in the bottom half.
Source: After Nilsson and Keddy (1988).

Reservoirs in Norway have been similarly studied (Rorslett 1984, 1985), showing that as well as abundance, the distribution of plants changes as the amplitude of water level fluctuations increases. In general, species shift into deeper water as the amplitude increases (Figure 4.25). Species respond to both the duration of flooding (causing lowered light availability) and the timing of drawdowns (causing patterns in ice scour) (Figure 4.26).

Downstream effects: altered hydrology

The impacts downstream from the reservoir may affect an even larger area of landscape than the reservoir itself. Downstream from the dam,

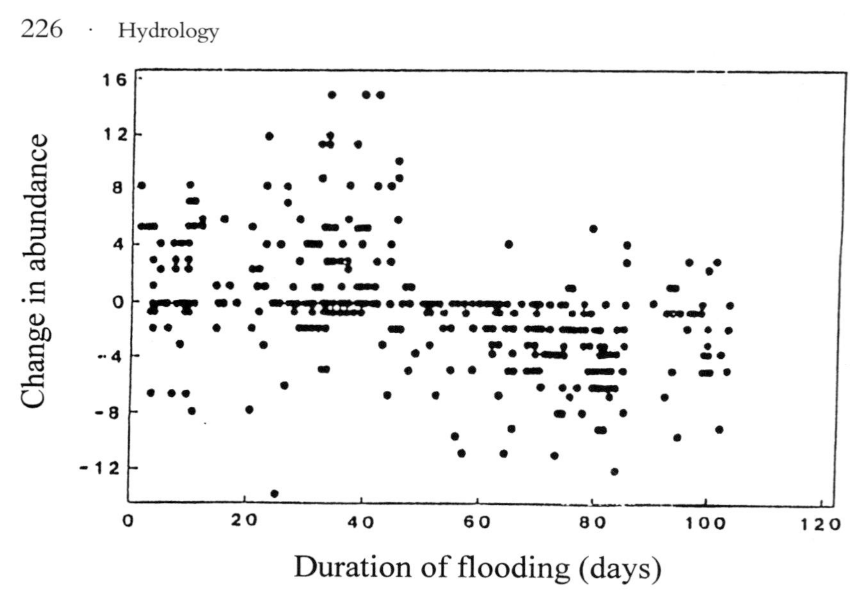

Figure 4.24 The effects of duration of flooding on the change in abundance of plants in a northern European reservoir (from Nilsson and Keddy 1988).

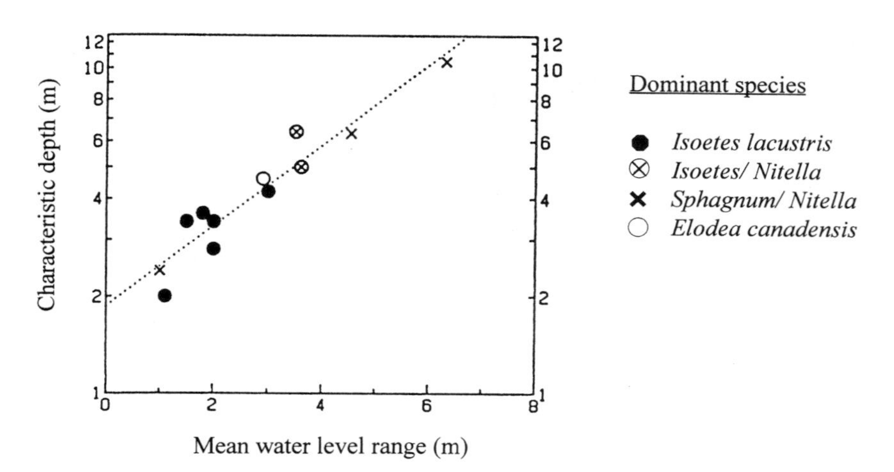

Figure 4.25 The characteristic depth of four aquatic plants plotted against range of water levels in a northern European reservoir. The depth of occupancy increases with amplitude of fluctuations (from Rørslett 1985).

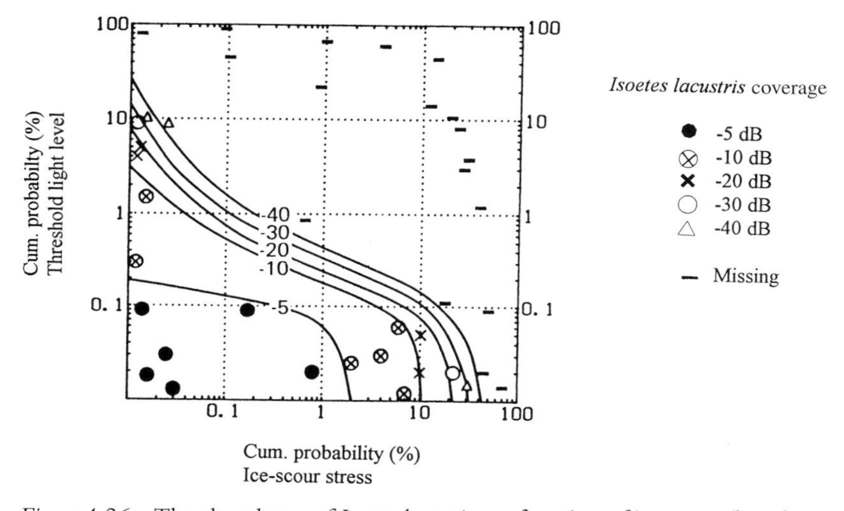

Figure 4.26 The abundance of *Isoetes lacustris* as a function of ice scour (largely determined by timing of flooding) and light levels (largely determined by duration of flooding) (from Rørslett 1984).

water levels are usually stabilized, particularly if the dam has been constructed to hold back spring flow and release water later in the year. If we roughly define the area of wetland in a watershed by the area inundated during spring flood, then it is obvious that, by reducing spring floods, dams eliminate immense areas of wetland in a watershed. Recent studies (Dynesius and Nilsson 1994) show that almost every watershed in the world has been altered by the construction of dams; there is therefore a global trend to reduce the maxima of spring run-off. Figure 4.27 shows the situation in Canada alone. We have yet to fully recognize, let alone tabulate, the magnitude of wetland losses resulting from this practice (but see Nilsson *et al.* 1991; Johnson 1994; Poiana and Johnson 1993; Toner and Keddy 1997).

Some of the severe negative consequences of altered hydrology are shown by the Peace–Athabasca Delta in Alberta, Canada (Rosenberg and Barton 1986; Rosenberg *et al.* 1995). It is part of the larger Mackenzie River system, the fourth largest discharging into the Arctic Ocean; the Peace–Athabasca delta is 'probably the most important northern delta in north America for nesting and is used by hundreds of thousands of birds' (Rosenberg and Barton 1986). The delta consists of 39 000 km^2 of wetlands formed where the Peace River flows into Lake Athabasca (Figure 4.28(*top*)). It is divided into three main vegetation types: shrub/forest

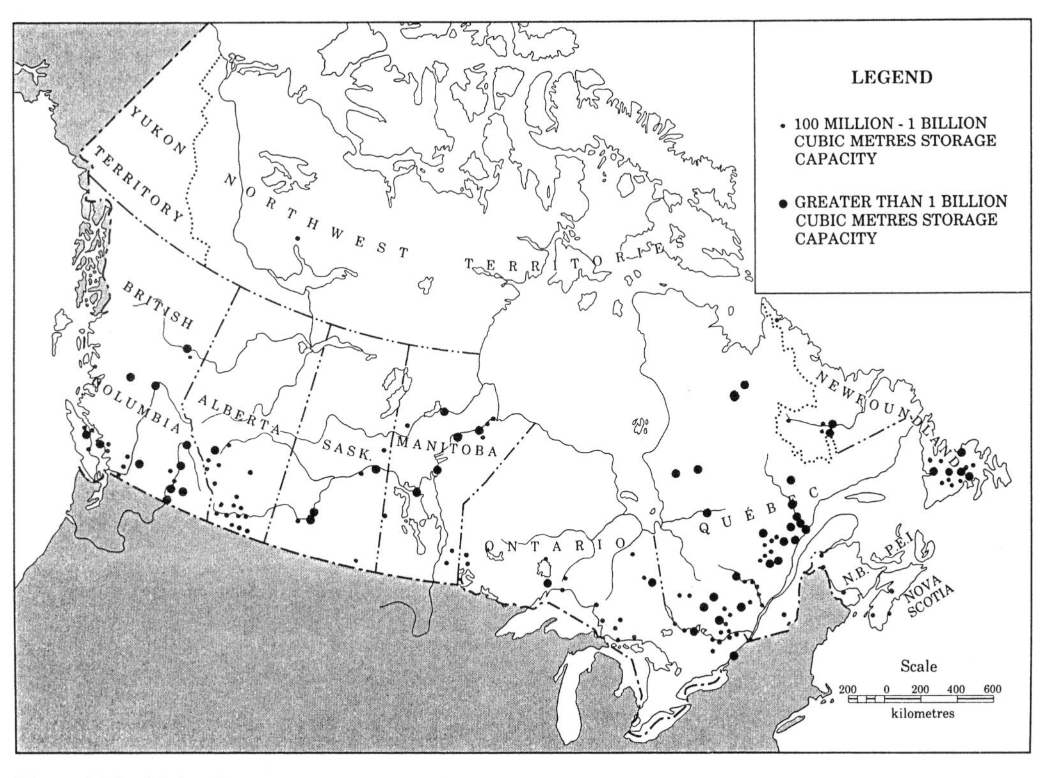

Figure 4.27 Major dams on rivers in Canada (Pearse *et al*, 1985).

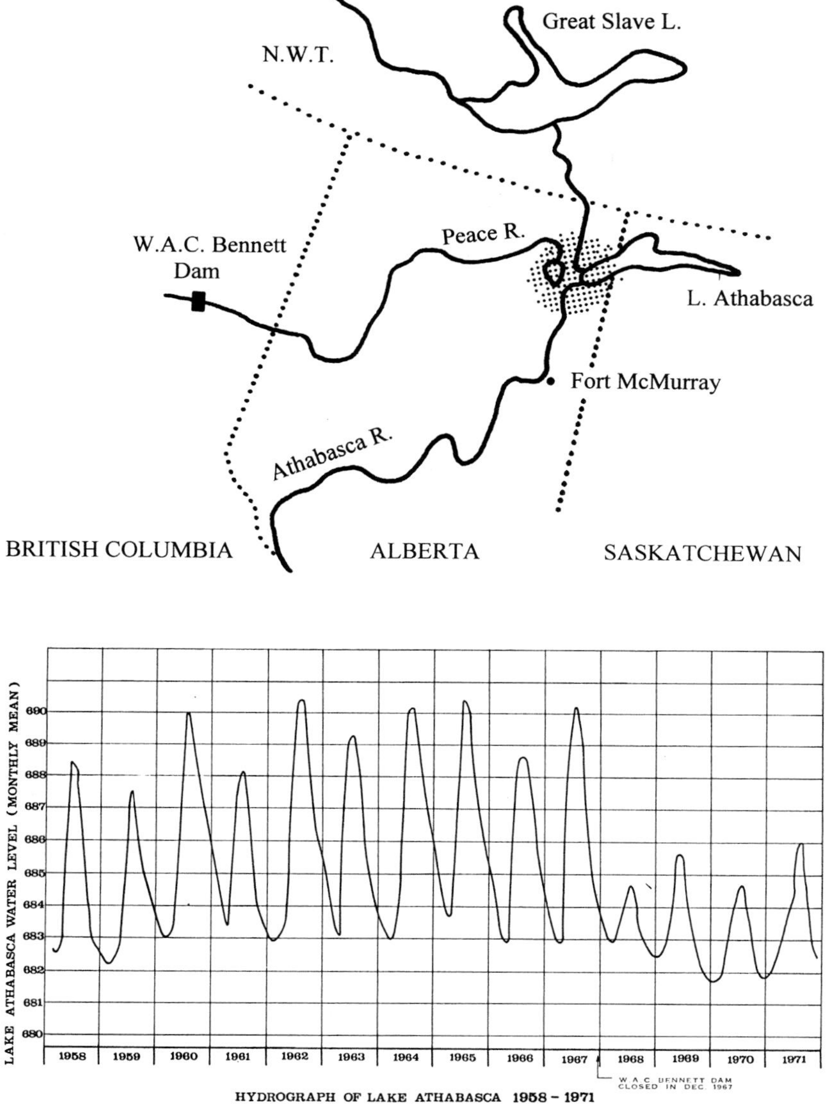

Figure 4.28 The location of the Peace–Athabasca delta and the W. A. C. Bennett Dam (*top*) and the changes in the hydrology of Lake Athabasca once the dam was completed (after Keddy 1990 and after Peace–Athabasca Delta Project Group 1972).

(40%), open water (36%), and 'productive' wetlands (24%) (Townsend 1984). The latter includes emergent vegetation, mud flat, immature fen and wet meadow – the approximate equivalent of the marsh and wet meadow vegetation types in Figure 4.9. In 1968 the Bennett Dam was completed 1200 km upstream on the Peace River. A major effect of this dam was to remove the annual peak in floodwaters that occurred in June (Figure 4.28(*bottom*)). This change in hydrology led to rapid changes in the vegetation of the delta, with many of the productive wetlands becoming converted to woody vegetation through short-term succession (Figure 4.29). The rate of change was remarkable: within 2 years (by 1970) the total area of the nine largest water bodies had decreased by 28%, and the numerous perched lakes and ponds of the delta were drying up at the rate of 12% per year. *Calamagrostis canadensis, Phragmites australis*, and *Salix* spp. spread rapidly. Because the delta is one of North America's most important waterfowl staging areas, and is part of Wood Buffalo National Park, The Peace–Athabasca Delta Implementation Committee conducted detailed studies into the problem (Peace–Athabasca Delta Implementation Committee, 1987), and recommended construction of weirs to hold back flood waters and recreate periodic flooding. Although this could not protect all the wetlands from the damage caused by the Bennett Dam, it made the best of a bad situation.

Similar changes can be expected wherever dams are built. Petr (1986) describes the effects of the Akosombo Dam on the Volta River in Ghana. This dam created 'Lake Volta' with a surface area of some 8500 km^2 and annual drawdowns of some 3 m. Prior to construction of the dam, yearly flows cycled across four orders of magnitude, between 10 and 10^4 $m^3 s^{-1}$; after closure of the dam, flow stabilized at 10^2 $m^3 s^{-1}$ throughout the year. Since river scouring was reduced, plant growth along the river banks exploded (primarily *Potamogeton octandrus, Vallisneria aethiopia*); a plant-associated snail became more common, and schistosomiasis is now endemic.

Similar disruption in vegetation downstream from dams has been observed in many other watersheds including the Colorado River in Arizona (Turner and Karpiscak 1980), the Milk River in Alberta and Montana (Bradley and Smith 1986), the Platte River in Nebraska (Johnson 1994), the Kissimmee River in Florida (Toth 1993), and streams in the Sierra Nevada (Harris *et al.* 1987). As in the Peace River, floodplains are being invaded by woody vegetation.

When riparian forests are surrounded by arid prairies rather than forested landscapes, the effects of reduced spring floods may be more

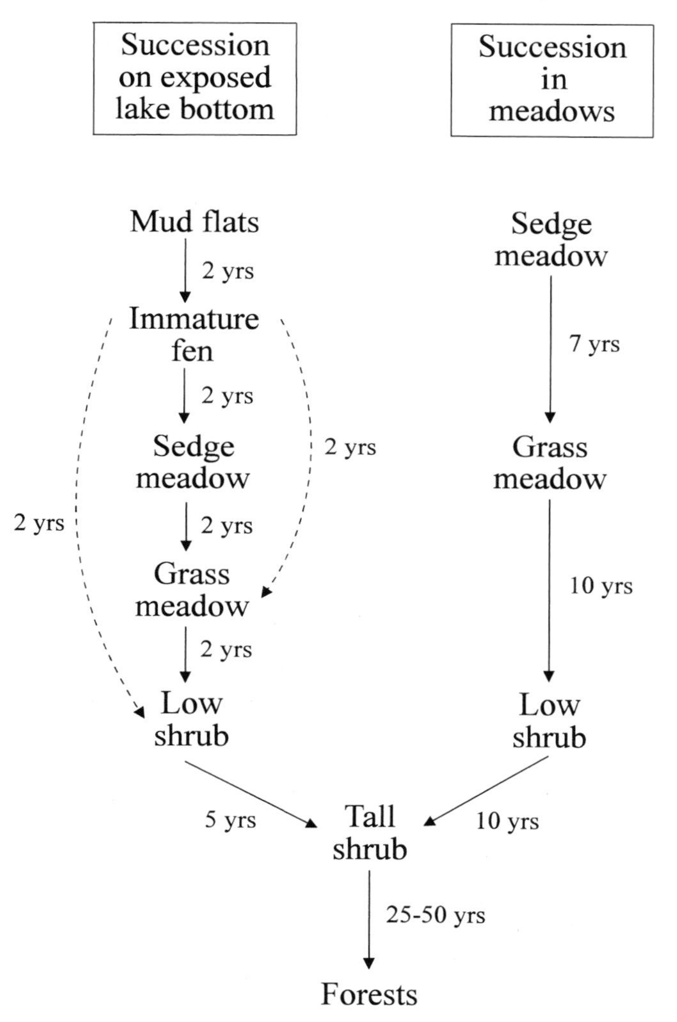

Figure 4.29 Changes in the vegetation in the Peace–Athabasca Delta after the dam in Figure 4.28 reduced spring flooding (after Keddy 1990b and after Peace–Athabasca Delta Project Group 1972).

severe because as the floodplain dries, it may be unable to support wetland at all. Rood and Mahoney (1990) describe the importance of poplar forests in river valleys (*Populus* spp., largely Prairie Cottonwood, *P. deltoides* and Balsam Poplar, *P. balsamifera*), which are 'thriving oases for wildlife' amidst the dry plans of western North America. In some cases, these riparian forests have complex vegetation and support a variety of

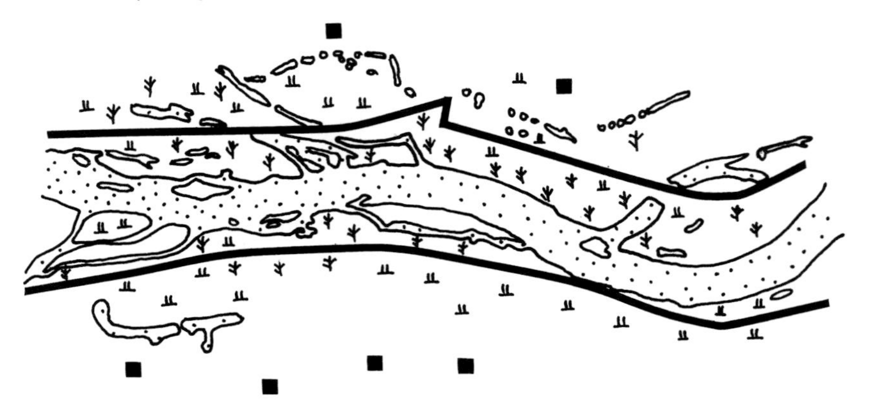

Figure 4.30 Rivers such as the Vistula are largely constrained within walls and dykes (after Kajak, 1993). Contrast this situation with Figure 1.6.

trees shrubs and smaller plants absent from the surrounding landscape (Johnson *et al.* 1976); in these cases, poplar is an early successional species. Further west, poplars may be the dominant or the sole tree in floodplains. 'As the poplars die, so dies the whole riparian forest. Wildlife habitat is lost, the forest canopy is lost, and the forest understorey dies.' Reduced river flows not only stress the older trees, but may prevent establishment of seedlings. Further, dams reduce erosion and the flows of sediment, thereby eliminating sites for poplar seedlings to establish. Patten (1998) has emphasized that in spite of differences in species composition, these changes are occurring throughout arid areas of North America; most likely, they can be extrapolated to other arid areas as well.

Dams are not the only way of reducing water level fluctuations in floodplains. Increasingly, rivers are contained within walls to prevent flooding, particularly in heavily populated areas. For example, the Vistula, which drains more than half of Poland, is more than 1000 km long, yet Kajak (1993) reports that it 'has side walls on almost its whole length' (Figure 4.30). There are now plans to wall in two of the largest rivers in the world, the Ganges and the Brahmaputra (Pearce 1991). While the emphasis of this chapter is upon the importance of water level fluctuations, it should be noted that there is an additional indirect effect of building dams and dykes: the lands temporarily protected from spring flooding are frequently converted to agriculture or urban uses. Because this secondary effect, the clearing for agriculture or cities, is so obvious and dramatic, it is sometimes easy to overlook the primary effect of dams and dykes, the loss of wetland area and conversion of meadows to swamps that would occur even if no further human activities had occurred. Dams, dykes and agriculture are a devastating combination for alluvial wetlands: in the Mississippi River floodplain, there were originally an estimated 8.5 to 9.5 million hectares of wetland forest; by the 1990s, only some 2 million remained. Although satellite images tell us that these 2 million remain (Figure 4.31), they cannot tell us how much these wetlands have been altered by reduction in spring floods, nor how many more may eventually be replaced by encroaching terrestrial species.

Finally, we have an exception. A nearly opposite situation occurs in the marshes of the Camargue at the mouth of the Rhone River in southern France. Here, riparian marshes that originally dried during the summer are turned into permanent water bodies by pumping in river water during the summer (Van Wijck and de Groot 1993).

A predictive model: changes in shoreline wetlands

It is one thing to describe changes in vegetation that occur after a dam is constructed. In some cases, this is the proverbial situation of closing the barn door after the horse has escaped. What we need to be able to do is predict the severity of the changes that will occur before a project is built so that we can fully establish the consequences beforehand. We have the general empirical pattern established around the world that dams reduce flooding, that reduced flooding causes wetlands to shrink, and that herbaceous wetlands are invaded by woody species. This pattern is important, although still perhaps not widely enough appreciated. Beyond this appreciation, we need quantitative models to predict not only the direction of change, but the amount that is probable. The logistic regression approach, used to construct Figure 4.14, may hold considerable promise. Simpler models may work equally well, as we shall now see using the water level changes in the Great Lakes.

The Great Lakes are among the largest bodies of fresh water in the world (Table 2.1). The water levels of these lakes have changed over both geological and historical time scales (Figure 4.3). There is a rich array of wetland types (Table 4.7). Rich wet meadow floras are particularly well developed on gently sloping sandy shorelines (Reznicek and Catling 1989). Great Lakes wetlands provide important habitat for fish, waterfowl, and rare plant species (Smith *et al.* 1991). Large areas of these wetlands have been drained, and humans have also already reduced the amplitude of water level fluctuations. More recently there was added pressure to control them further. The drawing in Figure 4.9 was used to

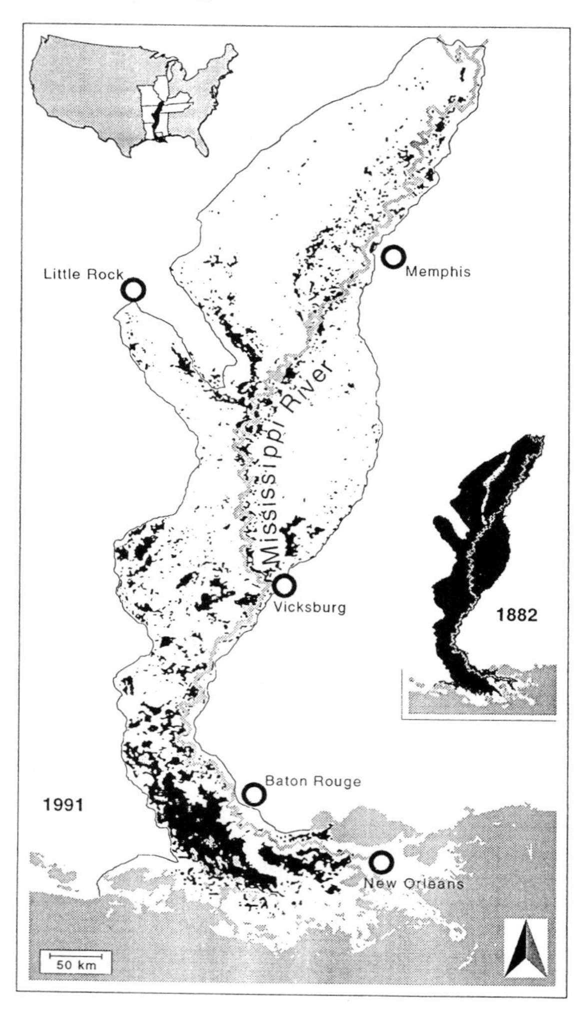

Figure 4.31 The remaining bottomland forests of the Mississippi floodplain (from Llewellyn *et al.* 1996). Once a river is constrained by dams and dykes, the floodplain is often cleared for agriculture. This can obscure the more widespread but less obvious effects of reduced spring flooding, such as encroachment by terrestrial species, conversion of wet meadow to woody plants, and changes in function and biological diversity.

Table 4.7. *Characteristics of wetlands in the Great Lakes*

	Lakes				
	St. Lawrence River	Lake Ontario	Lakes Erie and St. Clair	Lake Huron	Georgian Bay
Number of wetlands	27	64	28	13	10
Area of wetlands (ha)	3511	11555	19306[a]	1274	1829
% marsh	91.2	79.1	83.9	5.1	73.3
% swamp	8.8	20.3	16.1	27.8	24.6
% fen		0.2		16.7	2.1
% bog		0.1			
% palustrine and isolated	0.7	2.4	3.9	27	4.1
% riverine	95.8	28.9	3.9	7	27.4
% rivermouth	3.5	25.2	16.4	2.2	
% lacustrine on bay		27.5	13.8	5.2	
% lacustrine		16.6	61.3	58.6	68.5
% organic soil	6.2	45.4	50.3	31.9	0
% mineral soil	89.6	53.7	49.7	68.1	100
Number of vegetation communities	9.1	9.6	6.8	32.2	11.8

Notes:
[a] Walpole Island marshes account for at least another 10000 ha.
Source: From Smith *et al.* (1991).

provide preliminary estimates of the probable effects on wetland area. The objective was to predict the upper boundary of wet meadows, and the lower boundary of the marsh.

To model the landward edge of the wet meadow, it was necessary to consider the die-back and re-colonization by woody plants. Two assumptions were made. The first was that the die-back of woody plants was directly related to high water levels during the growing season, and that woody plants re-invade according to an exponential model. This allowed predictions of the lower limit of woody plants from projected water levels (Figure 4.32 top line). Lag times of 20 or 15 years as opposed to 18 years made little difference.

To model the lower boundary of the marsh required the assumption that marsh plants move downslope the same year as water levels fall; this would most likely be the result of germination from buried seeds. As water levels rose, the wetland plants would die back over several years

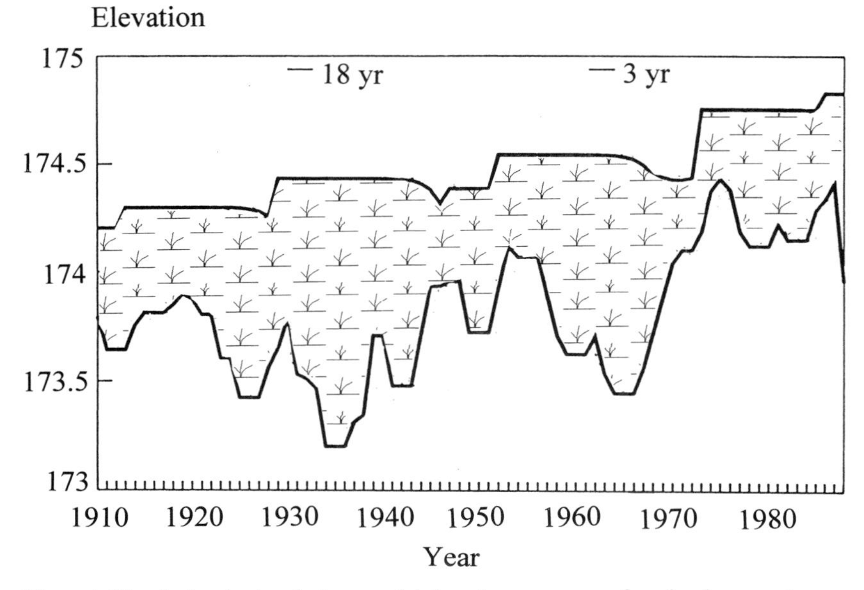

Figure 4.32 A simple simulation model showing response of wetland vegetation to changes in water levels in the Great Lakes (after Painter and Keddy 1992). Note that the area of wet meadow varies with water level history.

(Figure 4.32 bottom line). Lag times of 4 or 2 years as opposed to 3 made little difference.

The area between these lines is then the area of wet meadow and marsh as a function of time. The figure shows, for example, the great areas of wetland that occurred during the low water period of the mid-1930s. This model was then used upon projected water level scenarios to forecast the effects on wet meadow/marsh area in the Great Lakes. If further reductions in amplitude occurred, the model predicted losses approximating 30% of the wetlands in Lake Ontario alone.

A summary model: frequency and intensity of flooding

Two of the most important components of flooding are frequency and intensity, and these can be plotted on orthogonal axes to represent all possible pairwise combinations. We can then plot, for many sites, the frequency of flooding against the depth of flooding, or perhaps more conveniently, the frequency of flooding and amplitude of water level changes. There are several important properties we could then plot on

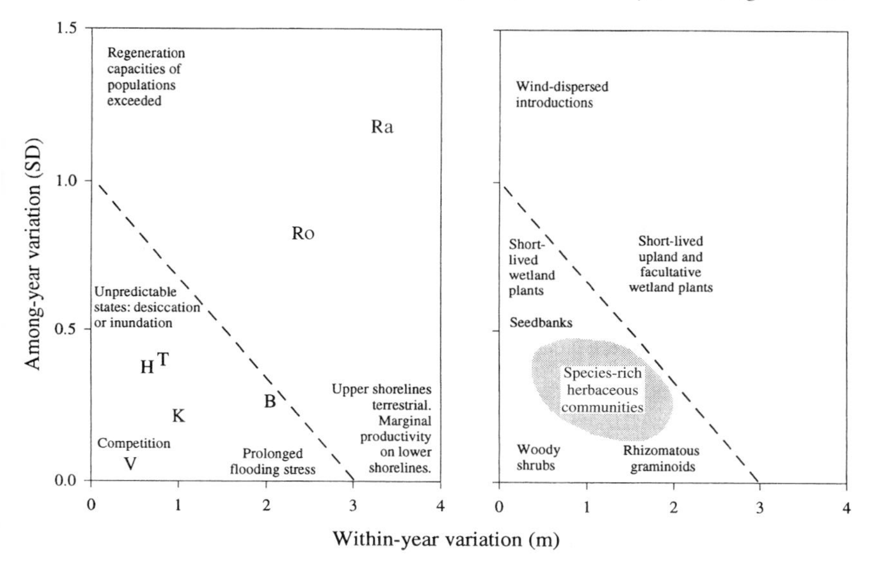

Figure 4.33 The effects of water-level variation upon shoreline vegetation. The stippled area contains lakes whose hydrological regimes produce rich floras, while hypo- and hypervariable zones represent impoverished systems. Rich floras with many rare species occur in unregulated lakes with high catchment areas such as Kejimkujik (K) and Bennetts (B), both lakes with immediate water level fluctuations (1 m changes in amplitude within years and 0.5 m variation among years). In contrast, hypovariable lakes such as unregulated lakes with small catchment areas an head ponds (e.g. V, Vaughan) lose many species through competitive exclusion by shrubs. Hypervariable lakes such as storage reservoirs (above the dashed line, e.g. Ro, Rossignol and Ra, Raynard) lose species and are subject to invasion by exotic species. The stippled area is therefore the desirable management target. Increased catchment area can push lakes into the region of high richness, but reservoir construction can push lakes into the hypervariable state. Data on Long Island pondshores (H, House Pond and T, Third Pond) (from Hill *et al.* 1998).

these axes. One could plot the many reservoirs or wetlands of the world in order to explore patterns. Are there, for example certain combinations that are rare and others that are common? One could plot important properties such as productivity or wildlife diversity, and explore how they respond to water levels. Notice the use of the word could. Unfortunately, the required data on water levels are scattered through a broad literature describing individual cases, and buried in reams of unpublished reports. Often the axes are not quite comparable. The data on vegetation, productivity or wildlife diversity are even more scarce. As a first step in this direction, Figure 4.33 shows such a plot for a few lakes, and identifies a

Figure 4.34 The globally significant wetlands of the Amazon River basin are threatened by dams altering annual flooding (from Sioli 1986).

region of high plant species richness. This is based upon a set of lakes in eastern North America, and there is currently no way to know how well we can extrapolate beyond this geographic region or to other properties. This chapter ends with a lament. Here we have possibly the most important environmental factor influencing wetlands, yet we lack the data to plot even the most basic properties against this factor. I have introduced some studies that suggest some steps in the right direction. We have our work cut out for us.

Back to reality: the Amazon river basin

The chapter opened with the Amazon River basin as a globally significant wetland ecosystem that is produced and maintained by annual floods (Junk 1983, 1986). The importance of these floods to the rich array of fish species was described in Chapter 2 (Goulding 1980; Lowe-McConnell 1975, 1986). A wide variety of other examples illustrates the importance of natural changes in water level upon wetlands and the impacts that dams have upon them. These impacts also can be expected along the Amazon, where dams are proliferating (Figure 4.34). This final figure reminds us that as our understanding of the effects of hydrology grows, its wise application becomes ever more pressing.

5 · *Fertility*

Water levels, and their patterns of variation, are the primary control factors in wetlands. But all organisms require more than just water (Table 5.1). If these basic elements are in short supply, growth and reproduction will be curtailed. The word fertility can be used to refer to the availability of resources for growth and reproduction. For plants, these resources include mineral nutrients such as nitrogen, phosphorus and potassium, as well as carbon dioxide. In flooded sites, resources like soil oxygen can limit growth; in others (but rarely wetlands), scarcity of water has similar effects. Factors such as high salinity or low temperatures can also reduce growth rates, and therefore are also considered to affect fertility. Although the term fertility is normally restricted to plant ecology, it may be usefully extended to other organisms such as animals and fungi, where rates of production are also dependent upon the availability of resources and constraints imposed by environmental extremes. Any environmental factor that constrains plant productivity is generally termed a stress; although not everyone agrees with the use of this term, there is a growing body of literature on its value for sceptics to consult (e.g. Grime 1977, 1979; Levitt 1980; Larcher 1995), and since this is a book about wetlands, I do not want to be distracted by terminology debates. Animal ecologists have used the word adversity to describe the same kind of constraints (Southwood 1977, 1988). Basic questions about fertility would include (i) What resources do wetland organisms require? (ii) What patterns occur? (iii) What are the consequences of these patterns? and (iv) How have humans modified these patterns?

Before exploring the impact of low fertility upon plant communities, it might be helpful to preview how shortages of nutrients for plants can have significant consequences for animals (a topic further considered in pp. 262–265 and in Chapter 8). Most plant tissues have relatively low ($<5\%$) nitrogen content; this may conserve an essential resource, but it

Table 5.1. *Atomic composition of four typical CHNOPS organisms*

Element	Mammal[a]	Vascular plant[b]	Arthropod[c]	Moneran[d]
C	19.37	11.34	6.10	12.14
H	9.31	8.72	10.21	9.94
N	5.14	0.83	1.50	3.04
O	62.81	77.90	79.99	73.68
P	0.63	0.71	0.13	0.60
S	0.64	0.10	0.14	0.32
Total	97.90	99.60	98.16	99.72

Notes:
[a] Humans; [b] alfalfa; [c] copepod; [d] bacteria.
Source: From Keddy (1989a) after Morowitz (1968).

simultaneously reduces their nutritional value for herbivores. Consider waterfowl. The small body size required for flight and migration may have prevented geese from developing the large gut required for efficient nitrogen extraction from low quality plant material. Geese therefore, appear to gain sufficient nitrogen by using high rates of throughput of superficially digested plant food (Fox and Kahlert 1999). Nitrogen supplies for waterfowl become particularly critical during moulting periods, when the birds are flightless and producing new feathers. Greylag geese, for example, select for protein-rich food, but during moult they feed only at night, thereby cutting foraging time in half (Fox and Kahlert 1999). This may reduce exposure to predators during a vulnerable period, but it also cuts food intake in half. During moulting, Fox and Kahlert found that food spent more time in the gut, 68% less uric acid was excreted, and the proportion of nitrogen extracted from food increased from a typical 18% to a more efficient 42%.

Fertility and primary production

The resources required by living organisms can be deduced from their chemical composition. Table 5.2 shows the basic elemental composition of different species of wetland plants (see also Gopal 1990 for similar data from India). These are not that different from the basic elemental composition of other living organisms shown in Table 5.1. These elements are the raw materials of primary production. Geographical patterns of primary production were described in Chapter 1. In temperate zones,

Table 5.2. *Atomic composition of different species of wetland plants*

	Temperate species		Tropical species	
Constituent	*N*	Mean	*N*	Mean
Ash (%)	40	14.03	75	14.23
C (%)	28	41.06	–	–
N (%)	27	2.26	75	1.99
P (%)	35	0.25	75	0.19
S (%)	25	0.41	–	–
Ca (%)	35	1.34	75	0.88
Mg (%)	35	0.29	75	0.29
K (%)	35	2.61	75	3.10
Na (%)	35	0.51	75	0.36

Source: After Boyd (1978) and Junk (1983).

most primary production is concentrated within a relatively short period of each year (Figure 5.1).

The fact that some wetlands have lower levels of primary production than others challenges us to find the resource that is constraining production. It is generally agreed that shortages of nitrogen are one of the major factors limiting plant growth (e.g. Good *et al.* 1978, Salisbury and Ross 1988), although depending on the environment, drought, cold and salinity can also be extremely important (Larcher 1995; Archibold 1995). The prevalence of nitrogen-limitation seems paradoxical, since all terrestrial plants are bathed in an atmosphere of roughly 75% nitrogen. But plants cannot use atmospheric nitrogen; it must be fixed into ammonium or nitrate for plants to use it, and this ability is possessed only by prokaryotic organisms.

The fundamental importance of ammonium is illustrated by the occurrence of an amino group (—NH_2) in every amino acid:

```
          H
H \       |     // O
    N  -  C  -  C
H /       |     \ OH
          R
```

Although there are more than 100 amino acids in nature, only 20 are commonly found, and they differ in the chemical units attached at R. For example, in glycine, R is replaced by H, in alanine by CH_3 and in serine

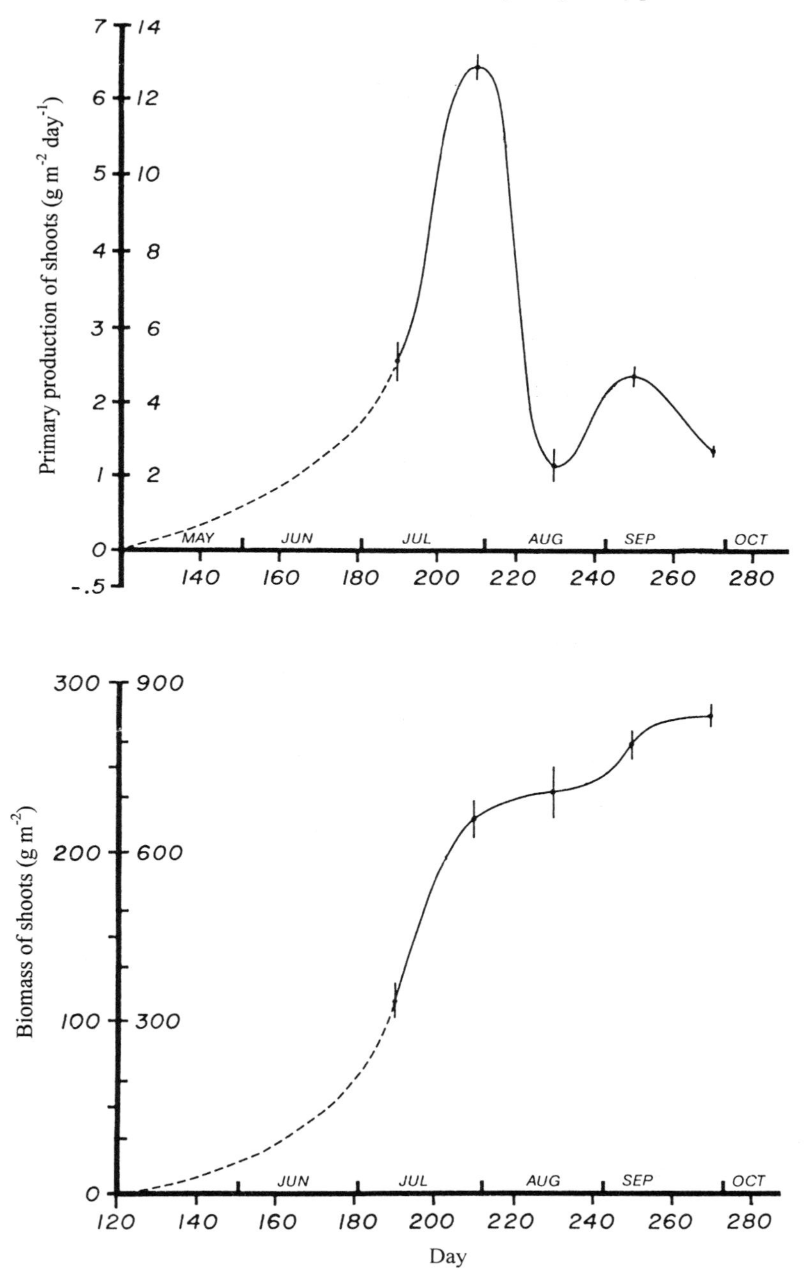

Figure 5.1 Primary production of shoots (*top*) and biomass of shoots (*bottom*) as a function of time of year (after Auclair *et al.* 1976b).

by CH_2–OH. Proteins are built up from long chains of amino acids, but each subunit of each protein requires the critical nitrogen molecule from ammonium. Ammonium and its counterpart ammonia are toxic at quite low concentrations; they are therefore usually absorbed and incorporated into organic compounds in the roots. In contrast, nitrate is easily transported in the xylem and can be stored in vacuoles of roots, shoots or storage organs. The costs of nitrate use lie in the necessity of reducing it back to ammonium to enable it to be further used in metabolic processes; this process consumes energy and is analogous to the process of reduction and assimilation of CO_2 in photosynthesis (Marschner 1995). These processes of nitrogen uptake and reduction are essential to our understanding of wetlands, not only because nitrogen is a key resource for plants, but because White (1993) argues that nitrogen is equally the key resource for animals. Nitrogen fixation, nitrogen absorption and nitrate reduction within plants may thus be the critical limiting steps in the production of the entire biota of wetlands.

With this background, we might at first assume that the inquiry should begin with nitrogen (or, more precisely, ammonia and nitrate availability) in wetland communities. Certainly, some ecologists believe that nitrogen is the key below-ground resource limiting terrestrial plant communities (Vitousek 1982; Tilman 1986; Berendse and Aerts 1987). However, such generalizations may be inappropriate, particularly for wetlands. Freshwater ecologists studying phytoplankton have concluded that phosphorus is the critical limiting nutrient (e.g. Vallentyne 1974; Schindler 1977; Smith 1982; Rigler and Peters 1995) This is why such an effort is made to remove phosphorus from sewage. So, were we to seek a general factor controlling plant production in wetlands, would it be best to start with the terrestrial situation and study nitrogen gradients, or to start with the aquatic frame of reference and study phosphorus?

The answer probably is that we should avoid creating dichotomies when nature is often rather complex (e.g. Dayton 1979; Mayr 1982); perhaps the dichotomy is not so clear cut as it first appears. Further, in marine situations, nitrogen rather than phosphorus appears to be the critical limiting resource (Valiela *et al.* 1992; Levine *et al.* 1998). Returning to lakes, while phosphorus levels are the primary control upon algal biomass, it seems that nitrogen does play an important role by modifying the relationship between algal biomass and phosphorus (Smith 1982). Further, as the ratio of nitrogen to phosphorus increases, cyanophytes (blue green algae) are replaced by other phytoplankton (Schindler 1977; Smith 1983). Thus N:P ratios can control both amounts and composition of phytoplankton.

One way to assess the relative importance of these elements in controlling plant growth in wetlands is to supplement the nutrient supplies in field experiments. Such experiments also suggest that, as in lakes, a simple dichotomy between N and P does not exist. In an early study of fertilization in species-rich dune slacks, a relatively dry type of wetland within the wet meadow category, Willis (1963) examined the effects of adding N, P, K and NPK fertilizer. The slacks were covered with sparse vegetation, in which the dominant species were *Agrostis stolonifera*, *Anagallis tenella*, *Bellis perennis* and three species of *Carex*. In field trials with different mineral salts added, complete nutrients produced three times as much biomass, and a reduced number of plant species, after only three years. The greatest deficiencies, concluded Willis, were for nitrogen and to a lesser degree, phosphorus. He then transplanted pieces of turf to the greenhouse and submitted them to different fertilization treatments. The results were similar to the field trials: complete nutrients yielded 151 g of shoots (fresh weight), nitrogen deficient treatments produced 34 g and phosphorus deficient treatments produced 44 grams. He concluded '. . . the sparse growth and open character of the vegetation of the Burrows are brought about mainly by the low levels of nitrogen and phosphorus in the sand . . .'. Further, treatment with complete nutrients resulted in grasses becoming dominant. He also observed that increased fertility is likely to result in declines in diversity as species intolerant of competition, particularly annuals, rosette forms and bryophytes, are stifled by grasses. Willis further observed that sedges and rushes appeared to be particularly successful in areas of phosphorus deficiency, suggesting that as with Smith's observations on N:P ratios on cyanophytes, the ratios of N to P may control which particular plant group dominates a site.

Hayati and Proctor (1991) did a similar fertilization study in mires in southern England, using sites including a dry *Calluna* blanket bog, a valley bog with *Sphagnum* and *Erica tetralix*, and an acid mire with *Sphagnum* and *Carex echinata*. A factorial design fertilizer application to *C. echinata* grown in pots of sieved peat from each site revealed three main effects: to N, to P, and to between site differences. N was more limiting on the wet heath peats, whereas P was more limiting on the blanket bog. There was a minor effect of K, suggesting that it was present in adequate supply everywhere except in the blanket bog. The use of peat in pots may have increased experimental control, but it is vulnerable to the criticism that it does not necessarily show the importance of these nutrients under more natural field conditions. Others have used sediment collected from waterways and put into pots to study growth of emergent plants in both saline

Table 5.3. *Limiting factors in seven habitat types, as determined by biomass response in fertilization experiments*

Habitat	N	P	K	N+P	N+K	P+K
Wet grassland	3	0	2	0	4	0
Wet heath	0	3	0	0	0	0
Rich fen	7	5	0	0	0	0
Poor fen	2	1	0	0	0	0
Litter fen	1	2	0	1	0	0
Bog	1	3	1	0	0	0
Dune slack	5	2	0	2	0	0
Total (45 cases)	19	16	3	3	4	0

Notes:
Figures indicate number of cases in which the element was shown to be limiting.
Source: From Verhoeven *et al.* (1996).

and freshwater conditions; the evidence suggested that N rather than P usually limited plant growth (Smart and Barko 1978; Barko and Smart 1978, 1979).

Koerselman and Meuleman (1996) have done a major service by reviewing 45 studies of fertilization in seven types of herbaceous mires. There was an almost even split between nitrogen limited sites and phosphorus limited sites (Table 5.3). Co-limitation, that is, a response only to combined fertilizations, was rare. Wet heathlands all had growth limited by phosphorus, whereas fens and dune slacks could be limited by either nitrogen or phosphorus. The wet grasslands were the most complicated, with cases of N, K and N + K limitation. Regrettably, there was no category for floodplain. How can we explain, ask Verhoeven *et al.* (1996), that one major subset was N limited and the other was P limited? The availability of a nutrient is the result of the sum of net nutrient inputs plus nutrient recycling. For P, the main source is weathering from rocks, and the main input to wetlands from water flow. In contrast, the main source of N is fixation from atmospheric nitrogen (and increasingly, deposition of pollutants in rainfall). There is therefore likely to be a shift from N to P limitation during succession, since early in succession P is available in ground water and there is little N stored in organic matter. Consistent with this view, Table 5.3 shows that early successional rich fens are generally N limited, while late successional bogs and moist heaths are P limited.

Finally, Verhoeven *et al.* examined tissue nutrient concentrations for the same set of studies. The N:P ratio of 15:1 (as measured from plants in control sites) clearly separated N-limited from P-limited situations. 'It can be concluded that the N:P ratio of above-ground biomass at the end of the growing season (August) provides a reliable indicator of the degree to which each of these elements has been limiting plant growth in herbaceous mires . . .'. They conclude that N:P ratios greater than 16 indicate P limitation, and N:P ratios less than 14 indicate N limitation.

Experiments usually provide results from one or a few sites. Larger-scale patterns can only be detected by description, and many field studies have therefore sought correlations between soil nutrient levels and plant growth, and often these similarly suggest important roles for both N and P. Figure 5.2 shows the interrelationships of soil nutrients, biomass, and primary productivity in one riparian wetland. In this case, biomass and productivity are positively correlated with nitrogen ($r=0.38, 0.39$) but negatively correlated with phosphorus ($r=-0.29,-0.23$). Yet, in a similar geographic region, but in a *Carex* dominated ecosystem, there were no significant correlations between production and either N or P (Auclair *et al.* 1976b). At a slightly larger scale, but still within the same region, Wisheu *et al.* (1990) found a positive correlation between biomass and soil phosphorus in three different wetlands (Figure 5.3), although in each case, soil phosphorus accounted for only about 50% of the variation. If we cannot even find consistency in gross patterns, it is difficult to proceed further. Perhaps some of these contradictions can be addressed by re-considering the scale at which we look for patterns in soil fertility. Let us therefore return to basics.

Fertility gradients

Fertility gradients in marshes and swamps

One of the most remarkable features of wetlands is their natural variation in fertility. At a global scale, fertility levels range between two extremes. At one end are ombrotrophic bogs, where the sole source of nutrients is compounds dissolved in rainwater. At the other extreme are large floodplains and deltaic swamps, such as those of the Mississippi, Rhine or Amazon, where the eroded nutrients from an entire watershed are deposited in rich deltaic sediments. Table 5.4 begins with the low levels of nutrients in rain water, and shows how nutrient levels increase as water flows over rock and through soils. At more local scales than Table 5.4 one can still find quite remarkable differences in fertility. Within a watershed, for example, the headwaters may have small sandy bottomed streams, and the river mouth may have fertile alluvial sediments. Within a lake, shores

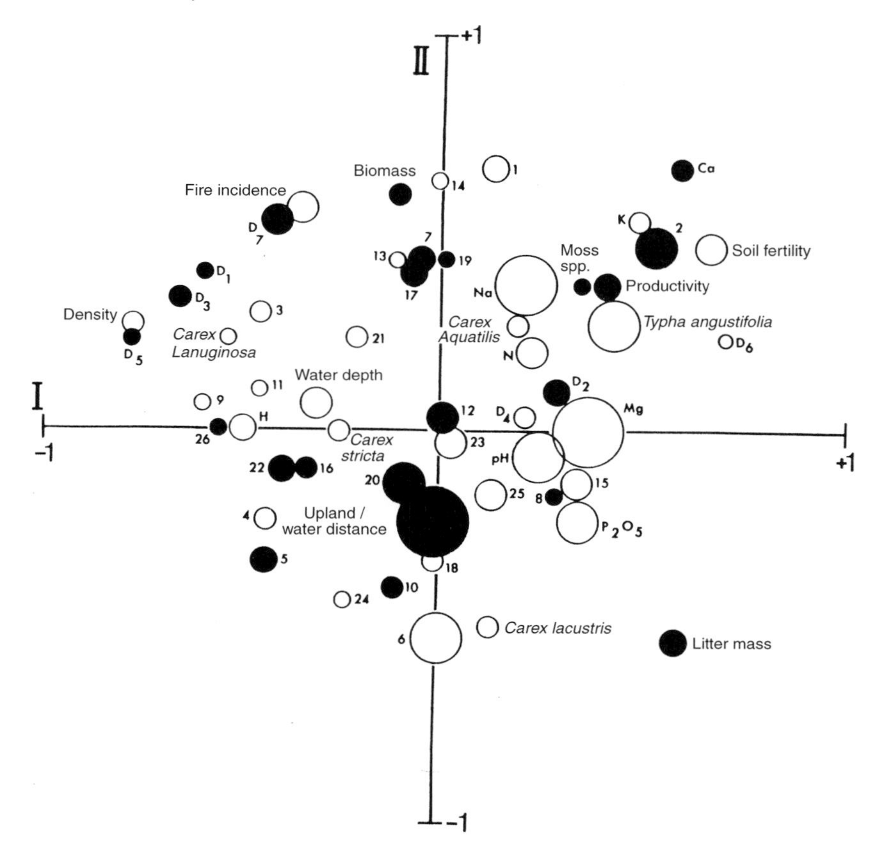

Figure 5.2 Interrelationships of fertility and productivity in a riparian wetland in eastern Canada (using a factor analysis model). The first and second components are shown on the horizontal and vertical axes, respectively. Factor loadings on the third component are represented by circle diameters (open = negative, closed = positive). D1–D7 are species diversity indices. Coded variables include: 1 soil organic matter, 2 soil (Ca + Mg)/(K + Na), 3 date since 1 May, 4 tussock incidence, 5 distance to upland, 6 distance to water, 7 biomass, 8 average stem height, 9 *Equisetum fluviatile*, 10 *Onoclea sensibilis*, 11 *Thelypteris palustris*, 12 *Potentilla palustris*, 12 *Viola pallens*, 13 *Hypericum virginicum*, 15 *Galium palustre*, 16 *Lysimachia thyrsiflora*, 17 *Lythrum salicaria*, 18 *Cicuta bulbifera*, 19 *Utricularia vulgaris*, 20 *Impatiens capensis*, 21 *Lycopus uniflorus*, 22 *Campanula aparinoides*, 23 *Carex diandra*, 24 *Calamagrostis canadensis*, 25 *Sparganium eurycarpum*, 26 *Sagittaria latifolia* (from Auclair *et al.* 1976a).

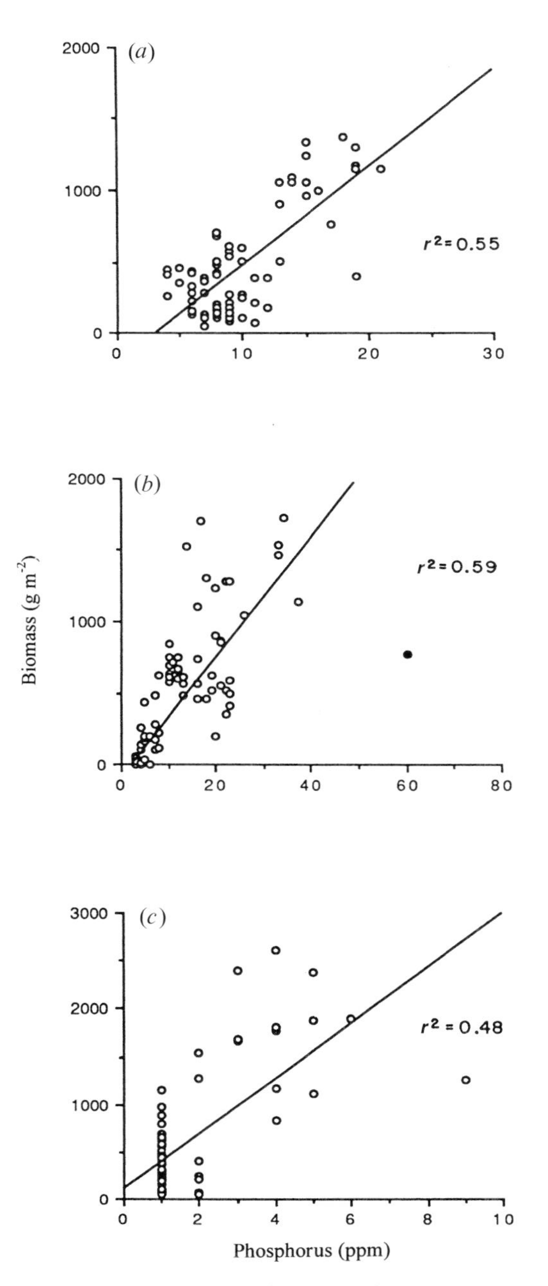

Figure 5.3 Biomass as a function of phosphorus in three temperate zone marshes ((*a*) Luskville, Quebec, (*b*) Westmeath, Ontario; (*c*) Presqu'ile Provincial Park, Ontario) (from Wisheu *et al.* 1990).

Table 5.4. *Ionic composition (ppm) of rain water and of run-off from rocks of different geological origin*

Site/rock type	Ca^{2+}	Mg^{2+}	Na^{+}	K^{+}	HCO_3^{-}	$SO4^{2-}$	Cl^{-}
Rain water							
Newfoundland	0.8	–	5.2	0.3	–	2.2	8.9
Wisconsin	1.2	–	0.5	0.2	–	2.9	0.2
Minnesota	1.0	–	0.2	0.2	–	1.4	0.1
Northern Sweden	1.2	0.2	0.4	0.3	–	2.5	0.7
Central Sweden	0.6	0.1	0.3	0.2	–	2.6	0.5
Guyana	0.8	0.3	1.5	0.2	–	1.3	2.9
Run-off water							
Nova Scotia							
Granite	1.0	0.5	5.2	0.4	*	5.9	7.7
Quarzite and slate	2.1	0.4	3.0	0.6	1.8	5.2	4.9
Carboniferous strata	3.0	0.6	3.6	0.5	6.1	5.3	5.4
Bohemia							
Phyllite	5.7	2.4	5.4	2.1	35.1	3.1	4.9
Granite	7.7	2.3	6.9	3.7	40.3	9.2	4.2
Mica schist	9.3	3.8	8.0	3.1	48.3	9.5	5.4
Basalt	68.8	19.8	21.3	11.0	326.7	27.2	5.7
Cretaceous rocks	133.4	31.9	20.7	16.4	404.8	167.0	17.3

Note:
* Not detected.
Source: From Gore (1983), after Gorham (1961).

exposed to waves have coarse infertile substrates, whereas shores in sheltered bays have rich accumulations of silt, clay and organic matter. It is little wonder, then, that wetland ecologists have studied the role of fertility in controlling vegetation composition.

Smart and Barko (1978) experimentally studied such substrate gradients by growing wetland plants on different kinds of sediment: sand, silty clay and clay. They found that measured nutrient levels were lowest in sand. N levels increased from 0.3 g kg^{-1} in sand to 5.2 g kg^{-1} in silty clay; similarly, P levels increased from <0.05 g kg^{-1} in sand to 1.65 g kg^{-1} in silty clay. By growing four different salt marsh plants (*Spartina alterniflora, S. foliosa, S. patens* and *Distichilis spicata*) in these sediments under a simulated tidal regime, they found that growth rates were an order of magnitude higher on clay than on sand, with intermediate values on silty clay. Smart and Barko concluded that growth was limited by nutrients in sandy substrates. They were, however, unable to explain why plants grew better in clay than in silty clay; their data show, however, that while N values were highest in silty clay, P values were significantly higher (1.94 g kg^{-1}) in clay. Perhaps on the finer sediments where both N and P were at relatively high levels, the plants were responding to the P gradient. Similar kinds of experiments have shown that growth on fine substrates is much higher than growth on coarse ones for a wide range of wetland plants including submersed aquatics (Denny 1972), emergent freshwater plants (Barko and Smart 1978) and plants of seasonally flooded shores (Sharp and Keddy 1985, Wilson and Keddy 1985).

Let us return to the apparent contradictions in the case studies mentioned on p. 244. Table 5.5 summarizes the fertility gradients that exist over a hierarchy of scales, scales ranging from among several ecoregions (*top*) to within a single vegetation type (*bottom*). Let us begin at the top. The upper panel shows correlations among major nutrients in a series of wetland types representative of eastern North America. Soil samples came from wetlands ranging from the highly fertile (e.g. *Typha* marshes and floodplains) to highly infertile sand or gravel shorelines where insectivorous genera such as *Drosera* and *Utricularia* were common. At this scale, all major nutrients are positively correlated, so one can reasonably talk about fertility gradients without distinguishing among the major nutrients. Both organic matter and the silt and clay content of the soil are also positively correlated with these nutrient levels.

At the intermediate scale, Table 5.5b shows correlations among major nutrients within a single lake. This gradient results from wave action and ice scour (Chapter 7). At one end are wave-washed sandy shores, and at the other end there are organic bays (Table 5.5). Again, major nutrients are positively correlated. Similar patterns occur within a single watershed (Table 5.5c).

Moving to the bottom of the table, at the most local scale, this pattern begins to break down. Table 5.5e shows correlations among nutrients within a single vegetation type – a *Carex* meadow. We note that nitrogen is still positively correlated with soil organic matter, but phosphorus now negatively correlated with nitrogen. (Similar results occurred in the more heterogeneous *Scirpus* – *Equisetum* wetland, Table 5.5d). In neither of these latter habitats is N significantly correlated with P.

Although natural fertility gradients appear to be a general feature of natural ecosystems, Table 5.4 suggests that, at small enough scales, nutrients become uncorrelated. It is therefore unfortunate that some ecologists find it convenient to maintain (contrary to Table 5.5) that variation

Table 5.5. *Fertility gradients in wetlands from large scale (top) to small scale (bottom). Note that patterns fade as the scale becomes smaller*

Table 5.5a. *Marshes in north eastern North America*

	% organic	P	N	K	Mg	pH
Standing crop	0.77	0.76	0.66	0.58	0.67	−0.28
% organic	1.00	0.77	0.57	0.50	0.51	−0.47
P		1.00	0.72	0.56	0.66	−0.13
N			1.00	0.53	0.63	−0.02
K				1.00	0.70	−0.28
Mg					1.00	−0.14

Source: Gaudet (1993), Table 1.2.

Table 5.5b. *The shores of a temperate zone lake*

	Mg	% organic	P	K	pH
Mg	1.00	0.52	0.86	0.73	0.22
% organic		1.00	0.51	0.48	0.16
P			1.00	0.64	0.20
K				1.00	0.09
pH					1.00

Source: Keddy (1984).

Table 5.5c. *One wetland complex in the Ottawa River watershed*

	% organic	P	N	K	Mg	pH
Standing crop	0.74	0.80	0.69	0.76	0.69	−0.45
% organic	1.00	0.80	0.61	0.66	0.62	−0.61
P		1.00	0.62	0.82	0.59	−0.46
N			1.00	0.68	0.53	−0.18
K				1.00	0.64	−0.35
Mg					1.00	−0.72

Source: Gaudet (1993), Table 1.4.

Table 5.5d. *One vegetation zone of the St Lawrence River*

	Standing crop	Organic	P	N	K	Mg	pH
Standing crop	1.00	0.34	−0.29	0.38	0.49	0.17	0.21
Organic		1.00	−0.27	0.37	0.75	0.59	0.18
P			1.00	−0.01	−0.48	0.33	−0.55
N				1.00	0.39	0.32	0.14
K					1.00	0.43	0.38
Mg						1.00	0.12
pH							1.00

Source: Auclair *et al.* (1976a), Table 1.

Table 5.5e. Carex *meadow, St Lawrence River*

	Standing crop	Organic	P	N	K	Mg	pH
Standing crop	1.00	0.13	−0.02	−0.02	−0.22	−0.23	−0.11
Organic		1.00	−0.39	0.30	0.52	0.17	−0.14
P			1.00	−0.26	0.18	−0.21	0.03
N				1.00	0.24	0.26	0.04
K					1.00	0.16	−0.01
Mg						1.00	0.52
pH							1.00

Source: Auclair *et al.* (1976b), Table 1.

in fertility is created only by plant competition (e.g. Tilman 1982, 1988). While it is undoubtedly true that, at the small scale (e.g. Table 5.5e), plants remove nutrients from the regions of their roots and create depletion zones, the nutrient variation attributable to root depletion zones is minor compared to that created by erosion and sedimentation. Perhaps it is part of our current fascination with small-scale processes that leads us to overlook natural habitat gradients. An overview of wetlands, and the data in Table 5.5, remind us that whatever models we create, they must explain how spatial variation in communities is connected to spatial variation in fertility.

Fertility gradients in peatlands

That fertility also has a major impact upon the characteristics of peatlands should not come as any surprise, because mires, like all wetlands, require

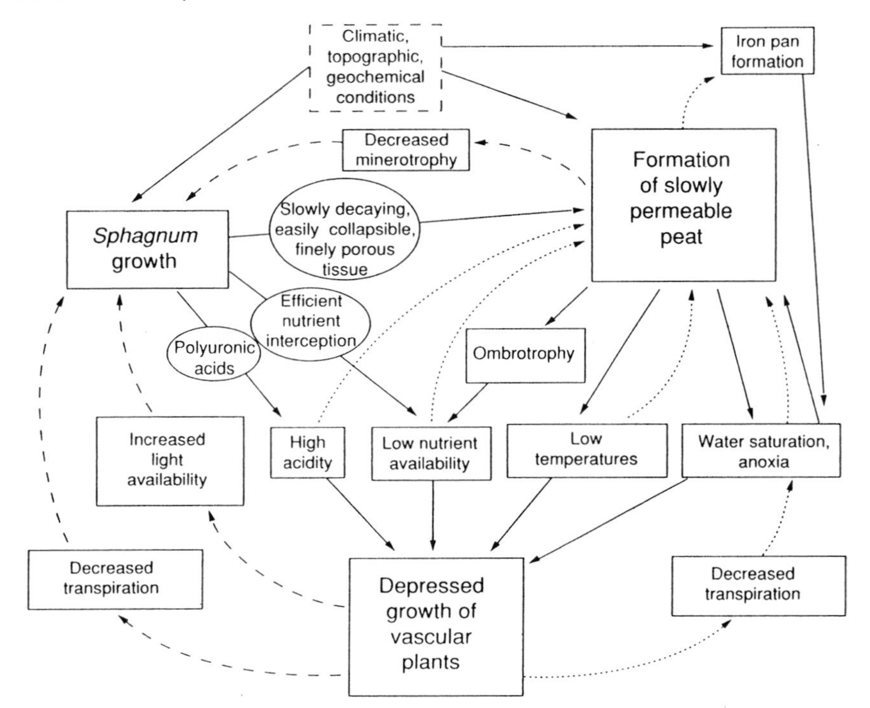

Figure 5.4 Pathways of peat formation (solid arrows) and feedbacks increasing the growth of *Spagnum* (dashed arrows). Dotted arrows show feedbacks involving physio-chemical processes and depression of decomposers. The ovals show the morphological, anatomical, physiological and organochemical properties of *Sphagnum* (from van Breeman, 1995).

plant growth, and plants in turn require certain elements. Fertility gradients in mires are, however, particularly important, because most mires are relatively infertile; elements such as N, P, K and Ca remain chemically bound in the peat instead of being released by decay (Figure 5.4). Dickinson (1983) writes 'Peat accumulations stand as impressive reminders of the high level of efficiency which is normally expected from decomposer micro-organisms', and re-emphasizes the important distinction between ombrotrophic and minerotrophic mires, with the latter characterized '. . . by luxuriant vegetation sustained by nutrient-rich water supplies'.

Slack *et al.* (1980) studied peatlands and nutrient gradients in the foothills of the Rocky Mountains using multivariate analyses, somewhat analogous to the work by AuClair *et al.* (in Figure 5.2). These peatlands were mostly patterned, pools of water (flarks) alternating with raised ridges (strings). The distribution of plant species was closely related to water level, with the flarks tending to be dominated by the moss *Scorpidium scorpioides* and the sedge *Carex limosa*. In contrast, nutrient gradients were thought to be less important, perhaps because the authors focused upon minerotrophically rich fens, thereby reducing variation in nutrients among study sites. Glaser *et al.* (1990) studied a larger mire in northern Minnesota where it was possible to identify an array of vegetation types from fen to swamp forest to raised bog. As the vegetation types would suggest, this site had a chemical gradient that covered the entire range reported for boreal peatlands. For example, the rich fens had a pH above 7 and Ca concentrations ranged from 20 to 45 mg l^{-1}. In contrast, the raised bog had a pH below 4 and a Ca concentration below 1.1 mg l^{-1}. Species composition changed dramatically along this gradient; species richness tended to increase with rising pH and Ca concentrations. A stratigraphic exploration of the raised bog showed that, under the roughly 3 m of accumulated peat, were the remains of a sedge fen and swamp forest, confirming the successional trends reported from other peatlands.

In European peatlands, similar patterns and gradients are evident, although distance from the sea is an added factor in controlling Na, Mg and Cl concentrations (Malmer 1986). Regrettably, tall sedge vegetation with *Carex* spp. or *Cladium mariscus*, which is emergent marsh vegetation, is treated as an aberration which is 'difficult to relate to any gradient' (Malmer 1986) rather than a distinctive type of ecological community.

One of the last extensive undrained valley mires in Central Europe occurs in the valley of the Biebrza River in north east Poland. Wassen *et al.* (1990) studied 58 stands of vegetation to explore correlations with environmental factors. The larger part consisted of fens with species such as *Carex lasicarpa* and *C. chordorrhiza*, but floodplain marshes occurred further downstream and along river margins dominated by *Glyceria maxima* (Palczynski 1984). Shrubs such as *Betula humilis* occurred on drier sites. Parts of the valley are mowed and grazed by either cattle or elk. Flood frequency and soil nutrients were apparently the primary factors controlling species composition. Species density was highest (33 species/10 m^2) in those fens receiving only groundwater that had low productivity (*ca.* 1 kg/m^2) and were typified by *Carex limosa*. Species density was lowest on the *Glyceria maxima* floodplains (13 species/ 10 m^2) where productivity was highest (*ca.* 6 kg m^{-2}). The flooding appeared to elevate potassium concentrations, whereas sites dependent upon groundwater discharge were low in phosphorus and rich in calcium. While these particular nutrient observations may be of significance, the productivity/diversity patterns are consistent with those from other fens in both Europe and North America.

Evolution along fertility gradients

Fertility gradients have likely existed as long as there have been wetlands along water courses (recall Figure 1.21). To explore the evolutionary responses of wetland communities to fertility gradients, we can consider just the two extremes in habitat type: chronically infertile, and highly fertile.

Infertile habitats

Many wetlands are chronically limited by nutrients (Figure 5.5). Ombrotrophic bogs are the best example, but sand and gravel shorelines are also low in nutrients relative to most other wetland types (e.g. Table 5.6). Such wetlands are an ideal place to study the impact of chronic infertility on plant traits and upon plant communities. Grime (1977, 1979) argues that infertility is a major force affecting plant evolution, and identifies a series of traits associated with plants occupying infertile habitats. Such traits include the uncoupling of growth from nutrient uptake, inherently slow relative growth rates, evergreenness, investment in anti-herbivore defence, and the occurrence of mycorrhizae. The postulated significance of these traits, the relationships among them, and their field distributions provide a rich terrain for physiologists, evolutionary ecologists and community ecologists to interact (Keddy 1994).

Here there is time to only briefly introduce some of the more obvious traits plants possess that allow them to tolerate low fertility (Figure 5.6). The evolution of carnivory is one of the most dramatic; although many people know that carnivorous plants occur in wetlands, it often is not understood that this is a dramatic example of the selective pressures imposed by chronic infertility (Givnish 1988). Indeed, a simple way to locate infertile habitats is simply to search for carnivorous plants. Other plants, such as the widespread *Myrica gale* of bogs and lakeshores, have root nodules in which nitrogen fixation occurs (Bond 1963). Others are evergreen; these include herbaceous genera such as *Lobelia, Eriocaulon, Xyris* and *Scirpus* as well as many trees and ericaceous shrubs (Richardson 1981; Richardson and Gibbons 1993). It is believed that evergreenness allows nutrient investment in leaves to be amortized over longer time periods.

Figure 5.5 Sandy lakeshores (*top*) and peatlands (*bottom*) are habitats with low nutrient availability.

Table 5.6. *Mean above-ground standing crop (g 0.25 m^{-2}) and mean macronutrient levels in soil (ppm) (n = 15) in each of five wetland sites sampled at three study locations in Ontario, Canada*

Location	Standing crop	NO_3	P	K	Mg
Presqu'ile					
1	17.8	7.33	1.13	17.3	43.3
2	67.0	9.07	1.33	34.5	69.9
3	86.8	7.80	1.00	35.2	53.9
4	173.1	8.93	1.00	52.7	63.1
5	421.6	9.20	4.20	15.5	32.9
Westmeath					
1	4.0	18.00	3.27	36.7	74.3
2	54.5	7.07	5.27	38.3	130.5
3	131.9	7.87	19.40	62.5	243.7
4	163.0	16.40	10.93	59.0	345.0
5	315.8	27.33	25.73	93.4	236.9
Luskville					
1	32.7	6.93	6.80	56.9	191.2
2	40.0	4.93	8.73	62.0	187.3
3	92.3	5.07	5.27	132.3	326.6
4	124.6	11.60	9.67	150.3	383.0
5	277.7	15.73	15.67	159.4	386.9

Notes:
Presqu'ile is a sand spit–lagoon complex in Lake Ontario, Westmeath is a sand dune-lagoon complex on the Ottawa River, and Luskville is a clay/sand bar system on the Ottawa River
Source: From Moore *et al.* (1989).

Some evergreen species also use CAM photosynthesis, in which carbon is stored at night for use during the day (Boston 1986; Boston and Adams 1986). CAM is thought to be associated with deserts, and its occurrence in wetland plants such as *Isoetes* is unexpected.

Fertile habitats

In highly productive habitats, nutrients are, by definition, not limiting to growth. One evolutionary response is likely to be rapid growth to exploit and store nutrients. In such circumstances, light soon becomes limiting. Competition for light has been a factor in plant communities since the first terrestrial plants began to form erect shoots and shade one another

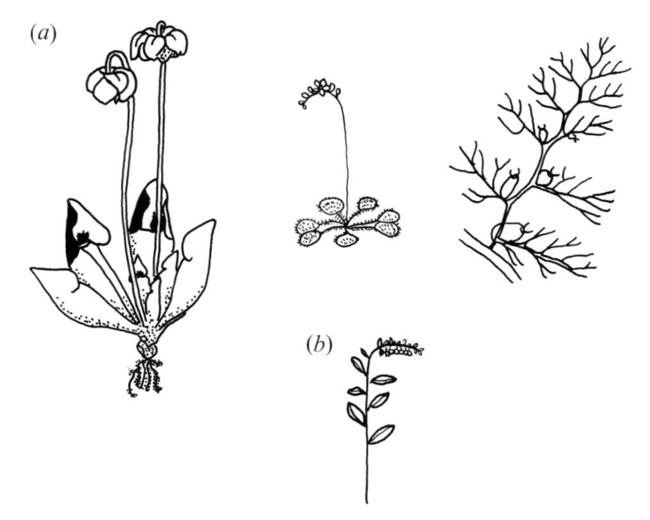

Figure 5.6 Plant traits associated with low soil fertility include (*a*) carnivory (*Sarracenia purpurea*, *Drosera rotundifolia*, *Utricularia* spp.) and (*b*) evergreen leaves (*Chamaedaphne calyculata*).

(Keddy 1989a, b). Thus, in herbaceous vegetation, tall shoots, broad canopies and deeply rooted rhizomes for support all can be regarded as traits that arose out of intense competition for light (e.g. Grime 1979, Givnish 1982). Woodiness, then, is an obvious ancient evolutionary response to light limitation; it has arisen independently in at least three major plant lineages (the Lycopodophytes, the Gymnosperms, the Angiosperms).

Of course, among all such traits there are inevitable trade-offs. For example, woody plants may be good competitors for light, but they are killed by ice scour and flooding, so many productive wetlands are dominated by herbaceous plants. Similarly, slow growth may be a way to tolerate infertile conditions, but grazing can then be very detrimental. But, we are getting ahead of ourselves. The theme of competition is covered in Chapter 7, and co-evolution of traits is tangentially considered in Chapter 9.

Mycorrhizae

Fungi associated with plant roots are thought to assist in taking up nutrients, particularly phosphorus, from infertile soils (e.g. Read *et al.* 1976; Marschner 1995; Smith and Douglas 1987). A recent review of the distribution of arbuscular mycorrhizae in 843 species in the British flora (Peat

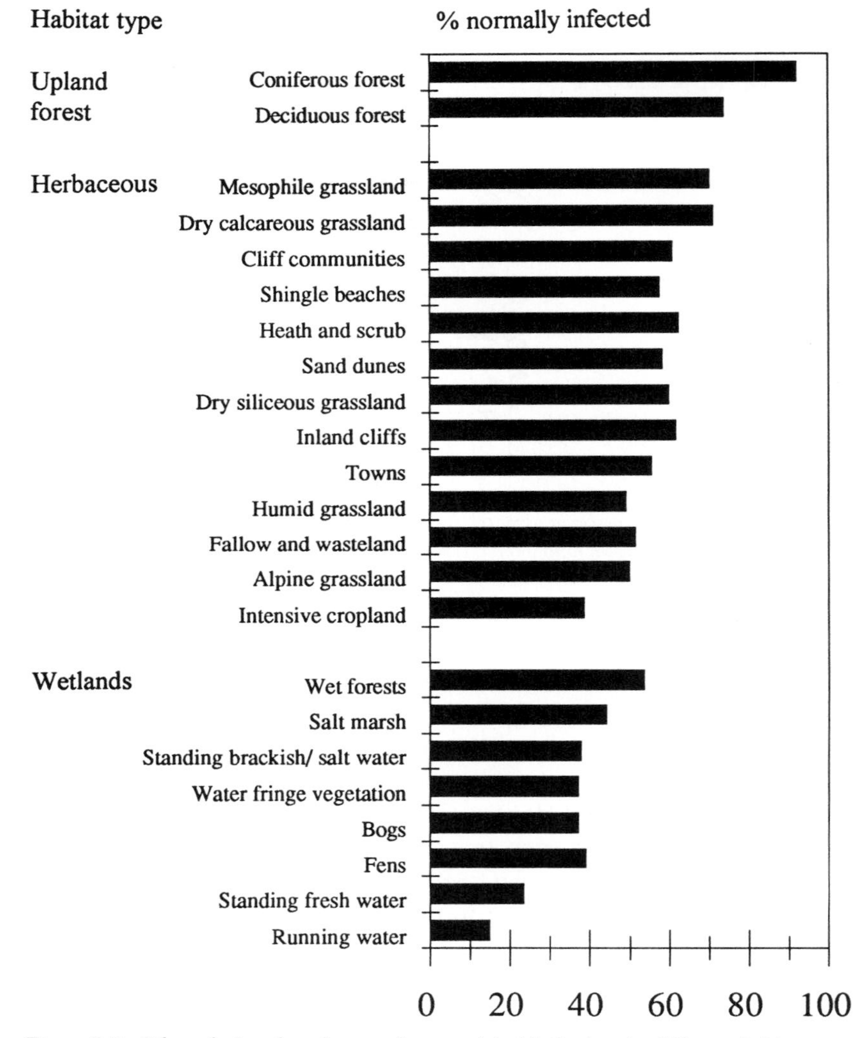

Figure 5.7 The relative abundance of mycorrhizal infection in different habitats (from data in Peat and Fitter 1993).

and Fitter 1993) yielded two important results for wetland ecologists. First, mycorrhizae are relatively uncommon in wetlands (Figure 5.7). Fewer than 50% of the plant species in wetlands have mycorrhizae, except for wet forests where slightly more than 50% of the flora is infected. In contrast, coniferous forest has more than 90% of the species infected.

In fact, there are few species of fungi in the Agaricales able to tolerate flooding, and most of these are saprophagous (Kalamees 1982). If mycorrhizae occur to augment nutrient uptake from infertile soils, then it would appear that this important evolutionary option is not available for many wetland plants. We may speculate that soil nutrient gradients are therefore even more important in wetlands than in terrestrial habitats. But Peat and Fitter also found little evidence to support the assumed relationships between mycorrhizae and soil fertility. Percentage infection was in the 80% range across all soil fertility levels and, if anything, the abundance of mycorrhizae increased with soil fertility. When the Cyperaceae (an important genus in wetlands which occupies infertile habitats, but is not mycorrhizal) were removed from the analysis, there was no significant relationship between occurrence of mycorrhizae and soil fertility. One cannot of course, generalize from the British flora to the world flora, but this is one of the largest data sets systematically explored.

Percentage infection is not likely to be a useful measure of the actual importance of mycorrhizae in a habitat. There are significant differences in the kind of mycorrhizal association, with the three main types (endomycorrhizae or VA mycorrhizae, ectomycorrhizae and ericoid mycorrhizae) being associated with different soil types (Read *et al.* 1985; Lewis 1987). In peatlands, the ericoid association is predominant. These are ascomycete fungi affiliated with ericaceous plants, and the fine individual hyphae ramifying near the soil surface appear to capture nitrogen (Read *et al.* 1985; Lewis 1987). Ericoid mycorrhizae apparently provide lower rates of N uptake than endo- or ecto-mycorrhizae, thereby supporting correspondingly low plant growth rates (Woodward and Kelly 1997). In actual assays in the field, Anderson *et al.* (1983, 1984) found that spore density declined with increased soil moisture (Figure 5.8). Rickerl *et al.* (1994) found that mycorrhizal infection declined with flooding (27% infection in dry areas vs. less than 1% in wet areas), with *Carex atherodes* and *Juncus tenuis* being entirely uninfected. *Scirpus fluviatilis* was non-mycorrhizal in wet sites, but slightly colonised (9% infection) in dry areas. In dry areas, but not wet ones, there was also evidence that infection was correlated with the amount of phosphorus in plant tissues. While the lack of mycorrhizae in wetlands therefore seems well established, exceptions exist. The aquatic plant, *Lobelia dortmanna*, is mycorrhizal. This may be related to the release of oxygen by *Lobelia* roots, which can create local aerobic pockets (Pedersen *et al.* 1995).

In general, then, it appears that the anoxic conditions in wetlands not only directly stress wetland plants, but also prevent the occurrence of most mycorrhizae. Another explanation for low infection rates could be

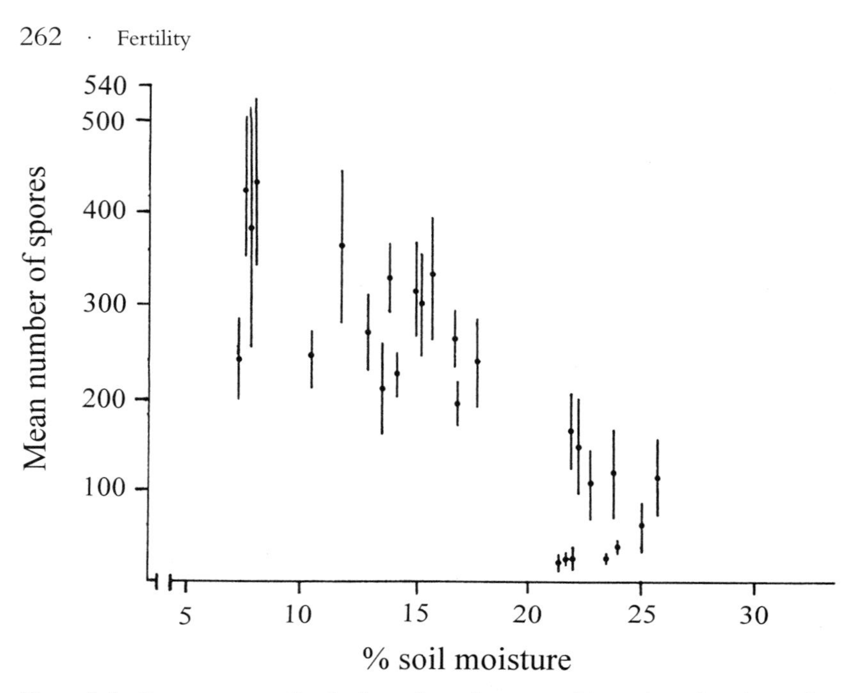

Figure 5.8 Spore counts of vesicular–arbuscular mycorrhizae plotted against soil moisture in prairie vegetation. One outlier has been removed (after Anderson *et al.* 1983).

that mycorrhizae exist, not as is commonly assumed, to increase access to soil nutrients, but to increase access to water; in the case of wetland plants, access to soil moisture is not a problem requiring an evolutionary solution. The question as to why mycorrhizae are least common in wetland plants, and how mycorrhizal and non-mycorrhizal species co-exist, still requires further study.

Animals and fertility

The very concept of fertility is usually seen to be an issue only for plant ecologists. It is not. Nitrogen availability may be a key problem for all animal life. We have already seen that (i) nitrogen is an important limiting resource for plants, and (ii) major changes in plant species composition and plant life form are associated with gradients in soil fertility. How do these two foregoing facts relate to animals? Of course, we can argue that, at the very least, plant community gradients produce wildlife habitat gradients. But, a more fundamental point is that plants provide the source of nitrogen for animals because animals must get their nitrogen in the form of ready made amino acids. Nitrogen is thinly spread throughout plant tissues. Because it is in such short supply, plants are very economical in their use of nitrogen. At very best, by concentrating on seeds, pollen, or cambium, animals may be able to harvest a food source that is about 5% nitrogen dry weight (White 1993). This has led to the suggestion that nitrogen, not energy, is the limiting resource for animal communities. Nitrogen, rather than kcal, may be the key currency for studying competitive interactions (Keddy 1989a) and trophic webs (White 1993). White provides extensive documentation of nitrogen limitation in the animal kingdom. Animals may therefore be more sensitive to soil fertility gradients than we might first assume.

This leads directly to two basic questions: (i) What is the nitrogen content of wetland plants? (ii) Does it vary along fertility gradients? Table 5.2 suggested an average N content of around 2%, less than half that of animal tissues (Table 5.1). Figure 5.9 compares N contents of a wide array of wetland plant types – and shows that, in spite of significant variation in life form, N content is rather similar, at least when the plants are grown under standard relatively fertile conditions. Even under these conditions, N values fall between 0.5 and 2%, well below the upper value of 5 to 7% described by White. These values would tend to support White's contention that animals are being strongly limited by nitrogen availability. We might naively expect fertile habitats to support plants with the ability to exploit and concentrate the higher nitrogen concentrations; in fact, Chapin (1980) suggests the opposite. Species from infertile habitats might store excess nutrients, thereby creating a N gradient in plant tissues running in the opposite direction of soil gradients. But the comparative data in Figure 5.9 show no difference in N concentration between plants typical of habitats with high and low fertility. Thus, while biomass and plant species composition may change with soil fertility, the per gram availability of nitrogen to animals may not.

Further work testing for correlations between soil nutrient levels and those in plant tissues might be helpful. However, if plants grown in fertile conditions in pots still only concentrate N at around 1% of dry weight, one might argue that tissue levels apparently are so low as to make any tissue gradients inconsequential. From the point of view of animals, then, wetland plant communities might present a consistent cover of tissue containing near-starvation levels of nitrogen. We have already observed that plants are bathed in atmospheric nitrogen, but often desperately short of nitrogen. It now appears that the situation may be even worse for animals.

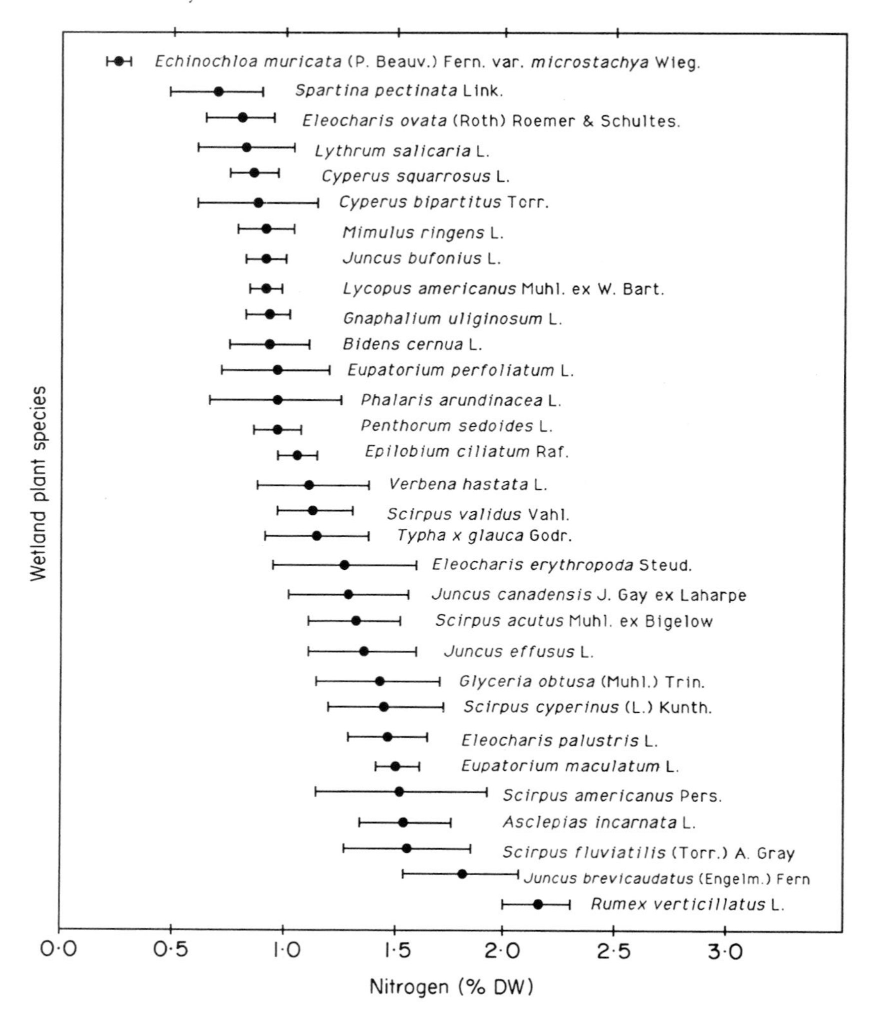

Figure 5.9 N concentration of emergent wetland plants grown under similar conditions with sustained fertilization. (Mean % dry wt. (DW), $n = 4$ or 5, error bars denote SD). Such concentrations provide an impoverished diet for animals.

White therefore suggests that animals have been strongly selected to counter these environmental shortages. He identifies six basic strategies:

(i) synchronizing life cycles with availability of food with higher nitrogen content;
(ii) selecting those tissues with higher N content;
(iii) eating quickly and digesting more efficiently;

(iv) supplementing plant food with animal protein;
(v) territoriality and social behaviour;
(vi) enlisting the help of micro-organisms.

Many examples of (i) to (v) can be seen throughout the animal kingdom. We could re-interpret a majority of dietary studies around the theme of extracting nitrogen from an impoverished environment. White's book is replete with examples.

Point (vi) may be worth an added observation. Over the past decades, plant ecologists have increasingly studied the role of mycorrhizae in plant nutrient budgets. Now White reminds us that animals too have associations with micro-organisms to cope with chronic nutrient limitations. For example, micro-organisms provide a large part of the nitrogenous food of ruminants. Some nitrogen fixing micro-organisms live in guts. Many detritivores may ingest micro-organisms along with partly decayed plant material. Micro-organismal associations with herbivory are found across the entire animal kingdom. Perhaps the use of micro-organisms to escape nitrogen limitation is an important evolutionary parallel in plants and animals.

Eutrophication: too much of a good thing

A major consequence of industrialized civilization has been the erosion of soils and heavy fertilization of agricultural ecosystems. As a consequence, nitrates and phosphates have increased in rainwater and run-off (e.g. Figure 5.10). Moreover, much of the nitrogen used in agriculture has not been obtained from natural sources, but extracted from the atmosphere industrially, thereby greatly increasing the biologically available nitrogen in the biosphere (Pimental *et al.* 1973; Freedman 1995). The amount of industrially fixed nitrogen applied to crops during the period 1980 to 1990 more than equalled all industrial fertilizer applied previously in human history (Vitousek *et al.* 1997). Wetlands are places where such nutrients tend to accumulate. Overall, there seem to be two general consequences. First, in marshes and wet meadows, fertilization alleviates nutrient limitation, causes shifts in competitive ability and leads to an increase in biomass (and therefore decreases species richness; see Chapter 3). Secondly, in aquatic situations, eutrophication can lead to the opposite effect – die-backs of macrophytes and consequent anoxic conditions that kill fish. Such effects have had a major impact on fish production in the Great Lakes (Christie 1974; Vallentyne 1974). Decreasing N:P ratios can also cause nitrogen fixing cyanobacteria to replace other species of planktonic algae (Schindler

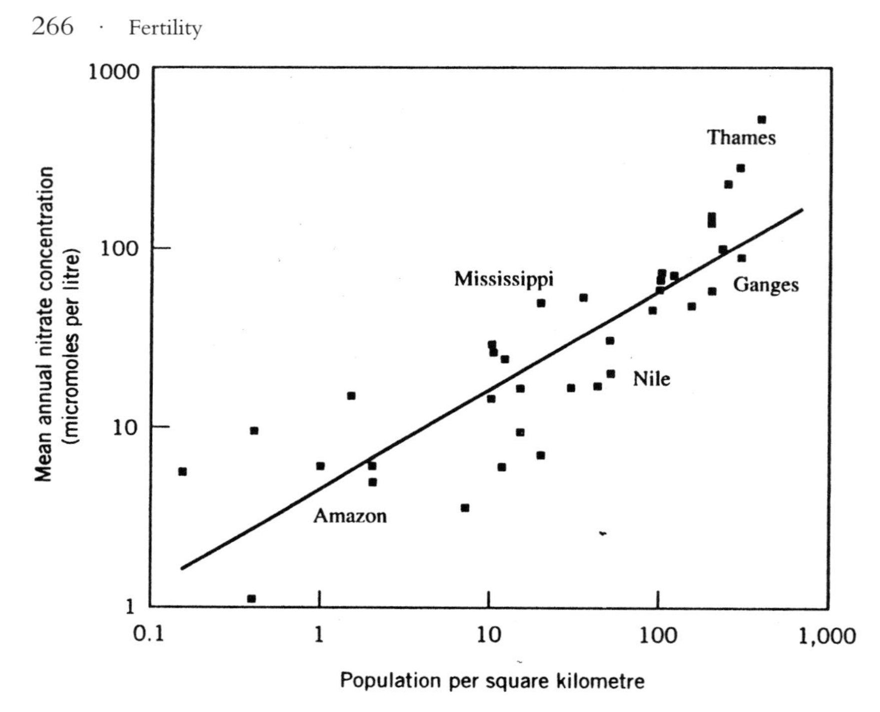

Figure 5.10 Nitrate concentration of water in 42 major rivers as a function of population density in the surrounding region (from World Resources Institute 1992).

1977; Smith 1983). Before leaving lakes and examining the consequences of eutrophication for wetlands, let us explore its causes.

Run-off

In this chapter, fertility has been treated as the independent variable, and we have briefly explored its effects on primary production and standing crop. Let us move back one step in causation by treating fertility as the dependent variable and then asking about the sources of nutrients that cause eutrophication. There is a large literature on this because the negative effects of eutrophication upon water quality were a matter of great concern more than a decade ago (e.g. Vallentyne 1974; Rigler and Peters 1995).

Since the Great Lakes, North America, are among the largest bodies of fresh water in the world, we will turn to them for an example (International Joint Commission 1980). Phosphorus is the major nutrient of interest, owing to the fact that these studies were carried out with an

Table 5.7. *Major sources of phosphorus (in metric tons) in the Great Lakes*

Source	Lake Superior	Lake Michigan	Lake Huron	Lake Erie	Lake Ontario	Initial St Lawrence River
Municipal sewage	268	2298	515	6828	2815	147
Industrial	135	279	122	347	102	42
Land use	2238	1891	2442	8445	3581	747
Atmospheric	1566	1682	1129	774	488	–
Load from upstream lakes	–	–	657	1070	4769	4545
Total	4207	6350	4857	17474	11755	5481

Source: After International Joint Commission (1980).

emphasis upon phytoplankton and fish production. Table 5.7 shows that the major sources of phosphorus in the Great Lakes arise from varying combinations of municipal point sources, rural land use and atmospheric deposition.

Let us first consider urban as opposed to rural sources of pollutants. Urban areas can be considered to be point sources, whereas rural areas are diffuse sources of nutrients. We can use forested areas, with the lowest levels of P loading (only 0.02 to 0.67 kg ha^{-1} yr^{-1}) as a natural reference point. Urban areas may contribute from 0.1 to 4.1 kg ha^{-1} yr^{-1}. Areas under construction have an even higher load as a consequence of soil erosion (see also Guy 1973 and Chapter 9). Apart from erosion during construction, sewage is an obvious factor in urban areas. Run-off into storm sewers is also important; the latter contains high concentrations of nutrients from lawns and pet faeces. Such urban nutrient sources are rather straightforward – one needs proper sewage treatment, and effective treatment of run-off from storm sewers. These can be dealt with technologically (waste water treatment) or culturally (reduced lawn area, reduced numbers of pets, etc.).

Rural land use, the diffuse source of loading, requires further exploration. Here run off contains from 0.1 to 9.1 kg P ha^{-1} yr^{-1}, with the lowest figures, as we would expect, for areas under forest. The amount of phosphorus in stream water (the dependent variable) has been related to a variety of predictors that describe land use. The best predictor is the amount of the watershed in row crops (International Joint Commission, 1980). In lakes with agricultural watersheds (Lake Michigan, Lake Huron, Lake Erie) two-thirds of the diffuse load of nutrients was attributable to run-off from cropland (Figure 5.11). In watersheds without agriculture, forestry becomes a more important source of nutrient loading to lakes. In an artificially deforested watershed, dissolved nitrogen jumped from *ca.* 1.0 to >40 kg ha^{-1} of nitrate, and from *ca.* 33 to >300 kg ha^{-1} of particulates (Bormann and Likens 1981). Sediment yield is also affected by the nature of forestry practices, with a commercial clearcut yielding turbidity values between 10 and 10^2 times larger than a cut conducted to protect water values (Lee 1980). Road construction is one of the greatest impacts of forestry upon water quality (Forman and Alexander 1998). Agriculture and forestry can therefore be modified to greatly reduce diffuse nutrient loading.

Much further south in Tampa Bay, the largest open water estuary in Florida, over two million people live in the adjoining watershed, contributing nitrogen, phosphorus and suspended solids (Greening 1995). Shading by algae has damaged seagrass (*Thalassia testudinum*) beds. Over half of the nitrogen loading was attributable to diffuse sources including residential run off (13%), rangelands (14%) and intensive agriculture (6%). A further quarter came from atmospheric deposition. Thus, a full three-quarters of the nitrogen load came from non-point sources. Although this study represented a different geographical region than the above study, the importance of land use in contributing to eutrophication was again emphasized.

A similar study was carried out in the nearby watershed of Lake Okeechobee. The lake's drainage basin covers 12000 km^2 and is connected to the Kissimmee River and the Everglades in one large surface water hydrologic system; at one time one-quarter of this land was wetlands (Rosen *et al.* 1995). Cattle ranching was the predominant land use for centuries, but improved cattle pasture and dairy farms are increasingly prevalent. The dairy farms are the major source of phosphorus loading, followed by cattle ranching. Land management activities were undertaken to reduce phosphorus loading by 40%; these included fencing cows away from streams, maintaining vegetation strips adjacent to streams, the construction of cattle crossings over streams, recycling dairy barn wash water, and even cash payments to buy out dairies, with the addition of a restriction on the property deed to prevent future use as a dairy. Changing the use of land is quite a different technical problem from constructing sewage treatment plants, but as the above studies show, it is equally important for the reduction of eutrophication of wetlands.

It has often been hypothesized that freshwater wetlands are nutrient sinks which trap nitrogen and phosphorus, reducing the eutrophication of downstream aquatic ecosystems (Richardson 1985). The value of wetlands as buffer strips between land and water is therefore often mentioned

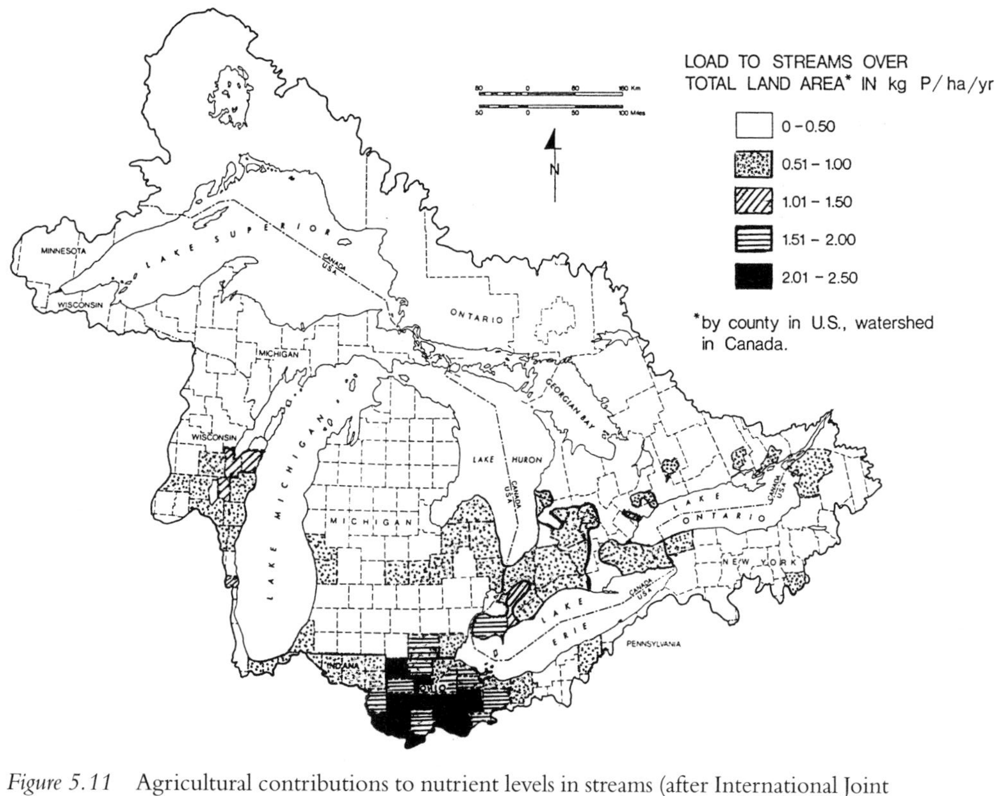

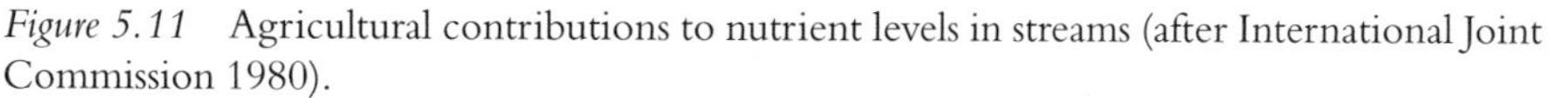

Figure 5.11 Agricultural contributions to nutrient levels in streams (after International Joint Commission 1980).

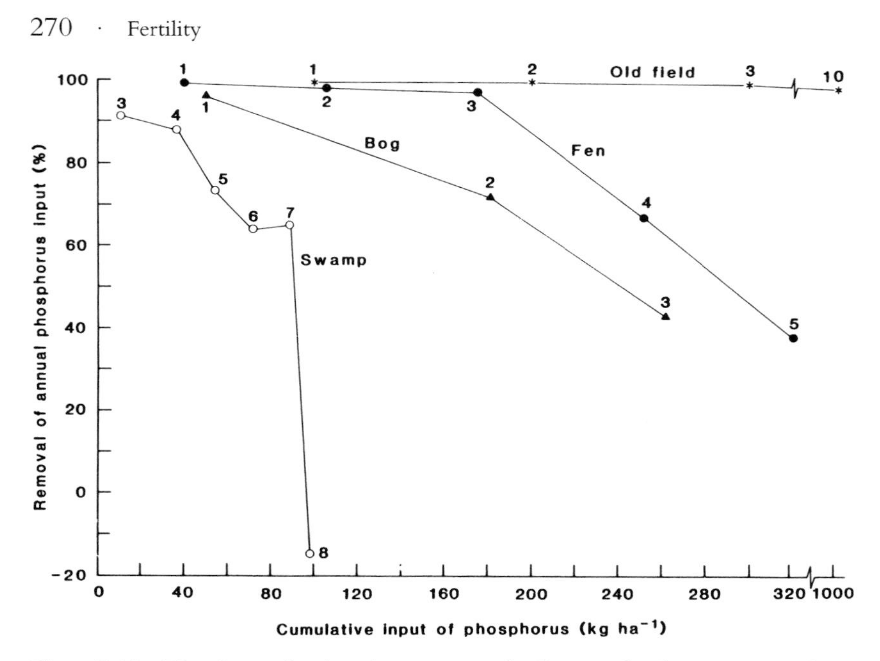

Figure 5.12 The change in phosphorus removal efficiency for the Houston Lake fen, a white cedar swamp forest, an Irish blanket bog, and an abandoned old field in Pennsylvania as a function of cumulative phosphorus imputs. The numbers along each line indicate the number of years of phosphorus addition (from Richardson 1985).

in functional assessments (Tables 1.6, 1.7, 1.12) (although such studies rarely, if ever, refer to the significant consequences of eutrophication upon the wetlands themselves). Comparative studies of phosphorus retention capacity among 20 sites showed, however, that it was extremely variable, with a phosphorus absorption index ranging from 163 in swamp forests to a mere 8 in pocosin peats (Richardson 1985). Richardson then compared four different sites that had received high loadings of phosphorus laden waste water (Figure 5.12). The abandoned old field continued to remove 96% of the added phosphorus in spite of a loading three times as high as the three wetlands. 'Collectively, these data indicate that high initial rates of phosphorus removal will be followed by large exports of phosphorus within a few years' (p. 1426).

Precipitation

Although sewage and run-off are the major sources of nutrient inputs to some wetlands, precipitation also plays a role (Table 5.7). The burning of fossil fuel, and to a lesser extent, biomass, adds both nitrogen and sulphur compounds to rainfall (Vitousek *et al.* 1997). Records over the last century from Europe show a steadily rising concentration of nitrogen-containing compounds in rainwater. Germany now receives in the order of 25 kg N ha^{-1} each year in rainfall (Ellenberg 1989), enough to produce significant changes in vegetation. More recently, pollution control measures may be having some positive effect. Over the period of 1980–1992, SO_4^{2-} concentrations fell at 42 of 58 monitoring sites in the United States (Lynch *et al.* 1995). Less than a fifth of the sites showed significant declines in NO_{3-}, however.

While monitoring stations are useful for detecting short-term changes in precipitation chemistry, much longer records can be obtained from ice cores taken from glaciers (Figure 5.13). Such work suggests that any recent reductions in nutrients in precipitation still leave us well above pre-industrial deposition rates. The scattered peaks of sulphate deposition can be traced back to volcanic eruptions such as Laki (1783) and Tambora (1815). Mayewski *et al.* (1990) also point out that the ratio of nutrients in rainfall has been changing. Pre-industrial concentrations of nitrate were roughly twice those of sulphate; in contrast, at the turn of the nineteenth century, they were nearly equivalent. Recent surges in nitrate emission have again caused nitrate levels to exceed sulphate.

Effects of eutrophication

Wet meadows and marshes

Chapter 3 discussed the relationship between biomass and species richness. We left, somewhat to one side, the question of what controls biomass. In general, nutrient levels or stress will control biomass, although in not such a tidy fashion as we might first expect (e.g. Table 5.6). Certainly, Figure 5.2 shows that soil fertility and productivity are related. Moreover, large rhizomatous species (such as *Phragmites communis* and *Typha angustifolia* (Figure 5.2)) are associated with high fertility.

Experimental fertilization provides the tool for studying effects of eutrophication. Recall the dramatic changes in salt marsh vegetation that Levine *et al.* (1998) produced with added nitrogen (p. 107). Figure 5.14 shows the results from a study where 12 freshwater wetland habitats were fertilized. The different habitats were created by various factors such as water depth, surface texture and litter loading (Moore 1990; Weiher and Keddy 1995). In each habitat, there were high and low fertility levels; high fertility levels led to much higher levels of biomass, irrespective of the habitat types. Auclair *et al.* (1976b) documented field patterns among

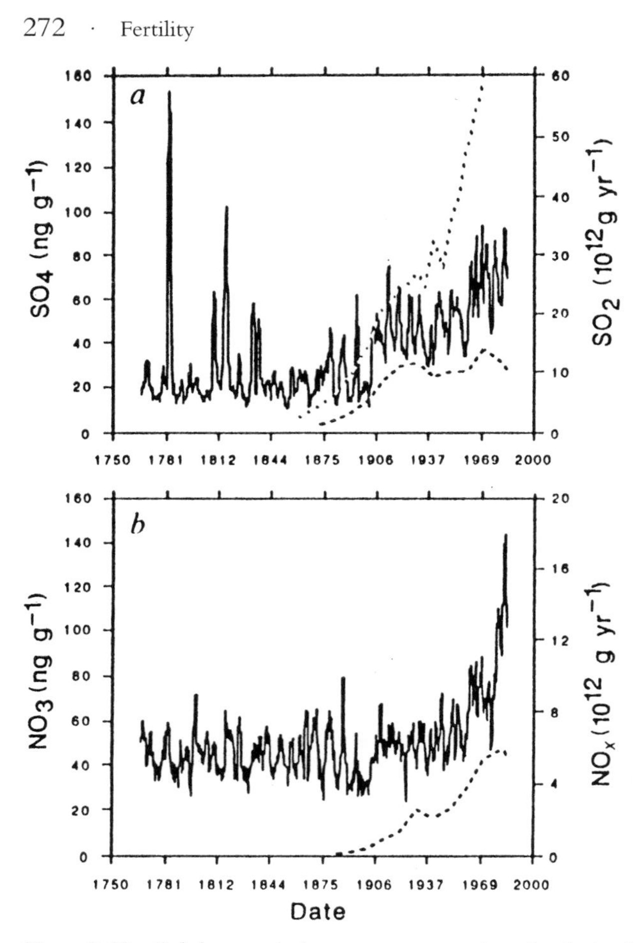

Figure 5.13 Sulphate and nitrogen concentrations in glacier ice taken from South Greenland (from Mayewski *et al.* 1990).

nutrients and plant communities. They propose that changing availability of nutrients, particularly phosphorus, can create competition gradients in wetlands (Figure 5.15). This hypothesis has been experimentally tested – as we shall see in Chapter 7.

The Everglades provide another example. Here there are some half million ha of freshwater marsh dominated by sawgrass (*Cladium jamaicense*), with sloughs, wet prairie and tree islands. Since the 1960s, nitrogen and phosphorus have been pumped into the northern Everglades from agricultural drainage. Reported changes include an increase in net primary production and invasion by cattail (*Typha domingensis*). Using nine different experimental treatments, Craft *et al.* (1995) showed that

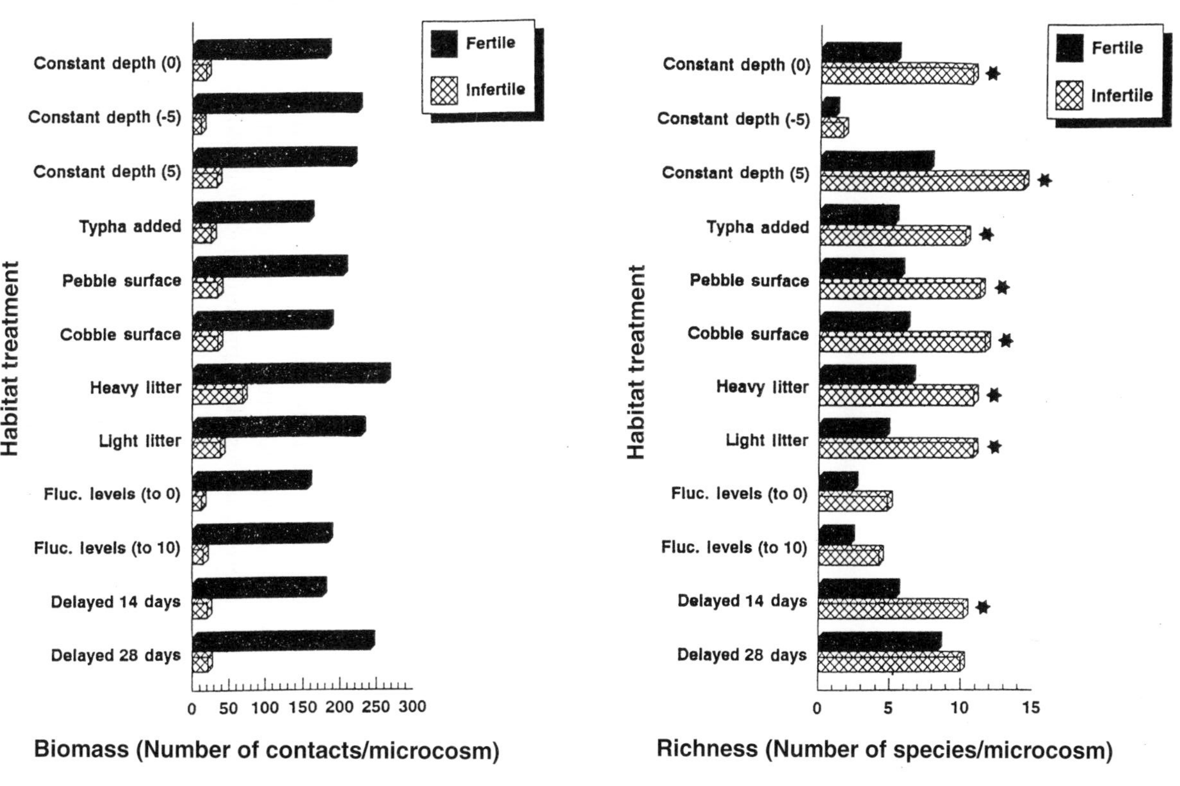

Figure 5.14 The effects of fertilization on biomass and species richness of wetland plants in 12 artificially created habitats (from Wisheu *et al.* 1990).

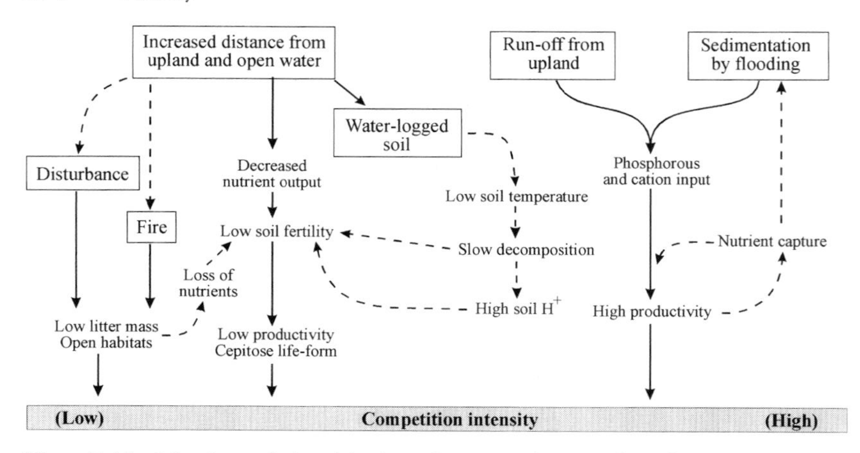

Figure 5.15 Nutrient relationships in a *Carex* meadow are thought to control both biomass and competition intensity (after Auclair *et al.* 1976b).

these wetlands are limited by phosphorus. There was no apparent shift in plant composition over the two years of the experiment, but it may take considerably longer for the consequences of competitive interactions between large clonal perennials to become evident.

The contrast between the above two freshwater examples and the salt marsh experiment on p. 107 suggests that freshwater marshes may be phosphorus limited, while saltmarshes are nitrogen limited.

Shallow water/aquatic habitats

The effects of eutrophication on wetland communities are studied rather less than effects on phytoplankton and fish. It often appears that eutrophication is associated with declines in the abundance of macrophytes. This at first appears counter-intuitive. It certainly is the opposite of what we see in marshes and wet meadows. The presumed explanation is that increased fertility leads to increased phytoplankton biomass. The plankton then absorb light and shade out macrophytes (Phillips *et al.* 1978; Moss 1983; Pieczynska 1986; Osborne and Polunin 1986). Figure 5.16 presents an hypothesis that relates the decline in macrophytes to eutrophication. In estuaries, eutrophication appears instead to stimulate macroalgae, which, through shading, or anoxia induced by decomposition, eliminate macrophytes such as *Zostera marina* (Valiela *et al.* 1992).

In order to understand the effects of eutrophication, it is necessary to know whether aquatic plants are able to use nutrients in the water column as opposed to the substrate. Carignan and Kalff (1980) used

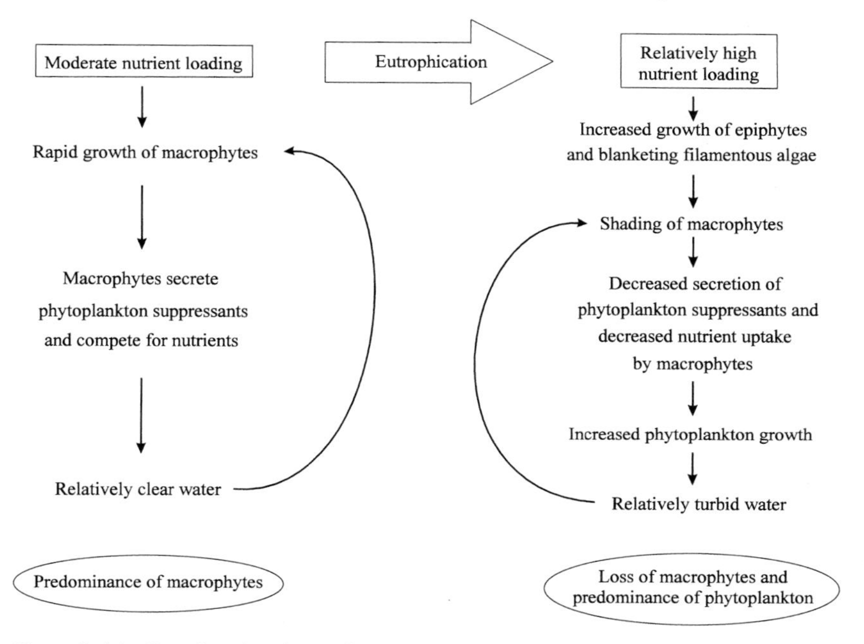

Figure 5.16 Postulated pathway for increased eutrophication causing a decline in macrophyte communities (after Phillips *et al.* 1978).

radioactive phosphorus to compare sediment vs. the water column as sources of phosphorus in nine common aquatic plants, and found that a majority of nutrients is removed from the sediments (Table 5.8). Even under hypertrophic conditions, the sediment contributed nearly three-quarters of the phosphorus taken up during growth. This has two important implications. First, from the point of view of this chapter, aquatic plants are not that different from other wetland plants. We can talk about fertility gradients in the substrate even among aquatic plants. Secondly, macrophytes may be visualized as pumps that remove nutrients from sediments and return them to open water (Barko and Smart 1980).

European wetland and terrestrial vegetation

Large-scale eutrophication may be expected to cause significant changes in landscapes. With its long history of human use, western Europe has significant eutrophication problems. Nitrogen levels in rainfall have reached very high levels. Ellenberg (1985, 1988) predicts that species normally found at low fertility will gradually disappear from the European landscape. Ellenberg has ranked species according to the nitrogen levels at

Table 5.8. *Uptake of phosphorus from sediments by nine macrophyte species*

Plant species	Percentage uptake from sediments
Myriophyllum alterniflorum	104.4
Potamogeton zosteriformis	107.4
Potamogeton foliosus	98.6
Callitriche hermaphroditica	94.2
Elodea canadensis	99.0
Najas flexilis	100.8
Myriophyllum spicatum	99.4
Heteranthera dubia	95.2
Vallisneria americana	103.1

Source: After Carignan and Kalff (1980).

which they normally occur, thereby identifying the subset requiring infertile habitats. The assumed mechanism behind losses of such species is that fertilization will increase plant biomass and competition intensity, so that smaller species and those that use infertile sites as refuges from better competitors, will be displaced. In Chapter 7 we will explore the effects of competition in more detail. Because eutrophication is occurring across the entire landscape, a significant proportion of the European flora is at risk, in habitats ranging from wetland and moorland to woodland (Figure 5.17).

In the heathlands of western Europe, species diversity has declined, and evergreen *Erica tetralix* heaths have been replaced by the grass *Molinia caerulea* (Aerts and Berendse 1988; Sansen and Koedam 1996). A nitrogen deposition rate above 10–15 kg N ha^{-1} yr^{-1} is sufficient to accelerate this change, and present rates in Flanders exceed 40 kg N ha^{-1} yr^{-1}. Sod cutting is used to reverse this process to retain species rich peatlands, with more deeply cut and regularly flooded sites allowing the growth and persistence of pioneer species such as *Drosera intermedia* and *Rhynchospora fusca*. Eventually, however, *Molinia caerula* achieves dominance, and at current N deposition rates, Sansen and Koedam (1996) believe that sod cutting frequencies will have to be increased from 50-year to 10-year intervals. Verhoeven *et al.* (1996) point out that if mowing removes N faster than it is deposited, it may be possible to retain a species composition more typical of pristine areas.

The experimental application of N fertilizer for agricultural purposes also illustrates the potential changes in vegetation with atmospheric deposition. In the moors of Somerset, which are wet grasslands on lowland

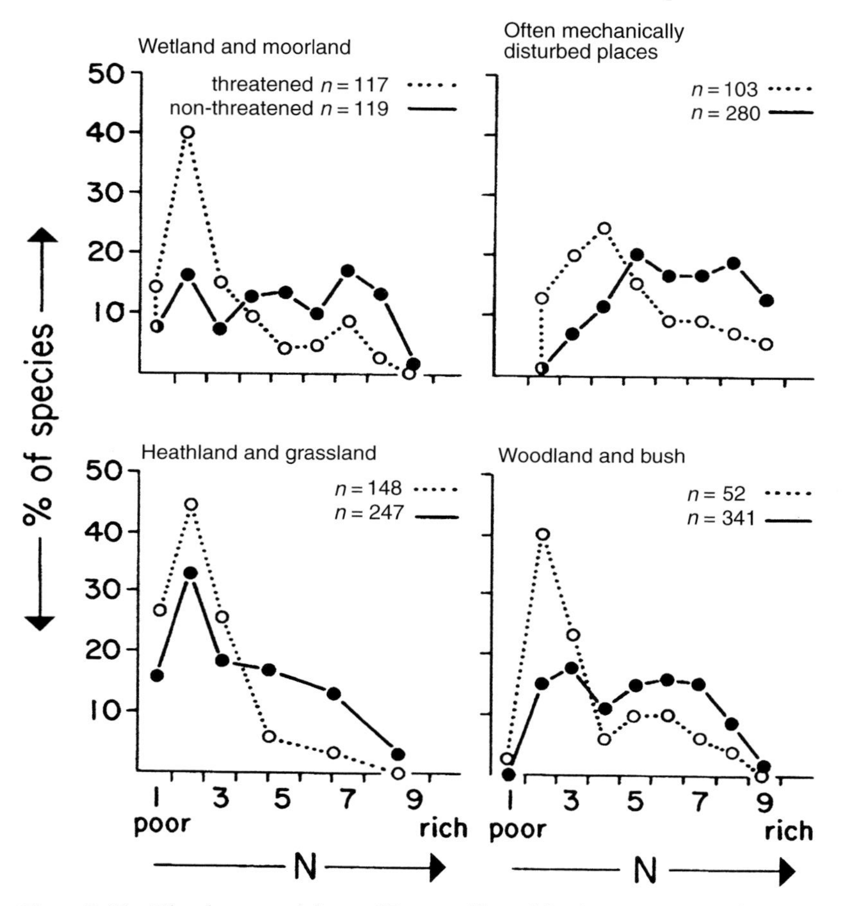

Figure 5.17 The threatened flora of Europe (dotted line) is concentrated in nutrient poor conditions of four different European habitats (from Wisheu and Keddy 1992 after Ellenberg 1985).

peat, experimental applications of N fertilizer at levels greater than, or equal to 25 kg ha^{-1} produced striking changes in the vegetation (Mountford *et al.* 1993). After only 4 years, sedges such as *Carex nigra* and rushes were replaced by widespread agricultural grasses such as *Holcus lanatus* and *Lolium perenne*. While there was an upward trend in species richness on control plots, richness declined in all high N treatments.

Returning to the global perspective, Townsend *et al.* (1996) present nitrogen deposition maps for the planet charting the consequences of increasing rates of industrial nitrogen fixation, showing that high rates of deposition already occur in eastern North America, and that there are early signs of a third area in eastern Asia. Losses of species from infertile habitats are not discussed, but Townsend *et al.* present some other implications. They note that the enhanced growth rates resulting from fertilization may remove extra amounts of carbon from the atmosphere, thereby reducing rates of global warming. Indeed, unexpectedly low CO_2 levels in the atmosphere present atmospheric scientists with a problem – where is the missing carbon? It may, in part, have been absorbed by fertilized plants. The possible beneficial effects of reducing rates of global warming have to be balanced against other negative effects such as damage from acid rain due to nitrogen and sulphur oxides, the role of NO_x emissions in ozone depletion, and other effects that may reduce ecosystem production. Whatever the atmospheric consequences, the examples in this chapter show that dealing with eutrophication and the problems posed by species requiring infertile habitats will be a long-term problem facing both scientists and managers.

Hydrology, fertility and wetlands

The wetland classifications introduced in on pp. 17–28 and pp. 189–194 assumed that hydrology and fertility were the two controlling factors in wetlands. Now that topics such as zonation, flooding and fertility have been explored in more detail, it is time to introduce one last summary diagram for wetlands. The elegantly simple scheme of Gopal *et al.* (1990) combines fertility on the horizontal axis (from nutrient poor to nutrient rich) with hydrology on the vertical axis (duration of waterlogging superimposed upon water level changes). Four wetland types are then produced from left to right: bogs, fens, marshes and swamps. The figure further discriminates between poor fens and rich fens. The addition of salinity as a factor produces both salt flats at the upper left and salt marshes at the lower right. Superimposed upon these is the zone of peat formation, showing the interaction between fertility and waterlogging. Finally, the region which woody plants can occupy is stippled. Many of the factors discussed so far are therefore neatly combined into one figure. Note too that there is a region between the peat formation and tree formation zone, showing that in some hydrological regimes, open meadows may be formed where the soils have limited amounts of peat, but are still too wet for trees to occur. Figure 3.15 presented the Hertzsprung–Russell star chart as an elegant example of useful simplification achieved from complexity; Figure 5.18 would appear to achieve many of the same benefits. Any particular wetland can be put in a context of other sites.

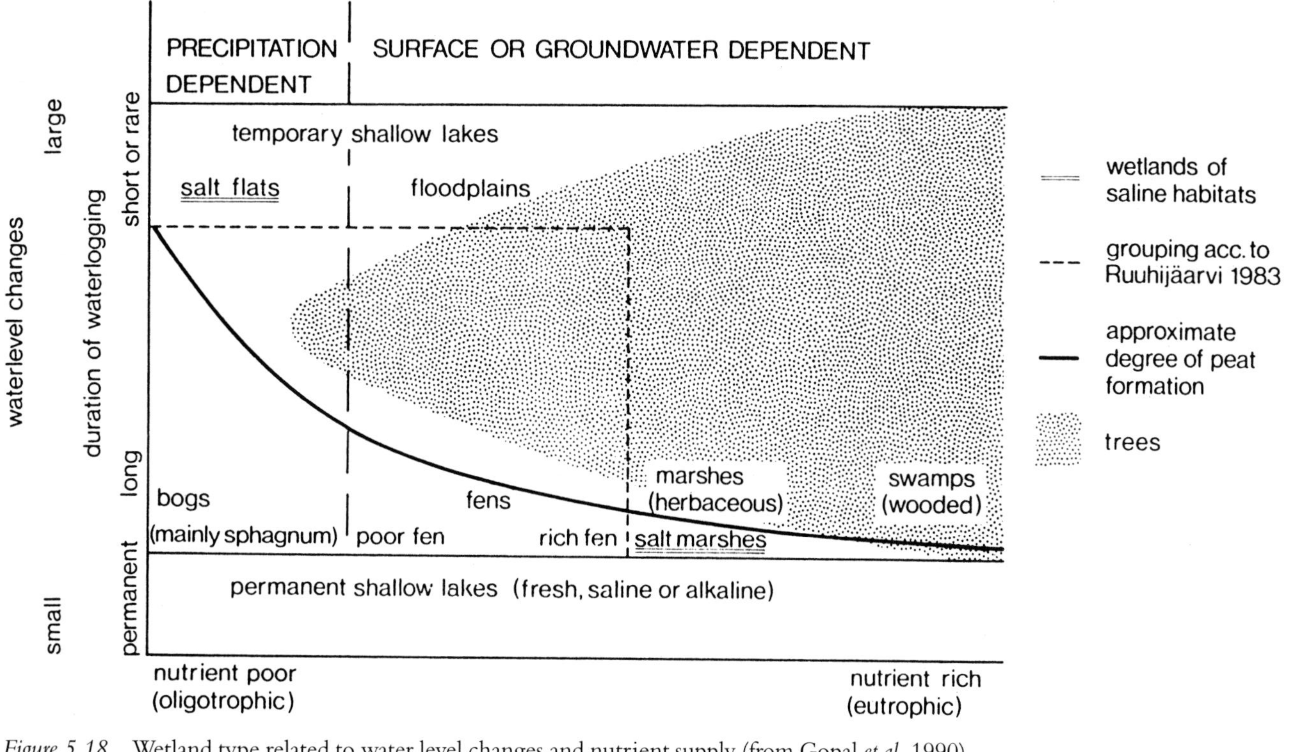

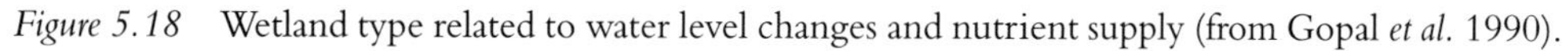

Figure 5.18 Wetland type related to water level changes and nutrient supply (from Gopal *et al.* 1990).

Two key controlling factors are given on the axes. Other factors such as peat production and the phase transition between woody and herbaceous species are also illustrated. Further, the diagram raises challenging questions. Where, for example would carnivorous plants occur? Is there a region where particular kinds of disturbance are most severe? Where might gap colonization predominate? Where would cylindrical growth forms as opposed to leafy ones tend to occur? Where would the most methane be produced? Where might above-ground competition give way to below-ground competition? Where would grazing be the most intense? Some answers appear obvious; others are not. A simple diagram such as this one challenges ecologists to put the pieces together into meaningful patterns.

6 · *Disturbance*

Introduction

Disturbance is an all-pervasive process in communities and ecosystems (e.g. Sousa 1984; Pickett and White 1985; Botkin 1990). But disturbance is perhaps a dangerous concept in ecology. It is dangerous precisely because the word is non-technical, and therefore many people assume they understand it when they do not. Moreover, in popular usage, disturbance encompasses so many effects that it hardly excludes anything in nature. It may encompass everything from a deer walking across a peat land, to an asteroid colliding with the Earth and eradicating entire genera. It may include everything from a botanist collecting plant specimens to an all-out nuclear war followed by nuclear winter. Words that mean everything end up meaning nothing.

According to the *Oxford English Dictionary* (Concise Edition) disturbance is 'interruption of tranquillity; agitation; tumult; uproar'. For ecologists, the key phrase is probably 'interruption of tranquillity'. We may re-phrase this and define disturbance as '*a short-lived event that causes a measurable change in the properties of an ecological community*'. This at first may seem vague. What is short-lived? Southwood (1977, 1988) suggests measuring duration in terms of organisms' life spans. By short-lived, we mean an event that occurs as a pulse with duration much shorter than the life span of the dominant species in the community. According to this definition, a fire or 1-year drought would be a disturbance; a slow and long-term climate change would not. By insisting upon measurable change we further require that the user of the word identify at least one property that is measurable (e.g. biomass, diversity, species composition) and show that it changes. No change, no disturbance (see Cairns 1980).

Grime (1977, 1979) has suggested defining disturbance as a factor that removes biomass. White (1994) explains: 'When the structural resistance and physiological tolerance of the vegetation is exceeded, substantial and sudden destruction of living biomass occurs; hence the recognition of . . . events as disturbances.' Disturbances include fire, damaging freezes, and floods; they tend to be discrete events in time. In general, destruction is fast and recovery is slow. Time lags are therefore important. This definition and explanation is useful, and it may work well in grasslands and forests, but we may find it somewhat limiting in its application to wetlands. Some disturbances have not only direct effects of biomass removal, but indirect effects of changing substrate texture and fertility. For example, wave disturbance on shorelines not only removes plant biomass, it also produces coarse textured infertile substrates. We may therefore subsume Grime's definition within the one above, since biomass is one of the properties, perhaps the most likely one, to change with disturbance.

Four properties of disturbance

Now that we have a possible working definition of disturbance, we must consider some of the properties it possesses. These would include (i) duration, (ii) intensity, (iii) frequency and (iv) area (Sousa 1984; Pickett and White 1985). Each of these is measurable. To some extent they are self-explanatory, but let us briefly consider them in turn.

Duration

Duration refers to how long the event lasts. A frost or fire may last only hours; floods or grazing may continue for years. For example, burial by litter can kill salt marsh plants in as little as eight weeks (Bertness and Ellison 1987), whereas flooding for 3 years is required to kill most emergent wetland plants in freshwater marshes (Figure 6.1). A deposit of wrack or pulse of flooding is therefore a disturbance. The duration of an event may reasonably be expressed in terms of the life spans of the organisms of concern (Southwood 1977, 1988).

Intensity

The intensity of a disturbance is best judged by the severity of its effects upon a wetland. The simplest measure of disturbance would be the proportion of biomass at a site which is killed or removed. A factor which disturbs one taxon (say, plants) might not disturb another (say, insects), so the change in abundance of several groups might be measured simultaneously.

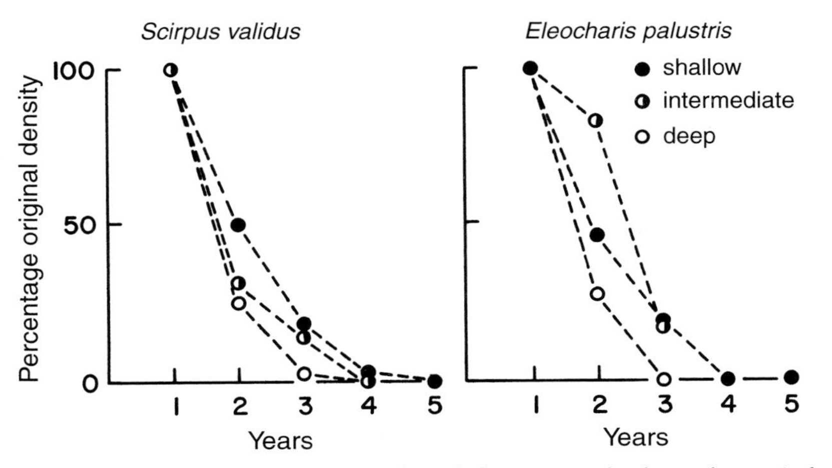

Figure 6.1 The effects of flooding to three different water depths on the survival of two emergent plant species (from Keddy and Reznicek 1986, data from Harris and Marshall 1963).

Change in species composition would be an alternative measure of intensity. There is a wide range of measures of similarity between samples (Legendre and Legendre 1983). Using a standard measure of ecological similarity, one could define a range of disturbance intensities from 0 (the community is the same before and after the disturbance) to 1 (the community is completely different after disturbance). Figure 6.1 shows that the deeper the flooding of a wetland, the more rapid the decline in abundance of two emergent plants. If we were measuring the degree of similarity each year, it is obvious that the dissimilarity through time increases most rapidly with the deepest flooding (that is, most intense disturbance). We would observe similar effects if, instead of using composition, we used biomass of these plant species as our measure.

Frequency

Some events, such as water level changes, happen on a yearly basis. Others, such as ice ages or asteroid collisions, happen rarely. Recall Figure 4.2, showing water levels in the Great Lakes for roughly a century, and illustrating the frequency with which different degrees of low water occur. In general, the greater the intensity of disturbance (extreme high or low water periods), the lower the frequency. It seems reasonable to argue that the more frequent an event is, the more likely organisms are to develop resistance to them. Again, in the short term, we may express frequency in years; in the long run we will probably find it necessary to measure it in life times.

Area

This is self-explanatory, but it bears mention because it reminds us of at least one way to rank disturbances in terms of their significance to ecology. Factors that affect huge areas (e.g. climate change, drainage, air pollution) will usually demand more attention than factors that influence only small areas (e.g. mining, trampling). Hurricanes have been observed to kill between 25 and 75%, and as much as 90% of low-lying mangal, suggesting that 'hurricanes may have a very large role in determining the ratios of species within vegetation types over large areas.' (Lugo and Snedaker 1974). Area of impact would be a first-order way of setting research priorities for conservation agencies.

Regeneration from buried seeds after disturbance

The episodic destruction of biomass in wetlands can be caused by many factors including fire, flooding, ice scour and grazing. When biomass is removed, resources such as light and nitrogen become available. Many plant species have traits that enable them to exploit these temporary patches of resources. Buried reserves of viable seeds, often called seed banks, allow plants to rapidly exploit disturbed patches. Seed densities in excess of 1000 seeds per square metre are common in both prairie marshes and freshwater coastal marshes, and densities in excess of ten thousand per square metre are common in wet meadows (Table 6.1). These high densities of buried seeds provide evidence of the importance of disturbance and regeneration in wetland vegetation. Possibly the most familiar example is the regeneration of freshwater marsh vegetation after disturbance by flooding and drying, well documented in both prairie wetlands (e.g. van der Valk and Davis 1976, 1978) and lake shores (Keddy and Reznicek 1982, 1986).

For many marsh and wet meadow species, regeneration in gaps provides the only opportunity for establishment from seed. Grubb (1977) coined the term 'regeneration niche' to describe the many ways in which plant seedlings and juveniles can exploit the different kinds of natural disturbance. Buried seeds appear to detect these natural disturbances in three ways: increased fluctuations in soil temperature, increased quantity of light, and changes in the quality of light (Grime 1979). Thus, most

Table 6.1. *Reserves of buried seeds are an important feature of wet meadows and marshes*

Study	Site	Seedlings m^{-2}
Prairie marshes		
Smith and Kadlec (1983)	*Typha* spp.	2682
	Scirpus acutus	6536
	S. maritimus	2194
	Phragmites australis	2398
	Distichlis spicata	850
	Open water	70
van der Valk and Davis (1978)	Open water	3549
	Scirpus validus	7246
	Sparganium eurycarpum	2175
	Typha glauca	5447
	Scirpus fluviatilis	2247
	Carex spp.	3254
van der Valk and Davis (1976)	Open water	2900
	Typha glauca	3016
	Wet meadow	826
	Scirpus fluviatilis	319
Freshwater coastal marshes		
Moore and Wein (1977)	*Typha latifolia*	14768
	Former hayfield	7232
	Myrica gale	4496
Leck and Graveline (1979)	Streambank	11295
	Mixed annuals	6405
	Ambrosia spp.	9810
	Typha spp.	13670
	Zizania spp.	12955
Lakeshore marshes		
Nicholson and Keddy (1983)	Lakeshore, 75 cm water	38259
Keddy and Reznicek (1982)	Waterline of lake	1862
	30 cm below waterline	7543
	60 cm below waterline	19798
	90 cm below waterline	18696
	120 cm below waterline	7467
	150 cm below waterline	5168
Wisheu and Keddy (1991)	Wilsons Lake	8511
McCarthy (1987)	Hirst Lake	24430
	Hirst Lake	16626
	Goose Lake	11455
	Goose Lake	3743
Beaver ponds		
Le Page and Keddy (1998)	Canadian shield	2324

Source: Adapted from Keddy and Reznicek (1986).

plants adapted to exploit natural disturbances are stimulated to germinate by a combination of high light levels and fluctuating temperatures (Grime *et al.* 1981).

In saline environments, colonization of gaps by buried seeds appears to be infrequent, seed densities being low (50 m^{-2} Hartman 1988 to 500 m^{-2} Bertness and Ellison 1987) and a majority of the revegetation results from expansion by neighbouring plants (Bertness and Ellison 1987; Hartman 1988; Allison 1995). This exception is probably a consequence of the constraints which salinity places on the establishment of seedlings; periodic flooding with fresh water may provide the only opportunity for some salt marsh species to establish from buried seeds (Zedler and Beare 1986). Species such as *Distichilis spicata* appear to persist by rapidly invading gaps with long adventitious runners, thereby avoiding stronger competitors such as *Spartina patens* (Bertness and Ellison 1987). The large viviparous seedlings found on some mangrove species (Tomlinson 1986) may also provide a means of circumventing constraints imposed by salinity. Mangrove seedlings are thought to play an important role in regeneration after damage by storms (Lugo and Snedaker 1974).

Examples of disturbance

The above generalizations may seem to be belabouring the obvious. But, so long as we use the word 'disturbance' in a careless manner, we will only end up confusing one another. Let us now turn to concrete examples of studies that have explicitly explored disturbance in wetlands. The treatment in this section is somewhat encyclopaedic in order to illustrate the wide array of types of disturbance in wetlands; even so, other examples including burial by wrack in salt marshes and deposition of fresh sediment along water courses are omitted and reserved for the next section.

Water level fluctuations

Because this is one of the most important disturbances in wetlands, an entire chapter (4) has already been devoted to this topic.

Erosion

Running water can create many kinds of disturbance in watersheds; these can range from splash erosion caused by individual rain drops, to soil wash on valley slopes, to bank erosion and slumping in valley bottoms (Strahler

1971; Rosgen 1995). The latter are certainly predominant disturbances in the lower reaches of watersheds, where rivers flow though valleys filled with alluvial sediments; these areas often have extensive flood plain forests (swamps) and to a lesser degree, marshes. The alluvial sediments are continually reworked by the river, thereby continually destroying the established vegetation and exposing new substrate for succession. The dynamics of meanders have received a good deal of attention from geologists (e.g. Strahler 1971). Meanders originate from the enlargement of bends in the course of a river. As the river undercuts and erodes a bank at one point, the material from this bank is carried downstream and deposited as a point bar; the point bar deflects the current to the other side of the stream, where the force of the water cuts into the bank, further producing another point bar on the opposite side. Once a bend is produced, centrifugal forces thrust the river flow towards the outside of bends, increasing rates of erosion until a meander loop is formed (recall Figure 1.6). Meanders in flood plains take on characteristic geometric forms in which meander wavelength is linearly related to the mean annual discharge. Once these meanders are formed, there is a gradual movement downstream (down valley sweep) so that, over time, all the alluvial sediments are eroded and reworked between the limits set by river bluffs.

These processes have important effects on vegetation because they continually remove established communities and allow the process of succession to begin anew. The result is a mosaic of vegetation types of different ages and species composition. Such processes have been described, for example, from lowland podocarp forests in New Zealand (Duncan 1993). These forests are dominated by conifers in the Podocarpaceae, along with the angiosperm *Weinmannia racemosa*. Two species of podocarp, *Dacrycarpus dacrydiodes* and *D. cupressinum* establish in large openings created by floods. In the absence of further flooding composition shifts towards the latter species. Over longer periods yet, the more shade tolerant *W. racemosa* would be expected to dominate, but stands of this tree are absent from this region of New Zealand, emphasizing the frequency of disturbance. Duncan concludes that Podocarp forests occur largely because of recurring disturbance. Similarly, in north western North America *Populus balsamifera* establishes on newly deposited sediments, whereas *Picea glauca* dominates older surfaces (Nanson and Beach 1977).

In a half million km^2 area of the Peruvian Amazon, Salo *et al.* (1986) report that 26.6% of the modern lowland forest shows characteristics of recent erosion and deposition, and that fully 12% of the Peruvian lowland forest is in successional stages along rivers. During one 13-year period, Landsat images showed that the mean lateral erosion rate of meander bends was 12 m yr^{-1}. The total area of newly created land available for primary succession was 12 km^2, representing nearly 4% of the present flood plain area. The new substrates were first colonised by herbaceous species in genera such as *Tessaria, Cyperus, Ipomea* and *Panicum*; smaller trees in the genera *Cecropia*, *Ficus* and *Cedrela* gradually formed a closed canopy, and eventually these became mixed with later successional species. Kalliola *et al.* (1991) have described the successional processes in more detail, documenting a pioneer flora of 125 plant species. Salo *et al.* conclude:

> According to the repetitive nature of river dynamics, the migration of the river channel course creates a mosaic of successional forests within the present meander plain. The mosaic forest is composed of patches of differentially aged sequential successional forest and patches of forests originating from a succession on the sites of former oxbow lakes. The annual floods further modify the mosaic pattern.

While the numbers are impressive, they do not convey the breadth of this phenomenon as powerfully as do aerial photographs (Figure 6.2). Although studies of rain forests have tended to emphasize gap dynamics, that is the regeneration of individual trees within light gaps created by fallen trees (Grubb 1977; Connell 1978), Salo *et al.* point out that the regeneration in flood plains is fundamentally different, since the disturbance is more intense, with entirely new substrates for colonization. Similar processes have been described in the Okavango Delta of Africa, where erosion and deposition produce a mosaic of habitats and vegetation types (Ellery *et al.* 1993). Such combinations of substrate heterogeneity and continual disturbance may partly account for the very high plant species diversity of tropical flood plain forests.

Animals

An entire chapter on grazing follows this one. The general conclusion from that chapter is that grazing can have major effects on wetlands, but the area of habitat affected appears limited. Such cases include muskrat grazing in the north American prairies, and goose grazing in salt marshes. In many other cases the effects of herbivores may be small relative to other environmental factors. But there are exceptions – animals that create wetlands are a significant cause of natural disturbance. We have already touched on the example of beavers (pp. 208–212). The southern

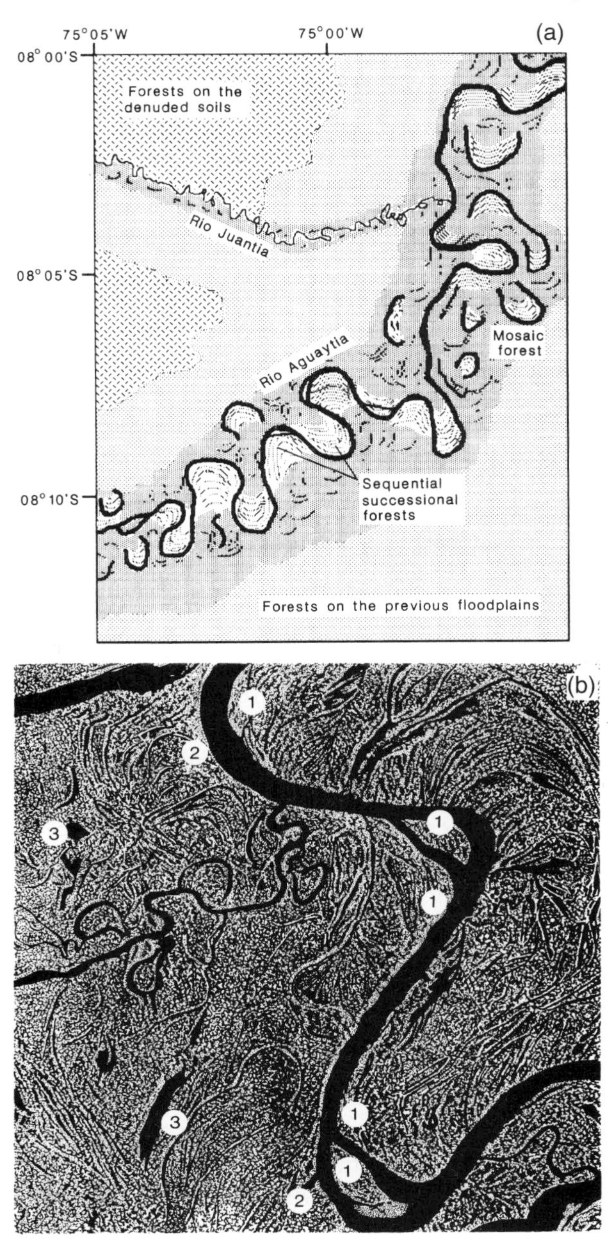

Figure 6.2 Meandering rivers generate variation in both the topography and the age of lowland forests. *Top*, a simplified map of a meander system in a tropical white-water river (the Ucayali at Pucallpa, Peru). *Bottom*, a detailed map of forests along a meander system downstream of the top figure, showing (1) areas of intense primary succession, (2) eroding forest at outer edges of meander and (3) isolated oxbow lakes (from Salo *et al.* 1986).

equivalent of the beaver pond may be the 'gator hole. These are depressions that are either made, or maintained by alligators. During winter dry periods, these holes may be the only ponds remaining in a wetland (Loveless 1959; Craighead 1968). The alligator maintains ponds by pulling loose plants and dragging them out of the pool. Thicker muck is either pushed or carried to the edges of the pond.

Gator holes were once a predominant feature of wetlands in South Florida. Craighead concludes that 'in the first two decades of this century every inland pond, lake and river held its quota of alligators'. He suggests a density approaching one alligator per acre in some regions. Historical records are always suspect, but the naturalist William Bartram, who travelled the St John's River in 1774–1776, described alligators massed around his boat. He reported that, when camping on beaches, it was necessary to keep a large fire burning all night for protection. As recently as 1929, 190 000 hides were sold, but by 1943 the figure had fallen to only 6800.

Gator holes are 'reservoirs for an amazing biological assemblage'. Within them live 'diatoms, algae, ferns, flowering plants, protozoans, crustaceans, amphibians, reptiles and fish' (Craighead 1968). The productivity of these ponds is enhanced by uneaten food. Larger animals, such as hogs and deer, are killed by drowning but may be left for several days for ripening. The aquatic flora includes widespread genera such as *Myriophyllum*, *Urticularia*, *Potamogeton*, *Nymphoides* and *Najas*. The shallow water near the banks has a marsh genera such as *Peltandra*, *Pontederia*, and *Sagittaria*. Indeed, the description of this flora is remarkably reminiscent of beaver ponds. Connecting the gator holes are well-developed trail systems. These trails may be eroded by heavy 'gators into troughs that are 6′′ deep and 24′′ wide'.

Craighead concludes that nearly all the animal life of the area revolves around the alligator, either directly or indirectly. In the context of this chapter, both beavers and alligators are tied to water level changes in wetlands. Both provide examples of animals acting as engineers of wetland structure. But there appears to be an important distinction. Beavers cause water level fluctuations by building dams and abandoning them. In contrast, in alligator wetlands, fluctuations in water level are caused by climate; the alligators then provide reservoirs where wetland life can find refuge from drought. From another perspective, both of these species may be regarded as agents of natural disturbance that maintain biological diversity in landscapes.

Fire

It may be difficult to imagine a fire ripping through a stand of water lilies, but some kinds of wetlands do indeed burn during dry years. Sutter and Kral (Table 6.2) postulate that fire frequency, along with hydrology, is the principal factor determining the kind of wetland community that forms in the south eastern United States. Fire is regarded as a major control of plant diversity in boreal circumpolar peatlands (Wein 1983), pocosin peatlands (Christensen *et al.* 1981) and the everglades (Loveless 1959). Loveless concludes 'The importance of fire and its influence on the vegetation of the Everglades can hardly be over-emphasized.' Fire becomes important during prolonged periods of drought. Low intensity fires can simply remove existing vegetation, shift composition from woody to herbaceous wetlands, and increase plant diversity (Christensen *et al.* 1981; Thompson and Shay 1988). Jean and Bouchard (1991) believe that fires set by aboriginal inhabitants prevented woody plants such as alders from invading wet meadows along the St Lawrence river in eastern North America. Figure 6.3 shows the effect of fire incidence on both litter accumulation and plant diversity in *Carex* dominated wetlands here. Hogenbirk and Wein (1991) measured vegetation responses to fire in two vegetation types of the Peace–Athabasca Delta (Figure 6.4). Fire reduced both the height and density of the dominant species, but had little effect upon species richness. However, during the longer droughts, more intensive fires can burn the organic matter in the soil, creating new depressions and pools (e.g. Loveless 1959; Vogl 1969).

Changes in vegetation will produce changes in other properties of ecological communities. Table 6.3 shows that, in one wetland, more birds were observed on a burned shoreline than on an adjacent unburned area. Laubhan (1995) studied intensively managed impoundments used to provide seeds, browse and tubers for waterfowl in Missouri. Using experimental 0.10 ha burns, he found that the composition of vegetation was affected by the time of year a site was burned. Sites burned in the spring had a greater cover of annual plants, such as *Bidens* spp. and *Leersia oryzoides*, but a lower cover of *Iva ciliata*, a species that tends to produce dense monospecific stands. Summer burned sites were more than three-quarters bare ground, with only *Ipomoea purpurea* and *Sesbania exaltata* showing slight increases relative to controls. No seeds were produced in summer burns; in contrast, the total seed biomass of two genera of annuals (*Bidens*, *Polygonum*) were 33.7 gm^{-2} in spring burns relative to

Table 6.2. *Summary of non-alluvial wetland communities of the south eastern United States*

Community	Canopy dominants	Soil	Hydroperiod/ water source	Fire frequency
Forested wetlands in basins				
Pond cypress pond forest	*Taxodium ascendens*	Mineral to organic	6–12 months rainfall	Infrequent, 20–50 years
Swamp tupelo pond forest	*Nyssa biflora*	Organic to peat	6–12 months rainfall	Rare, one fire per century
Cypress dome	*Taxodium ascendens*	Peat	6–9 months rainfall	20 + years
Basin swamp forest	*Nyssa biflora, Acer rubrum, Liquidamber styraciflua*	Organic	6–9 months groundwater	Infrequent, 20–50 years
Wetland complexes (from forested to open water) in basins				
Limestone pond complex (karst ponds)		Mineral	Deep groundwater	1–10 years/yellow sand, 36–60 years/white sand
Coastal plain small depression pond		Mineral	Variable	Dependent on surrounding forests
Coastal plain lakeshore complex		Mineral	Variable	Rare, one fire per century
Okefenokee swamp wetland mosaic		Mineral-peat	Variable	Infrequent, 20–50 years
Woodlands and savannas on flat coastal terraces				
Slash pine flatwoods	*Pinus serotina*	Mineral	<3 months groundwater	3–10 years
Wet longleaf pine flatwoods	*Pinus palustris*	Mineral	<3 months groundwater	3–10 years
Wet longleaf pine-slash pine flatwoods	*Pinus palustris, Pinus serotina*	Mineral	<3 months groundwater	3–10 years
Longleaf pine savanna	*Pinus palustris*	Mineral	3–6 months groundwater	1–5 years

Coastal plain pitcher plant flat	A diversity of graminoid and herbaceous species including *Sarracenia* spp.	Mineral	6 months groundwater	1–5 years
Woodlands and savannas in basins				
Pond cypress savanna	*Taxodium ascendens*	Mineral	6–9 months rainfall	20 + years
Pond pine woodland	*Pinus serotina, Cyrilla racemiflora*	Shallow organics and peats	6–9 months rainfall	10–20 years
Evergreen shrub wetlands				
Low pocosin	*Pinus serotina, Cyrilla racemiflora, Zenobia pulverulenta*	Deep peat >0.5 m	6–9 months rainfall	15–30 years
High pocosin	*Pinus serotina, Cyrilla racemiflora, Lyonia lucida*	Shallow peat <0.5 m	6–9 months rainfall	15–30 years
Small depression pocosin	*Pinus serotina, Cyrilla racemiflora, Lyonia lucida*	Shallow peat <0.5 m	6–9 months rainfall	15–30 years

Source: After Sutter and Kral (1994).

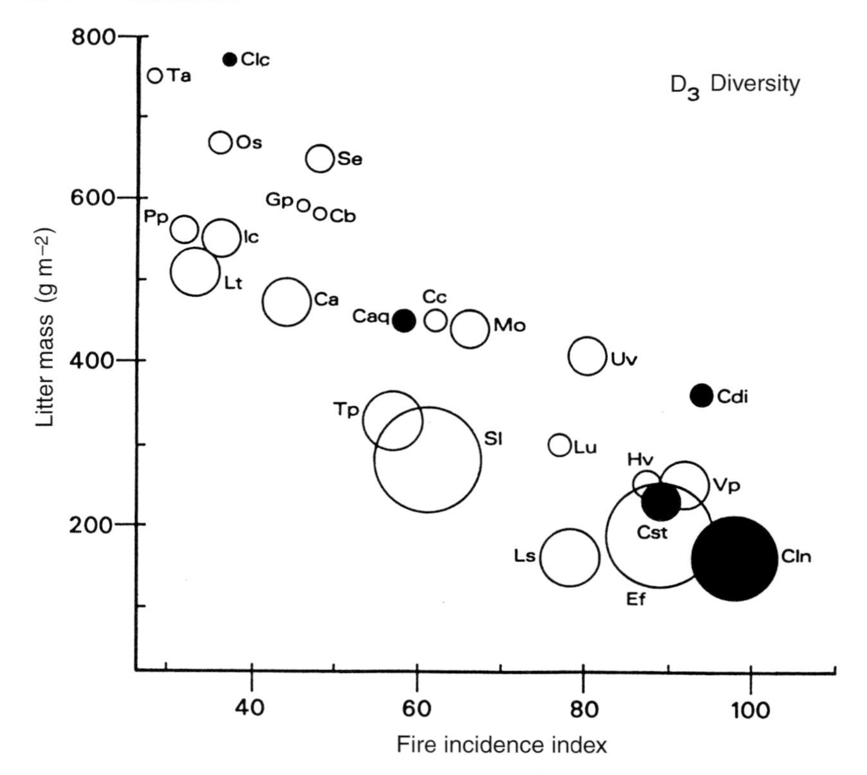

Figure 6.3 Relation of Shannon diversity index (D_3) and litter mass to fire incidence index in a *Carex* dominated wetland. Circle diameter is proportional to D_3 diversity (0.86–1.61 range). Principal *Carex* species, shown in closed circles, include: *Caq C. aquatilis, Cdi C. diandra, Clc C. lacustris, Cln C. lanuginosa, Cst C. stricta*. Other species include: *Ca Campanula aparinoides, Cb Cicuta bulbifera, Cc Calamagrostis canadensis, Ef Equisetum fluviatile, Gp Galium palustre, Hv Hypericum virginicum, Ic Impatiens capensis, Ls Lythrum salicaria, Lt Lysimachia thyrisflora, Lu Lycopus uniflorus, Mo* Moss species, *Os Onoclea sensibilis, Pp Potentilla palustris, Se Sparganium eurycarpum, Sl Sagittaria latifolia, Ta Typha angustifolia, Tp Thelypteris palustris, Uv Utricularis vulgaris, Vp Viola pallens* (from Auclair *et al.*, 1976b).

12.6 gm^{-2} in controls. There were no detectable effects on soil nutrient levels. Laubhan concludes that spring burns can create favourable conditions for waterfowl by stimulating seed production, whereas summer burns produce mud flats favourable to migrating shorebirds. Smith and Kadlec (1985a) also report that muskrats preferentially graze in burned areas of shoreline marsh.

Peatlands are particularly useful for the study of fire because, under certain circumstances, charcoal layers and macrofossils record both the

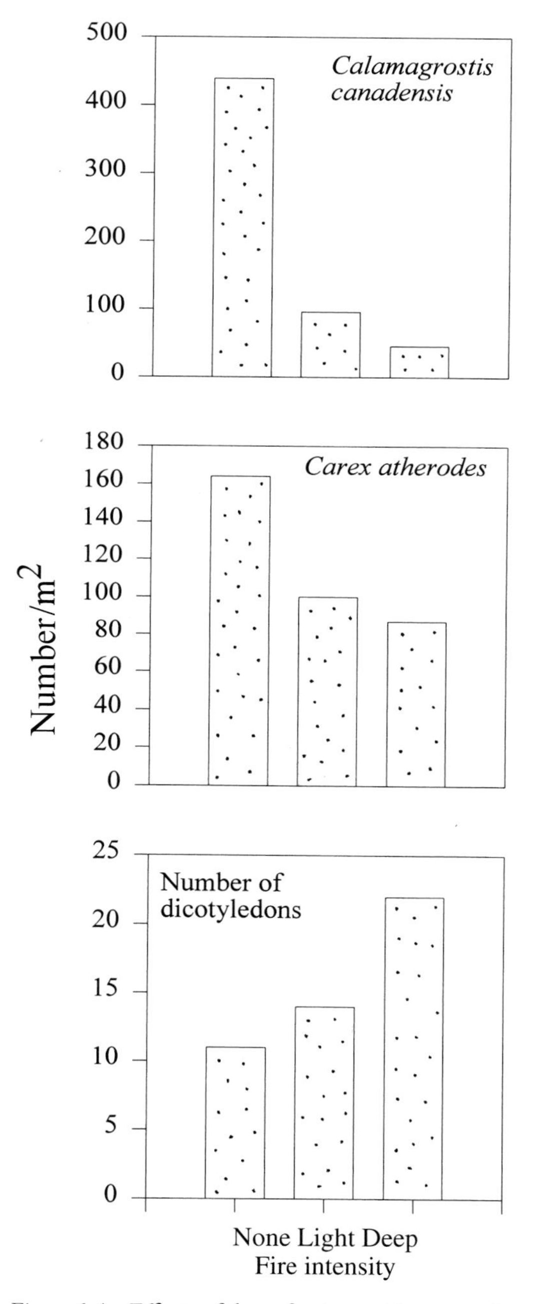

Figure 6.4 Effects of three fire intensities upon three properties of wetlands in the Peace–Athabasca delta (from data in Hogenbirk and Wein 1991).

Table 6.3. *Resident birds on unburned and burned portions of a shore line wetland near the Florida/Georgia border*

Species	Control	Burned
Common egret (*Casmerodius albus*)	5	22[a]
Bobwhite (*Colinus virginianus*)	14	1
Cardinal (*Richmondena cardinalis*)	2	14[a]
Common crow (*Corvus brachyrhynchos*)	0	10[a]
Common gallinule (*Gallinula chloropus*)	8	25[a]
Great blue heron (*Ardea herodias*)	0	8[a]
Little blue heron (*Florida caerulea*)	7	32[a]
Mockingbird (*Mimus polyglottos*)	0	6
Common grackle (*Quiscalus quiscula*)	0	15[a]
Red-winged blackbird (*Agelaius phoeniceus*)	66	150
Snowy egret (*Leucophoyx thula*)	0	7
Total	102	290

Notes:
The numbers represent totals obtained from 63 sampling trips in 1971.
[a] paired *t*-tests for repeated samples, but no replication of treatments.
Source: From Vogl (1973).

fire history and the vegetation responses to the fire. *Sphagnum*-dominated peatlands are probably the most abundant peatland type in western boreal North America. Kuhry (1994) studied a series of peat cores to reconstruct fire and vegetation history. He found that these peatlands had many macroscopic charcoal layers as a consequence of past fires. In the eight studied peat deposits, he estimated there had been one local surface fire approximately every 1150 years. While this may be a surprisingly high rate of fire frequency, it is still an order of magnitude less frequent than estimates of fire frequency in coniferous forests in western boreal Canada (e.g. Ritchie 1987). During the hypsithermal, a period of warmer and drier climate about 7000 years ago, fire frequencies in peatlands appear to have been twice a high as in the past 2500 years. These fires not only burned the vegetation, but they also burnt the superficial peat deposits. In spite of this, the cores suggest that the effect of peat surface fires on vegetation was short lived. This is apparently also the case in contemporary reports of peat fires. An interesting natural history story complements these findings; *Sphagnum* can apparently regenerate from stems at depths 30 cm into the peat deposit (estimated to be 25–60 years old) (Clymo and Duckett 1986).

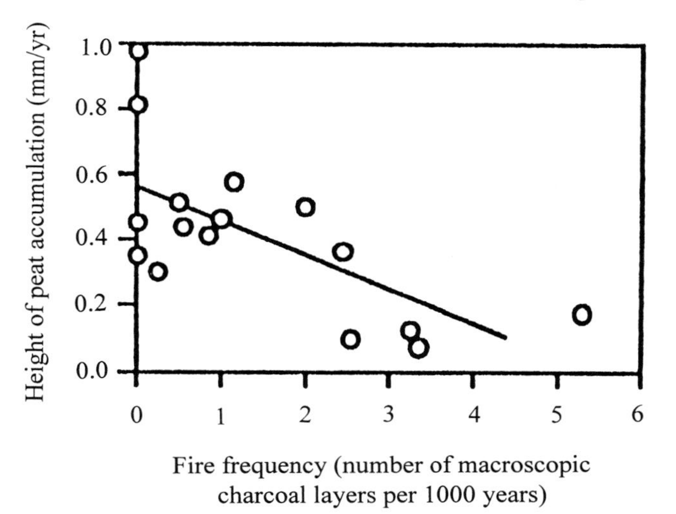

Figure 6.5 Peat accumulation as a function of fire frequency in western boreal Canada (from Kuhry 1994).

Kuhry could have stopped here with pictures of peat profiles, but in addition to these sort of qualitative observations, he went on to test for quantitative relationships among rates of peat accumulation and fire frequency. Fire frequency was estimated as the number of macroscopic charcoal layers per 1000 years, and peat accumulation rates were determined from radiocarbon dating. There was a negative relationship between peat accumulation rates and fire frequency (Figure 6.5). It appears, then, that the flush of nutrient-rich ash released by burning (and the presumed higher plant growth rates), does not compensate for the loss of peat consumed by the fire. Thus, fires significantly retard the growth of peatlands. This has important consequences for global warming because peatlands are an important reservoir for carbon storage. An increase in temperature would presumably lead to higher frequencies of burning, which, in turn, would lead to further releases of stored carbon in the peatlands (Gorham 1991; Hogg *et al.* 1992). This would then act as a positive feedback loop to increase rates of global warming.

The effects of severe fires in British peatlands became apparent in 1976; the period between May 1975 and August 1976 was the driest in England since at least 1727. In the North York Moors National Park, where there are 518 km^2 of open moorland, 62 uncontrolled fires burned in the summer of 1976 (Maltby *et al.* 1990). Some of the most severe fires

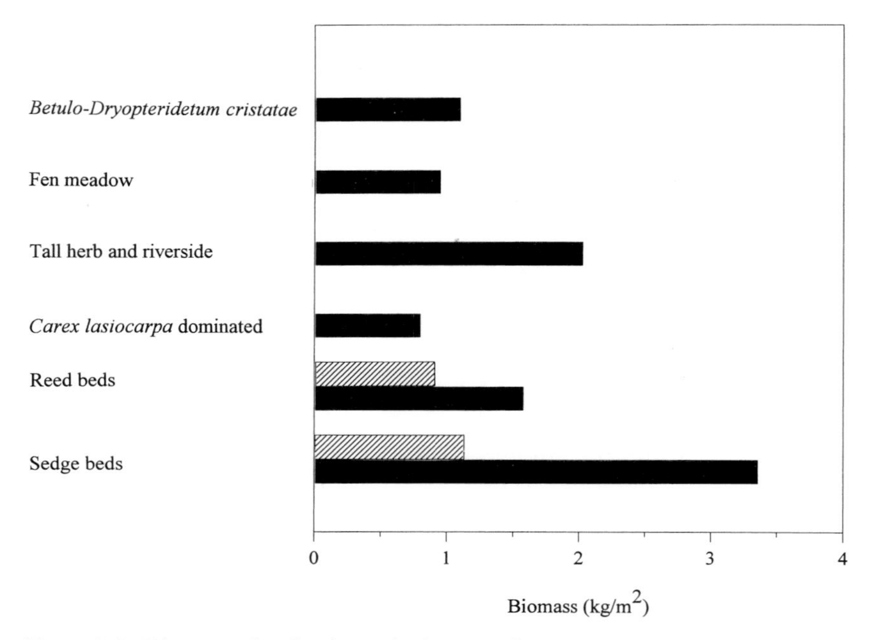

Figure 6.6 Biomass of wetlands in a herbaceous fen in western Europe; the hatched histograms are regularly mown (from data in Wheeler and Giller 1982).

affected 11 km^2 near Rosedale; here, fire burned deeply into blanket peats and largely removed the areas of thinner peat. Further alteration was caused by wind and rain erosion, leaching and freeze–thaw processes. Vegetation development was dominated by bryophytes, with *Ceratodon purpureus* most common on heavily burned sites. By 1982, *Polytrichum* spp. were widespread. By 1984, vascular plants such as *Calluna* were restricted to less than 2% of the area, 65% was still dominated by brophytes, and 30% was unvegetated. Perhaps, they suggest, other bryophyte dominated patches reflect past fires.

Mowing

In Europe there is a long cultural history of mowing wetlands (Elveland 1978, 1979; Elveland and Sjöberg 1982; Müller *et al.* 1992). Moreover, at some sites with a long history of mowing or grazing, a distinctive flora and fauna occurs as a consequence of the continual removal of biomass. Figure 6.6 shows differences in biomass among reed beds and sedge beds in a herbaceous fen in western Europe. Figure 3.9(*a*) showed changes in plant species richness and biomass associated with mowing. Peat cutting could be considered a more extreme kind of mowing, in that the substrate is removed along with plants and burned as a fuel. Disturbance by past peat cutting is thought to be an important factor generating plant diversity in European peatlands. Giller and Wheeler (1986), for example, studied nineteenth-century peat cuttings in the Norfolk Broadland, where some 50–70 cm of peat had been removed down to underlying clay. These had turned into ponds ('turf ponds') which now have unusual vegetation types, species rich fens, 'exclusive to the turf-pond sites'. To maintain such rich fen communities, they suggest that peat cutting will be a necessary management tool. Recall, too, that Sansen and Koedam (1996) proposed that peat cutting will be necessary in much of western Europe in order to remove nutrients accumulating from atmospheric deposition (p. 276).

Walker and Wehrhahn (1971) studied the environmental factors controlling prairie wetlands in Saskatchewan, Canada, and concluded that the most important environmental factor was disturbance ('grazing, mowing and natural disturbance'). This was followed by available nutrients, water regime and then salinity. This result occurred in spite of their intention of avoiding disturbed sites. Species such as *Eleocharis pallustris*, *Glyceria grandis*, *Alopecurus aequalis* and *Beckmannia syzigachne* occurred in disturbed areas. They note that, in spite of its importance, disturbance 'could not be quantitatively measured', a shortcoming that still hampers many ecological studies.

In some North American wetlands, invasion and dominance by cattail (*Typha* spp.)may reduce plant diversity and water fowl production. A variety of mowing implements, from machetes to 50 hp tractors have been used to cut cattail stands (Kaminski *et al.* 1985). While cutting temporarily reduced shoot density by 50 to 95%, success in limiting regrowth depends upon the duration of flooding after the mowing. Kaminski *et al.* argued that inundation after mowing would exhaust oxygen and carbohydrate reserves in rhizomes, and calculated that a depth of 40 cm in the spring would result in zero regrowth in their study sites.

Ice

Anyone who has watched great cakes of ice grind against the shoreline during spring flooding will be impressed by the power of ice scour to modify vegetation. In salt marshes or large lakes, one can find many metre square pieces of marsh, with 20 or more centimetres of substrate, chopped

Table 6.4. *Correlations among four environmental factors on a temperate zone lakeshore. This lake is frozen for approximately half of each year*

	Loss on ignition	Silt/clay	Ice[a]	Soil
Silt/clay	0.78	–	–	–
Ice	−0.47	−0.37	–	–
Soil	ns	ns	ns	–
Shrubs	−0.37	−0.43	0.31	ns

Notes:
$N = 121$ quadrats.
[a] Ice damage in each plot was measured with ten wooden pegs ; intensity of ice damage could range from 0 (no damage) to 20 (all pegs sheared off or removed).
ns, $P > 0.001$.
Source: After Wisheu and Keddy (1989b).

out of the ground and moved aside. At a smaller scale, there is the constant grinding of all freshwater shorelines as ice freezes to the shore and grinds the shore as water levels rise and fall. Although the effects of ice cakes grinding on shorelines are visible (and audible) during spring thaw, the processes beneath ice and snow during long winters are more difficult to study. Geis (1985) described how lake ice freezes onto the shoreline of lakes, forming an 'ice foot'. Sediments can become incorporated into this ice foot. Entire sections of shoreline are torn out of place when ice is lifted by rising water levels. According to Geis, plant biomass and diversity are reduced in the zone affected by the ice foot. Further north, ice pushing can create ridges, producing distinctive undulating topography along shorelines (e.g. Bliss and Gold 1994). As salinity declines and nutrient levels increase in the meadows, between the ridges, there is a transition from brackish wetlands with *Puccinellia* to freshwater wetlands with *Carex*.

One possible way to study ice damage is to put wooden pegs into wetland and measure the amount of damage accumulated over different periods of time. Figure 6.7 shows a typical vertical profile of ice damage. Note that the effects vary between an exposed shore and a sheltered bay. This technique could be extended to examine how ice damage correlates with other habitat and vegetation properties. Wisheu and Keddy (1989a) found loss on ignition and silt/clay content were both negatively correlated with ice damage (Table 6.4). Moreover, woody plants grew closer to the water on shores protected from ice damage. An alternative approach requires data on water levels, and the assumption that ice scouring can be estimated from knowing when the shoreline is frozen. Plots of water level

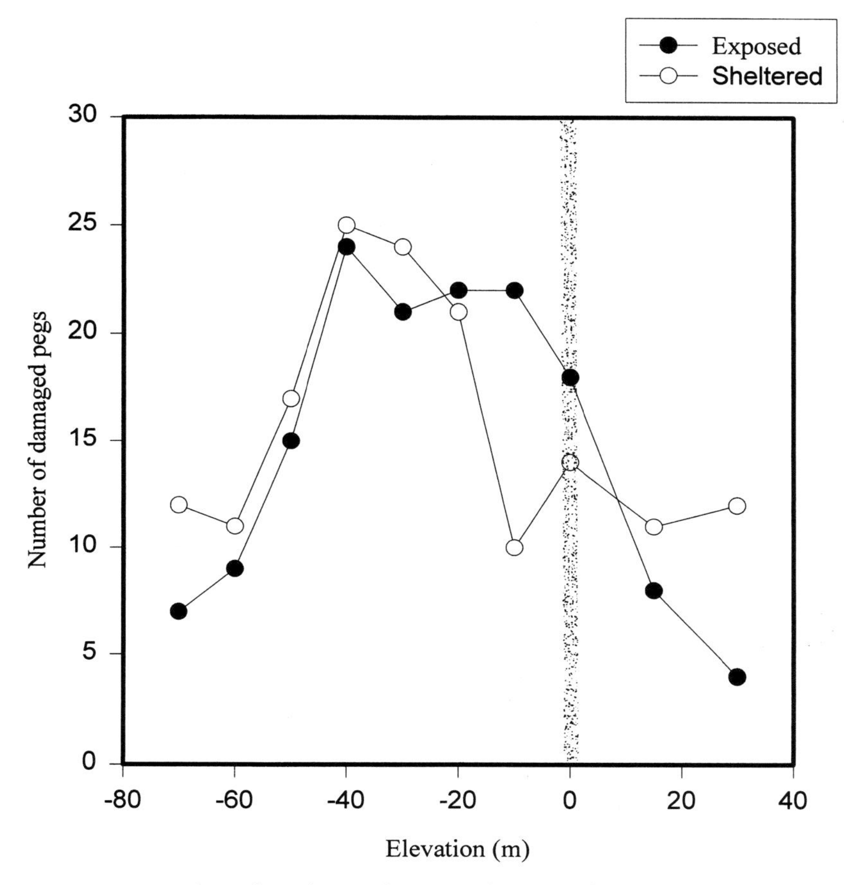

Figure 6.7 Number of ice-damaged pegs as a function of elevation on two contrasting shorelines; the stippled line shows typical late summer water levels (N= 25 1.25 cm wooden pegs, each 20 cm long, were pounded 10 cm into the ground in the summer of 1980 and damage was assessed the following spring. The study site is described in Keddy 1981).

during the freezing period can then be used to estimate the importance of ice scour on different shorelines (Rørslett 1985) (Figure 6.8).

Given the importance of ice scour on shorelines, it seems that there is much more that could be done with such simple techniques. For example, entire beds of pegs of different sizes could be used to map both the intensity and area of winter disturbance. These could be compared to known water levels during the winter. Both could be tested for their ability to predict vegetation patterns.

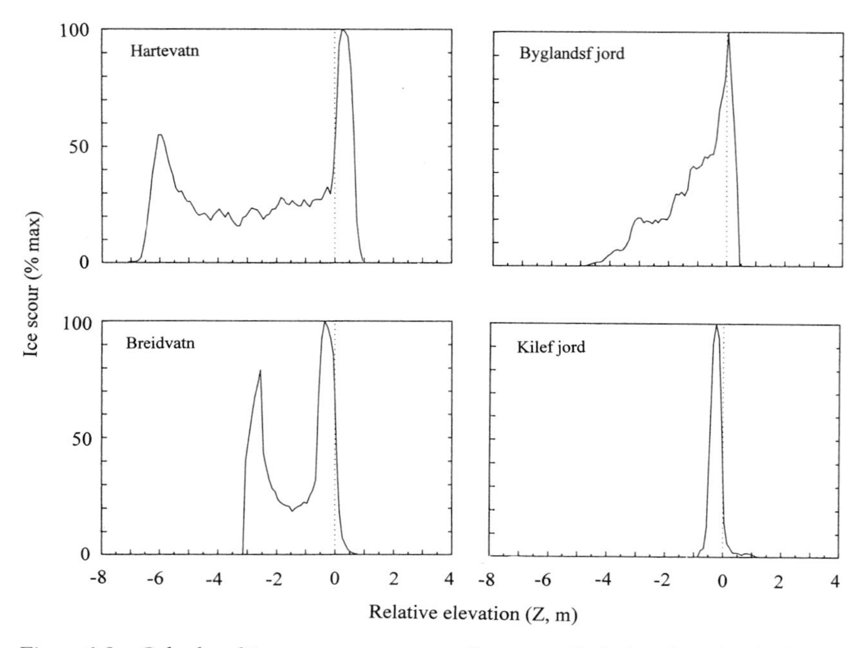

Figure 6.8 Calculated ice scour patterns as a function of relative elevation in four Norwegian lakes and reservoirs (after Rørslett 1985).

Waves

Waves are events of very short duration and high frequency. They provide an opportunity to study the effects of chronic disturbance. It has long been observed that vegetation varies with exposure to waves, and sketches such as Figure 6.9 are typical of many books on aquatic botany. The effects of chronic exposure to waves are complex. Pearsall (1920) noted that there were both direct effects (e.g. biomass removal from plants, uprooting, seed dispersal) and indirect effects (e.g. erosion of nutrients, sorting of substrates, litter transport). The indirect effects create fertility gradients, so as disturbance from waves increases, fertility decreases. Since many models (e.g. Southwood 1977; Grime 1977) assume that fertility and disturbance are the two fundamental axes that control life history evolution, the fact that both occur on shorelines, and that they are arranged in opposition to one another, makes this a particularly useful situation. In order to move from sketch maps to science, it is necessary to find ways to measure exposure gradients.

Here we are aided by waves having long had an impact on humans. Storms destroyed much of the Spanish Armada, thereby changing

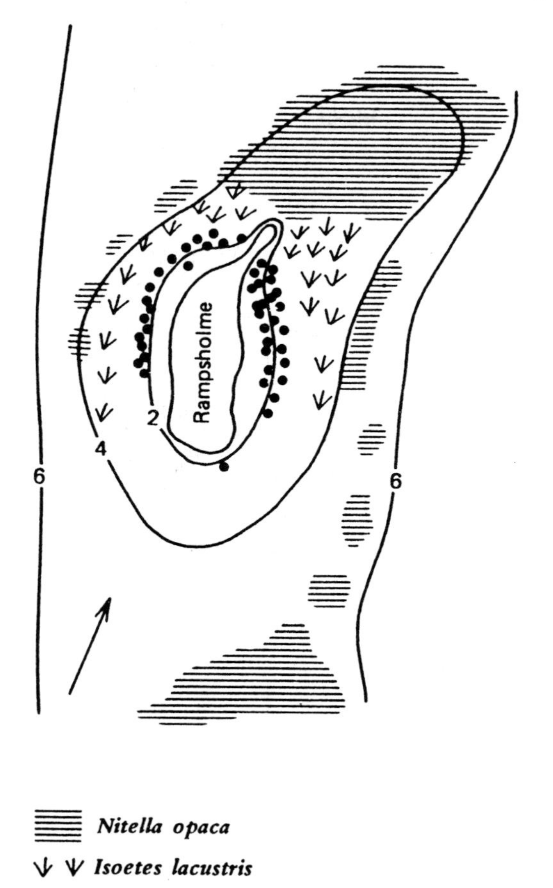

Figure 6.9 Distribution of rooted plants around a small island in Derwentwater. Arrow indicates wind direction. Note that *Nitella* grows in the lee of the island where it is assumed that sedimentation is more rapid (from Hutchinson 1975).

European history (Fernández-Armesto 1989). They also badly damaged artificial channel ports constructed for the Normandy landings in the Second World War, thereby nearly changing European history again (Blizard 1993). It is therefore natural that much of the work on waves can be found in manuals published by the US Army Corps of Engineers (e.g. US Army Coastal Engineering Research Centre 1977). Their equations have been recently adapted for use by aquatic ecologists (e.g. Keddy 1982, 1983; Weisner 1990). Using data on fetch and wind directions, one can calculate indices of wave energy expected at shorelines, and these can

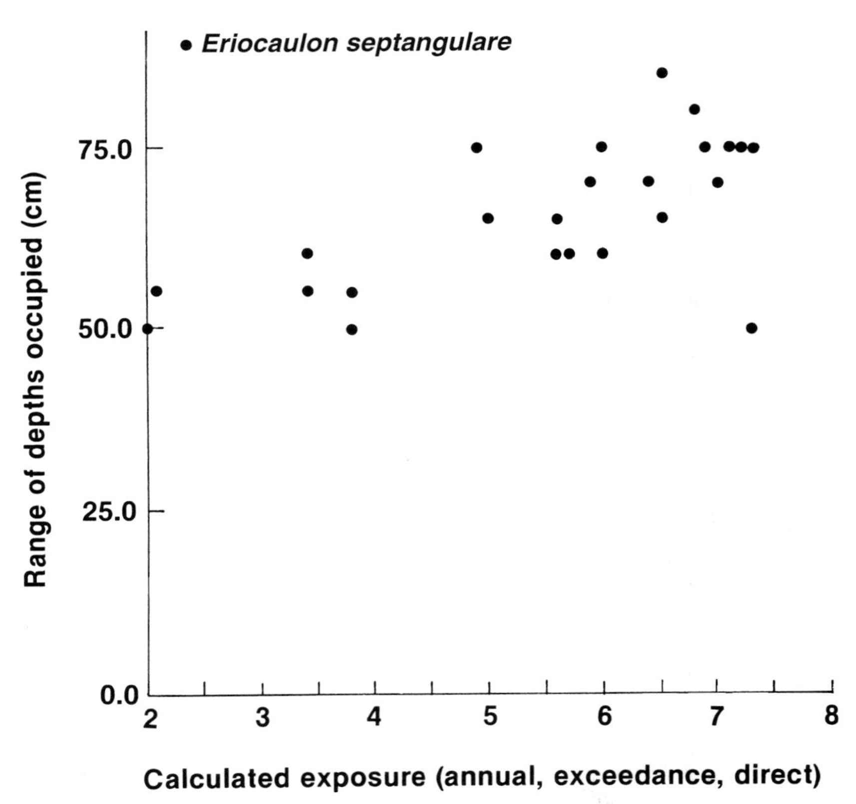

Figure 6.10 The range of elevations occupied by *Eriocaulon septangulare* increases with exposure to waves (after Keddy 1982).

be used to rank areas of shoreline in terms of relative degrees of wave energy that they experience. This allows construction of exposure gradients within lakes; they are generally capable of predicting both the proportion of silt and clay in sediments, and, in turn, the zonation patterns of common species. Figure 6.10 shows, for example, that the depth range occupied by *Eriocaulon septangulare* increases with exposure to waves. In contrast, the depth range occupied by shrubs decreases with exposure to waves (Keddy 1983). Weisner used similar methods and showed that the lower limits of emergent plants exhibit a striking pattern – as exposure increased, the lower limits of marsh plants extended into deeper water (Figure 6.11). In combination, the results of these two studies suggest that exposure to waves seems to spread out vegetation zones by simultaneously increasing both the landward and waterward (upper and lower)

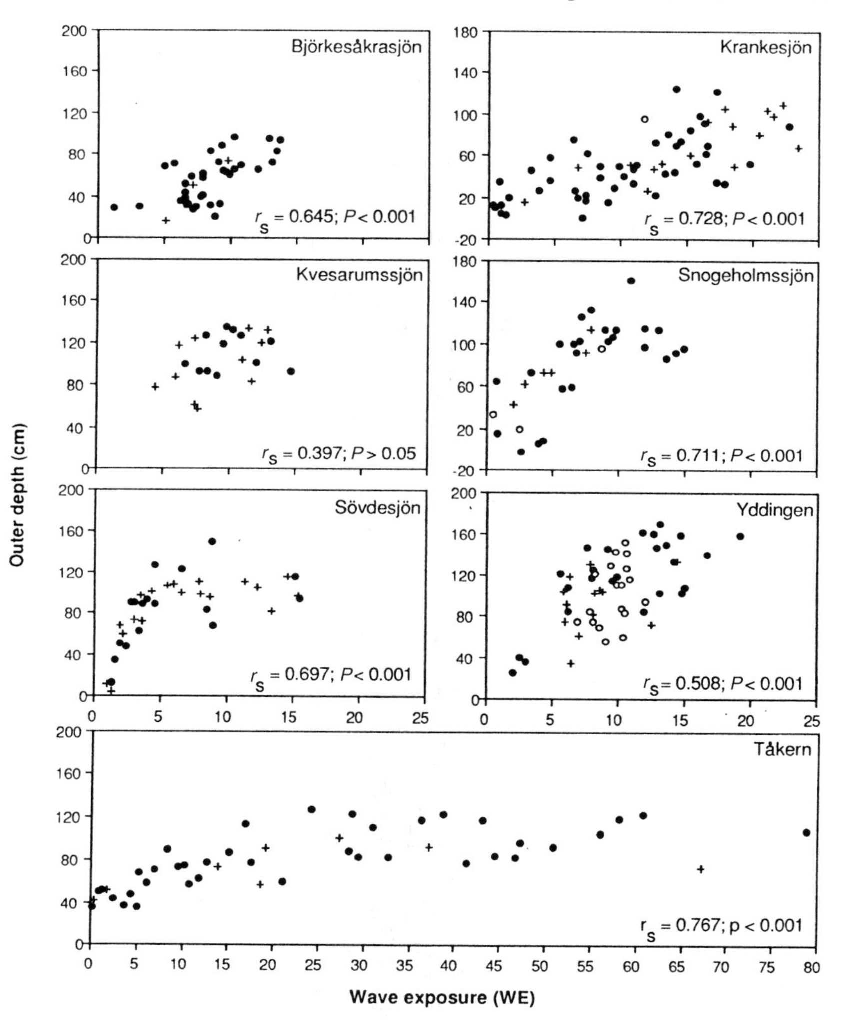

Figure 6.11 As exposure to waves increases, the outer depth of emergent plants extends into deeper water (from Weisner 1990).

limits of emergent marsh plants. Wet meadows and marshes may then be widest on shores with moderate levels of exposure to waves. Species composition also changes with exposure (Figure 6.12).

Similar processes have been reported from Australian floodplains (Roberts and Ludwig 1991). Four main vegetation types were found along the River Murray: (i) riparian grasses (*Cynodon dactylon*, *Paspalidium jubiflorum*), (ii) *Cyperus gymnocaulos* and riparian grasses, (iii) *Eucalyptus*

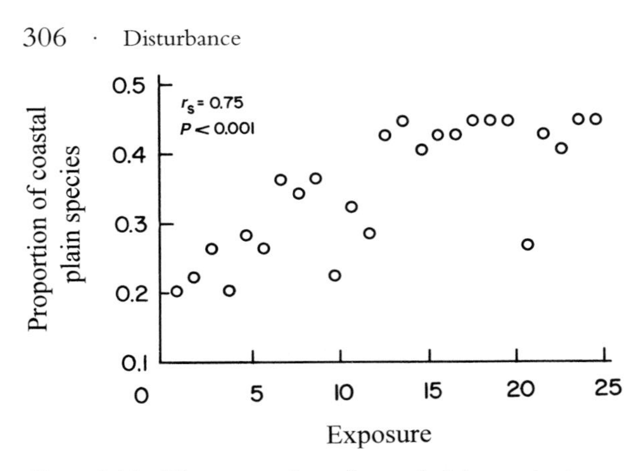

Figure 6.12 The proportion of coastal plain species in a transect as a function of exposure to waves (after Keddy 1985b).

camaldulensis (river redgum) with sedges and grasses and (iv) *Eucalyptus* with *Phragmites australis*. These vegetation types were thought to be produced by current speed and wave action. The riparian grasses, for example, occurred on steep banks with the swiftest currents.

Exposure gradients are difficult to study experimentally, although, in principle one could construct small breakwaters to protect sections of shoreline. In the absence of this, it is possible to vary individual components of the exposure gradient such as fertility or texture. It is well established that soil texture has a major effect upon germination (Harper *et al.* 1965; Harper and Benton 1966; Oomes and Elberse 1976; Vivian-Smith 1997). Given the conspicuous gradients in soil texture on shorelines, and the importance of regeneration from buried seeds (Salisbury 1970; Leck *et al.* 1989), we might anticipate that germination of marsh plants will be significantly affected by exposure. Figure 6.13 shows that most species have enhanced recruitment in fine substrates typical of bays.

Disturbance by waves and ice together can have a major effect upon adult shoreline plants. Of 840 ramets transplanted to seven sections of shoreline in Axe Lake in June 1984 (Wilson and Keddy 1988), fewer than one-third (265) were still alive in August 1985, in spite of the fact that competition was all but eliminated as a factor. Two species (*Viola lanceolata, Drosera intermedia)* were together nearly all dead, and the remaining species together had rates of mortality from 91 to 32% depending upon the exposure of the site to waves. When grown in pots in an outdoor garden, all of the species had showed a significant shared preference for organic as opposed to sandy soils (Wilson and Keddy 1985). But, under

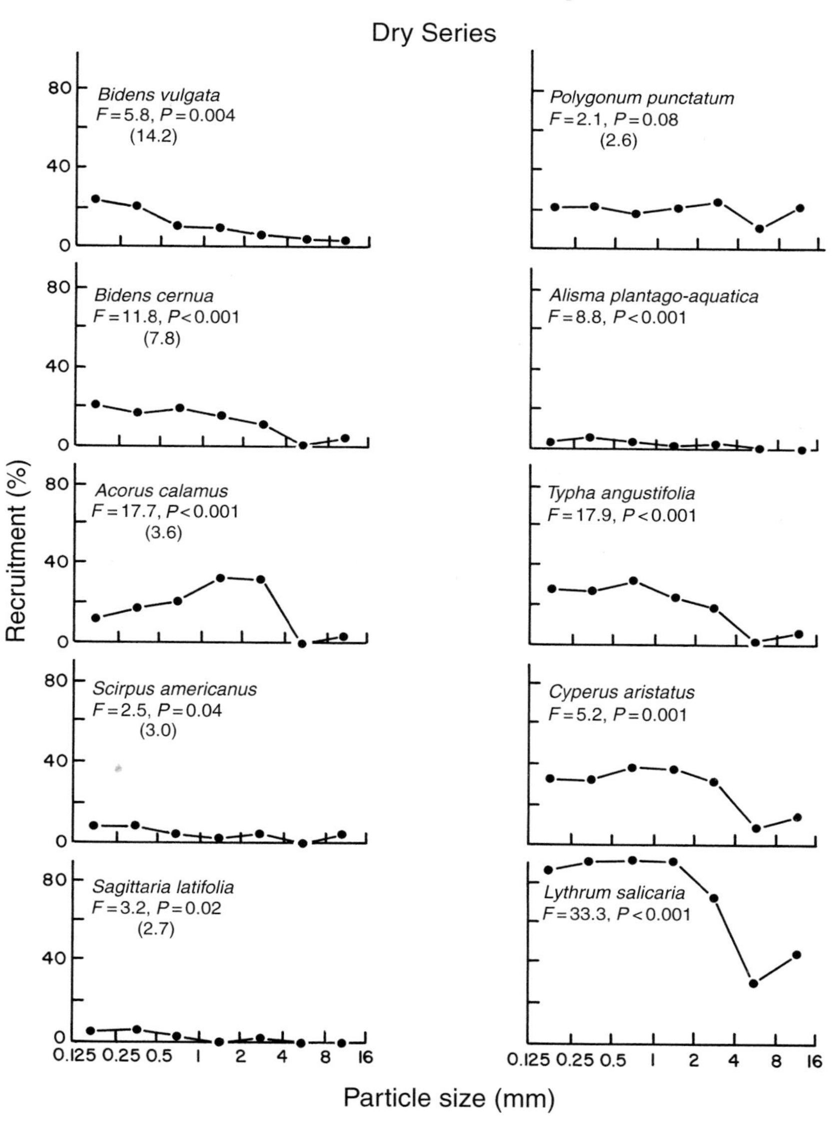

Figure 6.13 Recruitment (germination and early establishment) as a function of soil particle size for ten wetland plant species. The water table was 4 cm below the substrate surface (from Keddy and Constabel 1986).

field conditions, many of these patterns were obscured; only 3 of 12 species (*Dulichium arundinaceum, Juncus pelocarpus, Xyris difformis*) still showed significantly higher growth in organic bays. There was also a general tendency for higher survival in these organic sites. Since the effects of waves and ice may vary from year to year, depending, say, upon the timing of ice break up and the direction of winds at that time, these factors together can impose a great deal of disturbance that, while predictable in the long run, is local and patchy on a year to year basis.

Disturbance and gap dynamics in wetlands

The effects of disturbance can be studied using two different viewpoints. Many of the above examples have discussed disturbance gradients. In some situations, disturbance occurs instead as discrete patches (Sousa 1984; Pickett and White 1985). Much of the work on patch dynamics is concentrated in forests where effects of storms can create patches ranging from the size of a single fallen tree to entire stands (Urban and Shugart 1992). We have rather fewer studies of patch dynamics in wetlands, yet we might reasonably expect this process to be important. Examples might include patches burned by fire, eaten by muskrats, cut out by ice cakes, killed by floating mats of litter or buried by alluvial deposits.

Saltmarshes

The typical zonation of salt marshes in eastern North America, where there are well-developed zones, includes *Juncus gerardi* higher on the shore and *Spartina patens* lower on the shore. Bertness and Ellison (1987) report that bare space is commonly generated when floating mats of litter consisting mostly of dead *S. alterniflora* (wrack) are deposited by tides. Marsh species differ in the relative ability to tolerate cover by wrack, but if covered for more than eight weeks, they are generally killed (Bertness and Ellison 1987). Such patches are then recolonized in several ways. *Salicornia europea* invades by means of seeds. *Distichilis spicata* invades with adventitious runners. Eventually the patches are reinvaded by *Juncus gerardi* and *Spartina patens*. There is thus a continual process of patch creation and patch recolonization (Bertness 1991). Unlike freshwater marshes, seed banks appear to play a minor role in the re-establishment of vegetation; instead, vegetative expansion from adjoining plants predominates (Bertness and Ellison 1987; Hartman 1988; Allison 1995). To explore these dynamics, Bertness made artificial bare patches of three sizes

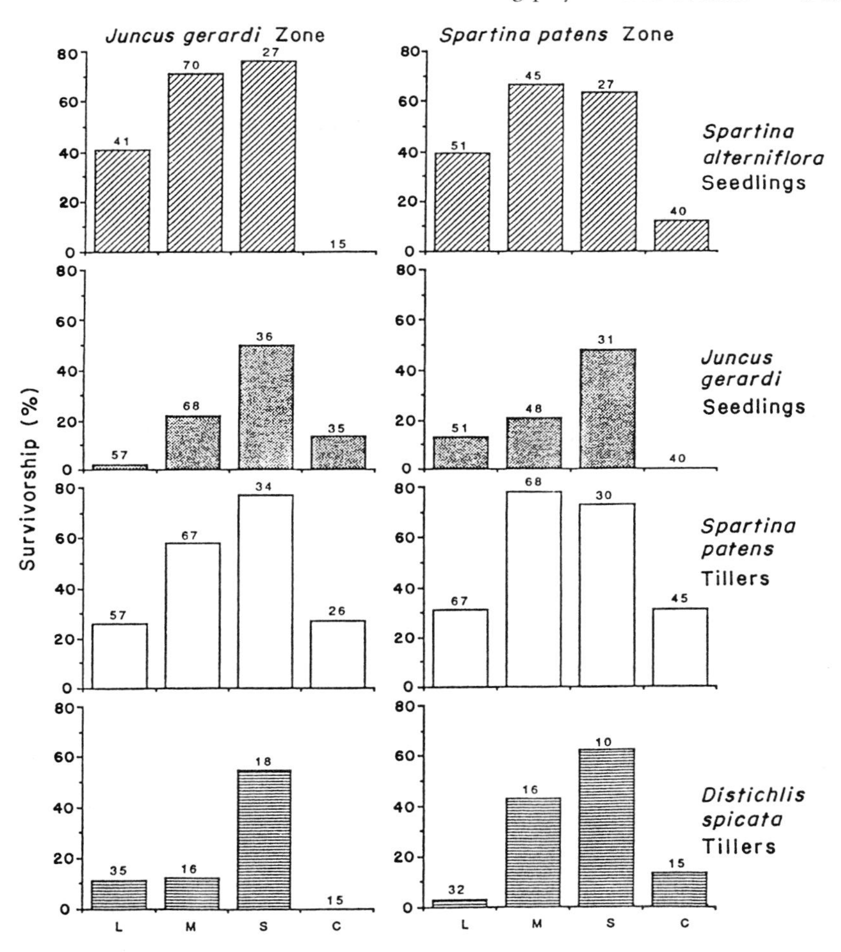

Figure 6.14 Survivorship of four saltmarsh plants in three different gap sizes (L = large, 1 m^2; M = medium, 0.5 m^2; small, 0.25 m^2; c = control, dense vegetation (from Bertness 1991).

ranging from 0.06 to $1m^2$. He then transplanted seedlings and tillers of four species into these patches and recorded their survival (Figure 6.14). He reports that larger patches (left) had higher salinity than the surrounding vegetation, and that survivorship therefore increased as patch size/salinity decreased.

Transplants into existing vegetation (Fig. 6.14 right) also had very low survival in spite of low salinity in the surrounding vegetation, showing that competition from existing vegetation also prevents the establishment of transplants (see Chapter 7). Apparently competition is temporarily reduced in gaps and the important factors then become relative colonization rates and relative tolerances to salinity. The relative amounts of *Salicornia europea, Distichilis spicata*, and *Juncus gerardi* or *Spartina patens* therefore depend upon the frequency of disturbance in these marshes. As the proportion of patches increases, the relative abundance of *Salicornia* and *Distichilis* increases.

Zedler and Beare (1986) describe a quite different process in salt marshes in south western North America, where Mediterranean-type rainfall patterns (wet winters, dry summers) occur. In addition to these seasons, there are substantial year-to-year differences in streamflow, and rare events appear to be particularly important in both destroying vegetation and allowing establishment of new plants. During dry years, hypersaline conditions develop; marsh species such as *Spartina foliosa* and *Typha domingensis* cannot tolerate these conditions and are slowly replaced by salt-tolerant species such as *Salicornia virginica.* This process is reversed during abnormally wet years, when higher streamflows and longer rainfalls flush accumulated salt from the soil. This creates a low salinity gap in which seedlings can germinate and establish. If the gap is short (3 – 6 weeks), only halophytic species such as *Spartina foliosa* can establish, but if it is extended, brackish and freshwater marsh species can establish as well. The duration and intensity of the next hypersaline period will then determine which of these species survive. There is thus a constant cycling through different vegetation types driven by changes in moisture supply (Figure 6.15).

Alluvial deposition

Patches are also formed where rivers deposit sediments, and an entire later chapter will explore the effects of burial on wetlands. However, let us consider just one example here, an example (from Schnitzler 1995), where newly deposited alluvial sediments provide a raw substrate for colonization and succession. Although the Rhine River has now been markedly manipulated with dams, dykes and channelization, there were once extensive alluvial forests associated with 2–3 m water level fluctuations. The forests were of two main types, a 'softwood' type (Alder/ Poplar/ Willow) and a 'hardwood' type (Maple/ Ash/ Oak/ Elm). Newly deposited alluvial sediments are generally colonized by the former 'softwood' group, but after a period of 20–30 years, the 'hardwood' species begin to establish. The human changes to the Rhine make

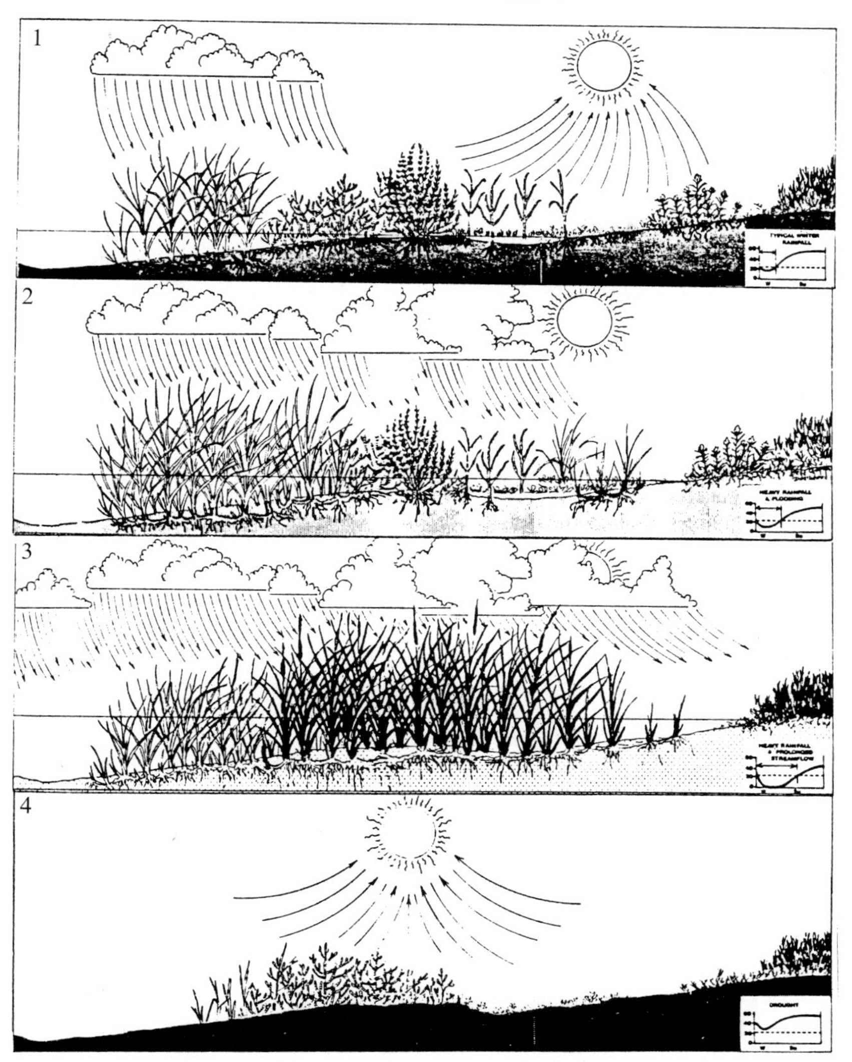

Figure 6.15 Cyclical changes of saltmarsh vegetation in arid climates. 1. Typical situation where brief periods of low salinity allow saltmarsh species to germinate and establish. 2. Floods reduce salinity and allow expansion of *Spartina foliosa*. 3. Prolonged flooding eliminates saltmarsh vegetation and allows brackish marsh species to establish. 4. Periods without rainfall or flooding create hypersaline conditions which kill all but a few highly salt-tolerant species such as *Salicornia virginica* (from Zedler and Beare 1986).

it difficult to sort out the natural processes that would once have predominated; there are, for example, two distinctive classes of trees (120–150 years, 30–40 years) that correspond to the two eras of canalization.

The overall diversity of plant communities along the Rhine can apparently be explained by two factors. First, there is the continual creation of new patches of alluvial deposits and successional development through time. From this perspective, the Rhine forests, like most alluvial forests (Mitsch and Gosselink 1986, Chap. 14; Salo *et al.* 1986), are a mosaic of different aged stands of forest on alluvial deposits. A second source of variation is superimposed upon this dynamic landscape, because all patches are not the same. Depending upon the habitat in the patch, a different community develops. For example, along the Rhine, areas of wet clay may remain as habitats for willow and poplar, whereas areas of drier gravels will develop stands of oak and elm. Much of the richness of alluvial plant communities can be explained by different kinds of gaps allowing different kinds of plant species to regenerate (Grubb 1977; Salo *et al.* 1986).

Freshwater marshes

In freshwater marshes, patches can be formed by flooding, fire or herbivores (e.g. van der Valk 1981; Weller 1978, 1994a; Ball and Nudds 1989). Figure 6.1 shows that as little as three years of flooding can kill stands of emergent plants. Such processes create a mosaic of different vegetation types; the simplest example may be dense stands of cattails interspersed with patches of open water (Figure 6.16). Experimental studies have shown that breeding ducks select a 1:1 ratio of these two patch types (e.g. Kaminski and Prince 1981).

Since many shallow water marshes are slowly dominated by cattails, mosaics can also be created by human manipulation. Ball and Nudds (1989) describe an experiment in which circular patches of 0.02, 0.09 and 0.15 ha were either cut or burned into cattail stands along Lake St Clair in Canada. They then measured food availability for water fowl by sampling the aquatic invertebrates. The mowed patches had higher invertebrate availability than the burned patches (Figure 6.17), but there was no detectable effect of patch size on invertebrate availability. They conclude that, if the objective is to increase food supplies for ducks, mowing is superior to burning; not only does it produce more invertebrates, but the clearings last longer.

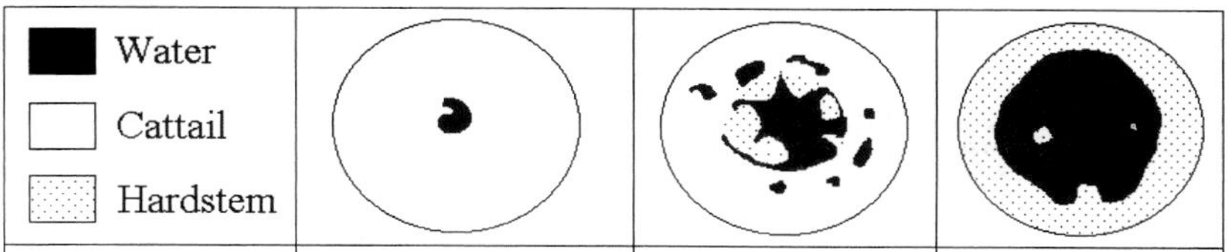

Water / Cattail / Hardstem			
Water depth	Shallow	Medium	Deep
Vegetation	Dense	Moderate	Sparse
Size of bird populations	Medium	Large	Small
Bird species richness	Low	High	Low
Number of muskrats	Few	Many	Few

Figure 6.16 Gaps create habitat interspersion in freshwater marshes (after Weller 1994a).

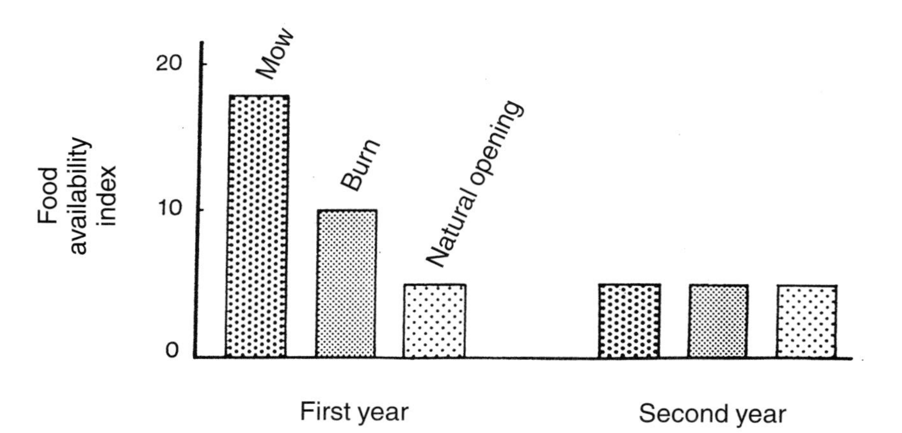

Figure 6.17 Burning and mowing can change the abundance of invertebrates in marshes (from Ball and Nudds 1989).

Szalay and Resh (1997) also experimentally manipulated fire and mowing to compare their effects upon plants and invertebrates. The study site was a 3500 ha brackish marsh in California, where over 100 000 dabbling ducks may over-winter. The dominant plant here is the grass, *Distichlis spicata*; stands of this plant were either subjected to hand mowing or burning in late summer, and then flooded. Flooding immediately after burning or mowing can eliminate *Distichlis* (Smith and Kadlec 1985b); here flooding was delayed until some weeks after mowing. During the following winter, invertebrate biomass was sampled. The dominant macro-invertebrates were a *Chironomus* larva (Diptera) and a water boatman, *Trichocorixa* (Hemiptera). Copepods were the dominant micro-invertebrate. The macro-invertebrates are known to be important food for dabbling ducks, and some ducks (e.g. Northern Shoveler) also feed heavily upon copepods. Burning increased the abundance of *Chironomus* and *Trichocorixa* by roughly a factor of ten relative to controls, but mowing did not have significant effects. Copepods were less abundant in the clearing created by burning. Invertebrate populations are thus sensitive to perturbations such as fire and mowing, and the timing, area and intensity of disturbance can affect their relative abundance.

White (1994) has synthesized the many kinds of disturbance active in the Everglades. This area has 'a great deal of structural and compositional variation, from open water sloughs with sparse macrophytes to sedge- and grass-dominated freshwater marshes, open pine stands and dense broad-leaved evergreen forest'. This variation arises because of a habitat template (see pp. 10–14) created by an elevation gradient. Superimposed upon this are nine physical driving forces, ranging from those that show gradual change (e.g. climate change, sea level rise, for which change is measured on time scales greater than 10^2 years) to natural disturbances (e.g. flood, drought, fire, storms, freezing temperatures, for which change is measured on scales of less than 10^2 years). These latter natural disturbances tend to be short-lived events, but the communities recover from them slowly. Hence there is a basic asymmetry: disturbance is fast, recovery is slow. These lags on recovery mean that periodic disturbances can generate a mosaic representing a particular degree of recovery from the last disturbance. The rates of recovery will depend upon the amount of vegetation (if any) that persists through the disturbance, the influx of new propagules from adjoining areas, and the productivity of the site.

In the Everglades, 11 of the 15 plant community types depend upon two key factors: hydrology and fire. Both of these key factors are in turn connected to relative elevation. In general, as peat accumulated, there is a

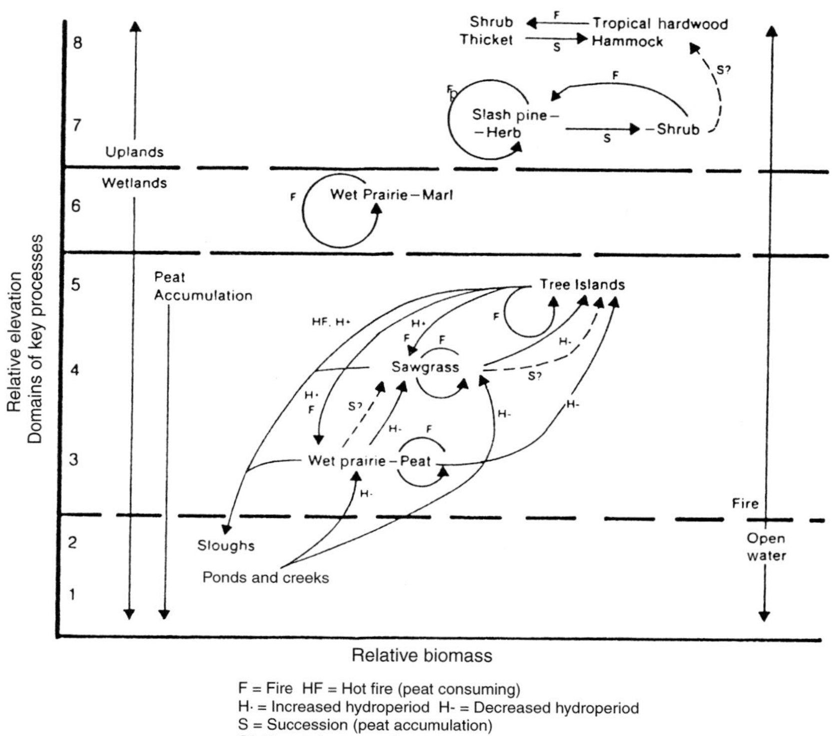

Figure 6.18 Plant communities of the central freshwater Everglades and the processes that produce them (from White 1994).

succession from sloughs to tree islands (the main sequence in Figure 6.18). Light fires will create patches in the vegetation, but severe fires can consume peat, thereby lowering the relative elevation of the site and returning it to a much earlier successional stage. That is, the more intense the fire, the longer the recovery time. Intense fire, and alligators, are apparently the only processes that can actually increase the hydroperiod of a site.

During the past 20 years, approximately 25% of wet prairie and slough has been replaced by stands of sawgrass, probably as a consequence of reduced flooding and decreased fire frequency. Wet prairies and sloughs have higher plant diversity, are major sites of periphyton production, and important habitats for crustaceans and fish. Drainage and fire control therefore has not only changed the vegetation, but the capacity of the area to produce and support other organisms. Restoration of the

Everglades will require restoration of flooding and fire as natural disturbances.

Measuring the effects of disturbance

If we are going to better understand the effects of disturbance on wetland communities as a whole and wetlands in particular, we must more precisely define, and then measure, the relative effects of disturbance upon community properties. Here is an example. Moore (1990, 1998) artificially created bare patches in five different riverine wetland habitats ranging from exposed sandy shorelines to sheltered organic bays. At each of the five sites, $1 m^2$ bare plots were created and the vegetation in them repeatedly compared with undisturbed controls over two growing seasons. There were two questions: (i) did the measured disturbance effects change among the particular ecological properties measured? and (ii) did the effects vary among the five wetland types? The properties measured included both community level properties (e.g. biomass, evenness) and lower order properties (the abundance of selected species). Moore found that a single growing season was sufficient for community level properties such as biomass, richness and evenness to return to control levels. The dominants removed at each site tended to remain depressed for the first growing season, although by the second year effects were negligible. At the guild level, recovery was also rapid, although there were minor changes, such as a modest increase in facultative annuals. The species level of organization tended to be the most sensitive to disturbance. Overall, it appeared that removing above ground biomass had a marginal effect on this vegetation type; this may not be a great surprise, given the dynamic nature of riverine wetlands.

Moore also tested whether removal effects varied among the five wetland vegetation types by measuring the magnitude of removal effects for each ecological property and wetland site. This is essentially the measure of disturbance proposed in the opening of this chapter, and is therefore worth examining more closely. This measure of removal effects for each property was $Z = (x_0 \star y_t) / (x_t \star y_0)$ where x_0 is the mean value for the property in the control sites during the pre-treatment survey, x_t is the mean value in the treatment sites during the post-treatment survey, y_0 is the mean value measured in the disturbance treatment during the pre-treatment survey and y_t is the mean value measured in the removal treatment during the post-treatment survey (see Ravera 1989). The value is thus independent of initial levels of the properties, and is independent of

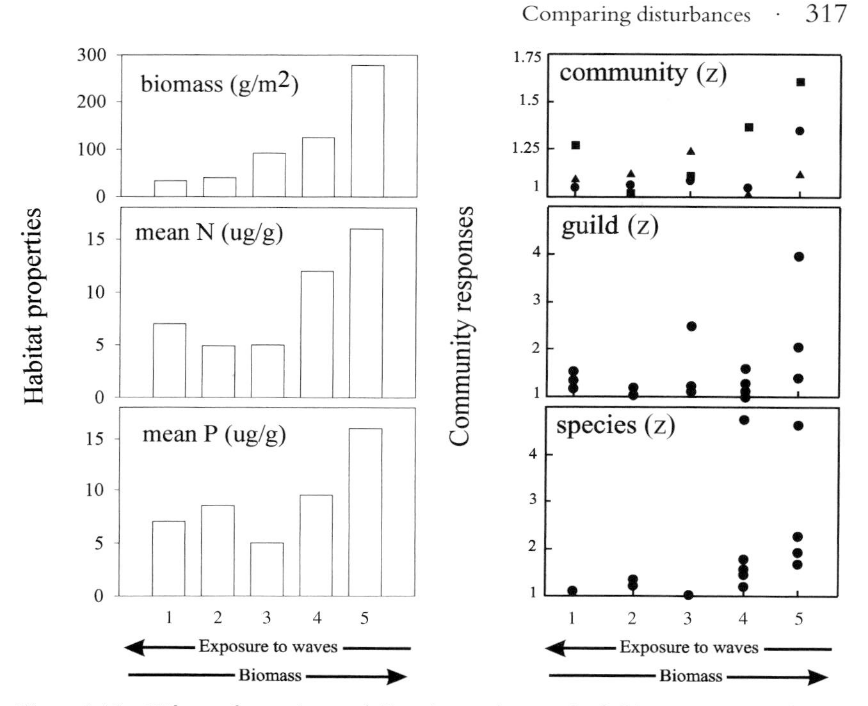

Figure 6.19 Effects of experimental disturbance (removal of all biomass) upon five different wetland communities of increasing biomass and fertility (*left*). Z is a measure of departure from control plot values, and is scaled so all change is greater than 1. The greater Z, the greater the departure from control values. Effects had largely disappeared by year two, and so are not included in the figure (after Moore 1998).

ongoing temporal trends in the community. A Z value of 1.0 indicates no treatment effects, while values above or below 1 indicate increase or decrease. Figure 6.19 shows that, in each case, the experimental disturbance had the greatest effects in sheltered bays. That is, sites with higher biomass and higher fertility tended to show the most response to disturbance. Perhaps this is because these are the riverine communities where disturbance is normally most infrequent.

Comparing disturbances

It is not my intention to catalogue and describe every disturbance that can occur in a wetland. What I have done so far is establish a basic terminology, and list some concrete examples. Two examples (water level fluctuations (Chapter 4) and herbivores (Chapter 8)) are important

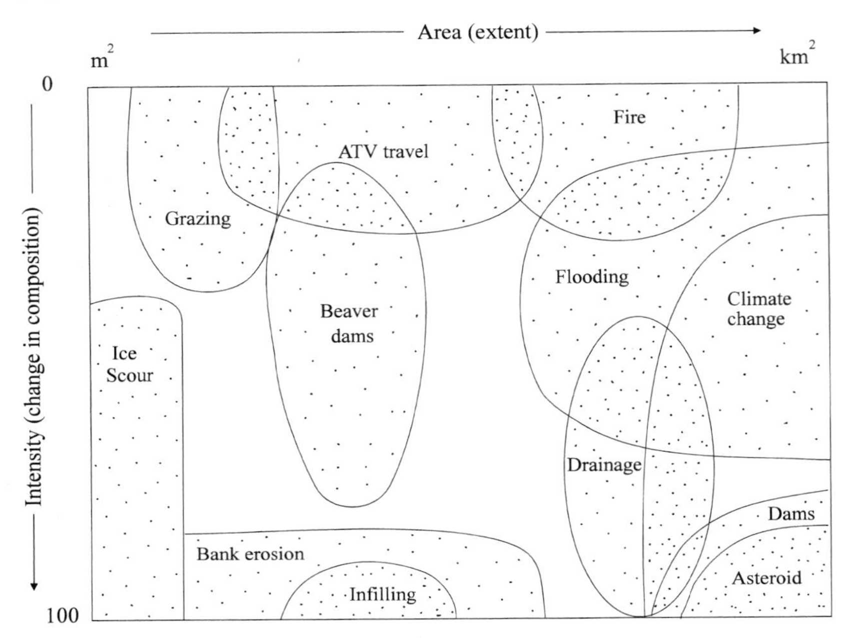

Figure 6.20 Intensity and area plotted for an array of natural disturbances in wetlands.

enough that they are given entire chapters. Let us close by examining the relative scale of different types of disturbance. To develop rigorous ecological theories, we must have ways of comparing very different processes. Comparing among sites and processes has been a theme throughout this book. Figure 6.20 compares a wide array of disturbances according to two of the foregoing measurable properties: intensity and area. At the upper left are disturbances that are generally of low intensity in small patches, such as grazing. Ice scour is a disturbance that still may occur in small patches, but with much higher intensity. Atmospheric deposition of pollutants (not shown) may produce only small changes in composition, but these may cover very large areas. Finally, there are effects of huge hydroelectric dams and asteroids which have high intensity and affect large areas. Most of the shaded regions are just best guesses, and I would welcome serious attempts to better define regions within this diagram.

7 · *Competition*

Introduction

Up to this point the theme has been largely the environmental factors that control the structure and function of wetland communities. The action of these controlling factors is most easily seen in the distribution of particular wetland organisms. For example, the presence of trees indicates the range of flooding in a particular wetland community. If the trees are Tamarack (*Larix laricina*), we may assume we have a peatland; if the trees are Silver Maple (*Acer saccharinum*), we may assume an infertile flood plain subjected to extremes of spring flooding; if the trees are mangroves (*Avicennia* sp.), we may reasonably assume tidal fluctuations and elevated salinity. Similarly, the occurrence of Bull Frogs as opposed to Chorus Frogs, or Sedge Wrens as opposed to Blue Winged Teal, allows the informed ecologist to draw inferences about the physical factors forming particular wetlands. There are limitations to these inferences, however; not every organism may be found in its preferred habitat. Some may be present only because they have been excluded from other sites by competing organisms. The fact that the distribution of an organism is correlated with certain nutrient levels, or certain water levels, by no means proves that the distribution is directly controlled by physiological responses to those factors. In fact, as we have already seen with zonation (Chapter 2), a species' field distribution may be only weakly controlled by physiological tolerance limits.

Let us therefore turn to a major biological factor, competition. This interaction between life forms may also control biological structure of wetlands. First we need a definition. Almost every textbook has its own. I prefer operational definitions and so will define competition as: *the negative effects that one organism has upon another by consuming, or controlling access to, a resource that is limited in availability.* This follows from Keddy (1989a), except for a slight improvement in grammar.

Table 7.1. *Major elements required by living organisms and their functions*

Elements	Function
C	Structure; energy storage in lipids and carbohydrates
H	Structure; energy storage in lipids and carbohydrates
N	Structure of proteins and nucleic acids
O	Structure; aerobic respiration for energy release
P	Structure of nucleic acids and skeletons; energy transfer within cells
S	Structure of proteins

Source: From Keddy (1989a) after Morowitz (1968).

Now consider these fundamentals. All living organisms require a rather limited number of elements to make up their bodies (Table 7.1). Most living organisms have relatively similar elemental compositions (Table 5.1). We may reasonably suspect that some organisms will accumulate these resources at the expense of others thereby reducing growth, survival or reproduction of their neighbours. If so, competition occurs.

Testing for competition

Is there any evidence that competition actually controls plant or animal distributions in wetlands? If not, we may safely move to another more profitable topic. (We have already seen a few examples of competition in vertical zonation, so to this extent we already suspect that some competition may be present in the field.) Further Sculthorpe (1967), who was a careful observer of the distinctive ecology of the monocotyledonae in wetlands, observed:

> Although . . . compositional changes may occur, it is apparent that numerous . . . reed-swamp plants tend to form extensive pure stands. These species assert their status early and attain a seasonal or permanent predominance. Of the numerous factors responsible, rates of vegetative reproduction and antagonism between species of similar or different life form are perhaps the most important. Vigorous vegetative spread, by means of rhizomes, stolons, and tubers, is a typical attribute of several reed-swamp dominants, notably species of *Carex*, *Cyperus*, *Glyceria*, *Phalaris*, *Phragmites*, *Schoenoplectus [Scirpus]*, *Typha* and *Vossia*. In a favourable site one species may gain an early initiative and increase much faster than any competitor . . . most mature reed-swamps are so dense that they resist infiltration by larger free-floating rosettes and severely reduce the amount of light reaching the water, thus indirectly inhibiting the growth of invading submerged species. (p. 426–427)

The most obvious test for the presence of competition is to remove one species, and test whether others respond significantly by changing either their distribution or abundance. Gopal and Goel (1993) describe a rich array of such studies documenting competition among wetland plants. My objective here is to present fewer examples organized to illustrate how competition produces the patterns that we see.

Let us begin with Silander and Antonovics (1982), who carried out such an experiment in south eastern North America. Certain plants were removed from wetlands, and the response, if any, of other neighbouring species measured. Their study included both low marsh and high marsh. In the high marsh, *Spartina patens* was removed, and in the low marsh, both *S. alterniflora* and *S. patens* were removed. The upper left panel in Figure 7.1 shows how to interpret such studies; what we are interested in measuring is the increase, if any, into the space that the test species normally occupied. Surprisingly, only one species, *Fimbrystylis spadiceae*, showed a significant response to the removal of neighbours! Apparently, competition was a very weak force here.

One problem with trying to examine so many pairs of interactions is that many non-significant responses may be observed simply because of small sample sizes combined with high levels of natural variation. Bertness (1991) examined fewer species and used both removals and transplants. In the transplant work, he moved monoculture turfs of two species and then measured their performance with and without neighbours. He found that neighbours had significant effects on performance (Figure 7.2). These effects were generally more dramatic than those reported by Silander and Antonovics. In general, competition severely reduced performance in the transplanted species, did so in both zones, and the effects increased with time. The one exception was *Juncus gerardi*, which performed well even when transplanted into other species. Bertness concluded that *J. gerardi* is competitively dominant to both *Spartina patens* and *Distichilis spicata*. This idea of competitive dominants and subordinates is one we will return to shortly.

The effects of floating leafed aquatics in wetlands are of particular interest because the competition they provide is inherently asymmetric; that is, floating plants can shade submersed ones, but the reverse cannot occur (Keddy 1976). Catling *et al.* (1988) studied the effects of a floating aquatic (*Hydrocharis morsus-ranae*, frog bit), a rapidly spreading exotic in eastern North America. They anchored 70 circular floating hoops 1m^2 in surface area in each of two study sites in eastern North America. Half the hoops had *Hydrocharis* added to a cover of 65%; it was removed in the

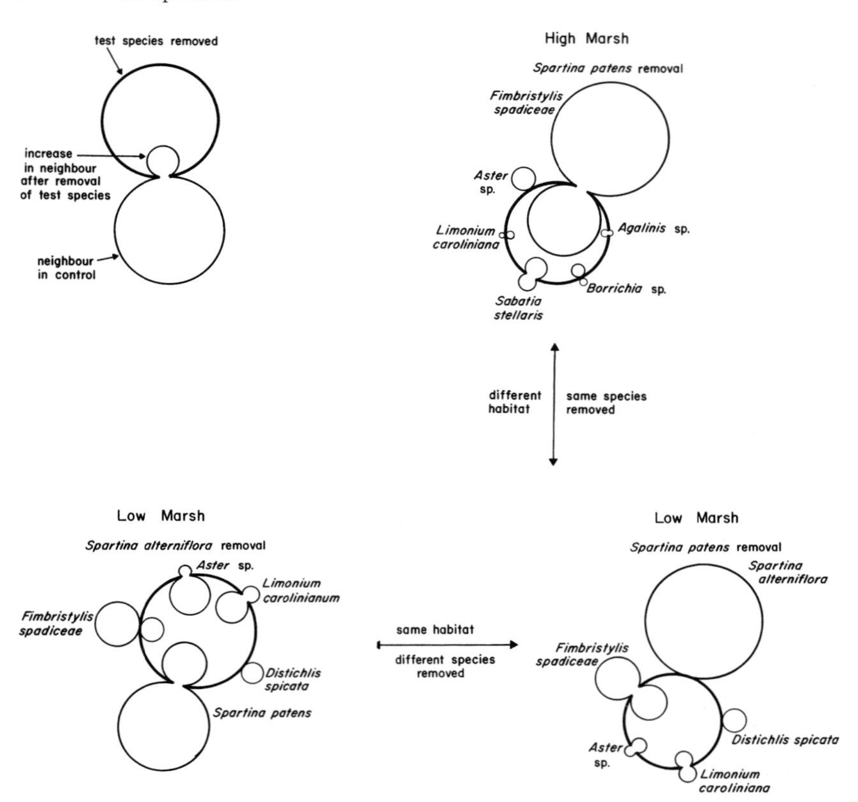

Figure 7.1 The results of a competitive release experiment in a community of coastal wetland plants. The panel in the upper left shows how to interpret the remaining figures; circle size is a measure of abundance either before or after the removals. Only the response of *Fimbrystylis* is statistically significant (from Keddy 1989a after Silander and Antonovics 1982).

other half. Thus this experiment both increased and decreased the abundance of the putative competitor. After one growing season, the cover of aquatic macrophytes was 72% in the exclusion plots compared with only 4% in the plots with *Hydrocharis* added. Species declining significantly with increasing *Hydrocharis* included *Elodea canadensis*, *Myriophyllum heterophyllum*, *Potamogeton pusillus*, *P. nodosus*, *P. zosteriformis*, *Sparganium eurycarpum* and *Utricularia vulgaris*. The effects on higher trophic levels are not known, although Catling *et al.* also suggest that snails, crustaceans and insect larvae appeared to be more numerous in beds of *Potamogeton pusillus* than in beds of *Hydrocharis morsus-ranae*.

Now let us consider two vertebrate examples. Further, instead of

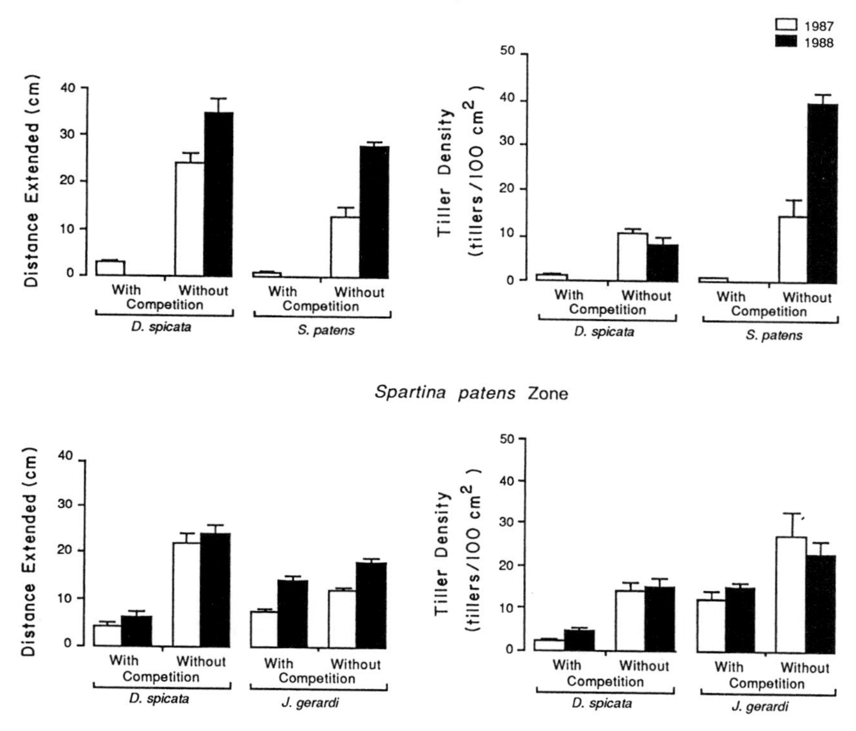

Figure 7.2 Performance of two salt marsh grasses with and without competition in two different vegetation zones (from Bertness 1991).

removal experiments, let us look at two examples where other kinds of manipulations were performed. Instead of removing selected species from existing communities, one can create artificial communities with different pairs of species.

Wilbur (1972), for example, explored competition in a community of North American amphibians which breed in temporary ponds in the spring. There were three species of mole salamanders (*Ambystoma* spp.); they return to water to breed, but normally lead terrestrial lives as adults. They are part of a larger amphibian community including Tiger Salamanders, American Toads, Grey Tree Frogs and Wood Frogs. Wilbur placed cages in the shallow water, and inoculated each cage with different numbers and kinds of amphibian eggs. The cages had one, two or three species. At the end of the summer, Wilbur measured three dependent variables: survivorship, body weight and length of time of the larval

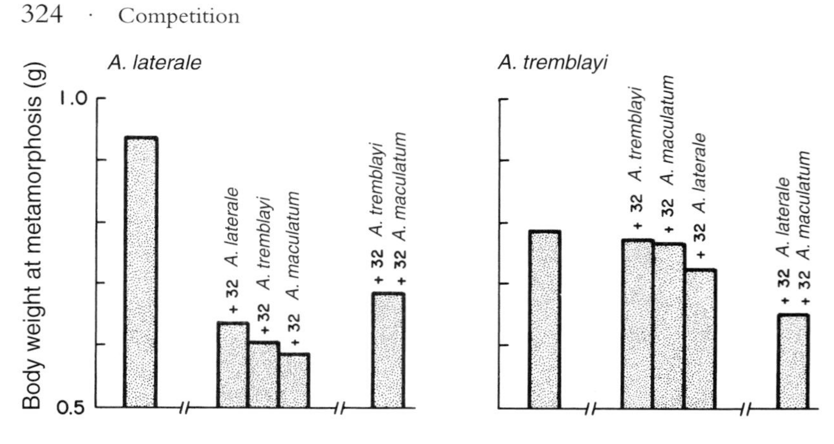

Figure 7.3 Body weight at metamorphosis in the salamanders *Ambystoma laterale* and *A. tremblayi* when grown in pens containing 32 individuals (control), and with experimental additions of either more conspecifics or two other salamander species. There is no evidence that intraspecific competition is greater than interspecific competition (data from Wilbur 1972).

period for all survivors. There was intense interspecific competition (Figure 7.3). For example, *A. laterale* (left) had body weights reduced by nearly two-thirds when 32 neighbours were added. Further, there was asymmetry; *A. laterale* (left) was far more sensitive to *A. tremblayi* (right) than *A. tremblayi* was to *A. laterale*.

Cetrachid sunfishes are a group of spiny-rayed freshwater fishes that dominate the fish faunas of small lakes over much of central North America. In Michigan, for example, there are 7–10 species, five of which are in the genus *Lepomis*. There are three principal habitat types: emergent vegetation, open water and near the substrate. The pumpkinseed (*L. gibbosus*) feeds near the substrate whereas the bluegill (*L. macrochirus*) is found higher in the water column, with further separation among species based upon food size. The green sunfish (*L. cyanellus*) is restricted to the shallow and vegetated inshore habitats (Werner 1984; Wootton 1990). In order to study this competition, fish were introduced into small experimental ponds, where habitat and diet could be measured for different mixtures of species (e.g. Werner and Hall 1976, 1979).

> Each species, when stocked in ponds alone, preferred the vegetation habitat where larger prey are found and evidently higher foraging rates are possible. In the presence of congeners, however, the bluegill and pumpkinseed underwent dramatic niche shifts to the plankton and sediment habitats respectively (Werner 1984, p. 365).

The green sunfish remained in the preferred vegetated habitat, the wetland vegetation providing the most profitable prey. Thus, to put it in other words, these fish had a shared preference for the wetland habitat, that is, the habitat providing the greatest rate of return per unit time spent foraging. The apparent resource partitioning into different habitats was actually the result of competitive displacement of subordinate species by the dominant species, that is asymmetric competition and inclusive niches (Keddy 1989a, b). Measures of niche overlap based upon diet or habitat (e.g. Wootton 1990) would therefore have been quite misleading descriptors of the kind of competition occurring among these fish.

Both Bertness and Catling found asymmetric interactions among wetland plants. Bertness found that *Juncus gerardi* was dominant over the other species. Similarly, it is obvious that floating plants can shade submersed plants, but that the reverse cannot occur. To what extent might such asymmetry be a general property of plant competitive interactions? This was one of the central questions posed in an experiment in which seedlings of twenty freshwater marsh and wet meadow plants were grown in pairs (Keddy *et al.* 1994). Each paired interaction was measured, and then a measure of asymmetry was calculated. The median of all interactions was strongly asymmetric. Moreover, the asymmetry levels were higher in flooded and fertilized conditions (Figure 7.4).

We have seen that marshes are often dominated by plants with deeply buried rhizomes, including, for example, *Papyrus* marshes on the Nile, reed marshes in Europe, cattail marshes in North America, and temperate marshes in China (Goodall 1983; Denny 1985; Whigham *et al.* 1992; Hou 1983). Most of these plants are also in the monocotyledonae, with the vast majority in the Cyperaceae (e.g. *Carex*, *Scirpus*, *Eleocharis*, *Cyperus*, *Cladium*, *Rhynchospora*) or the Poaceae (*Glyceria*, *Phragmites*, *Phalaris*, *Panicum*). Other families include the Typhaceae (*Typha* spp.), Sparganiaceae (*Sparganium* spp.) and Juncaceae (*Juncus* spp.). What intensity of competition is exerted by these stands, and does it vary according to the species comprising them? In order to answer this question, Keddy *et al.* (1998) created pure stands of seven marsh perennials, and then, after 3 years, introduced 48 other species of wetland plants for 4 months. The performance of these 48 species was calculated as $100(x_1-x_2)/x_1$, where x_1 was their weight grown alone and x_2 was the weight grown in the swards. Overall, the established species reduced growth and survival by more than one-quarter (Figure 7.5).

If organisms have different competitive abilities, we may ask what attributes they possess that might allow us to predict their competitive performance relative to other species. Gaudet and Keddy (1988) approached this

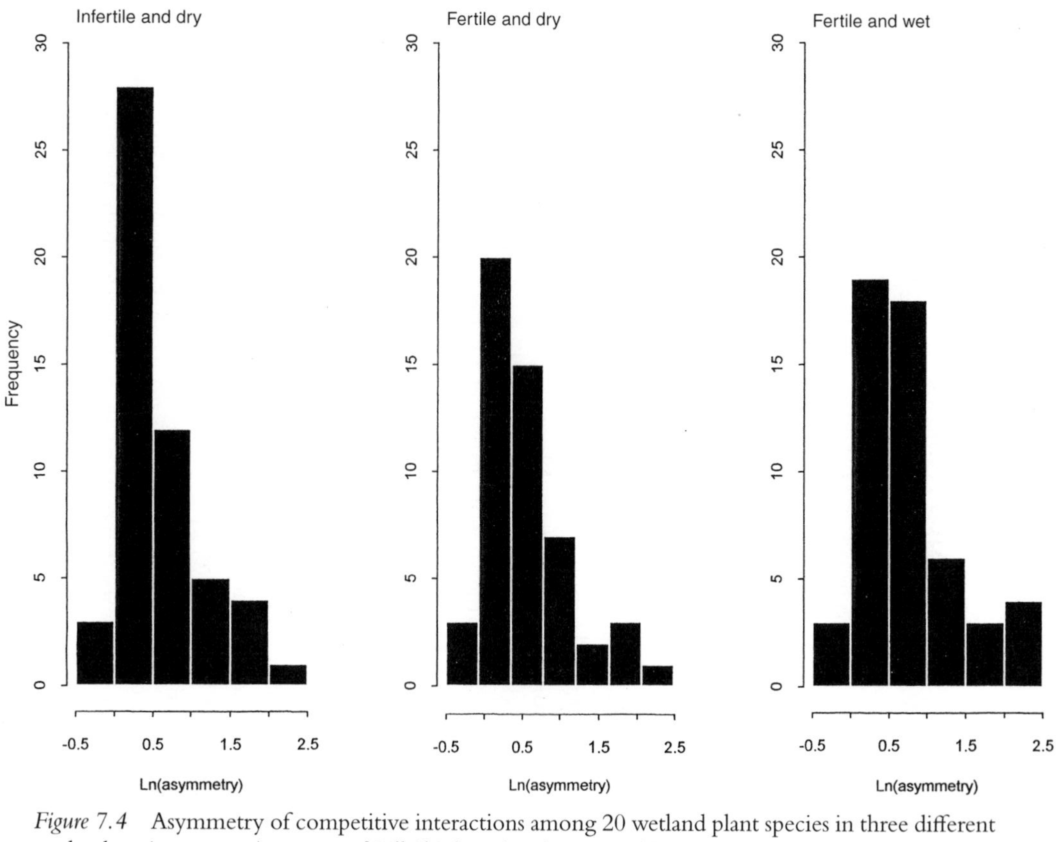

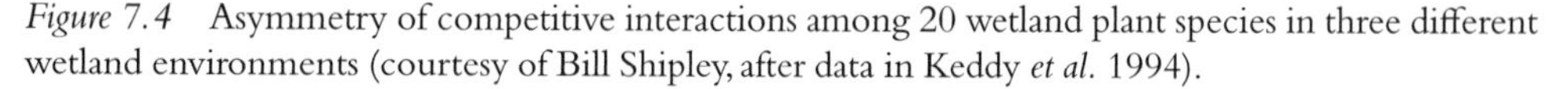

Figure 7.4 Asymmetry of competitive interactions among 20 wetland plant species in three different wetland environments (courtesy of Bill Shipley, after data in Keddy *et al.* 1994).

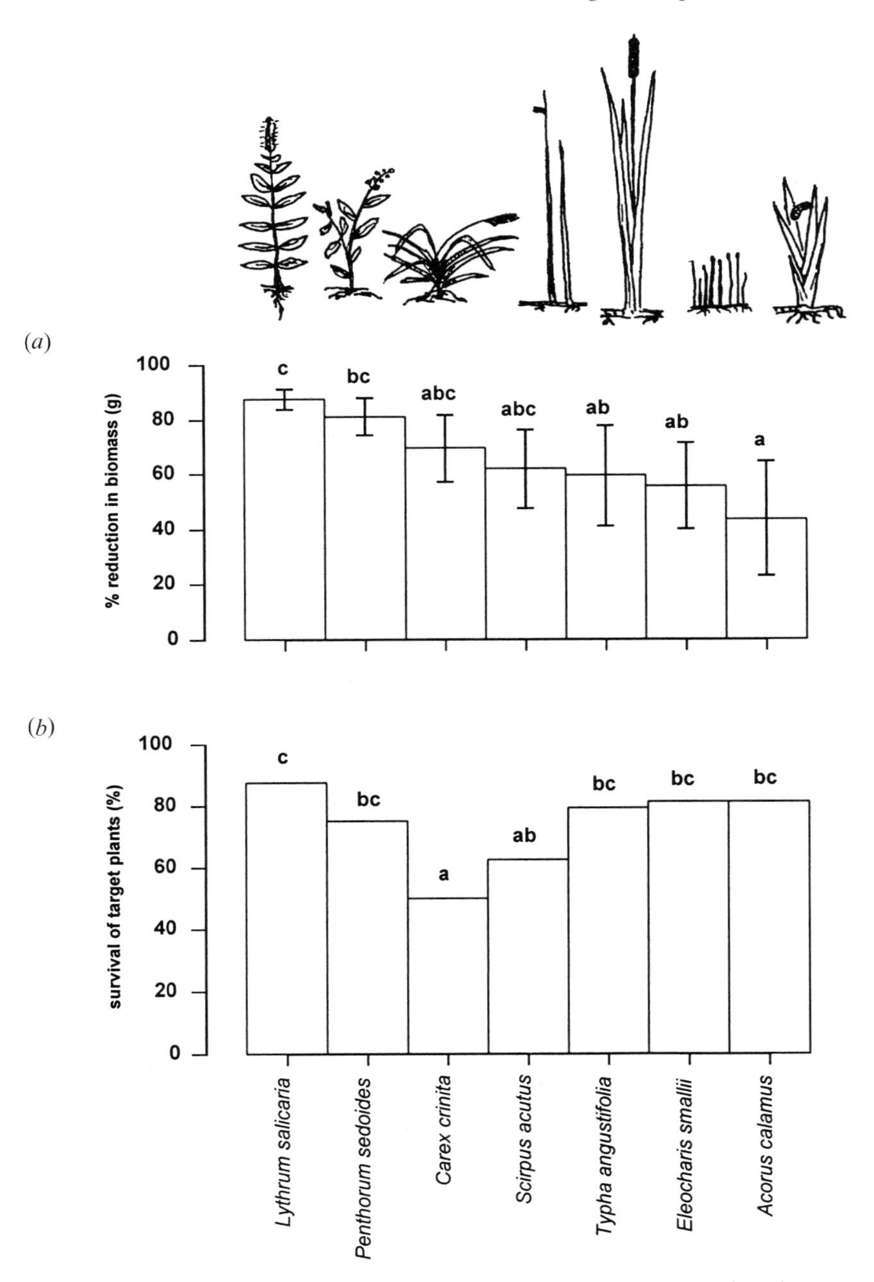

Figure 7.5 Effect of seven different perennial marsh plants upon growth and survival of 48 other (target) wetland plant species (from Keddy *et al.* 1998).

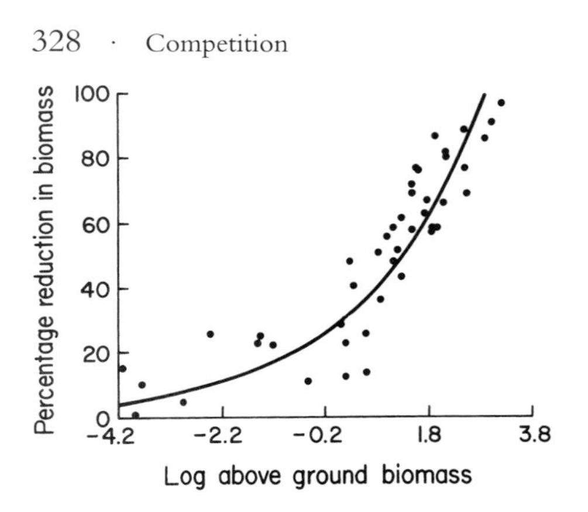

Figure 7.6 Competitive performance (measured as percentage reduction in biomass of *Lythrum salicaria*) plotted against the above ground biomass of 44 test species (from Keddy 1989a after Gaudet and Keddy 1988).

problem by measuring relative competitive performance of 44 freshwater wetland plants from a wide array of habitats. They estimated relative competitive performance by measuring the relative ability of each of these 44 species to suppress a common indicator species, the invasive *Lythrum salicaria*. The more the test plants were able to reduce the growth of *L. salicaria*, the better competitors they were judged to be. They then looked for plant traits that could predict this ability. Both height and above ground biomass were good predictors of competitive performance (Figure 7.6).

We may thus conclude that competition may be very important, and that species often tend to have unequal competitive abilities. Obviously, a further exploration of competition is warranted.

Competition and field distributions

If organisms differ in relative competitive performance, this may be expected to have a bearing on where they occur. In Chapter 2 we saw how one species of *Typha* excludes another to deeper water because of its higher competitive ability in shallow water. Let us consider one other example, a rare species restricted to vernal pools. Vernal pools form in small depressions during rainy periods, particularly in Mediterranean climates (p. 216). In California an unusual flora is distributed along the steep moisture gradient from the centre of the pool to the surrounding arid sites; over three-quarters of the species characteristic of these pools are endemic to the California floristic province. Many species are threatened with extinction by the rapid pace of development in vernal pool landscapes. The seasonal water level fluctuations are essential: if flooding lasts longer than 6 months, the endemic species are absent; without temporary flooding, typical grassland and chaparral occur. Restriction of the vernal pool species to depressions has been attributed to factors including drought, competition, a requirement for inundation to germinate, and even (improbably) low rates of dispersal. Bauder (1989) therefore tested for soil moisture and competition as controlling factors, selecting as test species a rare annual species of mint, *Pogogyne abramsii*. Near the upper limit of the *P. abramsii* distribution along a moisture gradient, 128 plots were established. Seeds were sown into the centre of each plot, with *P. abramsii*, *Hypochoeris glabra* or *Vulpia myuros* present as potential competitors. Eight densities of the competitors were created by different amounts of thinning, yielding three species × eight densities × four replicates for 96 plots. In an additional 32 plots, water was added. Surprisingly, the harvested biomass of the *P. abramsii* plants differed neither with type of neighbour nor with density of neighbours. Added water, however had strong positive effects upon survival, although it did not affect biomass or seed production. Bauder concluded that insufficient soil moisture is an important factor in restricting *Pogogyne abramsii* to vernal pools, while noting that the year the experiment was conducted was the second driest since 1850. Competition, she suggests, might be more important in mesic conditions near the elevation of the high water level in the pool, where neither drought nor inundation would restrict plant growth. It might also be more important in years with more rainfall. These possibilities remain to be tested.

Do such effects occur along other gradients besides water level? Fertility is an obvious gradient to choose because fertility gradients are so widespread in both wetlands and in vegetation in general. To test for competitive control of plant distributions along fertility gradients, Wilson and Keddy (1986a) first measured relative competitive performance of seven species of wetland plants by growing them in all possible pairwise combinations (Table 7.2). They were then able to predict the distributions of these plants in the field (Figure 7.7). Gaudet and Keddy (1995) repeated this at a much larger scale. They assembled data on field distributions for 40 of the 44 species shown in Figure 7.6, so they were able to test whether the simple measure of relative competitive performance was able to predict field distribution along complex natural gradients. Figure 7.8 shows that there was a striking positive relationship between relative competitive performance measured in the pot experiment, and the field distribution of the wetland species.

Table 7.2. *Relative increase in dry mass per plant (RIP) of seven plant species*[a] *grown in all pairwise combinations*

	Neighbour species							
Target species	*Dul*	*Jun*	*Lys*	*Hyp*	*Rhy*	*Dro*	*Eri*	Target scores
Dul	1.00	1.33	1.18	1.17	1.25	1.18	1.34	1.20
Jun	0.63	1.00	1.34	1.46	1.46	1.52	1.49	1.28
Lys	0.88	0.87	1.00	1.63	1.63	1.78	1.57	1.31
Hyp	1.09	0.99	0.91	1.00	1.22	1.29	1.23	1.11
Rhy	1.05	0.73	0.93	0.91	1.00	1.21	1.36	1.03
Dro	0.98	0.91	0.93	1.02	1.02	1.00	1.11	0.98
Eri	0.65	0.71	0.88	0.89	0.87	1.48	1.00	0.93
Neighbour scores	0.89	0.93	1.03	1.15	1.21	1.35	1.30	

Notes:

[a] *Dul: Dulichium arundinaceum.*
Jun: Juncus pelocarpus.
Lys: Lysimachia terrestris.
Hyp: Hypericum ellipticum.
Rhy: Rhynchospora fusca.
Dro: Drosera intermedia.
Eri: Eriocaulon septangulare.

Source: From Wilson and Keddy (1986a).

Competition intensity gradients

It is commonly believed that competition intensity does not change among habitats (e.g. Tilman 1982). In contrast, I personally would find it remarkable that any ecological property should turn out to be a constant. However, instead of arguing from theory, we simply need to do field experiments designed to test for competition intensity gradients. The design is simple; it requires growing test species (or phytometers *sensu* Clements 1935) in both cleared and uncleared plots of vegetation (Figure 7.9). If there is no difference in performance of the transplants, there is no evidence for competition. The greater the difference between treatment and control, the greater the intensity of competition. This is a relatively simple design and, when Wilson and Keddy (1986 b) repeated it at eight locations along an exposure gradient in a freshwater lake, they found that competition intensity increased as biomass increased. Obviously, additional studies are needed to test the generality of such results. Few exist. Shipley *et al.* (1991b) did a removal study in a different wetland – a freshwater riverine marsh. They showed that an equation fitted to Wilson and

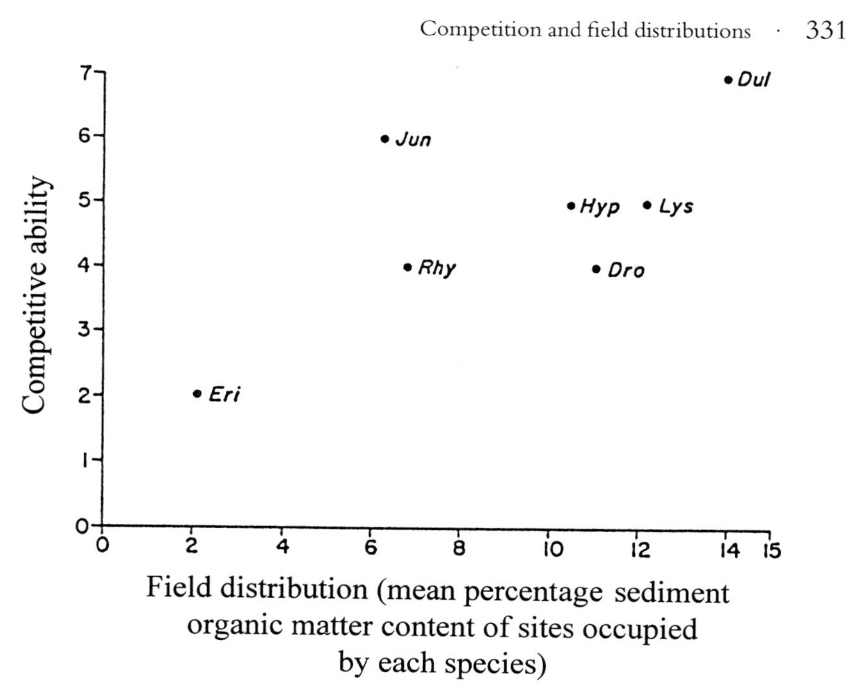

Figure 7.7 The relationship between competitive ability measured in pot experiments and field sediment characteristics can be used to predict the field distribution of seven shoreline plants (from Wilson and Keddy 1986a). Species names are abbreviated as follows: Eri: *Eriocaulon septangulare*, Rhy: *Rhynchospora fusca*, Jun: *Juncus pelocarpus*, Hyp: *Hypericum ellipticum*, Dro: *Drosera intermdia*, Lys: *Lysimachia terrestris*, Dul: *Dulichium arundinaceum*.

Keddy's data, predicts quite well the competition intensity measured in a completely different vegetation type (Figure 7.10).

Above and below ground competition

Although the above studies show that competition intensity increases with biomass and fertility, it does not separate the relative importance of above- and below-ground interactions. It might be tempting to invent explanations based upon presumed shortages of below-ground resources such as oxygen, but the best resolution is a field experiment. Only one such study has been done, but it compared a fertile wetland in a bay with an infertile wetland on an exposed sand beach. The design allowed for plants to interact above and below ground, or only below ground (Twolan-Strutt and Keddy 1996). The analysis showed once again that competition intensity increased with fertility, measured in this figure as

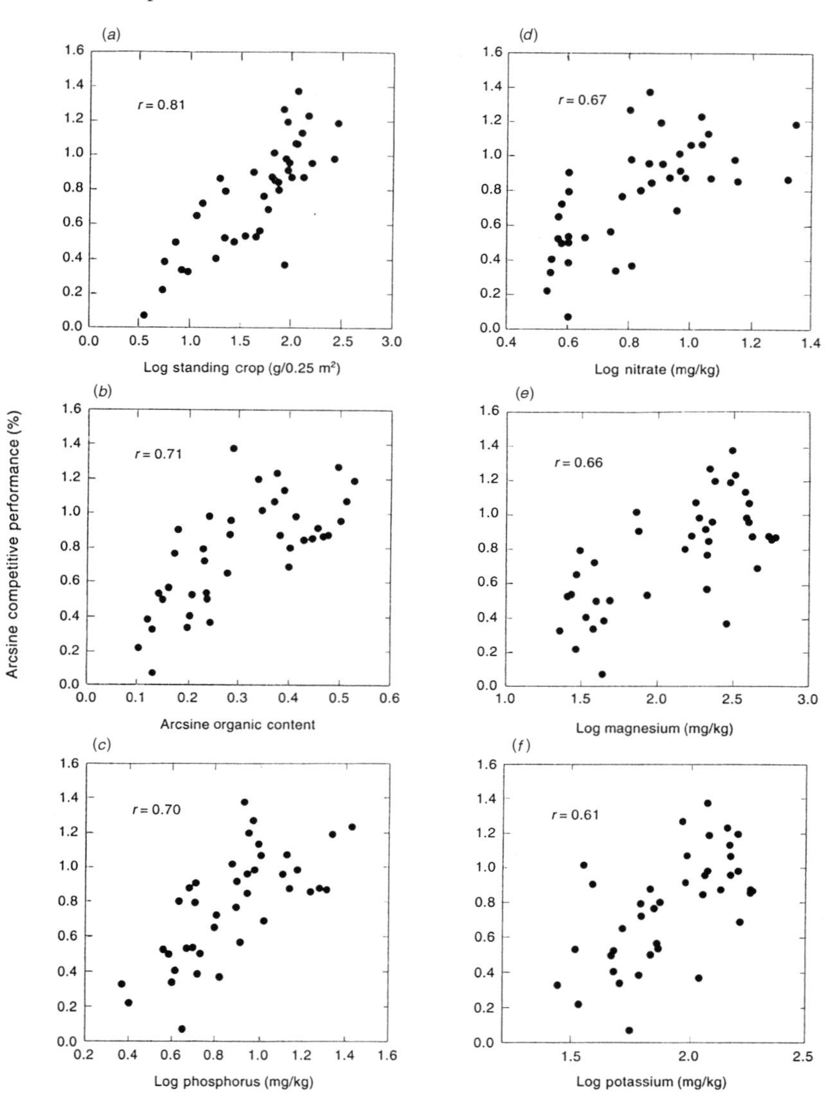

Figure 7.8 Competitive performance of 40 wetland plants (as measured in pots) plotted against their positions along six different field gradients (from Gaudet and Keddy 1995).

Figure 7.9 The use of two phytometer species to measure competition intensity at one point along a gradient. The ramets of the test species are transplanted into a clearing (*left*) and intact vegetation (*right*). The difference in biomass of the phytometers is directly proportional to the intensity of competition in the intact vegetation.

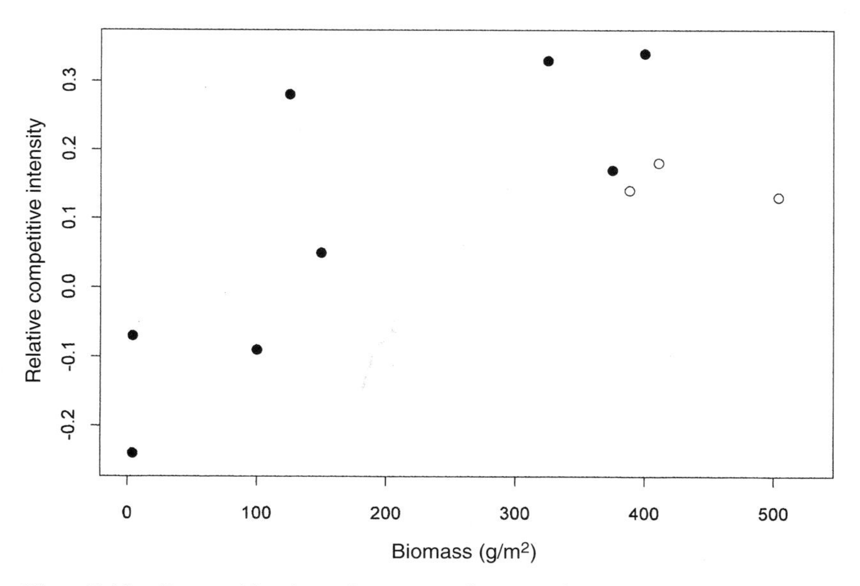

Figure 7.10 Competition intensity measured at seven locations in one lake (solid circles after Wilson and Keddy 1986a) and in a riverine marsh (open circles after Shipley *et al.* 1991b).

biomass (Figure 7.11). Below-ground competition did not change along the gradient, but above-ground competition was more intense in the fertile habitat with high standing crop.

Mechanisms of competition: exploitation vs. interference

According to our definition, competition occurs when organisms either consume resources that other organisms require, or when they deny neighbours access to resources. The former is called *exploitation* competition, whereas the latter is called *interference* competition (Keddy 1989a). There are good grounds for expecting exploitation competition to be important in wetland plants. The general logic is that there are only a few resources that plants need: light, water, carbon dioxide and mineral nutrients. It is commonly believed, at least with animals, that neighbours can avoid competition by using different resources, but this option is not open to plants since their requirements are so similar (Harper 1977); further, since water is normally available to wetland plants, there are even fewer possible resources that can limit growth. Figure 7.11 showed that both above- and below-ground competition were greater than zero. The most likely explanation is competition for limiting nutrients such as nitrogen or phosphorus below ground, and competition for light above ground. The increase in intensity of competition with above-ground biomass (Wilson and Keddy 1986a, b; Twolan-Strutt and Keddy 1996) is further evidence that light becomes limiting in fertile habitats, and therefore that the relative importance of below-ground competition declines in fertile conditions. We may presume that the dominance of tropical wetlands by large monospecific stands of high biomass plant species (Chapter 3) illustrates the consequences of intense above- and below-ground competition (although it still does not really answer the evolutionary question as to why we could not have many more such species co-existing with one another). We also know, from Chapter 5, that eutrophication generally leads to increasing dominance of wetlands by a few species, further suggesting that, as nutrients become more accessible, they are monopolized by a few species, which then produce dense canopies and shade out neighbours. There are fewer studies in which resource levels have been experimentally increased for wetland animal communities; one exception is the addition of food to experimental communities of amphibians.

Griffiths *et al.* 1993 studied the effects of added resources upon competition between natterjack toads (*Bufo calamita*). The toad is known to be

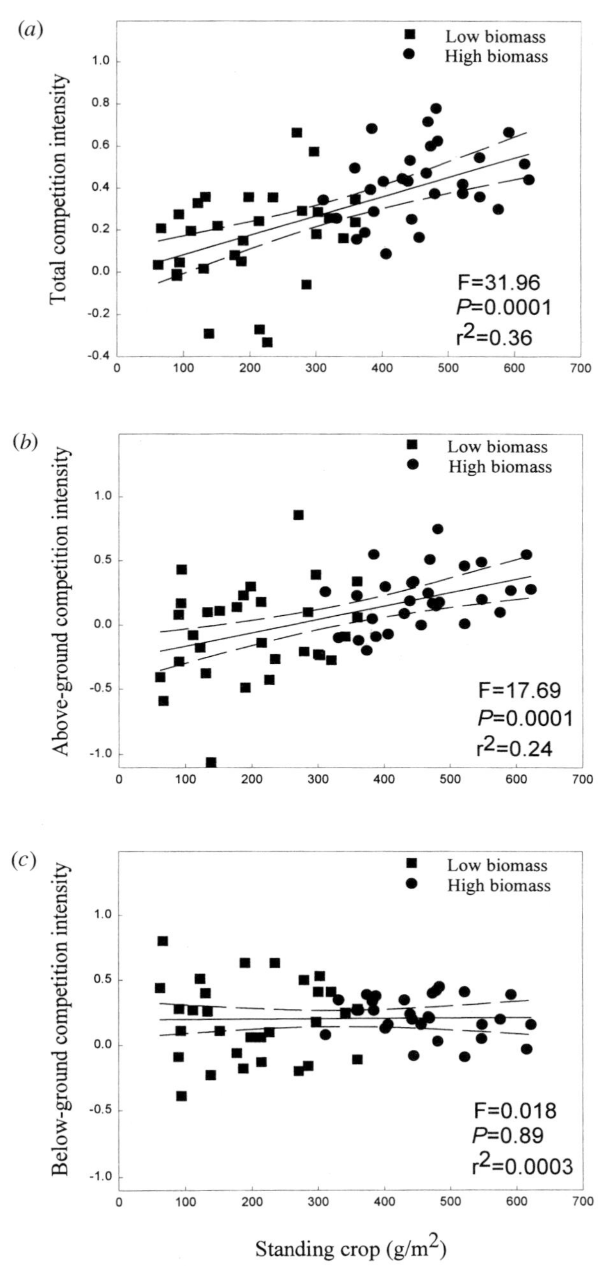

Figure 7.11 Competition intensity plotted against standing crop for a series of field experiments in two riverine wetlands (from Twolan-Strutt and Keddy 1996).

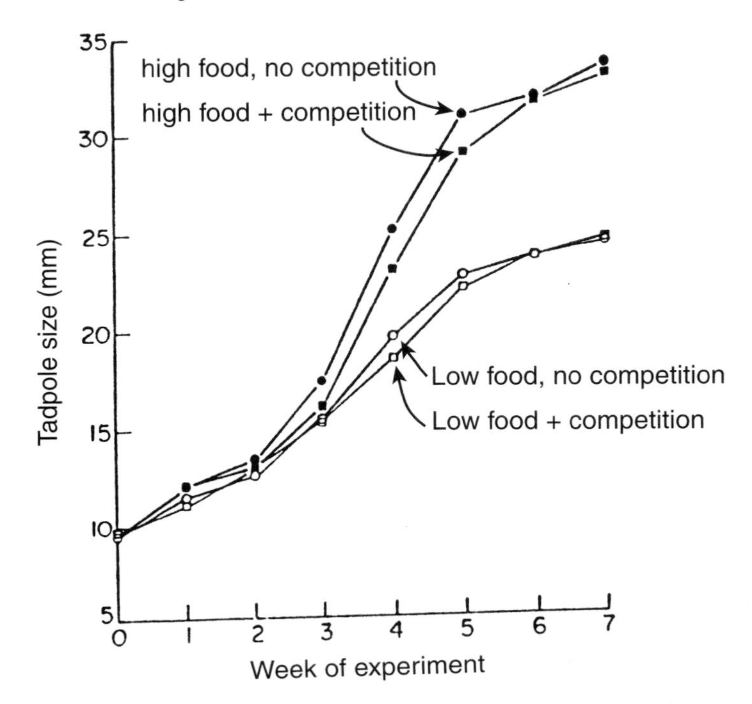

Figure 7.12 Growth of *Bufo calamita* raised under different conditions of food and competitors (after Griffiths *et al.*, 1993).

sensitive to effects of frog neighbours, and even to water conditioned by neighbours. The tadpoles were grown in experimental outdoor ponds constructed from fibreglass. All ponds received 150 toad tadpoles, and half of them also received 150 frog tadpoles. A subset of each received supplemental feeding of rabbit pellets. Figure 7.12 shows that the addition of food had a far greater impact on tadpole size than did the presence or absence of neighbours. We might expect the addition of food to reduce competition, but analysis of variance did not reveal any interactions between competition and food level, or among competition, food and time. In short, the effects of competition were more or less constant independent of resource levels, and, most significantly, much less important than food supply. Removing the supplemental food would reduce size by more than 100 g, whereas adding neighbours would reduce it by only some 10–20 grams.

Interference competition is much more poorly understood in wetland plants. We know of good examples of interference competition in animals, in part, because such interactions are relatively more easy to observe. For example, Marsh Wrens that nest in reed beds of deltaic marshes are known to smash the eggs of neighbouring birds, including both other wrens and Red-winged Blackbirds (Picman 1984). In contrast, interference between plants is more difficult to document. It is thought that plants may poison neighbors by releasing secondary metabolites, either through root exudates or through compounds leached from leaves or litter; Gopal and Goel (1993) provide many examples of such interactions in wetland plants, including common and widespread genera such as *Oryza*, *Eleocharis*, *Leersia*, *Nyphaea*, *Myriophyllum*, *Camboba* and *Brasenia*. Elakovich and Wootten (1989) illustrate the typical procedure: tissues of an array of 17 species from *Brasenia* to *Sparganium* were blended with distilled water, filtered, and then assayed for ability to inhibit the growth of lettuce seedlings in Petri dishes, or duckweed in sterile culture. All plant extracts were able to suppress lettuce seedlings, with greatest suppression produced by *Nymphaea odorata* leaves and petioles. Twelve out of 16 extracts were also able to reduce duckweed growth. Apart from damaging neighbours, there is also the possibility that plants may inhibit their own reproduction: reports that *Typha* can inhibit its own seedling establishment owing to auto-toxins have been contradicted by more recent work, but the possibility of auto-toxicity has since been offered in genera such as *Phragmites* and *Peltandra* (Gopal and Goel 1993). The problem with all such studies is that it is one thing to show that extractions from plants can reduce growth of test organisms in the laboratory; it is quite another to show that such interactions occur in the field. One could argue that the repeated flooding of wetlands is likely to dissolve and disperse water soluble allelochemicals, thereby making it even less likely that allelopathy will occur in such conditions. Williamson (1990) argues, however, that the demands of proof for allelochemical interactions are often impossibly strict compared to levels of proof accepted for other natural phenomena. In a noteworthy parallel, there is growing evidence that, in the laboratory, amphibian larvae in widespread genera such as *Rana* and *Bufo* produce secondary metabolites that reduce the growth of neighbours but there is again no consensus upon whether chemical interference occurs under field conditions (e.g. Griffiths *et al.* 1993; Biesterfeldt *et al.* 1993; Beebee 1996).

Constraints upon competition: three relevant models

The foregoing experiments might leave the impression that competition always leads to one small group of organisms dominating at the expense of others (see also Keddy and Shipley 1989; Shipley and Keddy 1994). This may well be true, at least for herbaceous plant communities, but we should not ignore other possibilities. There may be at least three ways to survive in the presence of competition.

Escape in space

Skellam (1951) showed nearly 50 years ago that weak competitors can survive as long as they can disperse better than strong competitors. The argument goes more or less like this (Pielou 1975). Imagine two competing species that reproduce once a year. Let A be the stronger competitor and B the weaker competitor. Wherever they co-exist, A invariably wins. Therefore, the only habitat in which B can reproduce is those sites in which it occurs alone (Figure 7.13). Assume that the region of interest contains N sites, or islands of habitat, and that at equilibrium the expected proportion of sites with a single A individual at the end of the growing season is Q. Since this means that NQ of the islands are dominated by species A, only $N(1-Q)$ remain available for B to occupy. If we call this remaining portion of islands q, then q must be greater than zero for the competitive subordinate to survive. We want to know how much better dispersal of B must be for this to occur. Therefore, let F and f be the number of seeds produced for species A and B respectively. For species B to persist, f/F must be great enough to ensure that $q>0$. It can be shown that, for this to occur, f/F must exceed $-Q\ /\ (1-Q)\ \ln\ (1-Q)$. Provided this condition is met, species B will continue to occur in the landscape in spite of its explicitly poorer competitive ability.

The many examples of disturbance in Chapter 6 may provide circumstances for this kind of process to operate. Tree species that grow in disturbed sites along floodplains (e.g. Salo *et al.* 1986; Roberts and Ludwig 1991; Duncan 1993) may illustrate Skellam's model, with the subordinant tree species being intolerant of shading, thereby requiring continual production of fresh sediments for reproduction. In the Podocarp floodplains (Duncan 1993), *Dacrycarpus dacryoides* could establish only in large openings. *Weinmannia racemosa* could reproduce in shade, however, it would slowly eliminate *D. dacryoides* from undisturbed floodplains. So long as new sediments are deposited, they can escape in space as competitive dominants exclude them from older sites. Similarly, many of the species in seed banks are relatively small and therefore probably weak competitors. They also produce large numbers of small seeds (e.g. *Juncus* spp., *Hypericum*

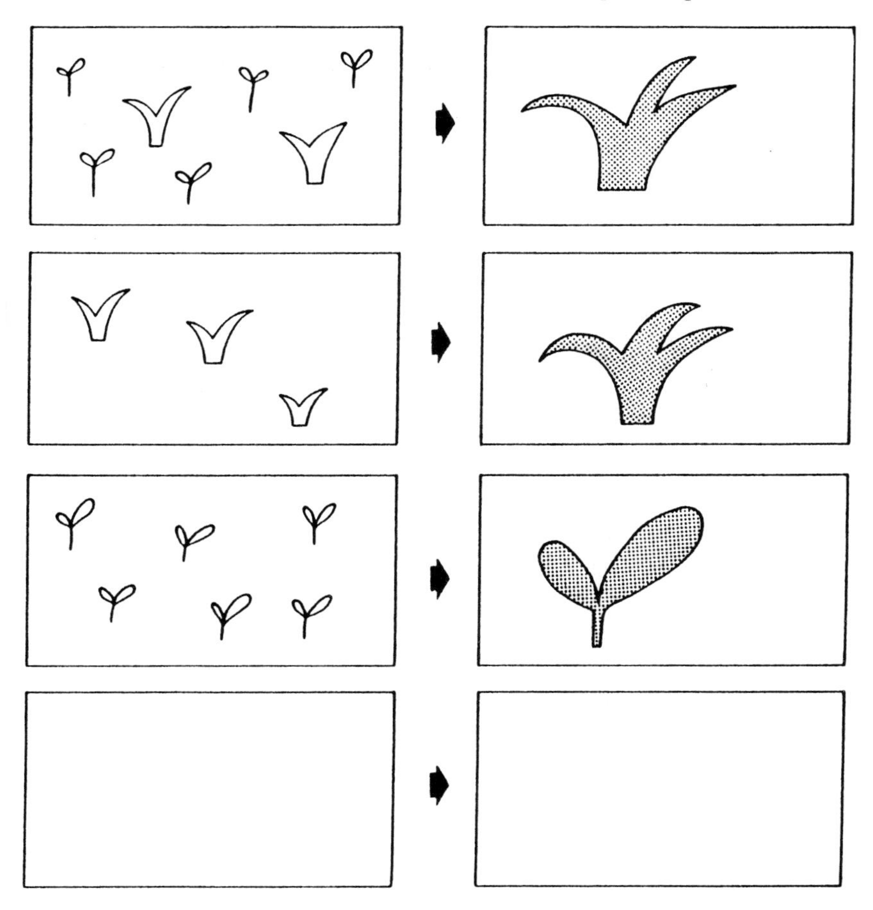

Figure 7.13 Weak competitors can survive by escaping to habitat patches that are not occupied by stronger species. Four possible combinations of seedlings are shown on the left, and the outcome of adults is given on the right (from Pielou 1975 after Skellam 1951).

spp., *Scirpus* spp.), which may similarly allow them to escape from superior competitors. The pioneer flora of 125 species listed by Kalliola *et al.* (1991) from the Amazon basin appears similarly to require escape in space. Similarly, in saltmarshes, weak competitors such as *Salicornia europea* are tolerant of high salinities; by being able to colonize bare spaces with high salinity, they are able to escape from competitive dominants such as *Spartina patens* (Bertness and Ellison 1987; Bertness *et al.* 1992).

Table 7.3. *Three possible kinds of community structure produced by competition*

Basis of competition		
Resources (collectivism)	Space (monopolism)	
Niche control	Dominance control	Founder control
Resource partitioning	Competitive hierarchies	Gap-creation
Symmetric competition	Asymmetric competition	Dispersal
Different fundamental niches	Inclusive fundamental niches	Colonization

Source: After Keddy (1989a) terminology from Yodzis (1989).

Founder control: an escape from competitive exclusion?

We may recognize three general ways in which competition organizes communities (Table 7.3). First we must distinguish between those organisms that collect a fraction of the resources from the whole area, and those organisms that harvest all the resources from a fraction of the area. The first case (collectivism *sensu* Yodzis 1989) is likely to occur with most motile organisms. In such circumstances, resource partitioning and symmetric competition may be common. The second case (monopolism *sensu* Yodzis 1989) is likely to occur with sessile organisms. Here, two types of competitive control are possible. In the first, dominance control, there are very different relative competitive abilities and the few dominants exclude the many subordinates. We have seen actual examples of this above. In the above model by Skellam, a subordinate escapes a dominant by dispersing ahead of it. But, there may be cases where the species all have relatively similar competitive abilities, in which case it is the first species to arrive that holds the site. Such communities are called founder controlled.

Founder controlled communities may have some quite unusual properties. If, indeed, there are communities composed of competitive equivalents, then competitive exclusion will take a very long time to occur. This is intuitively obvious, although it at first appears to contradict the common assumption that, because competition is most intense between similar species, it should rapidly cause the extinction of one of them. Models can help clarify this apparent contradiction. Assume a community in which each individual that dies is replaced by another randomly selected member of the community. Imagine, say, a forest in which a tree dies and is replaced by a random selection from among the remaining trees. In such cases, models show that many species will eventually become extinct, but it will require thousands of generations (Chesson and Warner 1981; Hubbell and Foster 1986). Such models may well apply to tropical riparian forests where large numbers of apparently similar tree species coexist (see Figure 3.11) and to morphologically similar submerged macrophytes (McCreary *et al.* 1983; McCreary 1991). Chesson and Warner (1981) suggest other factors that can extend co-existence. Environmental variability may control recruitment, thereby preventing any one species from gaining a dominant position. Adults may also survive prolonged periods inhospitable to recruitment; this has been called a 'storage effect'. Those unused to thinking in terms of simple models may be sceptical, but Grubb's (1977) thoroughly empirical review on regeneration niches provides more than enough examples to convince one that regeneration in small gaps differing in minor ways can generate a great deal of the diversity in monopolist communities.

Centrifugal organization

Introduction

There is another way in which competition can organize whole communities of organisms. We have already seen in Chapter 2 (Figure 2.10) that there is an important difference between ecological and physiological response curves (Zoologists tend to use the terms 'realized' and 'fundamental' niches for the same patterns.) In either case, inclusive niches are said to occur when all species perform best at the same end of the same resource gradient (Miller 1967; Mueller-Dombois and Ellenberg 1974; Colwell and Fuentes 1975). Plants, for example, all share a common requirement for a few basic resources: light, water and mineral nutrients. It therefore follows that most plants will grow fastest and will be most fecund where these basic resources are in greatest supply.

This concept of 'inclusive niches' or 'shared preference' is often overlooked, so let us examine this pattern further. Figure 7.14(*a*) illustrates inclusive niches for species A to F, the far left end of the gradient being used by all. Species A to F, however, vary in their competitive ability with A (*top*) being competitively superior and with F being subordinate. A trade-off is assumed here, between the ability to interfere competitively with neighbours, and the ability to tolerate adversity (*sensu* Southwood 1988) or stress (Grime 1979). The outcome of different competitive abilities and different tolerance limits is that species become differentially distributed along the gradient. This is illustrated in Figure 7.14(*b*) where the horizontal dashed lines show the region from which a species is excluded

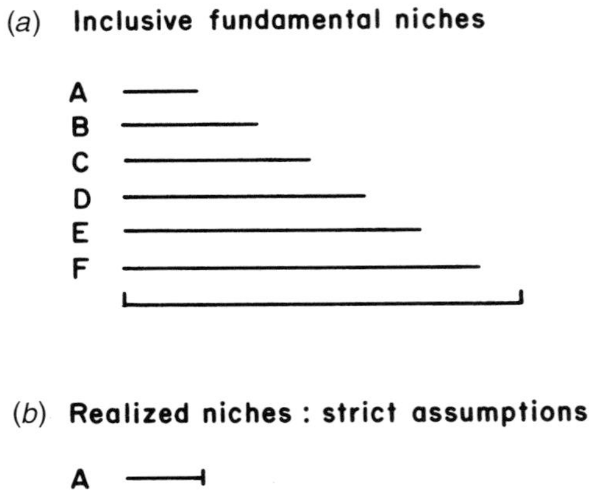

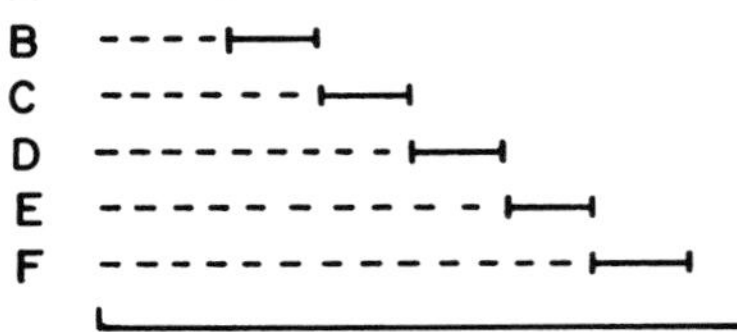

(c) Realized niches : relaxed assumptions

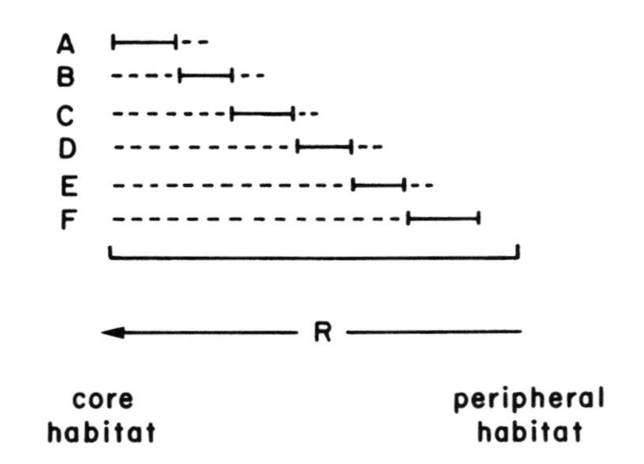

Figure 7.14 Species distributions along a gradient (*a*) where they have inclusive niches and (*b,c*) where there are competitive hierarchies (from Wisheu and Keddy 1992).

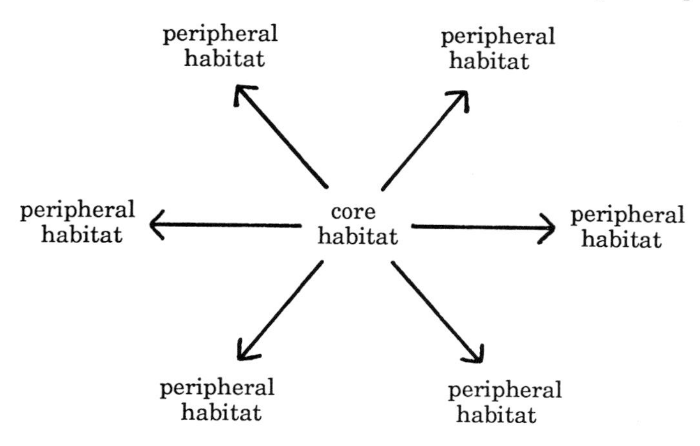

Figure 7.15 When many gradients radiate outward from a shared core habitat, the pattern is termed centrifugal organization (from Wisheu and Keddy 1992).

by one or more dominant species. Figure 7.14(*c*) shows a similar situation except that competitive exclusion is not complete. Near the limits of a species' tolerance, the dominant is excluded by the adjacent subordinate. The distribution pattern remains the same however, except that in Figure 7.14(*c*), each species is displaced slightly up the resource gradient. This displacement of species appears to be resource partitioning, although here, the pattern arises from quite different mechanisms than are usually assumed (Keddy 1989a). For whatever reason, ecologists observing zonation patterns far too often assume without testing that the ecological and physiological response curves (realized and fundamental niches) are very similar. Far too often they are wrong.

A second assumption is now introduced: that the same species prosper at the benign ends of the gradient, but that different species tolerate each of the adverse ends. That is to say, the benign ends of many gradients are similar enough that we can describe them as a 'core' habitat that can be dominated by the same species. At the peripheral end of each axis however, species with specific adaptations to particular sources of adversity occur. This pattern is termed centrifugal organization (Figure 7.15). Many peripheral habitats radiate outwards from the single, central core habitat.

In wetlands, the central habitat apparently has low disturbance and high fertility, and is dominated by large leafy species capable of forming dense canopies (Figure 7.16). In north eastern North America, *Typha* dominates the core region by reducing light levels for subordinate

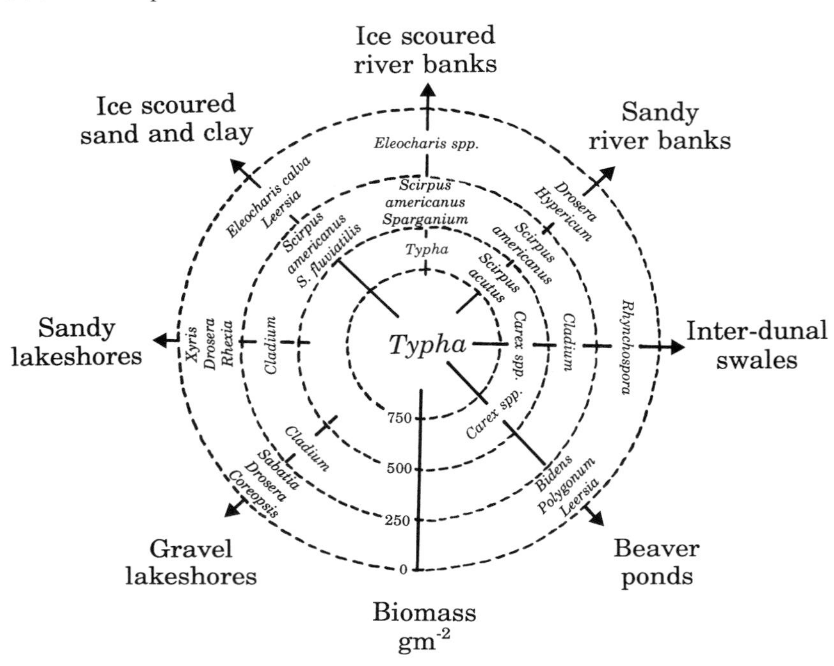

Figure 7.16 In herbaceous wetlands, it appears that the core habitat (fertile, undisturbed) is dominated by large canopy forming species such as *Typha* (from Wisheu and Keddy 1992, after Moore *et al.* 1989).

species. Beyond the core, different constraints create radiating axes along which different groups of species and vegetation types are arrayed. Along an axis of soil phosphorus, for example, that is highest in a fertile embayment and lowest at an infertile, open shoreline, the community composition changes from a high biomass *Typha* dominated wetland to a sparsely vegetated shoreline that is dominated by isoetid and insectivorous species (Moore *et al.* 1989), two groups of plants indicative of infertile conditions (Boston & Adams 1986; Givnish 1988).

Gradients of nutrient concentrations however, are only some of the many gradients that occur in wetlands. Others are gradients of disturbance and the species found along these gradients differ from those found along gradients of fertility. *Typha* would again occupy fertile, protected areas, but where ice scour or severe flooding occur, either reeds or annual species would be abundant (Day *et al.* 1988; Moore *et al.* 1989). The deeply buried rhizomes of reeds protect them from moderate ice scour, while fast growing annuals are able to set seed between periods of mud deposition (Grubb 1985; Day *et al.* 1988). Peripheral habitats that are formed by different kinds and combinations of infertility and disturbance support distinctive floras that reflect differing environmental conditions, e.g. shoreline fens (Charlton and Hilts 1989), interdunal swales and sand spits (Reznicek and Catling 1989), sandy coastal plain wetlands (Keddy and Wisheu 1989), and eroding river banks (Brunton and Di Labio 1989; Nilsson *et al.* 1989). In salt marshes, anoxic and salty patches may play a similar role (Bertness and Ellison 1987).

Evidence for inclusive niches and competitive hierarchies

The centrifugal organization model for wetland communities assumes that (i) species have inclusive niches, that (ii) there are competitive hierarchies, and that (iii) competition is most intense in the high biomass core habitat. Let us consider these assumptions in turn.

A number of experiments do provide evidence of inclusive niches and competitive hierarchies (Sharitz and McCormick 1973; Goldsmith 1973; Austin 1982) but only experiments that deal specifically with wetland habitats will be discussed here.

Testing for inclusive niches To test for the presence of inclusive niches, Wilson and Keddy (1985) grew 12 different wetland species by themselves in ten different substrates of increasing organic content. In the field these species occurred at different positions along this gradient, but when moved into experimental conditions where there was no competition, 11 of the 12 grew larger in more fertile soil (Figure 7.17). Actual field distributions differed, but there was an inclusive niche with preference for the organic end of the soil gradient. Similarly, transplant experiments by Snow and Vince (1984) and Bertness and Ellison (1987) demonstrated that saltmarsh plants grew best at higher elevations (pp. 99–100).

Testing for competitive hierarchies We have already seen that Bertness (1991) (Figure 7.2) and Gaudet & Keddy (1988) (Figure 7.6) have demonstrated that there are hierarchies of relative competitive ability in wetland plants. The presence of hierarchies appears to be a ubiquitous phenomenon of plant communities (Keddy and Shipley 1989; Shipley and Keddy 1994). It may be, however, that such hierarchies are fragile and change with minor environmental fluctuations. In contrast, they may be robust and relatively little affected by changes in environment. Gaudet

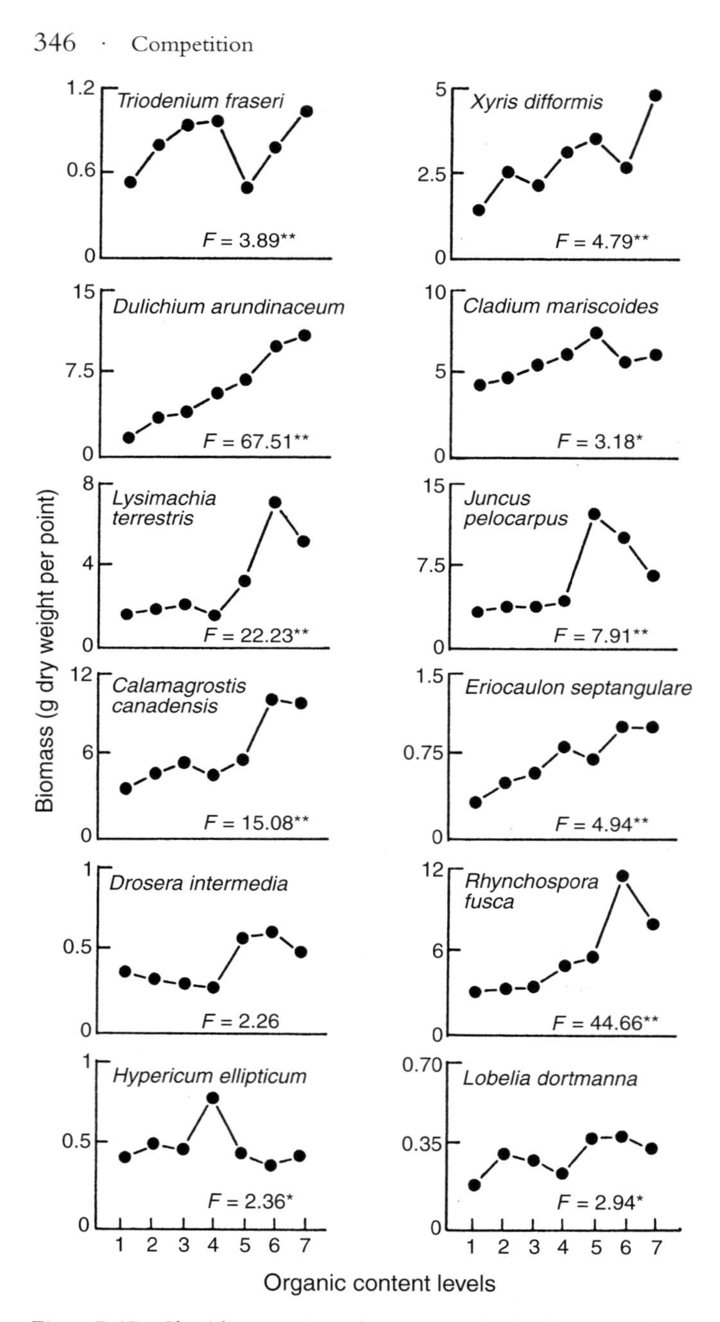

Figure 7.17 Shoreline species, when grown singly, show significant shared preference for the organic end of a shoreline gradient (from Wilson and Keddy 1985).

(1993) tested between these by measuring the relative competitive performance of ramets of more than 20 wetland plants in two different environments; fertile and infertile. The resulting hierarchies were surprisingly constant among environmental conditions (Figure 7.18 *bottom*), similar to patterns found with seedlings (Figure 7.18 *top*).

Testing for inclusive niches and hierarchies Zedler *et al.* (1990) grew *Typha orientalis* and *Juncus kraussii* along a salinity gradient. Four ramets of the two species were transplanted into either monoculture or mixed-species pots in each different salinity. When grown in mixed-species pots in the fresh, brackish and saline treatments, *Typha* outgrew *Juncus* (Figure 7.19). The relative yields of *Typha* were greater than 0.5, indicating that the mean total biomass of two *Typha* ramets in the mixed pots was more than half the biomass of four *Typha* ramets in monoculture. *Juncus*, with relative yields less than 0.5, was suppressed. Adult plants of *Typha*, however, were unable to survive in the hypersaline treatment, while specimens of *Juncus* could. The hypersaline treatment therefore exceeded the tolerance limits of *Typha* and provided a peripheral habitat where *Juncus* was no longer dominated by *Typha*.

Competition gradients Figures 7.10 and 7.8 show that competition intensity increases along the biomass/fertility gradient in freshwater wetlands. These two studies were carried out in different habitats with different species, which further supports the generality of the process. Further, the work by Twolan-Strutt and Keddy (1996) suggests that this gradient is largely produced by increasing intensity of above-ground competition as soil fertility increases. In salt marshes, higher elevations and lower salinities are conducive to faster plant growth and are also associated with plants having higher competitive performance (Bertness and Ellison 1987).

Evidence for centrifugal patterns Although preliminary work has shown that both wetlands (Keddy 1990b) and forests (Keddy and MacLellan 1990) can be arranged into centrifugal patterns, objective tests are now needed to determine whether vegetation gradients do converge in core habitats. One approach would be to test whether, for a given range of habitats, the number of vegetation types varies inversely with biomass. If they are centrifugally arranged, vegetation types would be relatively similar to one another at high biomass, but very different at low biomass. Thus we might test for a predicted increase in vegetation similarity with increasing biomass.

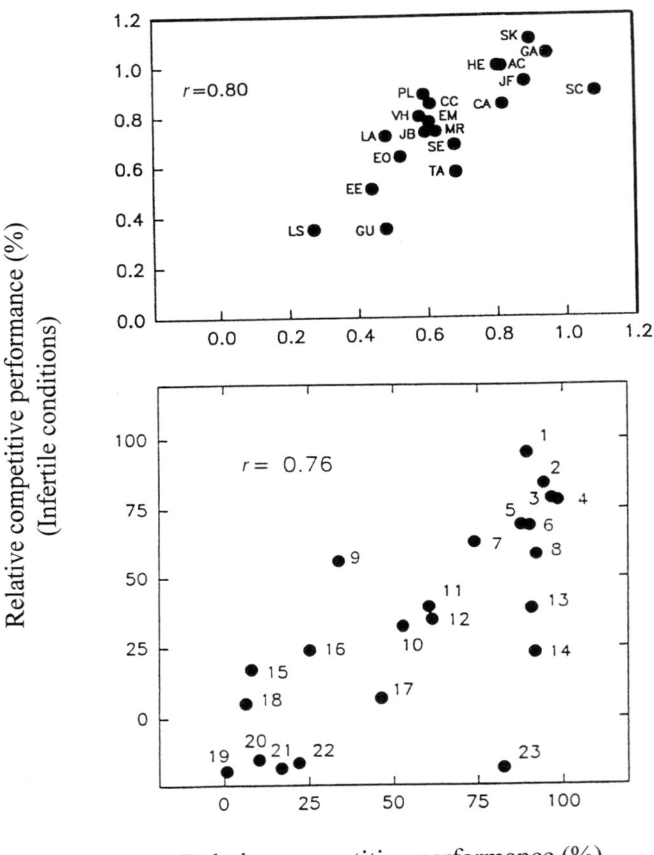

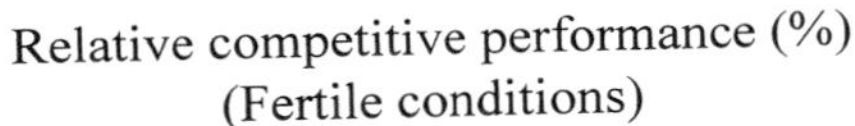

Figure 7.18 Relative competitive performance of wetland plants is little changed by the fertility of the experimental conditions (top; seedlings (from Keddy *et al.* 1994)); HE = *Hypericum ellipticum*, EE = *Eleocharis erythropoda*, JF = *Juncus filiformis*, AC = *Acorus calamus*, CC = *Carex crinita*, EM = *Eupatorium maculatum*, SC = *Scirpus cyperinus*, SE = *Sparganium eutycarpum*, TA = *Typha angustifolia*, GA = *Gratiola aurea*, PL = *Panicum longifolium*, SK = *Sabatia kennedyana*, EO = *Eleocharis obtusa*, LA = *Lycopus americanus*, LS = *Lythrum salicaria*, MR = *Mimulus ringens*, VH = *Verbena hastata*, CA = *Cyperus aristatus*, GA = *Gnaphalium uliginosum*, JB = *Juncus bufonius*; (bottom; adults (from Gaudet 1993)); 1 *Lythrum salcaria*, 2 *Phalaris arundinacea*, 3 *Phragmites communis*, 4 *Typha glauca*, 5 *Euthamia galetorum*, 6 *Mentha arvensis*, 7 *Acorus calamus*, 8 *Spartina pectinata*, 9 *Juncus filiformis*, 10 *Lysimachia terrestris*, 11 *Eleocharis palustris*, 12 *Eleocharis calva*, 13 *Sparganium eurycarpum*, 14 *Scirpus fluviatilis*, 15 *Triadenum fraseri*, 16 *Dulichium arundinaceum*, 17 *Carex crinita*, 18 *Viola lanceolata*, 19 *Scirpus torreyi*, 20 *Juncus pelocarpus*, 21 *Eleocharis acicularis*, 22 *Hypericum ellipticum*, 23 *Rumex verticillatus*).

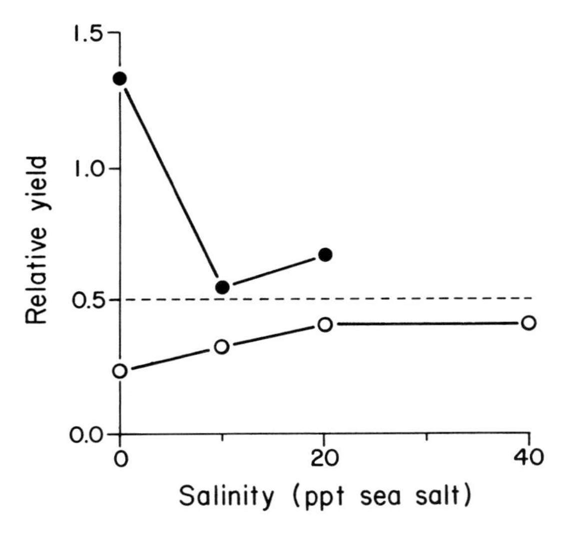

Figure 7.19 The relative yield of *Typha orientalis* (closed circles) and *Juncus kraussii* (open circles) along a salinity gradient (from Wisheu and Keddy 1992, after Zedler *et al.* 1990).

Moore (1990) tested for such a pattern using data from 15 different sites representing a wide array of biomass and vegetation types, including both lakeshore and riverine wetlands, high and low fertility wetlands, and different natural disturbance regimes. The raw data matrix consisted of 225 quadrats representing the abundance (on a scale of 0–9) of 141 species. These quadrats were then ranked in order of standing crop, and mean similarity of each quadrat to the ten nearest neighbours (nearest in measured biomass, not spatial location) was calculated. The Spearman Rank Correlation (Siegel 1956) was then used to test for a correlation between similarity and biomass. Similarity was calculated using the quantitative symmetrical formula $D_{jk} = 1 - (1/2)\{[\text{minimum}\ (X_{ij}, X_{ik})/(X_{ij})] + [\text{minimum}\ (X_{ij}, X_{ik})/(X_{ik})]\}$, which has been studied by Faith *et al.* (1987) and found to have a robust monotonic relationship with similarity even when species turnover between sites is high.

The similarity of quadrats among sites was positively correlated with biomass (Figure 7.20, $P<0.0001$). As Moore points out, there are provisional aspects to this test. First, it is based on only 15 sites; one could argue that more high biomass or more low biomass sites might change the pattern; at the same time, these sites already represent very different species composition and habitat types. Secondly, the patterns may be scale dependent; Moore measured similarity based upon $0.5 \times 0.5\ m^2$ quadrats;

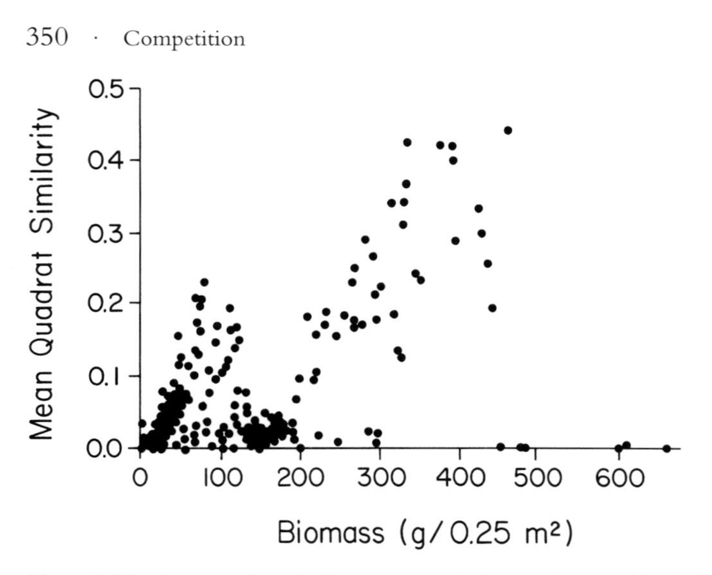

Figure 7.20 Interquadrat similarity is positively correlated with the biomass of the habitat (from Wisheu and Keddy, 1992, after Moore 1990).

different patterns may be found at larger scales. Thirdly, the measure of similarity may be inappropriate. In fact, species similarities may not be the best measure of similarity. For example, two high biomass quadrats differing only in that they have different kinds of *Typha* (*T. latifolia* and *T.* ×*glauca* (*T. angustifolia* × *T. latifolia*)) would be classified as having little 'similarity' when, in fact, the species are similar in morphology and ecology. Yet they are treated as being as dissimilar as a pair of quadrats with species of widely contrasting ecology – say, *Utricularia cornuta* (a carnivorous species) and *Nymphoides cordata* (a floating leaved aquatic). Thus, the best test might incorporate measures of similarity of form and function as well as taxonomic composition.

Predictions of the model

One prediction of the centrifugal organization model is that rare species will be restricted to peripheral habitats. An outstanding test of this prediction is shown in Ellenberg (1985); high numbers of rare plant species occur in the nitrogen-deficient regions of four very different habitats (recall Figure 5.13). The restriction of rare plant species to low biomass, infertile wetlands has also been shown in North America (Figure 7.21). The centrifugal organization model also addresses changes in sets of ecological communities in response to eutrophication. It is well known that eutrophication produces changes in individual plant communities, as

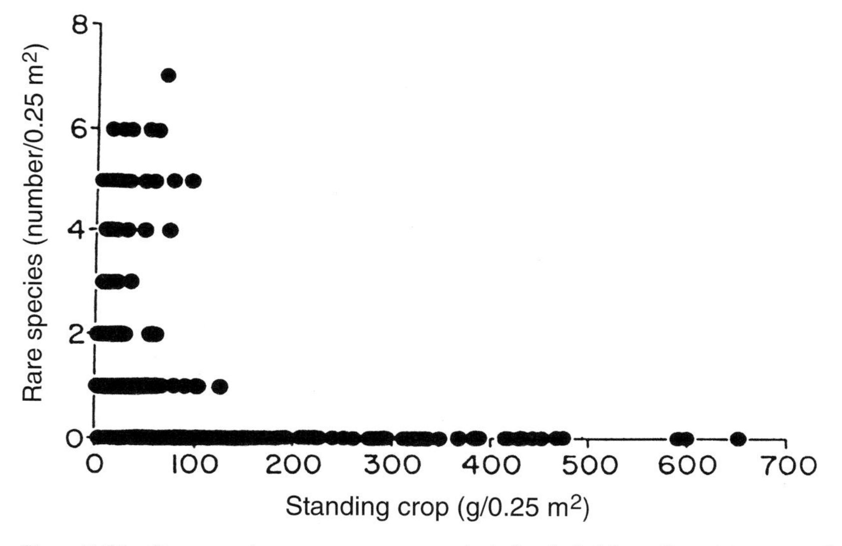

Figure 7.21 Rare species are most common in infertile habitats (from Moore *et al.* 1989).

documented in early work on the fertilization of interdunal communities (Willis 1963). Chapter 5 has already surveyed the effects of eutrophication; it appears that shifts in life form and species composition are towards those forms and species associated with core habitats. That is, small plants are replaced by large canopy forming species, and diverse communities are replaced by simpler ones. Over the landscape as a whole, eutrophication removes peripheral habitats and replaces them with core habitats, leading to an overall decrease in the diversity of wetland vegetation types in the landscape.

Competition and co-operation

Closer inspection of the competition experiment in Figure 6.8 shows some unexpected results: at one location (bottom left), competition intensity is negative! That is, plants grew better with neighbours than without them. Wilson and Keddy (1986a, b) suggest that this is an example of mutualism, where neighbours help protect other plants from the damage caused by waves. The same pattern can be seen in Figure 6.9 – once again, sandy infertile shorelines with very low biomass yield many points with negative values for competition intensity. It may well be that mutualistic interactions are relatively common in such stressful and disturbed sites.

Bertness and Yeh (1994) studied recruitment in the marsh elder (*Iva frutescens*), and found it was influenced by both negative and positive interactions with adult plants. In undisturbed areas, perennial vegetation prevented seedlings from establishing. Adult elder plants accumulated plant debris which killed the other perennial species, and then acted as nurse plants, protecting seedlings from radiation and hypersaline soil conditions. Marsh plant recruitment, they conclude, is dictated by a balance between co-operative and competitive interactions.

The oxygen transported by aerenchyma into hypoxic sediments appears to be an important kind of facilitation (recall pp. 37–39). Consider two more examples. Callaway and King (1996) showed that, under cool conditions (11–12 °C), *Typha latifolia* transported oxygen into sediments, raising oxygen levels from 1 to near 4 mg l^{-1}. The enhanced survival of *Salix exigua*, and enhanced growth of *Myosotis laxa*, was attributed to the beneficial effects of oxygen. At higher temperatures, however, the dissolved oxygen levels fell to control amounts, suggesting that there is a shift from facilitation to competition as temperature increases. Similarly, Pedersen *et al.* (1995) report that the evergreen rosette plant, *Lobelia dortmanna*, already known to absorb CO_2 from the rooting zone (Chapter 1), also releases O_2 into the soil zone during photosynthesis. This is slightly different from the above case, because, in *Lobelia*, the oxygen is a byproduct of photosynthesis, rather than atmospheric oxygen transported by aerenchyma. The release of oxygen from roots should lead to aerobic degradation of organic matter, enough oxygen to support mycorrhizae and an aerobic fauna associated with the roots. Apparently *Lobelia* roots have vesicular–arbuscular mycorrhizae, and are devoid of root hairs; the mycorrhiza may assist with phosphorus uptake, since the solubility of phosphorus declines under aerobic conditions. Thus, Pedersen *et al.* observe, sediment oxygenation may have the beneficial effects of allowing aerobic decomposition to occur, and mycorrhizae to operate, but the mycorrhizae may be necessary because of the low solubility of phosphorus and loss of nitrogen by denitrification caused by the enhanced oxygen levels in the substrate. None the less, it appears that oxygen release from roots can be an important factor in interactions among marsh plants, particularly when species lacking aerenchyma grow near those with aerenchyma. Further, the roots of such plants may allow aerobic fauna to thrive within otherwise anoxic sediment, raising the possibility of further mutualisms between microbes and plants. The effects of such positive interactions combine with those of competition to produce typical saltmarsh zonation (Figure 2.13).

In general, ecologists may have paid insufficient attention to the importance of co-operative interactions in nature (Keddy 1989a, b). There is growing evidence that positive interactions are important (Bertness and Shumway 1993; Castellanos *et al.* 1994; Bertness and Leonard 1997) particularly in stressful habitats (Bertness and Hacker 1994). The data in Figures 6.8 and 6.9 show that it would be possible to test for co-operation and competition in the same experiment, and it increasingly appears that the relative importance of positive interactions as well as negative interactions needs to be simultaneously assayed in field experiments.

Conclusions

Competition appears to be a ubiquitous process in wetland plant communities. That is, we may assume that many of the species distributions we see are the result of stronger competitors excluding weaker ones. Competition gradients and asymmetric interactions provide two useful conceptual tools for understanding and predicting such patterns. The process of competition is closely linked to several preceding chapters. For example, there is an obvious relationship between the patterns of zonation (Chapter 2) and competition. Facilitation (pp. 351–353), disturbance (Chapter 6) and grazing (Chapter 8) may be thought of as counter-balancing factors that reduce the degree to which competitive dominants can monopolize habitats, or as factors that allow weaker competitors (including invasive aliens) an opportunity to invade wetland communities. Fertility gradients (Chapter 5) appear to control both the intensity of competition and the degree to which above-ground interactions predominate; many of the effects of eutrophication can be understood as a consequence of these changes in the nature of competition.

We know rather less about the role of competition among animals in wetlands. Competition has been documented among, for example, birds (e.g. Miller 1968; Robertson 1972; Picman 1984), amphibians (e.g. Wilbur 1972; DeBenedictis 1974; Griffiths *et al.* 1993; Pehek 1995; Beebee 1996), fish (e.g. Werner and Hall 1977; Werner 1984; Wootton 1990; Todd and Davis 1995; Sakakura and Tsukamoto 1998), zooplankton (Smith and Cooper 1982; Gilbert 1988, 1990; Rothhaupt 1990), and benthic invertebrates (Hart 1983; Hemphill and Cooper 1983; McAuliffe 1984). As I have discussed elsewhere (Keddy 1989a), it is difficult to assemble general principles from studies of selected pairs of species, and so, at present, our knowledge of competition in wetland plant communities is greater than that for animals. It does appear generally true for animals as well as plants that habitats with the best supply of resources will be seized by the strongest competitors (Wisheu 1998), weaker competitors being excluded to less desirable habitats. Further animal studies would be helpful, particularly if they test for general patterns rather than assessing selected pairs of organisms on a case-by-case basis.

8 · *Herbivory*

Introduction

In ecological communities, some processes add biological material, and other processes remove it. The former include photosynthesis, growth and reproduction; the latter include fire, decomposition and herbivory. Processes that remove biomass are generally considered to be disturbances (Chapter 6). Disturbances can be considered either abiotic (flooding, fire, ice scour, landslides) or biotic (herbivory, burrowing, trampling). In some ways these disturbances are similar; in other ways they are different. They are similar in that standing crop is temporarily reduced, and light penetration is increased; they are different in that herbivory has the potential to be far more selective than other disturbances. I have decided that this difference is important enough to treat herbivory in its own chapter. At the risk of rendering this chapter irrelevant, it is important to emphasize that herbivory by large animals is remarkably inefficient at removing biomass from plant communities. Study after study over the past 50 years has demonstrated the same startling result: a vast majority of plant biomass goes directly into the decomposer food web, where it is processed by small invertebrates and micro-organisms. This generalization ranges from arid tropical grasslands (Deshmukh 1986) to temperate saltmarshes (Adam 1990), although aquatic algae are an apparent exception (Cyr and Pace 1993). Further, fire often removes a substantial portion of biomass not consumed by decomposers; in tall grass areas like the Serengeti plains, more than half of the plant biomass is burned (Deshmukh 1986). Therefore, while the study of herbivores and their impacts on vegetation may seem appealing, we should remember that most such studies often deal with animals that are processing less than 10% of the biomass in the vegetation.

Given that animals remove such a small proportion of the biomass, it seems reasonable to question whether it has enough effect in a community to produce measurable changes in properties of that community such as biomass, life form, production or diversity. If so, one next wants to determine what those changes are. To answer such questions, there are apparently three lines of investigation open to us:

(i) Comparative and evolutionary: do plants possess traits that confer resistance to herbivory (e.g. thorns, spines, secondary metabolites)?
(ii) Empirical: are there patterns in the importance of herbivory measured across a wide array of ecosystem types?
(iii) Experimental: if the density of herbivores is manipulated, do properties of the plant communities change?

A majority of the published studies of herbivory in wetlands does not fit into any of these categories. Rather, they are observations upon the diet of commercially valued animal species. Such studies generally do not allow us to determine what, if any, effects herbivores have upon wetlands. Therefore, we cannot use these studies to determine the importance of herbivores in controlling the structure or function of wetland communities, or to compare herbivory with other factors such as flooding or competition. None the less, such studies provide the historical foundation for future work on herbivory. Further, they provide some information on the natural history of wetlands. Therefore, let us begin this chapter by looking at a few historical examples of herbivory by commercially valued species. Then we can turn to that smaller subset of studies that provide evidence about the impacts of herbivores on wetlands. We will explore them under the three classes of evidence listed above: (i) comparative/evolutionary, (ii) empirical and (iii) experimental.

Field observations on wildlife diets

Bottom-up or top-down? A brief digression

This chapter largely side-steps the issue of top-down and bottom-up control. The terms refer to the issue of whether the composition of ecological communities is controlled from the top of food webs down, by predators, or whether the control is exercised from the bottom up, with the consumers being controlled by plants (e.g. Hunter and Price 1992; Power 1992). There are well-known cases of top-down control. In 1925 for example, Tansley and Adamson were able to show that rabbits controlled the vegetation composition of British grassland. Certainly, herbivores have significant impacts upon individual plant species (Louda *et al.* 1990; Marquis 1991). In lakes, control can extend across four trophic levels (Carpenter and Kitchell 1988), with secondary carnivores suppressing primary carnivores, thereby unleashing herbivores to decrease the standing crop of plants. One could therefore reasonably ask whether wetlands, like lakes, can be controlled from the top by predators. I will avoid this debate here.

At the very least, we can be certain that there is bottom-up control, for the very simple reason that, without plants, the consumers disappear (Hunter and Price 1992). It is therefore quite reasonable to start off with the assumption that the vegetation in wetlands controls wildlife, both through habitat and food. But, as for the second issue, whether the consumers also influence or control the producers, this turns out to be much less clear-cut. Resurrecting Hairston *et al.* (1960) we can naively observe that most wetlands are green; since the plants are not eradicated by herbivores, something else must be controlling herbivore abundance. So far, it seems plausible. But then, as White (1993) argues, a good deal of this green matter has such low nitrogen content that it hardly qualifies as food anyway, and the growing literature on secondary metabolites (Rosenthal and Berenbaum 1991) suggests that much visually apparent green food is well protected from herbivores. We shall see that both of these constraints upon herbivory have been demonstrated in wetlands. Therefore, the issue of whether herbivores control the abundance of plants, and the composition of wetlands, is open for evaluation.

Secondly, apparently, clear-cut dichotomies like this, while attractive, often turn out to be misleading (Dayton 1979; Mayr 1982; Keddy 1989a). It is possible that both operate simultaneously, that neither operates except for rare exceptions, or that other factors such as habitat productivity (Oksanen 1990), habitat heterogeneity (Hunter and Price 1992) or omnivory (Power 1992) may override the apparent dichotomy.

Oksanen *et al.* (1981) propose that, rather than a simple dichotomy of 'bottom-up' and 'top-down', there are at least three different possible kinds of grazing systems, and the kind of system found depends upon the primary productivity of a site, including the supply of soil resources to plants. According to their model, herbivore pressure should be most severe in relatively unproductive environments. As primary productivity increases, the impact of herbivory should decline because growing abundance of the herbivores allows predators to survive and regulate herbivore populations. In very productive systems, herbivory again becomes important owing to the occurrence of predators upon the predators, which releases the herbivores from regulation. Oksanen *et al.* present a model, building upon work by Fretwell (1977) that shows how such transitions in herbivore–plant relationships might occur, and they present some data that is qualitatively consistent with these kinds of changes. There are, in fact, many possible complex feedbacks, such as animals increasing the rates of nitrogen cycling, fertilizing plants with their waste products, and even altering competition between plants and soil microbes for nitrogen (McNaughton *et al.* 1988). Hence, generalizations about interactions between herbivores and plants, while highly desirable, await further experimental testing of such models. In the meantime, simplistic dichotomies should be viewed with caution.

Finally, the emphasis in the last four chapters has been upon the environmental factors that control the composition of wetlands. In this chapter, I wish to focus upon the role of herbivores as another possible control agent upon plant communities in wetlands, assuming, in turn, that if herbivores alter the abundance of plants, the effects will spill over into other species dependent upon those plants. For all these reasons, therefore, this chapter will assume that plants control the consumers by producing food and shelter; having seen some examples of this, we will move on to the unresolved issue: is there any evidence that herbivores can control composition in wetlands and if so, when, where and how do they do it?

It has long been observed by naturalists that animals feed on wetland plants. Most of us will have seen one or more examples: a beaver lodge made of willow trees, a muskrat house made of *Typha* and *Sparganium*, a moose munching on water lilies at sunset, or a duck feeding on *Potamogeton*. Wildlife biologists have investigated this in two principal ways: they have observed feeding by wild animals, and they have studied faeces to reconstruct diets. They have then tabulated the biological utilization of wetland plants for the use of managers, as we saw in Tables 1.8 and 1.9. Let us consider four other examples.

Waterfowl consume both plants and invertebrates. Egg-laying females and young tend to emphasize invertebrates in their diets, presumably because of the higher food quality of animal protein. Even so, Figure 8.1 shows that some species such as the Northern Pintail and Gadwall consume plants as one-quarter of their diet. Most such studies focus on the food quality of plants for waterfowl, however. Whether the waterfowl, in turn, affect the plants is much less explored.

Small mammals such as muskrats have long been studied because of their importance to the fur industry. Fritzell (1989) and Murkin (1989) have reviewed some aspects of muskrat grazing in prairie wetlands. They report that muskrats not only consume large amounts of fresh plant material, but the amount of cattail destroyed and not consumed may be two to

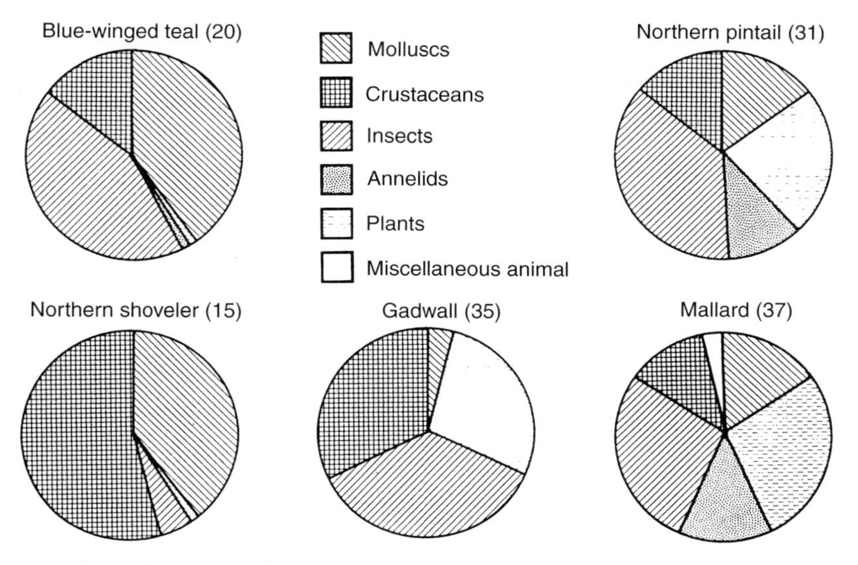

Figure 8.1 Plants can make up a significant proportion of the diet of waterfowl (n = sample size) (from van der Valk and Davis 1978).

three times that. Around their lodges, muskrats may remove 75% of the above ground standing crop in areas 4–5 m in diameter. The oft-reproduced drawing in Figure 8.2, shows the observed interactions between muskrat grazing, water level fluctuations and marsh dynamics (see also Weller 1978).

Although muskrats are a dominant herbivore in prairie marshes, with a large number of descriptive papers published, Fritzell notes that 'the details of muskrat ecology in prairie wetlands are poorly quantified'. A fourth factor, not shown in Figure 8.2 is fire; Smith and Kadlec (1985a) found that grazing intensity was particularly high in burned areas, where it ranged from 48% for *Typha* to 9% for *Scirpus maritimus.*

Many fish are also dependent upon wetland plants. A striking example is the fish that feeds upon fruits and seeds in floodplain forests (Goulding 1980). The Amazon basin has some of the largest areas of flooded forest in the world – some 70000 km^2. Some trees are flooded to depths of 15 metres and for up to 10 months of the year. Plant germination and growth appear to be restricted to the few months when the floodplain is drained. Up to 3000 species of fish may inhabit this region. Of the more than 1300 described to date, about 80% are either catfishes or characins. The latter group has radiated extensively in the Amazon lowlands, and

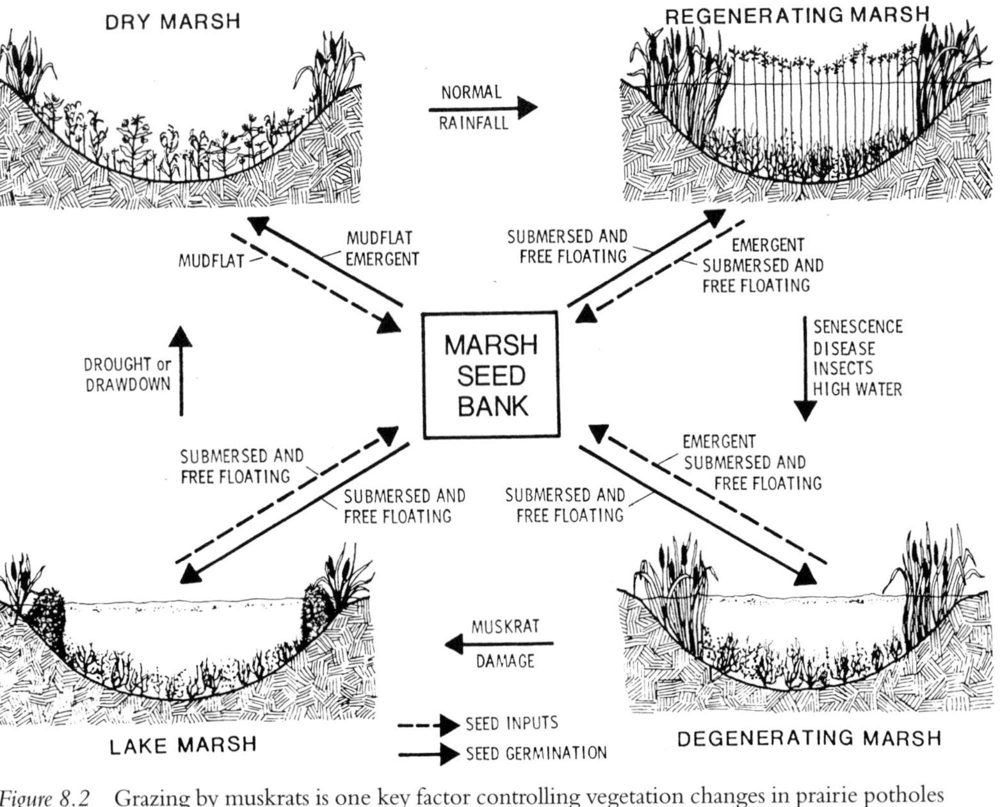

Figure 8.2 Grazing by muskrats is one key factor controlling vegetation changes in prairie potholes (from van der Valk and Davis 1978).

includes carnivores, frugivores, detritivores and planktivores. There are some 38 genera and 21 families of plants exploited by fish. It is clear that some (many?) of the species are dispersed by the fish. Goulding closes by suggesting that this is very important for human welfare too – some 75% of the commercial catch may originate in flooded forests.

Herbivores in African grasslands

As the three examples above already indicate, much of the literature on herbivory in wetlands is focused upon year round inhabitants of wetlands, and, in particular, upon wetter habitats such as macrophyte communities in shallow water. By contrast, peatlands have received comparatively less attention, as have seasonally flooded grasslands. To keep things in perspective, let us therefore return to a topic introduced in Chapter 1: the seasonal use of wetlands by large African mammals (Western 1975; Sinclair and Fryxell 1985), and then explore peatlands. Recall that many of the large ungulates on the African plains graze in wetlands during the dry season, and then use the surrounding grasslands in the wet season. As a consequence, each vegetation type receives a period free from herbivory, and as well, by using the combined productivity of this range of habitats, many more animals can be supported (Sinclair and Fryxell 1985). The ungulate populations in Africa are large and diverse; for example, Sinclair (1983) points out that the one family, the Bovidae (in the order Artiodactyla), containing the buffalo and antelope, has as many species (78) as the most diverse rodent family, the Muridae. Some of these bovids are adapted to wetlands, such as the kob and lechwe. Ungulates, as a whole, have four main habitats: forest, savanna, desert and wetland (Sinclair 1983); and the wetlands range from forested swamp to *Papyrus* marshes to seasonally flooded wetlands (Thompson and Hamilton 1983; Howard-Williams and Thompson 1985; Denny 1993a, b). Most large mammals use these wetlands at some time of year (Table 8.1), and distance from water is a good predictor of biomass of herbivores (Figure 8.3), but the shortage of water has placed constant selective pressure upon herbivores. There have been two main evolutionary responses. Independence from water requires a shift from herbivory on grasses to browsing upon shrubs; browsers are less dependent upon water and wetlands (Figure 8.3). Further, reproduction is timed to coincide with the rainy season when the habitat is as productive as possible; this is found in species such as the elephant, white rhinoceros, zebra, hippopotamus, warthog, buffalo, giraffe, and kudu. The importance of seasonal surges in

Table 8.1. *Seasonal habitat changes in the large herbivores of the Rukwa Valley, Tanzania*

Animal species	Time of year											
	Jan.	Feb.	Mar.	April	May	June	July	Aug.	Sept.	Oct.	Nov.	Dec.
Elephant	Acacia and escarpment woodlands						*Flood plain*				Woodlands	
Buffalo	Woodlands – *Lakeshore and delta grasslands*						*Flood plain*				Woodlands	
Hippo	*Fringe river and delta grasslands*				Wander widely along drainage						*River fringe*	
Puku	*Delta and lake shore grassland all year*											
Topi	Perimeter grassland woodland			*Lakeshore and delta grassland*				*Vossia pasture*		Acacia		
Zebra	Acacia woodland		Perimeter grassland	Acacia parkland			*Flood plain grassland*			Acacia woodland		
Bohor Reedbuck				*Flood plain grassland*								
Eland	Dry perimeter plains woodland			*Delta grasslands and Vossia pasture*						Acacia		
Giraffe and Impala					Acacia grassland							
Warthog				Acacia grassland and forest edge								
Waterbuck, Duiker, Baushbuck and Steinbuck					Woodlands							

Source: After Vesey-FitzGerald (1960).

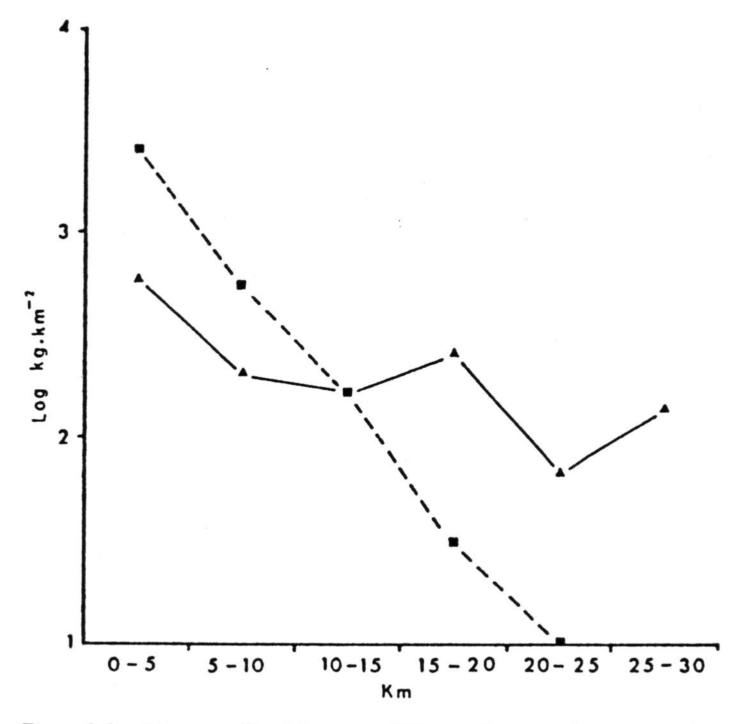

Figure 8.3 Biomass of herbivores at different distances from water during the dry season, Amboseli, Kenya. Water independent browsers (triangles) are less restricted than water dependent herbivores (squares) (from Western 1975 in Sinclair 1983).

production is illustrated by an exception to the above rule (Sinclair 1983). 'Lechwe live on riverine floodplains that are seasonally flooded ... Optimum food conditions occur when water is at the lowest level exposing the greatest area of floodplain, and it is then that the peak of births occur.'

The use of wetlands by many animals not normally considered 'wetland' animals is an important conclusion from such work. It is further evidence of the importance of wetlands to wildlife.

Mires, grasslands and sheep in Great Britain

In contrast to the African plains, in the peatlands of the British Isles, herds of slugs and sheep may be the most important herbivores. Overall, there are more than a million ha of moorland in Britain (Miller and Watson 1983). The principal habitat gradients are soil moisture, soil nutrient supply, and sheep grazing intensity. These areas have been extensively modified by humans. The original oak forests were cleared during Roman and Mediaeval times, and eventually replaced by scrub and grassland following the use of the mountains for grazing. The density of the main vertebrate herbivores is estimated as 50 sheep, 65 red grouse, 10 red deer and 16 mountain hare km^{-2} in the highlands of Scotland. Even so, less than 10% of the primary production of *Calluna vulgaris* is actually consumed by herbivores (Miller and Watson 1983).

Grazing in moorland was studied as part of the International Biological Program. One study area, Snowdonia, northern Wales, contained a mosaic of vegetation types including grassland, mire and heath; unfortunately for our purposes, the grasslands received particular attention, although *Juncus* and *Eriophorum* mires occurred intermixed with the grasslands (Perkins 1978). The slug populations are very variable in time and space, but species such as *Agriolimax reticulatus* and *Arion intermedius* can reach densities exceeding ten per m^2. They consume approximately one gm^{-2} per month (Lutman 1978), based on data emphasizing grasslands within the peatland complex. Sheep are the dominant herbivore vertebrate, and their effects have profoundly affected British vegetation. Brasher and Perkins (1978) report that herbivore densities can range from 5–19 animals ha^{-1}. The sheep show a preference for grassland areas (*Agrostis–Festuca* swards), and reject sedges, rushes and herbs, many of which are typical of wetter sites. Red Grouse are often studied as well because of their hunting value, but they feed primarily upon *Calluna vulgaris* shoots, eating only a negligible proportion of the primary production on their territories (Miller and Watson 1978); they therefore have rather little effect themselves upon peatland vegetation. The principal effects of grouse are likely to arise from the human practice of burning to improve the uplands for hunting grouse. This changes plant species composition, stimulating the growth of *Calluna* in particular, and may have deleterious effects upon the development of wet blanket bog (Rawes and Heal 1978). Further, the burning leads to volatilization of nitrogen and leaching of potassium from the remaining ash (Miller and Watson 1983).

At a second study site, in the Pennines, similar observations were made. Sheep again are dominant, have grazed this landscape for some 1000 years, and concentrate upon the smaller preferred grasslands within the mire and peatland complex. Rawes and Heal (1978) conclude that sheep grazing '. . . probably exerts more influence on the vegetation and fauna than any other biological factor'. Sheep grazing has increased the amount of grassland at the expense of the bog, as well as suppressing re-growth of woody species. An enclosure experiment showed that after 7 years of excluding sheep, the biomass increased by 50%, and the number of plant species declined from 93 to 67. In contrast, grazing on the blanket bog itself is so low that the sheep appear to have 'little noticeable effect', although comparison with a bog that has been grazed continually for many years shows that grazing reduces *Calluna vulgaris* and increases *Eriophorum vaginatum*.

Comparative studies

Food quality and nitrogen content

Nitrogen content is thought to be the most important factor determining food value of plants (Lodge 1991; White 1993). The distribution of nitrogen in plants could produce fertility gradients similar to those experienced by plants, as we have seen in Chapter 5. We have also already seen that nitrogen content of aquatic plants is frequently well below 5% (Table 5.2; Gopal 1990), and Lodge (1991) shows that emergent, floating and submersed macrophytes, as well as algae, all have similar nitrogen contents, usually of 2 to 3% (with extremes from 1 to at least 5%). These are very low values for supporting grazing animals. White describes attempts to control the *Salvinia molesta*, an aquatic fern from Brazil that has become a serious weed in many tropical regions. Initial attempts to import and establish insects from Brazil to control *Salvinia* in Australia and Papua New Guinea had variable success; at concentrations of nitrogen of 1% or less dry weight, the imported pyralid moth could not establish. 'However, increasing the level of nitrogen in the fern to only 1.3% dry weight by simply adding urea fertilizer to the water can cause an explosive increase in the abundance of the moth and severe damage to the plants' (p. 77). The species of weevil introduced from Brazil to Australia to combat *Salvinia* was also limited by nitrogen availability. In contrast, when Lodge (1991) studied herbivory preferences of the crayfish *Orconectes rusticus* among 14 submersed macrophytes, he found clear preferences for certain species, but he was unable to detect statistically significant differences in nitrogen content among the plants. Simple comparisons of plant tissue may conceal real differences in nitrogen content if herbivores are only consuming selected tissues; Sinclair (1983) and White (1993) have described many examples of herbivory where new growth occurring after damage is preferentially consumed, and White is of the opinion that the new growth has the highest tissue nutrient concentrations. He gives the example of Green turtles (*Chelonia mydas*), marine herbivores that feed on the aquatic vascular plant called seagrass (*Thalassia testudinum*). These turtles maintain areas of cropped seagrass and feed upon the flush growth in the cropped area, ignoring adjacent stands of tall grass.

Morphological defence

The presence of morphological defences against herbivores might at first be taken as unambiguous evidence that grazing is having a major effect upon a community. But, there are at least two critical assumptions in such arguments. First, we must assume that we have correctly and unambiguously interpreted the function of the alleged anti-herbivore structure. For example, do thorns on vines that scramble along shorelines (e.g. *Smilax* spp.), act to deter predators, or are they a morphological device for climbing on other plants? This first problem of interpretation always complicates comparative evidence.

Spines, thorns and prickles are generally thought to deter herbivory (e.g. Crawley 1983; Marquis 1991; Raven *et al.* 1992) and thus their presence would appear to offer relatively convincing evidence that herbivores are important in wetlands. Although many plant taxa are well protected by such structures, it would appear on first inspection that all such traits are relatively uncommon in wetland plants. Figure 8.4 shows a few selected examples. I know of no study which systematically compares the abundance of anti-herbivore defences in wetland plants with plants in other habitats, but desert plants, for example, appear far better protected than wetland plants. The absence of anti-herbivore structures might be a significant attribute of wetland plant communities.

Where anti-herbivore traits are present, there is evidence that suggests herbivory is less important under water than above it. *Pontederia cordata*, for example, has showy flower stalks, but once the flowers are pollinated, the stem bends to hide the flower stalk under the water (Figure 8.4(*c*)). Similarly, *Nymphaea odorata* has conspicuous flowers on the surface of many northern lakes and slow-moving rivers, but once the flowers are pollinated, the peduncles coil like a spring, pulling the fruits down to the bottom of the lake (Figure 8.4(*d*)).

The second problem with comparative data is that it confuses time scales of evolutionary and ecological processes. The presence of an anti-herbivore structure tells us that herbivores shaped the evolution of a taxon, but it in no way demonstrates the active occurrence of herbivory in present day communities. This caveat is by no means trivial or pedantic.

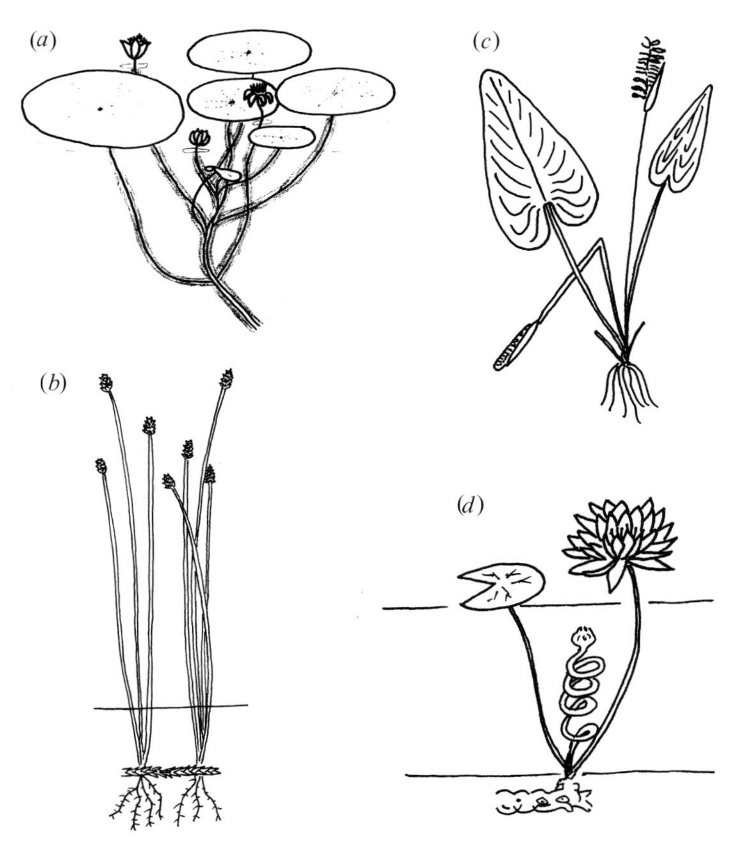

Figure 8.4 Some visible traits that may confer resistance to herbivory (*a*) gelatinous coating on stems and foliage (*Brasenia schreberi*): (*b*) buried rhizomes (*Eleocharis palustris*); (*c*) peduncle that bends to insert fruits under the water; (*d*) peduncle that coils to withdraw fruits below the water (*Nymphaea odorata*).

We know that, in relatively recent times, only about 10 000 years BP, both North America and Australia lost an entire megafauna (Figure 8.5). It has been argued that many plants possess adaptations to dispersal by large mammals that are now extinct (Janzen and Martin 1982). It seems equally plausible that many plants could have adaptations to protect themselves from herbivores that no longer play a role in determining wetland community structure. Further into the past, in the late Mesozoic, we also find many herbivorous dinosaurs, and some of these are thought to have been semiaquatic. The effects of herbivory on wetlands may thus extend back hundreds of millions of years. But the evolutionary history of wetland

Figure 8.5 Some examples of the megafauna that became extinct at the time humans arrived in North America (top) and Australia (bottom). **North America**: 1 *Platygonus, Castoroides*. 2 *Bison latifrons, Noth rotheriops*. **Australia**: 3 *Diprotodon optatum, Zygomaturus trilobus, Euowenia grata, Thylacoleo carnifex*. 4 *Procoptodon goliah, Sthenurus maddocki, Sthenurus atlas, Protemnodon brehus, Macropus ferragus*. 5 *Progura gallinacea, Genyornis newtoni, Megalania prisca, Wonambi naracoortensis, Zaglossus ramsayi*. Human included for scale (adapted from Martin and Klein 1984).

species does not tell us very much about extant forces shaping today's wetland communities.

Secondary metabolites

Chemical traits are less obvious than morphological ones, but may be equally important in deterring herbivory. While some plant compounds have obvious roles to play in photosynthesis, growth and reproduction, others do not. These latter 'secondary metabolites' were thought to be just waste products. It has now become clear that many of these compounds play active and important roles in defending plants against herbivores (Marquis 1991). There are three main groups of anti-herbivore compounds: terpenes, phenolics and nitrogen-containing secondary products (e.g. Taiz and Zeiger 1991).

There is a striking lack of information on anti-herbivore defence compounds in wetland plants in standard references such as Rosenthal and Berenbaum's (1991) compendium. This could be a consequence of either one of two causes: the actual rarity of defence compounds in wetlands (a phenomenon of real ecological interest) or the lack of interest in plants in wetlands (a phenomenon of interest only to those who study the behaviour and sociology of scientists). There are passing references to glucosinolates (Louda and Mole 1991), coumarins (Berenbaum 1991) and possibly iridoid glycosides (Bowers 1991) in protecting wetland plants from herbivorous invertebrates. Coumarins have been found in more than 70 plant families, and these include important wetland families such as the Cupressaceae, Araceae, Cyperacae, Poaceae and Juncaceae (Berenbaum 1991).

In contrast with these sources, McClure (1970) documents a prominent role for secondary metabolites in aquatic plants; going from wet to dry, he found that flavenoids are predominant in free floating species, phenols and flavenoids are found in submerged and emergent taxa, and alkaloids dominate among rooted floating-leaved species (e.g. the Nymphaeceae). In contrast, terpenoids are apparently more common in waterlogged soils and seasonally flooded areas (e.g. Cyperacae, Poaceae, Acanthaceae). Ostrofsky and Zettler (1986) also comment upon the lack of morphological defences in submerged aquatic plants. They therefore examined 15 species of aquatic plants including *Cabomba caroliniana*, *Vallisneria americana* and nine species of *Potamogeton* to assay for alkaloids, finding between 0.13 and 0.56 mg g^{-1} dry weight, values that are 'low, but certainly within a range which is pharmacologically active, and consistent with a potential role as herbivore deterrents'. The actual kind of alkaloid varied greatly among species, with the *Potamogeton* species being no more similar to each other than to other genera. Gopal and Goel (1993) list other examples such as fatty acids, allomones, mustard oils and steroids, but in general the role of such secondary metabolites is still poorly documented and even more poorly understood. The compounds may provide defence against herbivores, but there may be other functions such as antimicrobial activity and allelopathic interactions with competing neighbours including planktonic algae.

Simply screening for the presence of possible defence compounds in wetland plants, while helpful, still leaves important unanswered questions.

We need to know whether these compounds are actually able to reduce impacts of herbivores, and whether the production of defence compounds varies among habitats. McCanny *et al.* (1990) not only evaluated the anti-herbivore defences in 42 wetland plant species, but they also tested whether anti-herbivore defences were increased in infertile habitats where the costs of grazing to plants should be greater (Coley 1983). First they extracted secondary metabolites from the test plants, and mixed them into the diet of an insect herbivore. The larvae showed reductions in growth of up to 50%, thereby showing some evidence of anti-herbivore compounds in wetland plants. There was no difference in toxicity of forbs and graminoids. The food quality index (as measured by the performance of the insect herbivore) was then plotted against the fertility of the habitat typical of each plant species. There was no relationship between the food quality index and soil fertility, plant biomass (Figure 8.6 *top*) or plant relative growth rates (Figure 8.6 *bottom*). These results tend to agree with the above observations on morphological traits: there is only limited evidence for anti-herbivore defences in wetland plants, and these do not vary much among species or habitats.

In conclusion, while there is some evidence that morphological traits or secondary plant metabolites play a role in defence against herbivores, the evidence is far from conclusive. The study of effects of grazing upon existing communities requires evidence outside the comparative realm.

Empirical relationships

One of the most fundamental properties of grazing is the proportion of the primary productivity that is consumed. This proportion can be considered a measure of the 'importance' of herbivory in a particular habitat. Cyr and Pace (1993) compiled estimates of this property for a wide array of aquatic and terrestrial habitats: the producers were phytoplankton ($n = 17$), reef periphyton ($n = 8$), submerged macrophytes ($n = 5$), emergent macrophytes ($n = 14$), and terrestrial plants ($n = 67$). Figure 8.7 shows the importance of herbivory when these are lumped into three groups: aquatic algae, aquatic macrophytes, and terrestrial plants. A striking result from this figure is that aquatic macrophytes are much more like terrestrial plants than aquatic algae. This echoes earlier themes in fertility, where we were challenged to decide whether wetland plants were limited by phosphorus (as with algae) or by nitrogen (as with many terrestrial plants), finding that both phosphorus and nitrogen could be important depending

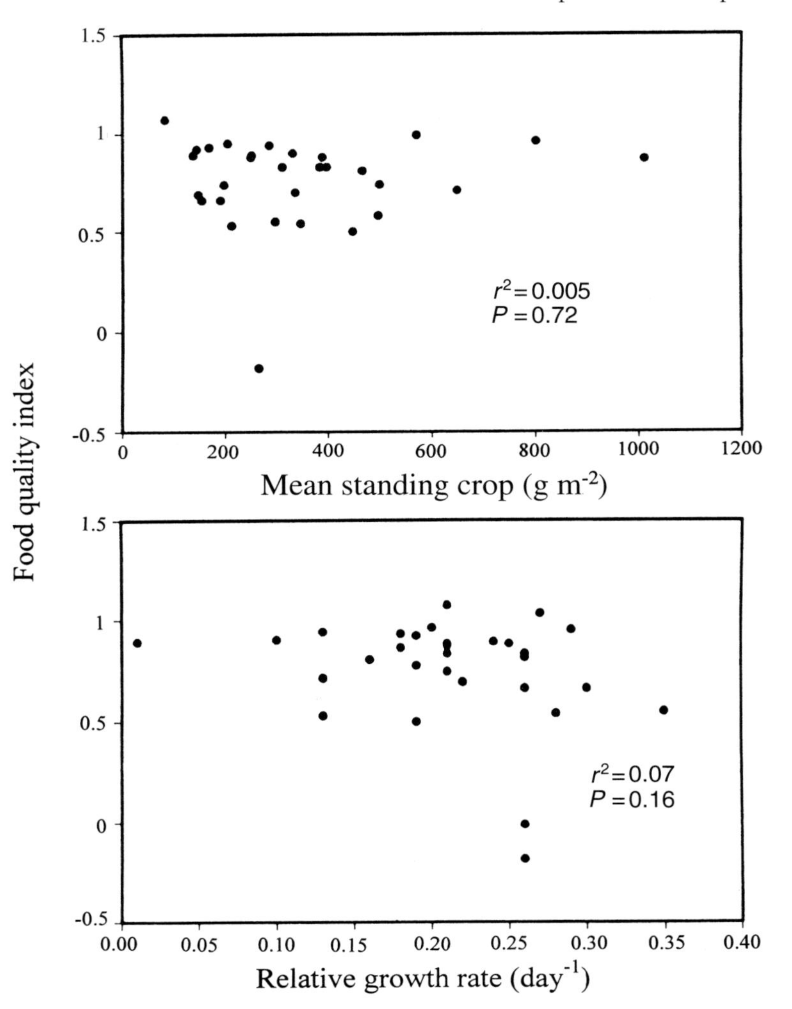

Figure 8.6 The food quality of 30 kinds of wetland plants is correlated neither with the biomass of the habitat (*top*) nor with the relative growth rate of the species (*bottom*) (after McCanny *et al.* 1990).

upon the type of wetland. The median proportion of productivity removed by herbivores of aquatic macrophytes is some 30% (compared to 79% for algae and 18% for terrestrial plants). Plotting the rate of removal of biomass by herbivores against primary productivity (Figure 8.8 *top*) gives a linear relationship with a slope not different from one, suggesting that herbivores remove the same proportion of primary productivity

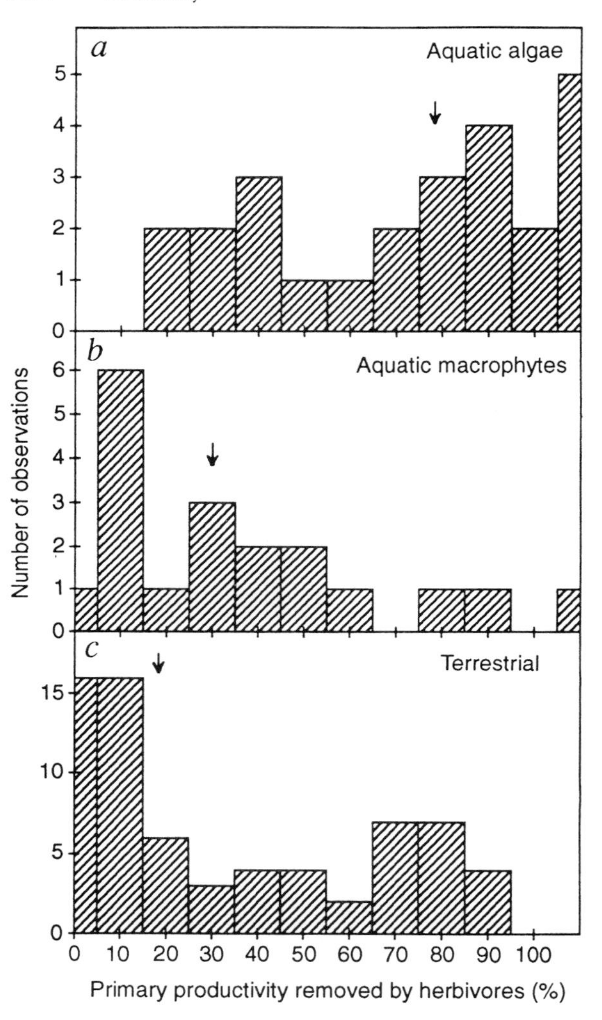

Figure 8.7 Frequency distributions of the proportion of annual net primary productivity removed by herbivores in (*a*) aquatic algae (phytoplankton, $n = 17$, and reef periphyton, $n = 8$); (*b*) submerged ($n = 5$) and emergent ($n = 14$) vascular plants; and (*c*) terrestrial plants ($n = 67$). Arrows indicate median values (aquatic algae, 79%; aquatic macrophytes, 30%; terrestrial plants, 18%) (from Cyr and Pace 1993).

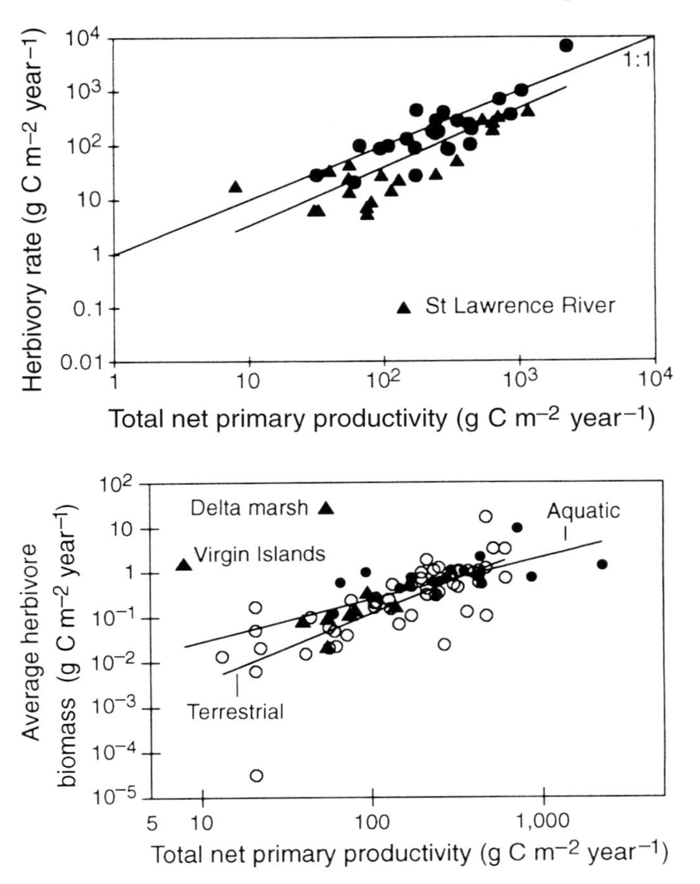

Figure 8.8 Rate of grazing (*top*) and herbivore biomass (*bottom*) both increase with net primary productivity (from Cyr and Pace 1993).

across a wide range of fertility levels. The top of Figure 8.8 also shows that consumption rates are apparently an order of magnitude lower in macrophytes (triangles) than algae (circles).

In the rest of their analyses, Cyr and Pace regrettably combine algae and macrophytes into one 'aquatic' category for comparisons with terrestrial plants. However, certain general conclusions about herbivores in wetlands can be extracted. Figure 8.8 (*bottom*) plots the biomass of herbivores against net primary productivity in all habitats. The two triangles at the upper left are submerged macrophyte beds where herbivore biomass was strikingly high. (The circle at the lower left is a terrestrial tundra site.) Excluding the two outlying triangles, herbivore biomass increases significantly with productivity, and, also excluding the outlying circle, there is no significant difference between the lines for aquatic and terrestrial habitats. Therefore, for a given level of net primary productivity, herbivores reach similar average biomass in aquatic and terrestrial ecosystems. Important questions about wetlands remain unanswered, and Figure 8.7 suggests that much could be learned by treating wetlands as a separate category in future work of this sort. This criticism aside, Cyr and Pace have provided an important introduction to the study of herbivory in wetlands.

Field experiments

Here is where we may expect to get a more complete answer on the effects of contemporary herbivores on wetlands. The principle is simple (Bender *et al.* 1984): exclude herbivores from some areas, and compare the vegetation in these areas with other control plots. One could also add other treatments that remove only selected herbivores, or even add certain species of herbivores. (Remarkably, while Crawley's 1983 monograph on herbivory devotes an entire chapter to plant populations, it hardly mentions the importance of properly designed exclosure experiments to the study of herbivory!) We again have rather few examples of such studies in wetlands; studies of large herbivores are particularly lacking.

Lodge (1991) has reviewed some 25 studies of herbivory in wetlands, covering examples from invertebrates grazing upon submersed macrophytes to mammals and birds grazing upon emergent macrophytes (see also Brinson *et al.* 1981). Herbivore impact, as estimated by the difference in biomass between grazed and ungrazed plots, ranged from 0 to 100%, with many values in the 30 to 60% range. He concludes that many herbivores can therefore have a substantial effect upon macrophytes. Further, these experimental studies are supplemented by observational reports for many other herbivores including manatees, fish, isopods and insects. Let us now turn to a few specific examples.

Effects of snails in lakes

Sheldon (1987) explored the effects of snails on aquatic macrophyte communities. She manipulated snail density indirectly by manipulating the abundance of predatory fish. When snail density increased, plant

Table 8.2. *Snail density and plant community responses in Christmas Lake fish exclosures, controls, and non-experimental areas*

	Controls	Exclosures	Christmas Lake
Mean number of snails/kg plant	8.95	93.9	4.2
Mean diversity (*H*)	0.758	0.063	0.815
Mean biomass (g/m^2)	678	457	1823
n	11	12	8

Source: From Sheldon (1987).

biomass declined by about one-third, and species richness dropped by a factor of four (Table 8.2, middle column). Moreover, there was a direct relationship between species richness and snail density (Figure 8.9). A possible criticism of these results, apart from the indirect way in which the snails were removed, is that while this shows that artificially increasing snail density will have dramatic effects on vegetation, it does not show that existing densities of snails have much of an effect. This latter problem is an inherent weakness in any experiment in which experimental treatments only increase densities of grazing animals.

Brönmark (1990) queried whether the effects of the exclosures reported in Table 8.2 actually provided good evidence of herbivory by snails; intricate indirect effects and chemical interactions may have occurred instead. Snails, he said, tend to graze on epiphytic algae, bacteria, and detritus rather than on macrophyte tissue itself, leading to suggestions that grazing snails may benefit plants by removing the cover of epiphytes (Thomas 1982). Certainly, a laboratory experiment in which snails (*Lymnaea peregra* and *Planorbis planorbis*) were allowed to graze on shoots of the submerged macrophyte *Ceratophyllum demersum* showed that growth rates were actually higher on the grazed *C. demersum* (Brönmark 1985). The removal of the epiphytes growing on the *C. demersum* might have increased light availability or reduced competition for nutrients including bicarbonate. If plants equally provide snails with habitat and shelter from predators, then the plant–snail interaction might actually be one of mutualism (Carpenter and Lodge 1986). Brönmark (1990) continues that aquatic plants do seem to be rather well protected by secondary metabolites. As well, other aquatic invertebrates, such as trichopteran or lepidopteran larvae, are well-known consumers of aquatic plants and, along with crayfish, may be implicated in producing the observed effects.

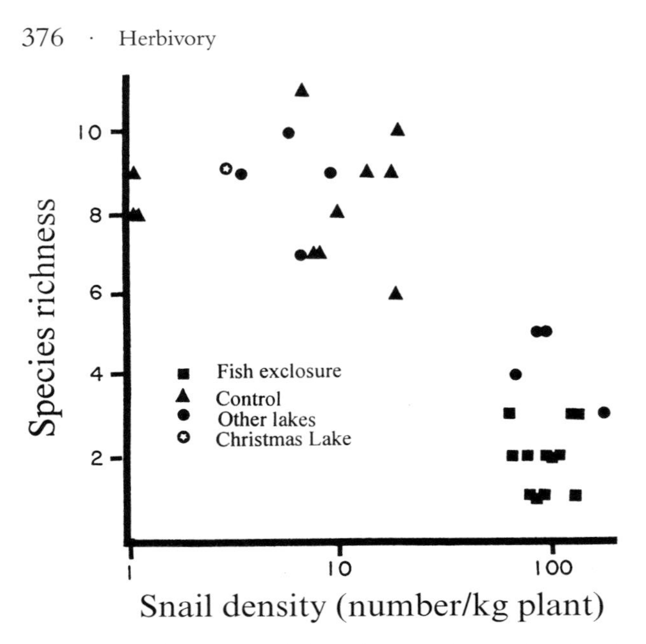

Figure 8.9 Plant species richness is inversely correlated with snail density in Minnesota lakes (from Sheldon 1987).

Sheldon (1990) agreed that, when there is a choice, snails prefer epiphytes to macrophytes, and may select plants with green algae or diatoms as opposed to those with cyanobacteria. The newly growing shoots of macrophytes may, however, be particularly vulnerable to snail damage. All the same, concedes Sheldon, the experiment probably measured the effects of herbivores rather than of snails alone.

Effects of snow geese on boreal salt marshes

The effects of foraging by Lesser Snow Geese on coastal wetlands have also been extensively studied (e.g. Jefferies 1988a; Bazely and Jefferies 1989; Belanger and Bedard 1994). Jefferies reports 'dramatic' effects on vegetation from two feeding activities: spring grubbing of roots, and summer grazing of above ground vegetation. Table 8.3 shows, however, that grazed plots are nearly identical to control plots, suggesting that, in fact, herbivory itself actually has minimal effects on the vegetation composition. In contrast, grubbing for rhizomes, as one might expect, significantly reduces the number of shoots of both graminoid and dicotyledonous species. Jefferies (1988b) suggests that the effects of herbivory are small in part because the geese also fertilize the plants and enhance the

Table 8.3. *Effects of herbivory by geese. Total number of shoots of graminoid plants and dicotyledonous plants in plots (10 × 10 cm,* n = *10) on intertidal flats immediately after grubbing or grazing in mid-June and early August 1986*

	Graminoid species		Dicotyledonous species	
	June	August	June	August
Ungrubbed plots				
Grazed plots	45.5 (5.0)	45.0 (7.5)	4.0 (2.0)	4.8 (1.8)
Exclosed plots	45.5 (5.0)	45.8 (7.8)	4.0 (2.0)	4.1 (1.7)
Grubbed plots	7.0 (1.0)	15.0 (5.2)	2.2 (0.7)	1.0 (0.8)

Note:
Standard error is given in parenthesis.
Source: From Jefferies (1988a).

rate of nitrogen cycling (Figure 8.10). Similar processes have been postulated for tundra swans (Monda *et al.* 1994).

Low marsh consists of *Puccinellia–Carex* swards, which slowly changes to *Calamagrostis–Festuca* swards as elevation increases from isostatic uplift. Geese can delay this process by heavy grazing, but when small exclosures (0.5 × 0.5 m) were built, there was an increase in biomass from about 50 to 200 g m^{-2}, and eventual dominance by *Calamagrostis deschampsoides* and *Festuca rubra* (Hik *et al.* 1992).

Effects of small mammals in marshes

Given Fritzell's (1989) observation on the paucity of quantitative data on the effects of grazing by mammals in marshes, we shall have to draw upon an example of an introduced exotic, *Myocaster coypus*, a large (up to 10 kg) South American rodent that has been introduced to both North America and Europe. Typical of the problems in wetland terminology, this animal is called coypu in the European literature (Moss 1983, 1984) and nutria in the American (Shaffer *et al.* 1992; Taylor and Grace 1995). Moss describes how animals, introduced to fur farms in England about 1929, escaped and multiplied to an estimated 200 000 animals by the 1960s. He observes that coypus 'are extremely destructive grazers, uprooting reed and other swamp [marsh] plants to eat the rhizomes', and attributes the loss of fringing reed marshes to herbivory by *M. coypus* (Moss 1984). *Myocaster coypus* has also reached high population levels in Louisiana deltaic marshes (Atwood 1950). Shaffer *et al.* (1992) built four 50 × 40 m exclosures to test

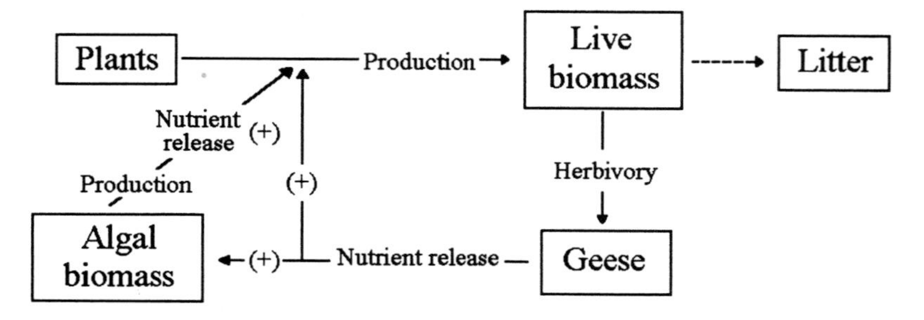

Figure 8.10 In addition to damaging vegetation through grubbing for rhizomes, snow geese may stimulate plant growth with nutrients released in their faeces (from Jefferies 1988b).

whether herbivory was changing the vegetation. Their data in Table 8.4 show that, relative to controls, the exclosure plots had much higher cover and more plant species. Plants that were preferred food of *M. coypus* (e.g. *Sagittaria platyphylla, S. latifolia*) dominated the exclosures, while species presumably less preferred (*Justicia ovata, Leersia oryzoides*) dominated the control sites. Although *Sagittaria latifolia* was one of the most flood-tolerant species, it was restricted to higher elevations, which Shaffer *et al.* (1992) attribute to the grazing by *M. coypus* at lower elevations. Shipley *et al.* (1991b) similarly found that damage (in this case from muskrats) to emergents such as *Acorus calamus* was much greater at lower elevations in riverine marshes. More recent work in Louisiana by Taylor and Grace (1995), using smaller exclosures, showed that the biomass of dominant plant species such as *Panicum virgatum*, *Spartina patens* and *S. alterniflora* increased if *M. coypus* was excluded, but they were unable to detect changes in the number of species.

The secondary effects of herbivory may be even more dramatic. Shaffer *et al.* suggest that to the direct effects of herbivory, one must add the indirect effects of increased environmental sensitivity in plants that have been damaged. Not only do shoots transport oxygen to rhizomes (pp. 37–39), but below-ground organs allow regeneration after burial by fresh sediment (Chapter 9). Grazers which feed only on shoots will still indirectly damage rhizomes by reducing the activity of photosynthetic shoots, but those like *Myocaster coypus* that feed on both above and below ground parts may be particularly destructive. *M. coypus* is thought to be a major factor reducing vegetation in the Atchafalaya Delta in Louisiana,

Table 8.4. *The effects of grazing by* Myocaster coypus *on deltaic wetlands as illustrated by four* 40×50 *m exclosures and paired control areas*

	I		II		III		IV	
Species	Exclosure	Control	Exclosure	Control	Exclosure	Control	Exclosure	Control
Amaranthus tamariscina	–	–	–	–	16	–	–	–
Alternanthera philoxeroides	12	–	–	–	14	–	6	–
Justicia ovata	27	19	31	11	62	40	24	35
Leersia oryzoides	2	–	3	–	51	7	87	27
Paspalum distichum	–	–	–	–	3	3	5	–
Polygonum punctatum	14	1	2	–	52	1	33	12
Sagittaria latifolia	95	1	128	–	82	59	73	22
Sagittaria platyphylla	18	1	11	–	18	4	52	5
Scirpus americanus	–	–	–	–	4	1	9	–
Scirpus validus	1	–	–	–	5	–	6	2
Spartina alterniflora	–	–	–	–	1	–	6	–
Typha domingensis	9	–	–	–	–	–	–	–
Total cover	178	22	175	11	308	115	301	103
Total species	8	4	5	1	11	7	10	6

Note:
Numbers are cover value sums for 30 plots.
Source: Shaffer *et al.* (1992).

not only through consumption of plants, but because the remaining plants are easily killed by flooding and sedimentation (Shaffer *et al.* 1992).

Effects of cattle: herbivory in the flooding Pampa

Unlike African grasslands, Pampean grasslands in South America developed under low intensities of natural herbivores (Facelli *et al.* 1989). Cattle and horses were introduced by the Spanish settlers in the 1500s, and in the mid 1800s, fences were built, so that herbivory was further intensified. As agriculture replaced ranching, natural grasslands were ploughed, except for areas subjected to regular flooding, the Flooding Pampa. Such trends are similar to those found in the Pantanal (Chapter 1) and the North American prairies. The Pampas of Argentina cover some three quarters of a million km^2; the main wetland area is in the Salado Basin, a flat area approximately 60 000 km^2 with mild winters and warm summers. Facelli *et al.* (1989) compared a 1 ha plot that had been grazed steadily at a stocking rate of roughly one head per 2 ha with a 1 ha plot from which cattle had been excluded for 9 years. Grazing had major effects on species composition. The ungrazed site had cover that was 95% monocots, particularly large tussock grasses; *Paspalum dilatatum* and *Stipa bavioensis* dominated. The tall grasses form a dense canopy which probably shades out shorter species. In contrast, the grazed community was almost 60% dicots, many of which were exotic species; *Mentha pullegium* was by far the most common species. Further, species composition was more variable in the ungrazed site, and suggest that in the absence of herbivory, the monocots may be reaching different competitive equilibria according to subtle environmental differences; in contrast, in the grazed areas, where continuous grazing damages plants taller than 5 cm, the effects of the cattle may be so severe as to override environmental heterogeneity.

Effects of multiple herbivores along a biomass gradient

Bonser and Reader (1995) used transplants of a test species (*Poa compressa*) to compare the effects of herbivory in herbaceous vegetation across a range of sites with different biomass levels. Two of the three high biomass sites were wetlands, a streamside with *Phalaris arundineacea* and a riverside with *Aster puniceus* and *Glyceria striata*. A combination of cage designs, combined with a plastic barrier, protected the enclosed test plants from herbivores including slugs, snails, voles, rabbits and deer. Insects, however, could enter the cages. The test species in wetlands had losses to herbivores that were greater than 90%. Terrestrial sites with similar biomass had nearly identical rates of loss to herbivores. Overall, losses to herbivores increased along the biomass gradient. Presumably, they conclude, sites with more biomass supported more herbivores.

Effects of Rhinoceros on a tropical floodplain

Asian lowland forests contain several large herbivores including the Asiatic elephant, greater one-horned rhinoceros and Javan rhinoceros. Tree diversity is relatively low, but large browser biomass approaches some of the highest values reported from Africa (Dinerstein 1992). More than 300 of the greater one-horned rhinoceros (*Rhinoceros unicornis*) occur in Royal Chitwan National Park in Nepal. Two woody species, *Litsea monopetala* (Lauraceae) and *Mallotus philippinensis* (Euphorbiaceae), together comprise one-third of the woody stems, and only six species comprise >95% of the stems. All of the understorey *Litsea* showed signs of moderate to heavy browsing and trampling by Rhinoceros. Exclosure experiments showed that *Litsea* growth was enhanced when it was free from browsing for 3 years. Rhinoceros also distribute the seeds of floodplain trees such as *Trewia nudifloa*, which Dinerstein (1991) considers to be the most common riverine forest tree in Chitwan. Dung piles in floodplain grasslands appear to be important colonization sites for this species. Thirty-seven other plant species have been recorded from Rhinoceros latrines, although the flora includes a total of 77 fleshy-fruited species that are dispersed by vertebrates (Dinerstein 1991). Dinerstein (1992) also notes that, at the time of his studies, the rhinoceros population was recovering from heavy poaching, so more natural population levels would be expected to have greater impact than those he measured.

Effects of humans upon fens

Humans sometimes harvest wetland vegetation to feed livestock or gather thatching for roofs. Although such land uses are often considered quaint by non-European scientists, they are considered important in European work because many European systems have been produced by the constant pressure exerted by human populations. Wheeler and Giller (1982), for example, report on the effects of different mowing practices in British fens. Unmanaged sedge beds (*Cladium mariscus*) had higher biomass, more litter, and lower species richness than unmanaged beds (Figure 8.11). As well, bryophytes were largely restricted to managed beds. In contrast, effects of

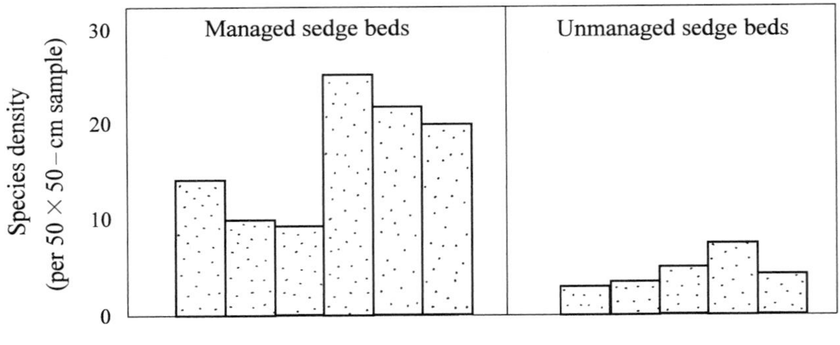

Figure 8.11 Mowing by humans can change species density in English sedge beds. Figure 7.5 provides the data on biomass (after Wheeler and Giller 1982).

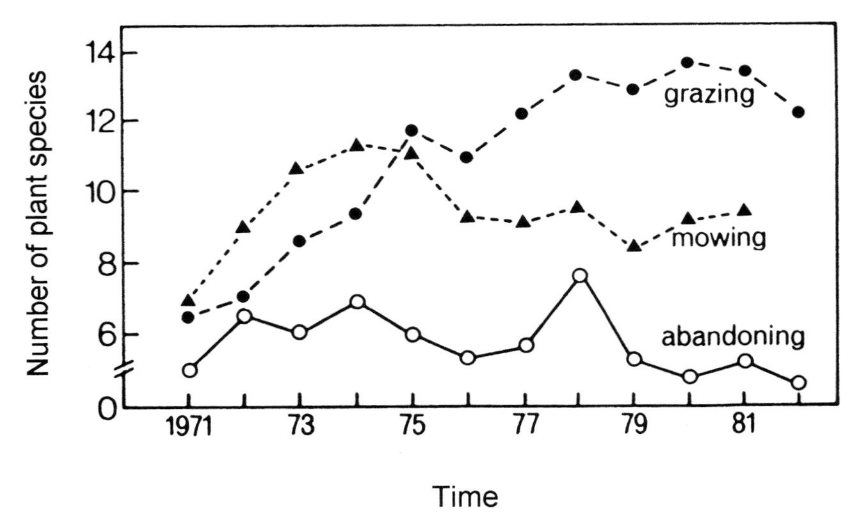

Figure 8.12 Species richness plotted against time in European saltmarshes with three contrasting types of management ($n = 5, 2 \times 2\ m^2$ quadrats) (after Bakker, 1985).

mowing on reed beds (*Phragmites communis*) were much less noticeable. Bakker (1985) similarly reports that mowing and cattle herbivory increase the number of species in European saltmarshes, with grazed areas also having more species than mowed areas (Figure 8.12).

Since atmospheric deposition of nitrogen is fertilizing both fens and wet grasslands (Chapter 5), managers face the challenge of maintaining species diversity in these ecosystem types (Ellenberg 1985, 1988; Verhoeven *et al.* 1996). This problem may increasingly face North American managers since habitats with low soil nutrients here have high plant diversity (Ehrenfeld 1983; Moore *et al.* 1989). Mowing provides a tool to remove dominant plants that may be excluding weaker competitors from multi-species communities. In the longer run, it also removes N and P from the site if the mown plants are harvested. In more natural wetlands in North America, disturbance by waves may produce a similar fertility gradient by removing nutrients from the substrate along with plant biomass (Keddy 1983) thereby creating refuges for species of infertile habitats (Keddy 1981; Wisheu and Keddy 1989b). When such natural gradients do not occur, mowing or grazing may be one of the few remaining options.

Müller *et al.* (1992) studied wet meadows along the Oste valley in north western Germany. When traditional management was abandoned, the marsh marigold meadows (e.g. *Caltha palustris, Senecio aquaticus*) developed into stands of reeds and tall forbs (e.g. *Glyceria maxima, Phalaris arundinacea, Urtica dioica*). These latter species produce dense shade and thick accumulations of litter, which (recall Chapter 3) reduce diversity in plant communities. Along the Oste, plant species richness declined from *ca.*30 species to *ca.*10 species. Muller *et al.* (1992) showed that, when mowing twice a year was re-established, diversity increased rapidly within 3–5 years. Similar problems occur in the management of wet grasslands scattered along slow-flowing rivers in Belgium. Dumortier *et al.* (1996) studied 40 7 × 7 m plots in which they varied the time of harvesting (one of 6 months, June to November) and two harvests (July and October). Overall 63 plant species were recorded. Harvesting once in either June or October appeared to produce rapid increases in numbers of species, as did the twice yearly harvesting. Richness declined with time in the control plots. Harvesting also reduced biomass below ground as well as above ground. Depending upon the characteristics of individual species, different mowing times may select for different species composition. Dumortier *et al.* (1996) consider germination characteristics, and rhizome production to be two key traits that interact with harvesting. Rhizomatous species (*Carex acuta, C. disticha* and *Lysimachia vulgaris*) are most damaged by mid-summer harvesting, when shoots would normally translocate internal resources back to roots and rhizomes; consequently, rhizomatous plants are favoured by late autumn harvesting.

Before mowing is advocated in other wetland types, it is essential to appreciate that many of the wetlands in western Europe have been produced by, or at least shaped by, herbivory, for hundreds if not thousands of years. The conservation problems arise either because traditional herbivory practices are reduced, or because eutrophication increases the intensity of competition or the rate of succession. Reintroducing harvesting may not have a deleterious effect upon such a flora. Other vegetation types, however, may not have a history of mowing or herbivory, particularly the infertile peatlands and alluvial wetlands in less populated regions of the earth. The floras in such regions may be stress tolerators (*sensu* Grime 1977, 1979) and mowing or herbivory could have negative effects upon them. The maintenance of species diversity in these habitats in the face of increasing eutrophication will remain a challenge to managers. Certainly, in such cases, natural disturbance by waves, ice, flooding and fire may be important. Dyking is likely to increase the problems of eutrophication by reducing these natural disturbances that may assist in the export of nutrients from a site. In the next chapter we will look more closely at such natural disturbances and what is known of the effects upon wetland communities.

These examples suggest that herbivory can, indeed, have a major effect upon community structure. Indeed, in the case of the Hudson Bay marshes, Jefferies' work (pp. 376–377) shows that herbivores can almost remove the wetland plant community.

Some relevant theory

Effects of Selective Grazing

The effects of herbivory can be measured with many dependent variables: biomass, species composition and species richness are three of the most obvious to choose. The latter is of particular interest, because it reminds us about one of the essential distinctions between herbivory and mowing: the potential of herbivores to be selective in the removal of biomass. Selectivity of herbivores can lead to a wide array of consequences for the plant community.

Imagine the following circumstances, a plant community with a mixture of species, some with high competitive ability, and some with low competitive ability. Their abundance in nature, in the absence of herbivory, tends naturally towards exclusion by the competitive dominant. Over time, then, species richness will decline. Now introduce a herbivore. What will happen? The effects of this herbivore upon the diversity of the plant community are impossible to predict without further knowledge about the feeding habits of the herbivore. Consider the two extremes. At one extreme, the herbivore feeds upon the weaker competitors, and avoids the dominant. In this case, the herbivore will actually reduce the diversity of the community. At the other extreme, imagine that the herbivore feeds solely upon the competitive dominant (although cannot entirely eliminate it) and avoids the weaker competitors. In this case, the herbivore will increase the biological diversity of the community. Third, imagine that the herbivore feeds on species in direct proportion to their occurrence in nature; in this case, the effects will be small, and largely determined by the species' relative degrees of resistance to the damage of herbivory. Yodzis (1986) provides a mathematical exploration of these situations. Such investigations illustrate that the effects of introducing exotic herbivores, or reintroducing extirpated herbivores, may be difficult to predict.

Mowing provides a good example of the application of such ideas. In one sense, mowing is just a form of herbivory, carried out by a relatively unselective herbivore: human beings. However, as Figure 8.12 showed (see also Chapter 3) it is generally found that mowing increases biological diversity. This increase in diversity appears to occur because mowing is actually somewhat selective: it tends to preferentially remove larger species with dense canopies, thereby allowing smaller species such as rosette forms to persist. Although mowing has its disadvantages, particularly when we are trying to re-establish natural processes in plant communities, it may be one of the few ways to manage for increased diversity in the presence of eutrophication.

A simple model of grazing

The effects of grazing upon vegetation, and the response of herbivores to vegetation, can both be explored with simple mathematical models. One of the simplest models adapts the logistic equation, which is widely used by ecologists to describe the growth of populations (Wilson and Bossert 1971). The logistic model assumes that, when there are few organisms and abundant resources, growth is (almost) exponential, but that, as population size increases, and resources becomes scarce, the population growth slows and reaches a level known as the carrying capacity, K. This equation can be equally used to describe plant populations (Noy-Meir 1975; Starfield and Bleloch 1991) as:

$$\frac{\mathrm{d}P}{\mathrm{d}t} = gP\frac{(K-P)}{K}$$

where P is the amount of plant material (e.g. biomass/unit area), g is the growth rate, and K is the maximum amount of plant material that a unit

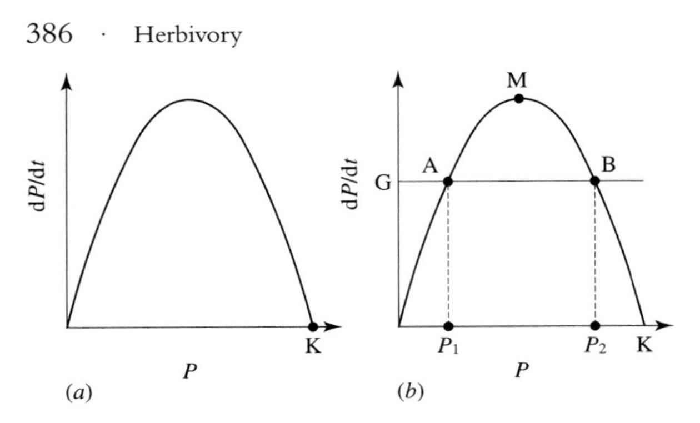

Figure 8.13 Vegetation growth rate dP/dt plotted vs the vegetation biomass P for the logistic model: (*left*) no herbivory and (*right*) constant herbivory pressure G (after Starfield and Bleloch 1991).

area can support. Another way of thinking about this that is more similar to familiar animal population models is to consider P to be the number of plant cells and K the carrying capacity of plant cells for a particular area of landscape.

To explore the behaviour of vegetation without herbivores, we can plot growth rate (dP/dt) against biomass (P), which produces an inverted parabola (Figure 8.13 *left*). The growth rate of the population of plant cells therefore at first increases as more and more cells are available for photosynthesis, and then slowly declines as the resources available to each cell become restricted. The botanical logic behind this seems to make sense: when plant biomass is low, each new cell will improve the photosynthetic capacity of the vegetation, but as biomass increases, more and more cells will be needed to provide structural support for photosynthetic cells, and others will be shaded so that photosynthesis is below the maximum potential. If we compare short turf, for example, with young forest, the number of plant cells allocated to support tissues (trunks, branches and stems) becomes a considerable proportion of the biomass in a forest. Further, the lower leaves on the trees are shaded by the upper leaves. Yet another way to think of this is the compounding effects of competition for resources such as light and nutrients; growth ceases when resources become severely constrained. In any case, when the mean photosynthetic yield of all cells just balances their mean respiratory demands, growth will come to a halt; the level K on the horizontal axis will have been reached. Half-way between 0 and K the growth rate is at a maximum. This is the familiar pattern of logistic growth; the novelty lies solely in applying it to plant biomass. The level of biomass K will depend upon environmental factors such as flood duration, growing season and soil fertility. In the absence of herbivores, all vegetation will tend towards point K.

Now, add in a constant grazing pressure from an herbivore. Assuming that the herbivores removed a fixed amount of biomass per unit time, designated G, the equation becomes:

$$\frac{dP}{dt} = gP\frac{(K-P)}{K} - G$$

Since the grazing rate is set to be independent of biomass, we can plot G as a horizontal line across the parabolic model of plant growth (Figure 8.13 *right*). There is no need to solve the differential equation to learn a good deal about the behaviour of such a herbivory system; a good deal can be deduced simply from the structure of the equations and the resulting graph (Starfield and Bleloch 1991). Returning to the growth of vegetation, it is apparent that the growth rate is positive only between points A and B, where the growth parabola lies above the herbivory rate, and biomass therefore accumulates. On either side of this range, the herbivory rate exceeds the growth rate. At points A and B, growth just matches herbivory.

The next step is to examine stability by considering what kinds of changes might occur through a period of time. Let's consider point B, where the corresponding amount of plant biomass is indicated as P_2. If growing conditions improve, pushing the amount of biomass to the right, the growth rate will fall below the herbivory rate, and the vegetation will decline back to level P_2. If, on the other hand, drought or flooding were to reduce biomass below P_2, then simultaneously, the difference between the herbivory rate and the growth rate increases, so that biomass accumulates, pushing the system back towards point P_2. Since the system returns to point B when it is lightly perturbed, this is called a stable equilibrium point.

Point A, in contrast, is unstable, because the same procedure shows that, if the system is perturbed, it slides even further away from Point A. If it is perturbed to the left of P_1, say, by a drought, then growth rates fall further and further below the herbivory rate until the plants disappear; the system slides to the bottom left and collapses. Conversely, if there is a surge of growth above P_1, then the vegetation temporarily escapes from herbivory, and continues to move to the right, because as biomass increases, the difference between herbivory rate and growth rate increases as well. Eventually the entire system slides over to point P_2. In this simple system, then, the only stable point is one where plant biomass is P_2. Over a broad range of biomass levels, this model herbivory system will return to this point after perturbation.

These dynamics can be deduced slowly from the structure of the equations. If, further, the growth rate of plants were actually measured to establish the maximum growth rate (point M), then one can see that if the herbivory rate were increased above this level M (equivalent to sliding the horizontal line above the parabola) the animals would graze faster than the vegetation grew, which is an unstable situation.

Other models could be used to describe herbivore/plant interactions, by, for example, allowing for growth rates to fluctuate in response to rainfall or flooding, or using a different model for plant growth (Starfield and Bleloch 1991). Others have addressed the interactions between plants competing for light (Givnish 1982) and their responses to added herbivory pressure (Oksanen 1990). If grazing pressure is not constant, but varies with plant biomass, then a variety of outcomes is possible, depending upon the functional responses of the herbivore (Yodzis 1989).

Conclusions

The effects of animals on evolution of wetland plants

Some thought may suggest an inherent contradiction within this chapter. It appears that wetlands contain few plants with morphological defences against herbivores, and that the secondary metabolite picture is still unclear but far from compelling. Yet Tables 1.5 and 1.6 and the examples above revealed a strong dependence of wildlife species upon wetland plants! How is it possible to reconcile the apparent importance of wetland plants as a food source with the absence of conspicuous anti-herbivore defences? There is no convincing answer, although several possibilities deserve consideration.

One possible reconciliation is to argue that the principal plant–herbivore interaction in aquatic and wetland plants is not predation, but rather mutualism. Animals are thought to play an important role in the dispersal of aquatic and wetland plants (Savile 1956; Sculthorpe 1967). The requirement for dispersal of propagules may be so important that wetland plants have tended to remain palatable precisely to attract the herbivores that will disperse their seeds. This logic seems weak, since many other plants have adaptations that protect foliage while making sure that fruits are attractive and conspicuous; in theory, there is no reason why this could not have happened in aquatic plants. One could imagine thorn-covered aquatic plants with bright flowers and edible fruits.

There seem to be three other possibilities. One is that herbivory just does not occur all that often. There are some contradictory examples and their generality is unclear. Certainly, evidence from Chapter 5 showed that wetland plants may, in general, be a very poor source of food because of their low nitrogen content. A second possibility is that the costs of herbivory to wetland plants are so low that there has been minimal natural selection to deal with them. But this argument seems less than compelling, because surely, if anti-herbivore defences arose in some wetland plants, then over time, they would be expected to replace more palatable neighbours under even low or moderate levels of herbivory. Perhaps the cost of such structures or compounds is more than the cost associated with occasional losses of tissue to herbivores. The third possibility is that there are always trade-offs among plant traits, and while herbivore defences do have costs, the advantages of attracting herbivores are overwhelming. Wetland plants perhaps have been selected more to attract herbivores to disperse their fruits than to defend their foliage. Consistent with this are the high levels of productivity in many wetlands. The availability of nutrients and water, and the limited investments of N and P per gram of plant, may make it relatively cheap to replace damaged tissue.

The effects of herbivory on wetland communities

To what extent are grazing animals, just like flooding or fire, able to control the composition and functions of wetland plant communities? The mere fact that herbivores are listed as a potential control factor inclines one to assume that the answer is yes, but actual examples, particularly those based upon properly designed exclosure experiments, are too few and far between to draw any firm conclusions. The evidence to date suggests that herbivores are far less important than flooding, fertility or competition are in creating the types of wetland communities we see. It appears that wetlands are generally controlled from the bottom up, in that the composition and structure of the vegetation determines the abundance and distribution of herbivores rather than vice versa. But, there may be important exceptions, such as alligators (Chapter 6), beavers (Chapter 4), and snow geese (this chapter). Wetland ecologists thus face two tasks: the first to determine what generalizations about herbivores are possible, and the second to discover the noteworthy exceptions. There is much to be done.

9 · *Burial*

Introduction

Our fear of being buried alive is illustrated by its frequent occurrence in our literature, from Sophocles' (*ca.* 495–406 BC) play *Antigone*, in which King Creon condemns Antigone to entombment 'in a hollowed cave living', to Edger Allen Poe's (1809–1849) macabre stories such as *The premature burial*. Yet being buried alive is a common, one might even say routine, occurrence for many plants and benthic animals found in wetlands.

This constant exposure to burial is one way in which wetlands differ from most terrestrial ecosystems. Many of the other factors that control wetlands are similar in kind to those acting on terrestrial communities: disturbance, competition and herbivory, for example, are more or less ubiquitous. In contrast, terrestrial communities are rarely subject to burial, an exception being catastrophic events such as volcanic eruptions or landslides (e.g. del Moral *et al.* 1995; Grishin *et al.* 1996) or chronic deposition of wind-deposited sand (e.g. Maun and Lapierre 1986; Brown 1997). Such events may be dramatic and conspicuous, but they are also infrequent enough that they are rarely significant factors in controlling community structure over large areas. In contrast, rivers continually erode the land's surface and carry sediments that are deposited in wetlands as water movement slows (Figure 9.1). It is estimated that the world's rivers deliver in the order of 10^{10} tons of sediment per year to their mouths, leading to the formation of large flood plains and deltas (Figure 9.2). Burial is clearly a routine experience for riparian wetlands.

The importance of sedimentation as an ecological factor varies among watersheds (Figure 9.3). The Ganges/Brahmaputra River apparently carries the largest load of river sediment in the world (Milliman and Meade 1983). In fact, south eastern Asian rivers, in general, are among the most prodigious producers of sediment. Taiwan, for example, an island of

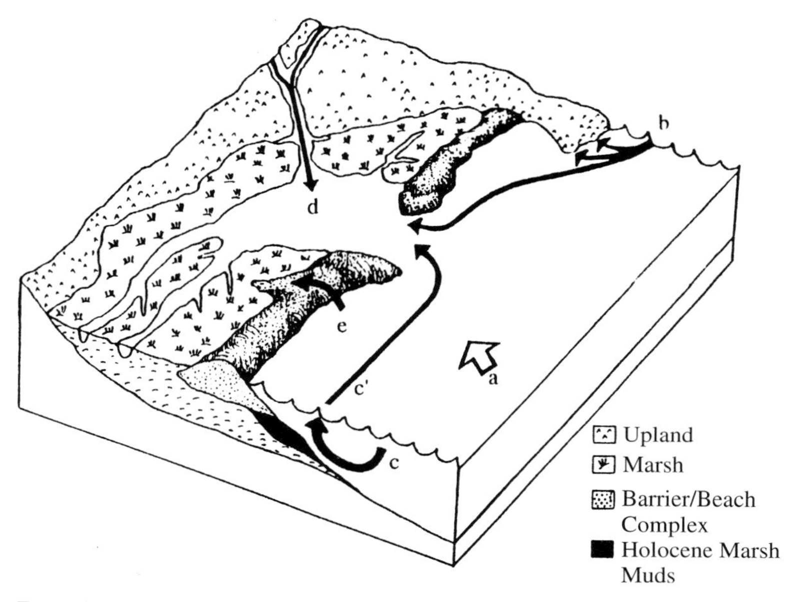

Figure 9.1 Sources of coastal marsh sediments and processes of delivery. (a) resuspension of offshore shelf or lagoonal muds with landward transport during storms; (b) erosion of Pleistocene headlands or abandoned deltas with transport to marsh via longshore currents; (c) wave attack of Holocene marsh muds exposed in lower shore face with transport to the marsh via longshore currents c'; (d) riverine input; (e) overwash redistribution (from Michener *et al.* 1997).

36 000 km^2 (roughly half the size of Ireland or the same as Indiana), produces nearly as much sediment as the entire coterminous United States (Milliman and Meade 1983). The Yellow, Ganges/Brahmaputra and Amazon have the highest annual suspended sediment loads in the world (Figure 9.4). On the coast of China's Jiangsu province, sediment from the Yellow River has accretion rates exceeding 40 cm per year (Lu 1995).

It is frequently useful to distinguish between *autogenic* burial (burial by locally produced organic matter such as occurs in peat bogs) and *allogenic* burial (burial by externally produced sediment or organic matter such as occurs in river deltas). Much of this chapter is concerned with allogenic burial, if only because rates of burial are generally much higher in this category. Both can cause changes in plant and animal communities, but with autogenic burial, this may occur on timescales of 10^3 to 10^4 years, whereas allogenic burial typically requires 10^0 to 10^2. Brinson (1993a, b) uses the terms *biogenic accumulation* and *fluvial deposition* to distinguish between these two types of sedimentary accumulation. He proposes that

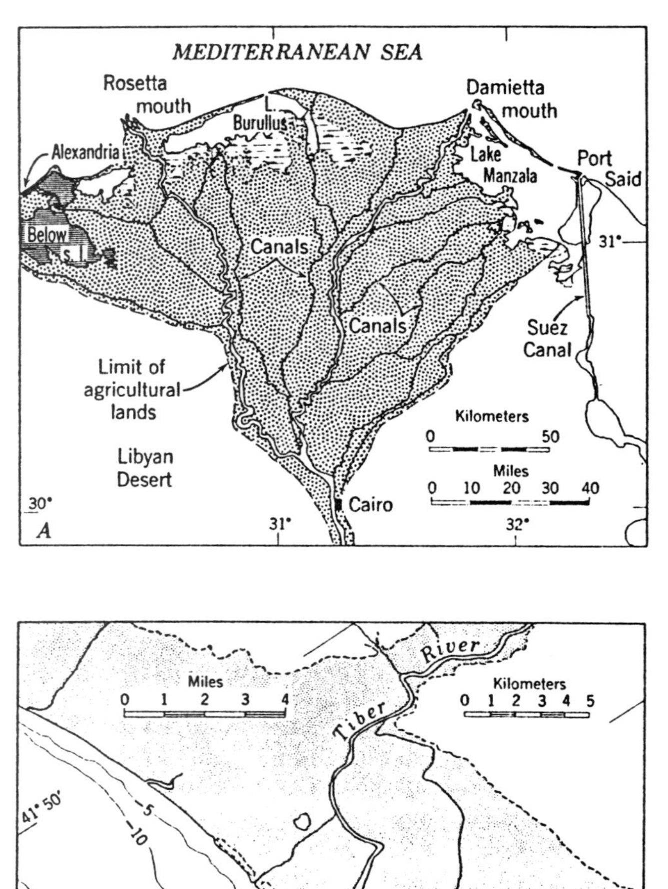

Figure 9.2 The world's large deltas illustrate the amounts of sediment transported and deposited by rivers (from Strahler 1971).

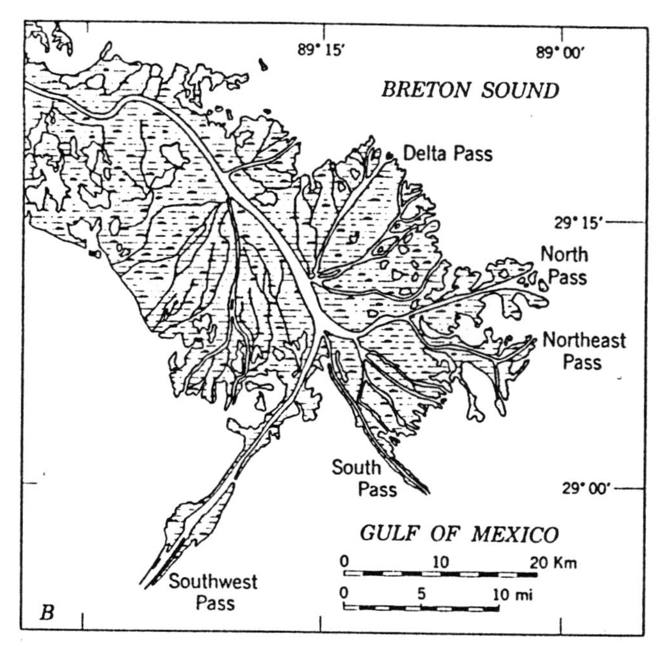
89° 15′
89° 00′
BRETON SOUND
Delta Pass
29° 15′
North Pass
Northeast Pass
South Pass
29° 00′
GULF OF MEXICO
0 10 20 Km
0 5 10 mi
Southwest Pass
B

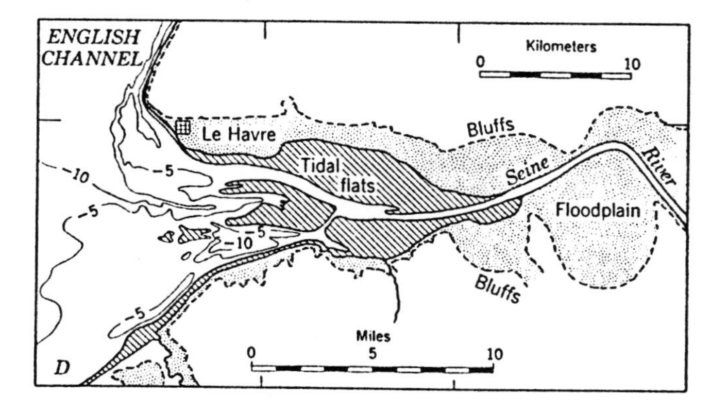
ENGLISH CHANNEL
Kilometers
0 10
Le Havre
Bluffs
Tidal flats
Seine
River
Floodplain
-10
-5
-10
Bluffs
Miles
0 5 10
D

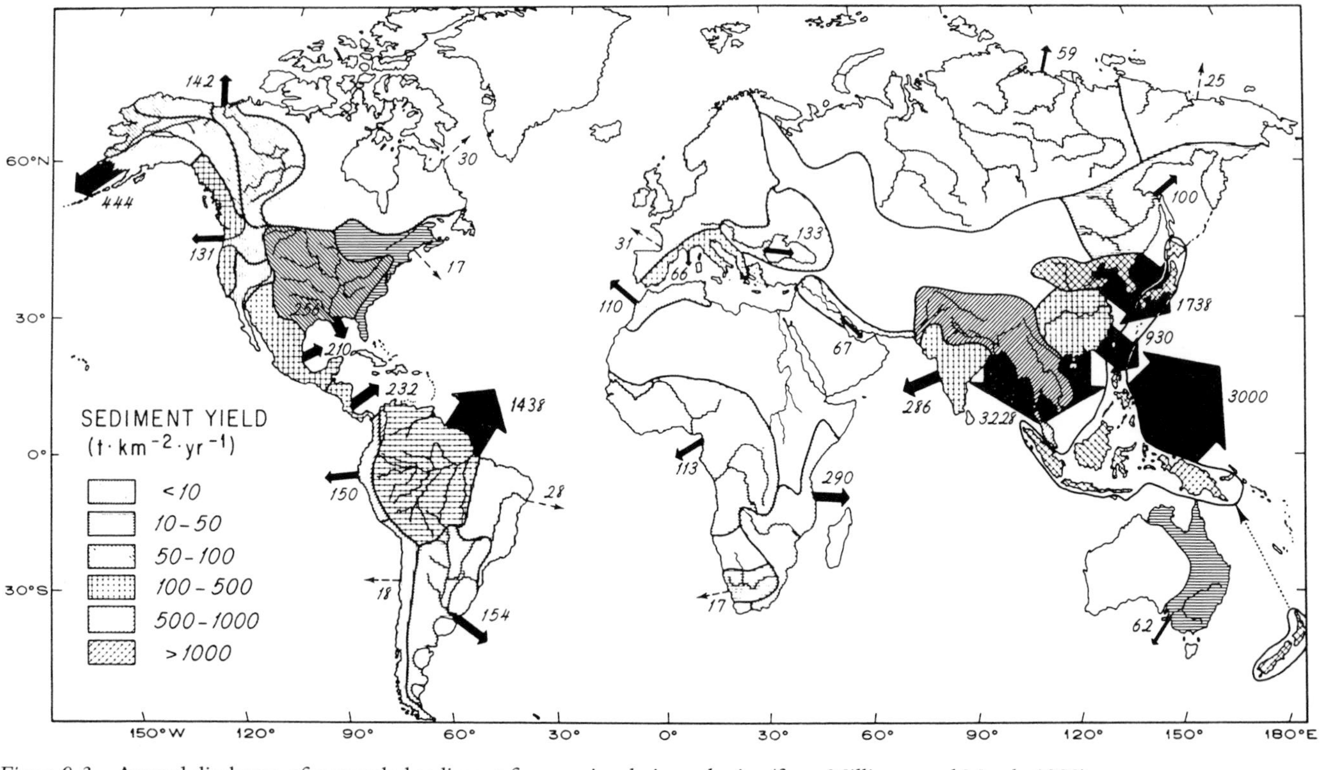

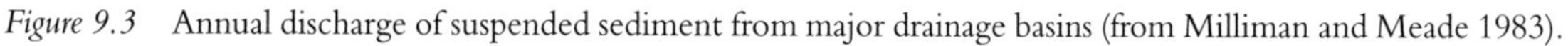
Figure 9.3 Annual discharge of suspended sediment from major drainage basins (from Milliman and Meade 1983).

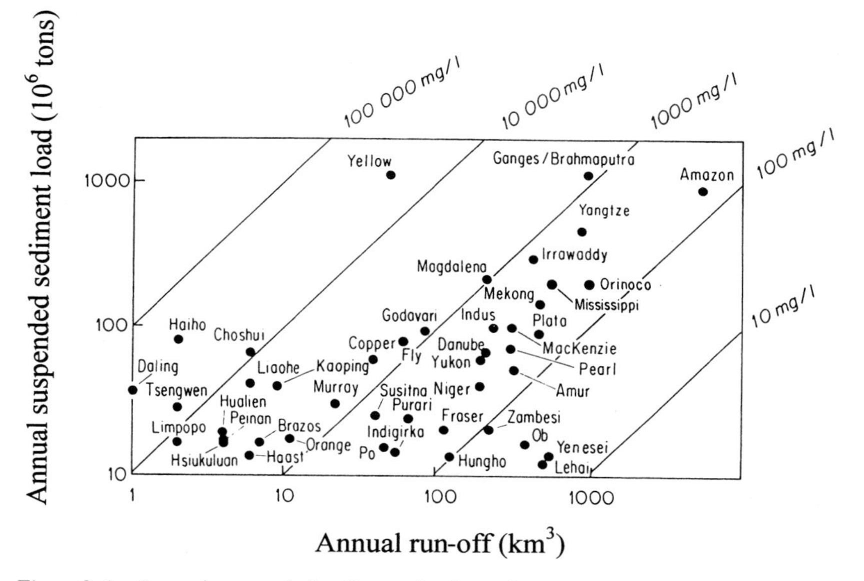

Figure 9.4 Annual suspended sediment load as a function of annual run-off in major rivers (from Milliman and Meade 1983).

the two sedimentary regimes are controlled by one gradient, water turnover (as measured by the ratio of flux to storage) (Figure 9.5).

Rates of sedimentation

Deposition rates in the order of 0.2 to 3 mm/year (2 to 30 cm/100 years), with a majority in the lower range, are suggested by palynological studies in the English landscape (e.g. Walker 1970), although Moss (1984) reports rates of 1–2 cm per year in the eutrophic Norfolk Broadlands. Compiled data from boreal and subarctic peatlands yield peat accumulation rates in the range of 0.3 to 0.8 m per year (Gorham 1991). Higher rates of deposition, some 3–6 mm per year appear to be more typical of saltmarshes (Niering and Warren 1980; Stevenson *et al.* 1986; Orson *et al.* 1990) and mangrove swamps (Ellison and Farnsworth 1996). Annual rates of deposition in a floodplain in eastern north America were below 0.1mm prior to this century, but then accelerated to approximately 1 cm/year (Rozan *et al.*1994). Deposition rates closer to 1 cm/year appear more typical in coastal deltas (Boesch *et al.* 1994) and also interdunal ponds (Wilcox and Simonin 1987). In contrast with the above figures, floods or storms can deposit 10 or more cm of sediment in a single year

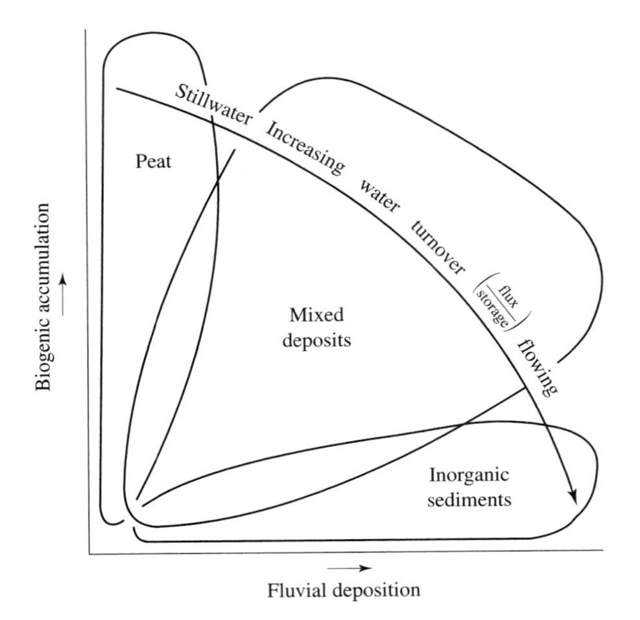

Figure 9.5 Biogenic accumulation and fluvial deposition are the two causes of burial in wetlands. These orthogonal axes may be represented (and caused by) one gradient, water transfer (from Brown *et al.* 1979 in Brinson 1993b).

(e.g. Robinson 1973; Zedler and Onuf 1984; Rybicki and Carter 1986). In rapidly eroding watersheds of Asia, deposition rates can exceed 40 cm per year (Lu 1995).

In general, sediment loads in rivers are determined by rainfall and vegetation cover, with cultivated watersheds having sediment loading rates orders of magnitude higher than forested watersheds (Figure 9.6). This is consistent with the results of studies on eutrophication (Chapter 5) where the clay content of the soil and amount of land in row crops are the best predictors of phosphorus loadings to watercourses. Although larger rivers can be expected to carry larger volumes of sediment, rainfall and human disturbance to vegetation can play equally important roles in determining sedimentation rates in watersheds.

Autogenic burial

Although most of this chapter will deal with allogenic burial, let us first digress to consider autogenic burial in peatlands. We have already seen (Chapters 1, 4) how peat, composed largely of *Sphagnum*, may accumulate

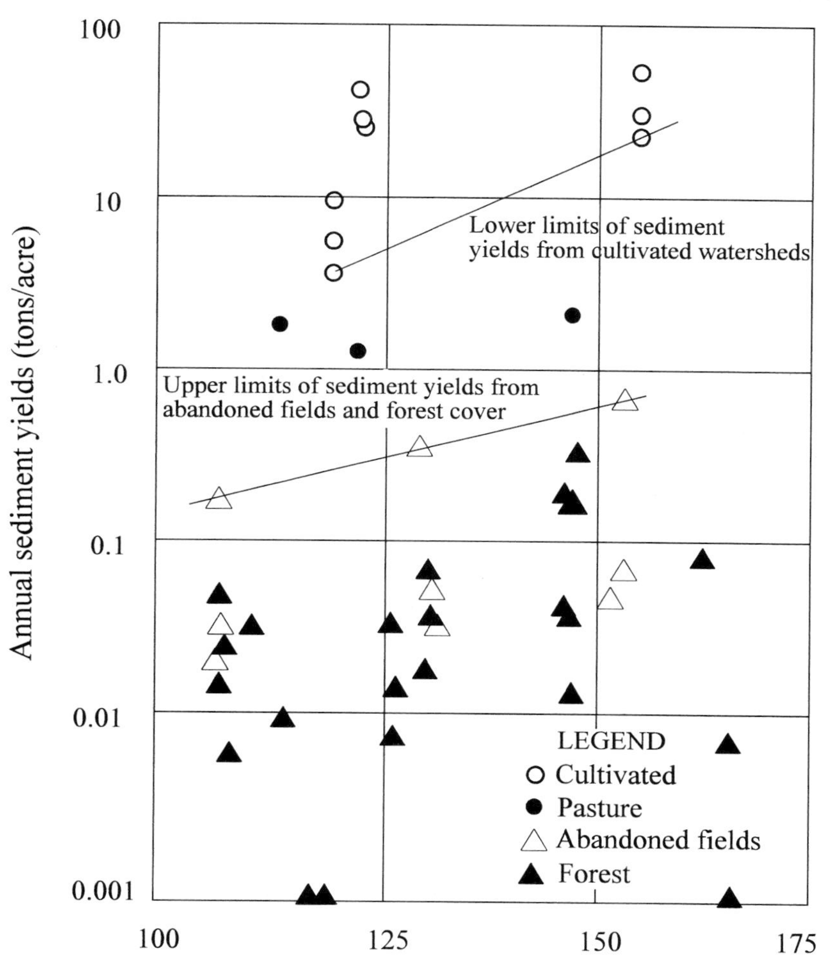

Figure 9.6 Annual sediment yield is a function of annual precipitation and land use (after Judson 1968).

and in doing so cause changes in the water table. Further, as peat accumulates, plants become increasingly isolated from the mineral substrate, so that distributions are controlled both by water levels and nutrient gradients produced by the peat itself (Chapter 5). The general outline of how *Sphagnum* buries the underlying substrates has been understood for at least a century (Gorham 1953, 1957; Gore 1983; Zobel 1988), and Figure 9.7 shows how the underlying substrate becomes blanketed in peat, with small depressions becoming forested, and larger depressions

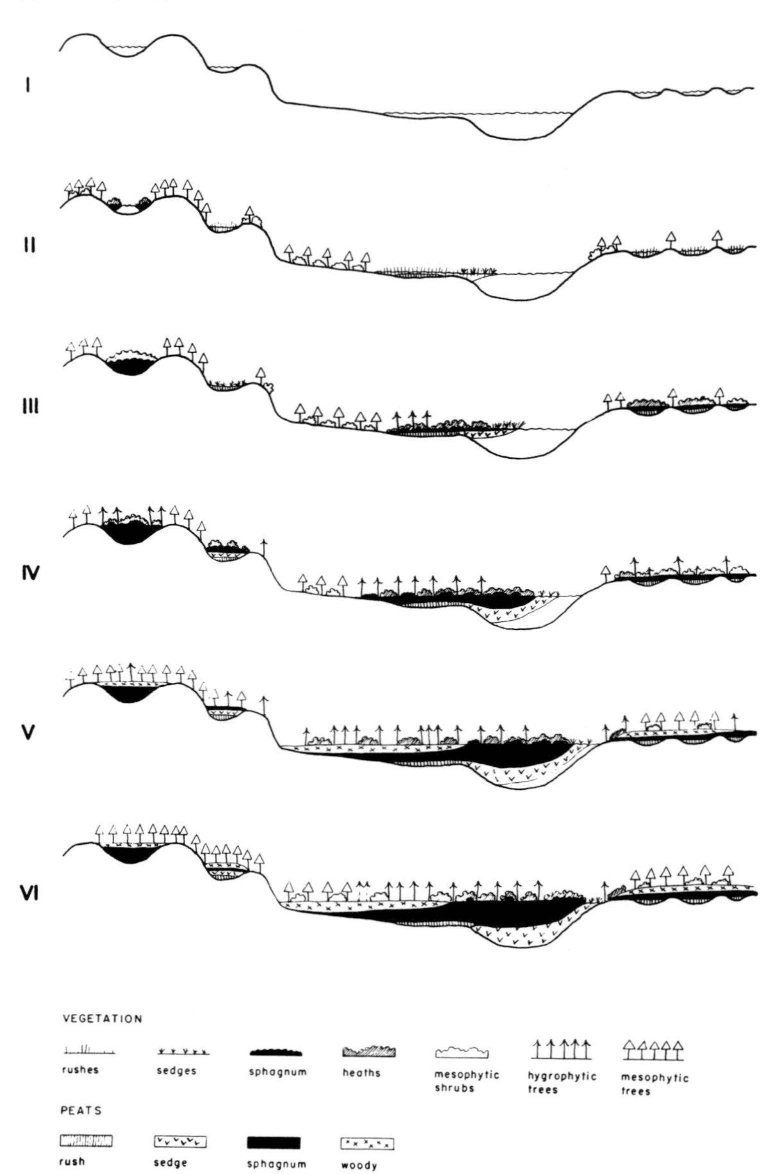

Figure 9.7 The development over time of peatlands on landscapes on the Precambrian shield, showing both the events in closed drainage (A,B,F,G,H) and situations where there is more seepage or water movement (C,D,E) (from Dansereau and Segadas-Vianna 1952).

going through a longer period ringed by floating bog vegetation. Eventually, the peat accumulates to such a depth that the vegetation is little affected by the underlying topography, and instead becomes largely controlled by climate (Foster and Glaser 1986; Zobel 1988). If, however, the topography has sufficient relief, run-off can then continue to control the peatland, with areas of comparatively rapid drainage remaining as fens, and those isolated from moving water developing into ombrotrophic raised bogs. Some idea of the time required for the transformations in Figure 9.7 is available, since many areas now dominated by peatlands were deglaciated less than 10 000 years ago.

With the increasing use of radiocarbon dating, and intensive study of individual bogs, much more is now known about the process by which a landscape becomes buried in peat. One can recognize three hypotheses which might explain how large ombrotrophic bogs form. There could be initiation of peat accumulation across a broad area, with steady accumulation of peat but no lateral expansion, in which case the area of the bog would remain unchanged but the depth would increase steadily through time. Another possibility is that peat could begin to accumulate at a number of individual sites followed by expansion and fusion of the separate peat islands into one large bog. Peat might also begin to accumulate at one site and gradually increase both in depth and area. This process has been explored in the Hammarmossen bog in the Bergslagen region of central Sweden; this bog developed on a broad flat outwash plain and has been well-studied by European scientists. To discriminate among the three models for bog formation, Foster and Wright (1990) took peat cores from a series of locations in this bog and obtained radiocarbon dates from the bottom of each core near the mineral soil. Figure 9.8 shows the general outline of this bog, with the open-water pools covering its surface; the adjoining sketch gives contours of bog age as determined by radiocarbon dating. The bog began forming some 6000 years BP, with growth initiated near the centre under what is now the deepest peat. It seems clear that in this case, the bog has not only grown upward by peat accumulation (the peat depth near the middle is some 4 m, for a rate of accumulation of 0.67 mm per year), but it has also expanded laterally at a rate of some 200 m per thousand years.

The careful dating of pools also allowed Foster and Wright to study the process by which pools form on the surface of raised bogs. They conclude that 'pool development is the result of biological process under hydrological control'. Pools apparently begin as small hollows on the relatively steep slopes covered by shallow peat. As the peat accumulates,

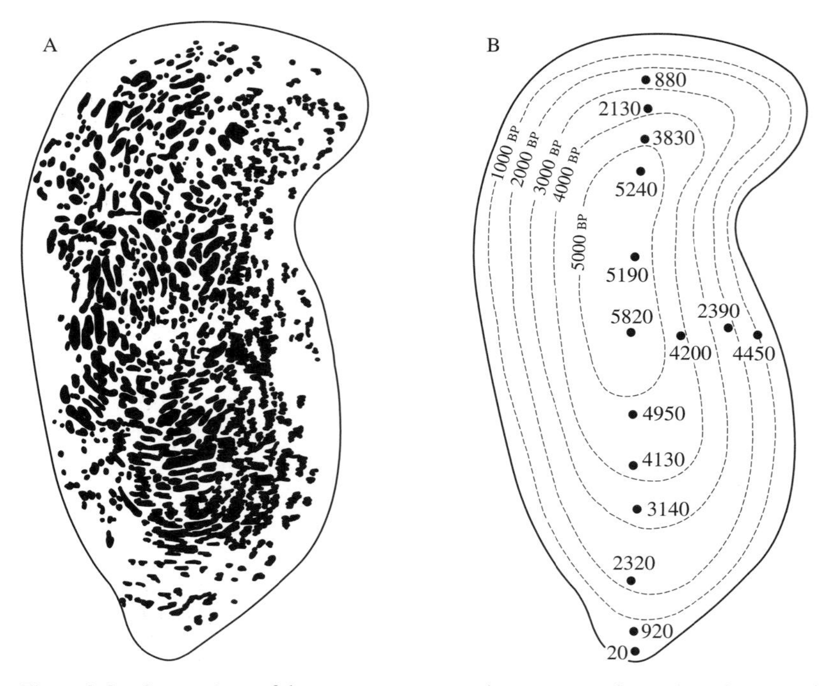

Figure 9.8 A top view of the Hammarmossen bog in central Sweden, showing the distribution and size of open water pools (*left*) and basal radiocarbon dates with interpolated isochrones for bog expansion (*right*). The peat is 4m thick at the centre of the bog (from Foster and Wright 1990).

these turn into pools. Presumably the rate of peat accumulation in the hollows is less than that of the adjoining ridges, so that over time the peat rises around the depression. At the same time, the water table rises. The plants near the centre of the depression are gradually killed and replaced by open water. Adjoining pools may coalesce to produce larger pools.

Peat cores taken from five peatlands in boreal Canada were also examined to study processes of peat formation (Kuhry *et al.* 1993). All five cores were initially dominated by wetland plants such as *Typha* and *Carex*. These were replaced by fen mosses, leading to inferred pH of about 6.0 and a water table at 5–15 cm below the vegetation surface. Subsequently, *Sphagnum*-dominated peatlands developed at each site, in which case pH levels apparently fell to 4.0–4.5. This transition from fen to bog was rapid (Figure 9.9). The overall sequence from marsh to rich fen to poor fen to *Sphagnum* bog took place over >2000 years in southern sites, but <1500 years in northern sites.

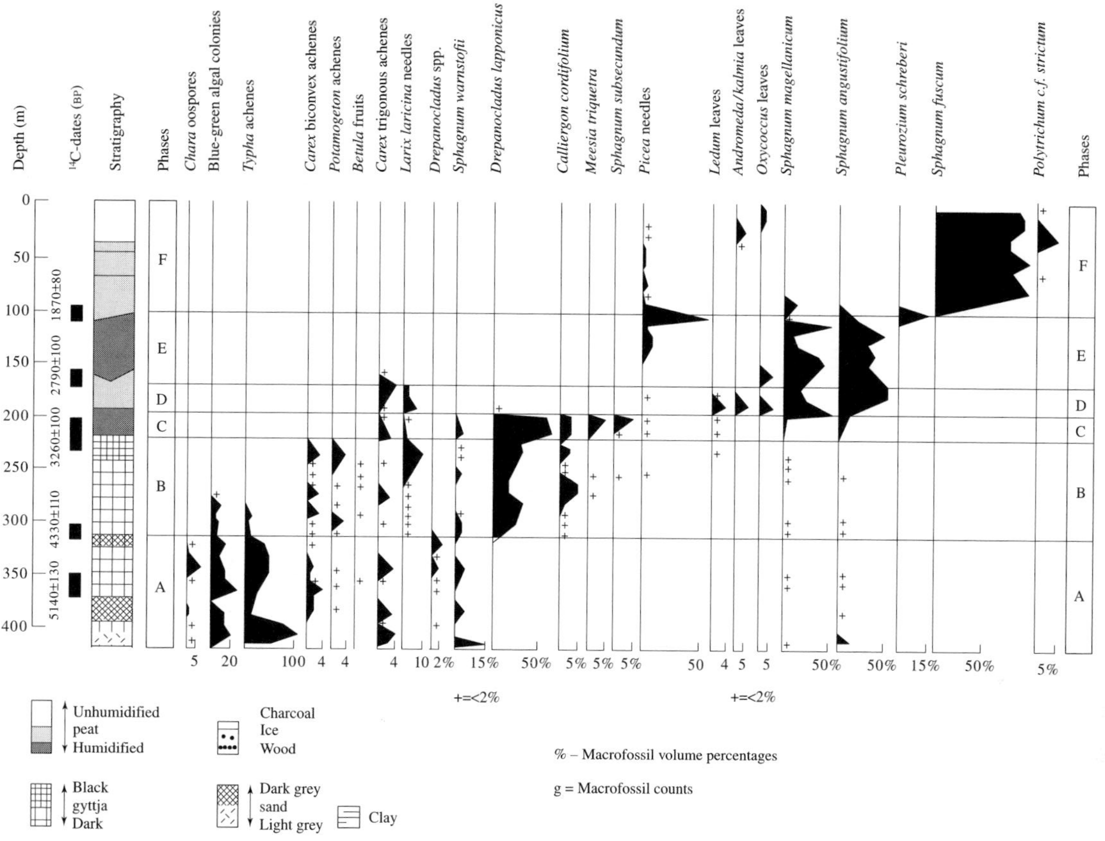

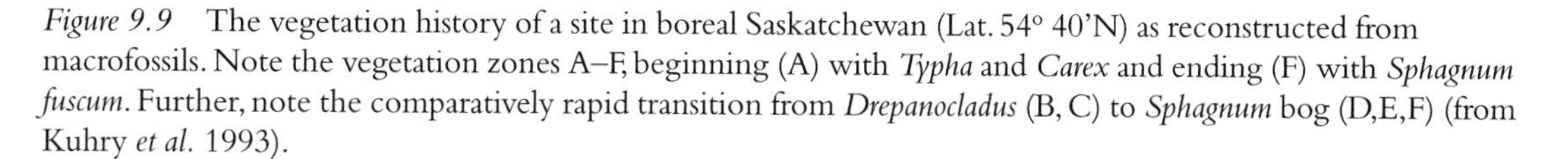

Figure 9.9 The vegetation history of a site in boreal Saskatchewan (Lat. 54° 40'N) as reconstructed from macrofossils. Note the vegetation zones A–F, beginning (A) with *Typha* and *Carex* and ending (F) with *Sphagnum fuscum*. Further, note the comparatively rapid transition from *Drepanocladus* (B, C) to *Sphagnum* bog (D,E,F) (from Kuhry *et al.* 1993).

Effects of burial

Evidence from plant traits and field patterns

There is an unfortunate contradiction in the scientific literature at this point. One often reads that wetlands are important as filters or sinks that prevent suspended solids from entering water courses. One assumes therefore, that these solids must be accumulating in the wetlands. But, when one looks for literature on the effects of these solids as they bury plants, the attention is minimal. For example, Hutchinson's (1975) treatise on limnological botany has only one relevant reference 'rate of accumulation, supposed effect' which refers to Pearsall's views in the 1920s. One recent compendium (Sharitz and Gibbons 1989), 1265 pages dealing with wetlands and wildlife (roughly twice the length of Hutchinson), has not a single main reference to sedimentation effects. The single sub-reference (Richardson 1989) occurs in a section titled 'wetlands as filters' and refers to a series of studies which document the effects of wetlands as filters of suspended solids. One can only look at Figures 9.2 and 9.3, and wonder.

Given the gaps in the literature, we can start by considering the morphology of wetland plants. Many wetland plants have well-developed rhizomes and pointed shoots (Figure 9.10). Examples include genera such as *Typha, Juncus, Scirpus* and *Carex*. Pointed shoots and underground storage structures are considered to be adaptations for penetrating accumulations of leaf litter (Grime 1979), and it is likely that the same traits also are adaptations for penetrating accumulations of sediment. In contrast, evergreen rosette life forms will probably be extremely intolerant of burial, and this may in part be why they are largely restricted to eroding shorelines (Pearsall 1920). At a larger scale, this may also explain, in part, why such plants are often restricted to oligotrophic lakes. Eutrophic lakes and bays with high sedimentation rates are generally occupied by larger rhizomatous plants. While we can explain such patterns in part by differences in relative competitive abilities (Chapter 6), differing tolerances to burial may also play a role.

Sediment deposition will often be correlated with litter deposition. We have already seen that litter can control the species composition of a wide array of plant communities (Chapter 3) and that patches of litter can kill the established plants, in saltmarshes (Chapter 6). Auclair *et al.* (1976b) further suggest that burial and competition intensity are positively correlated (Figure 5.13), with sedimentation leading to increased soil fertility, and increased soil fertility leading to increased competition intensity.

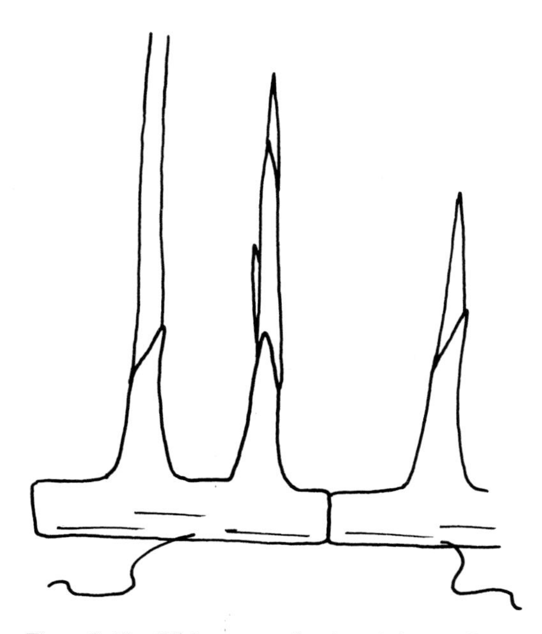

Figure 9.10 Rhizomes and pointed shoots allow buried plants to re-emerge.

Experimental studies of burial

The above observations on plant morphology allow us to infer that burial has long been a factor in wetlands. Three recent experimental studies support this inference.

Van der Valk *et al.* (1983) did a thorough study of the effects of burial upon three wetland vegetation types: alpine, lowland freshwater and coastal. In general, the alpine wetlands were most sensitive to burial, and the freshwater lowland wetlands the least so (Figure 9.11). This, in part, was a consequence of differences in plant growth form, with the alpine wetlands having more short species (e.g. *Oxycoccus microcarpus, Parnassia palustris*) and the freshwater lowland wetlands having mainly tall, erect species (e.g. *Eleocharis palustris, Equisetum fluviatile*). After a further year of growth, the coastal wetlands showed most recovery and the alpine wetlands the least. In general, regeneration from buried seeds was marginal; most recovery was from buried rhizomes.

Although tubers and rhizomes provide reserves for shoots to re-emerge after burial, as little as 20 cm of sediment can kill more than half the tubers of *Valisneria* (Figure 9.12). Burial by sand was more damaging than burial by silty clay; only 15 cm of sand caused as much mortality as 20 cm of silty clay. Rybicki and Carter (1986) conclude that, since

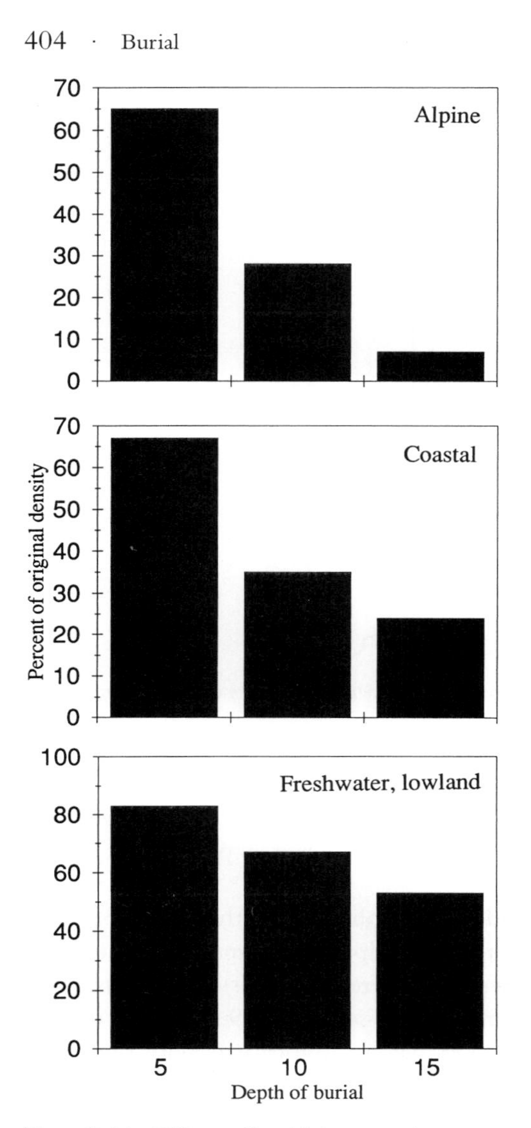

Figure 9.11 Effects of burial (measured as percent of original shoot density) plotted against depth of burial in three wetland vegetation types (from data in van der Valk *et al.* 1983).

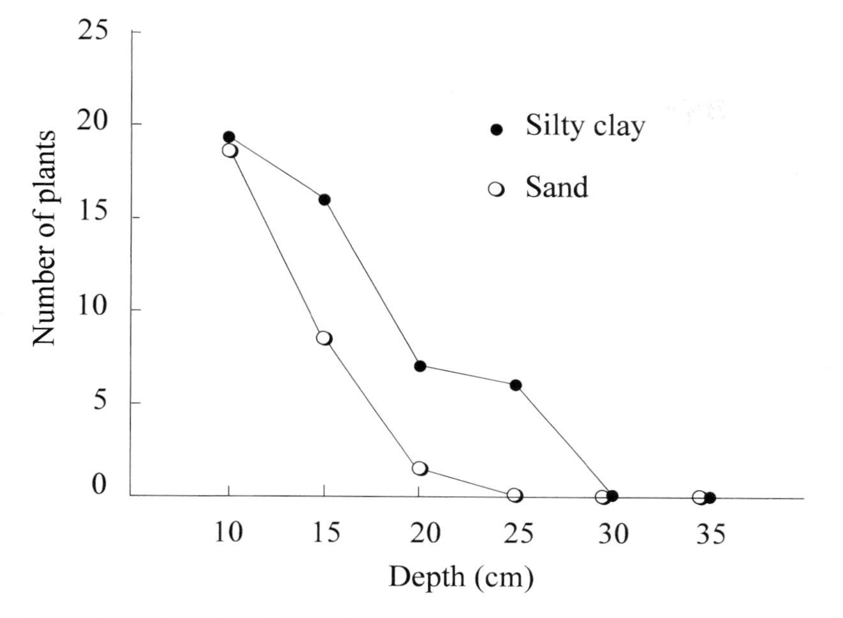

Figure 9.12 Number of viable *Valisneria* plants plotted against depth of burial (from data in Rybicki and Carter 1986).

Valisneria tubers are already covered by 10 cm of sediment under field conditions, storms carrying as little as 10 added cm of sediment can damage stands of aquatic plants.

Allison (1995) covered saltmarsh vegetation near San Francisco with 10 cm of sediment dug out of nearby tidal channels. He then followed the recovery of the plots for 4 years. For all species combined, vegetation cover returned to control values after only 2 years. Species such as *Salicornia virginica* and *Distichlis spicata* recovered quickly; other species such as *Frankenia grandifolia* and *Jaumea carnosa* recovered only when the disturbances occurred early in the growing season. In general, plots were revegetated by ingrowth from adjoining plants, or else from buried rhizomes. There was very little seedling establishment. Recovery was relatively rapid because the disturbed areas were only 1 m^2 circular plots; since most recovery was from adjoining areas, larger areas of spoil or sediment would presumably take much longer to recover. In contrast with freshwater wetlands, it would seem that seed banks play a minor role in recolonizing disturbed areas in salt marshes, since both Bertness and Ellison (1987) and Allison (1995) have found vegetative expansion of

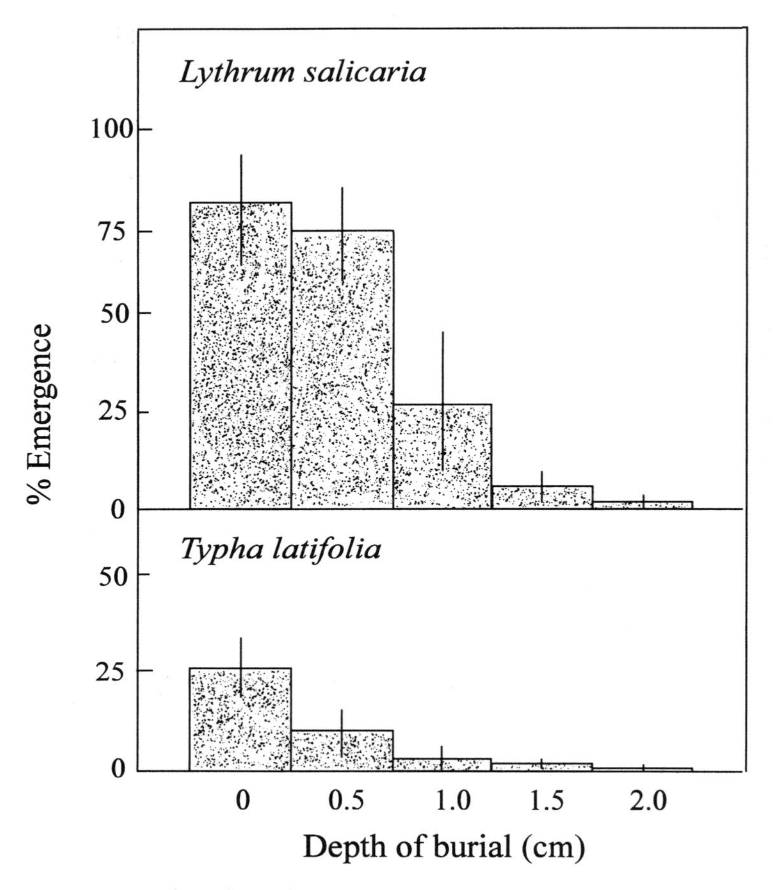

Figure 9.13 The effects of burial upon the emergence of *Lythrum salicaria* and *Typha latifolia* from seed (F. Terillon and P. A. Keddy unpublished data).

adjacent plants is far more important than buried seeds in re-establishing saltmarsh vegetation.

Effects of burial upon seedlings

Seedlings are likely to be much more sensitive to burial. They lack both the storage organs and the stiff pointed shoots of adult plants. A survey of 25 wetland plant species revealed germination was frequently above 80% in the light but many of the same species had 0% germination in darkness (Shipley *et al.* 1989). One could therefore assume that relatively small amounts of sediment would therefore prevent many species from even germinating. Figure 9.13 shows that even 1 cm of sediment is sufficient to

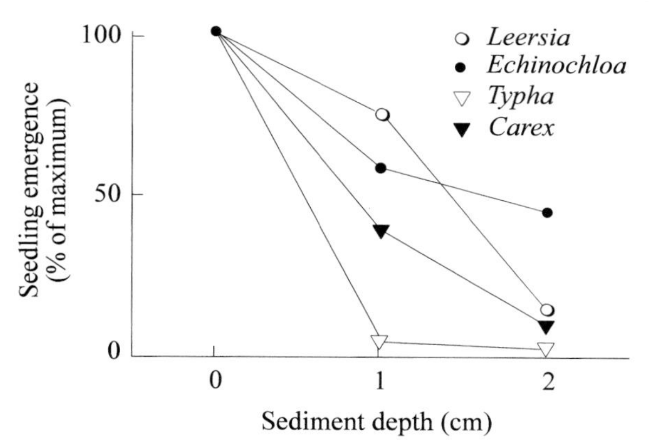

Figure 9.14 Emergence from natural seed banks is reduced by burial (after Jurik *et al.* 1994).

reduce emergence by more that 50%; 2 cm of burial reduces emergence to negligible levels. Similar results are reported by Galinato and van der Valk (1986) and Dittmar and Neely (1999).

These studies of natural seed banks covered by extra sediment also show that even small amounts of sediment significantly reduce germination, and can therefore change the species composition of wetlands (Figure 9.14). Not only does germination of individual species decline, but diversity as a whole decreases significantly with depth (Jurik *et al.* 1994).

Although Figures 9.13 and 9.14 show that *Typha* seedlings are very sensitive to burial, Wang *et al.* (1994) found that depths of even 4 cm of sediment have no effect upon the density of adult *Typha* plants. These results should not surprise us given the large stored energy reserves in *Typha* rhizomes, the pointed shoots, and the data in Figure 9.11 from a species with smaller energy reserves. *Typha* can often be seen emerging in spite of burial beneath litter and sediment. This tolerance of burial may partly explain why increasing numbers of temperate zone wetlands appear to be dominated by *Typha*. This is occurring in spite of the sensitivity of *Typha* seedlings to burial. This is a practical example of the generalization in Shipley *et al.* (1989) that juveniles and adults of the same wetland species can have contrasting life history strategies.

An unfortunate consequence of using natural seed banks in studies of burial (e.g. Jurik *et al.* 1994) is that many of the less common species cannot be evaluated because of very small sample sizes. This suggests that further comparative studies might benefit from using individually sown species. Perhaps, however, detailed comparative studies are unnecessary,

Table 9.1. *Effects of contaminated melt water upon percentage germination of five wetland plant species*

	Snowmelt concentration (%)		
Species	0	20	100
Aster umbellatus	5.8	2.0	0
Dulichium arundinaceum	11.6	3.4	0
Scirpus cyperinus	14.2	10.2	0
Typha latifolia	13.2	7.2	1.0
Lythrum salicaria	30.0	19.2	9.0

Note:
$n = 5$ replicates of 36 seeds each.
Source: From Isabelle *et al.* (1987).

since one might reasonably expect that larger seeds will be less sensitive to burial. This has been confirmed experimentally (Jurik *et al.* 1994).

There is a confounding factor in such studies. Sediments that bury plants may also contain a variety of toxic substances, particularly if the sediments originate in agricultural fields or urban areas (e.g. Reynoldson and Zarull 1993). The foregoing studies by Jurik *et al.* (1994) used sediment collected from a sediment trap in a ditch draining several soy bean and corn fields. This has the advantage of being a relatively natural treatment, since these sorts of habitats are a major source of sediment for wetland ecosystems. However, these sediments may also have contained herbicides or fungicides which could affect germination quite independently of burial. Sediments washed from urban areas are likely to contain contaminants, particularly salts from road de-icing (Field *et al.* 1974; Scott and Wylie 1980). In cold climates, contaminated snow is routinely dumped directly into rivers, or else allowed to melt in vacant lots which drain directly in storm sewers. To test for effects of such contaminants upon the establishment of wetland plants, Isabelle *et al.* (1987) watered pots containing standard seed mixtures of five wetland plant species with melt water from snow removed from urban streets. Both the biomass and richness of the experimental plant communities were reduced by increasing concentrations of snow melt. Table 9.1 shows that melt water alone significantly reduced germination; it is noteworthy that the only two species growing at high concentrations were *Typha latifolia* and *Lythrum salicaria*, two widespread plant species that are now common in ditches and roadside wetlands.

Effects of burial on other wetland life

Sedimentation is regarded as one of the three leading threats to freshwater aquatic ecosystems, the other two threats being exotic species and impoundments (Richter *et al.* 1997). As a consequence of these threats, Richter *et al.* observe that there is 'a quiet crisis taking place beneath the surface of the world's rivers and lakes', conservative estimates suggesting, for example, that 20% of the world's freshwater fishes are extinct or in serious decline. Aquatic organisms seem to be disproportionately at risk of extinction; in the United States of America, for example, between 14 to 18% of terrestrial vertebrates are considered to be at risk, whereas the figures for aquatic life are two to four times higher (some 35% for amphibians and fishes, 65% for crayfish and 67% for unionid mussels.) Richter *et al.* report that the primary cause of the altered sediment loads is agricultural non-point pollution, a factor already seen to be major cause of increases in nutrient levels in wetlands (pp. 266–269). Road construction is another major source of sediment in watersheds (Forman and Alexander 1998).

Burial by sediment has two main consequences for wetland animals. First, aquatic invertebrates and fish eggs are smothered by fine layers of silt and clay (e.g. Cordone and Kelley 1961; Ryan 1991; Waters 1995). Secondly, aquatic plants can be stimulated by the nutrients in the sediment, and when these plants decompose under the ice during the winter, they can reduce oxygen to levels where aquatic life is killed (e.g. Vallentyne 1974; Wetzel 1975). Lemly (1982) studied the effects of both nutrient loading and sedimentation upon aquatic insects in an Appalachian mountain stream (Figure 9.15). The Plecoptera, Trichoptera and Ephemeroptera all declined in richness, density and biomass with increased sedimentation. Many of the insects collected had their respiratory structures clogged with soil particles. Moreover, fine sand and organic silt adhered to their body surfaces. The most sensitive groups to sedimentation were the filter feeding Trichoptera and Diptera. Sediment loading appeared to be more detrimental than simple eutrophication. Other effects of sediment may be more indirect; in aquatic communities, suspended clay may shift competitive dominance from cladocerans to rotifers by interfering with feeding by the cladocerans (Kirk and Gilbert 1990).

Sediment yields for forested watersheds typically are 3–12 tons km^{-2} yr^{-1}. This leaps to *ca.* 300 for a clearcut watershed and to *ca.* 3600 for clearcutting followed by farming and pastures; a construction site yielded *ca.* 49000 tons km^{-2} yr^{-1} (Bormann and Likens 1981, Table 2–4). These

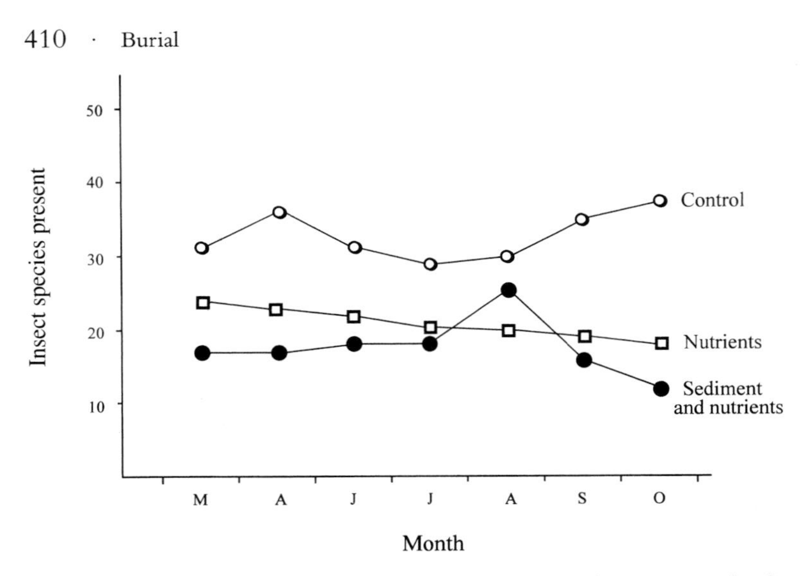

Figure 9.15 Changes in the number of insect species with nutrient and sediment loading to a stream (after Lemly 1982).

changes are also evident in Figure 9.6, and appear to have been a consequence of human activity for millennia (e.g. Hughes and Thirgood 1982; Binford *et al.* 1987); even Plato complains, in his dialogues, that humans have caused extensive soil erosion around Attica.

In the above study by Lemly, logging, residential construction and grazing were the sources of inorganic siltation, and cattle were the source of nutrients. In the case of streams, the effects of deforestation are even more serious because in addition to the above, there are two further consequences. First, the water becomes warmer, thereby reducing concentrations of dissolved oxygen. Secondly, tree leaves are the base of stream food webs. When trees are removed, the stream food web is removed along with them (e.g. Lemly 1982; Steedman 1988). For these reasons, the amount of riparian forest is considered to be an important predictor of the biotic integrity of streams (Figure 9.16). As the amount of urban land use in a watershed increases, increased amounts of riparian forest are needed to compensate. In Figure 9.16, excellent biotic integrity values are only possible if riparian forest exceeds 75% and urban land use is less than 20%.

Sedimentation and hydrosere succession

We explored the possible connections between wetland zonation and ecological succession in Chapter 2. One important procedure for testing

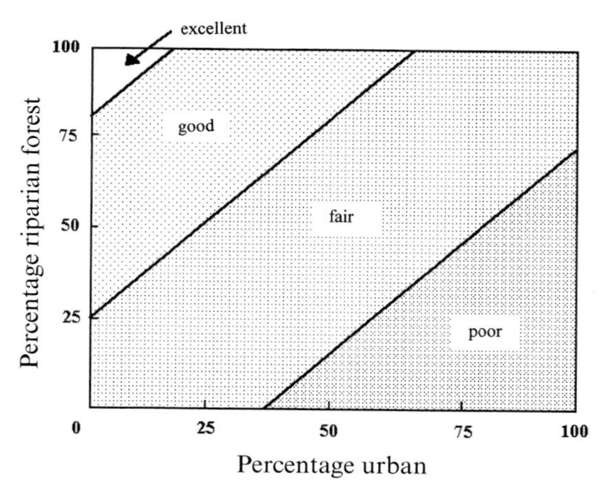

Figure 9.16 Biotic integrity of streams can be predicted from two watershed properties: percentage of land that is urban and percentage of riparian forest remaining (after Steedman 1988).

the connection between these is through the examination of pollen and macrofossils in sediment cores. Walker (1970) therefore studied sedimentation rates in a set of 20 sediment cores from across England, trying to reconstruct changes in wetland vegetation type through time. Although the accumulation of sediment was associated, as would be expected, with a gradual change from open water to floating leaved plants to reed swamp to bog, Walker emphasized that the sequence of changes in vegetation was not so constrained as one might expect. In all, he recorded 71 vegetation transitions. Of these, 17% showed reversal of this sequence, most short-lived, which he attributes to local changes in lake level, temperature or trophic status of the lake water. In a second stage, he extracted 159 transitions, and concluded 'The most impressive feature of these data is the variety of transitions which have been recorded and which must reflect the flexibility of the succession.' For example, '. . . significant numbers of transitions to bog take place directly from reed swamp, fen and swamp carr'. Many of these vegetation types or 'seral stages' last 1000 years or longer. While such palynological data are valuable, they have at least two problems. The first is the potential for external factors to interfere. Water levels or climate changes could cause reversals or delay transitions. Erosion of adjoining areas associated with fire or agriculture could drive increases in the rate of succession. These externally driven changes should be removed from any study of autogenic succession; it is not clear that this can be done. Secondly, while compilations such as Walker's are valuable, they lack a necessary statistical rigour, a problem which Walker himself acknowledges. While both individual cores (e.g. Yu *et al.* 1996) and compendia (e.g. Walker 1970) may cast doubt upon simple models of succession, the importance of unidirectional change in wetland vegetation deserves continued study. Of particular interest is the role of locally produced (autogenic) sediments, as opposed to erosional (allogenic) sediments in driving vegetation succession. As well, the relative strength of succession as opposed to counter forces (fire, flooding) could be better quantified. The persistence of individual vegetation types for 1000 years or longer emphasizes that ecological communities may possess a certain resilience when faced with either allogenic or autogenic forces of change.

A simulation model for sediment production

The sources of sediments in watersheds must be understood in order to manage watersheds wisely. Models can help scientists and managers clarify their thinking about environmental problems, and if the models are successful, perhaps even simulate the consequences of different management scenarios (e.g. Starfield and Bleloch 1991). In a watershed study of the heavily populated Hudson River watershed in eastern North America, Howarth *et al.* (1991) used the following equation to estimate soil erosion:

$$X_t = 0.132\ (RE_t)\ (A)\ (K)\ (LS)\ (C)\ (P),$$

where 0.132 was a dimensional conversion factor, RE_t was rainfall erosivity on day t, A was the area of the land type, K was the soil erodibility factor, LS was a topographic factor, C was vegetation cover and P was an agricultural practice factor. The general structure of this model, developed by Haith and Shoemaker (1987), may therefore help clarify thinking about the sources of sediments: sediment yield will depend upon storm intensity, soil erodibility, slopes, and area without forest. Each of these terms can then be estimated from technical manuals (Haith and Shoemaker 1987; Howarth *et al.* 1991). For example, rainfall erosivity (RE_t) includes assessments of storm energy and intensity, modified for dormant periods as opposed to the growing season. For the terms (C) (P), a value of 0.06 was used to represent abandoned fields whereas 0.001 was used for forests. In pp. 267–268 on eutrophication, the importance of row crops as sources of phosphorus and sediments has been described. While the general structure of this model has wide applicability, the specific parameters will, of course, vary with climate, soil type and other features of the landscape.

Using such a model, it is possible to explore the sediment yields that might arise out of different land use practices. Overall, however, the results were not surprising, given the structure of the equations and the parameters entered: forests and pasture had minimal contributions to sediment loading. The principal sources of sediment were the area of intensive agriculture, and in heavily populated watersheds, area of urban/suburban landscapes. While the output from such models is entirely dependent upon the model structure and the validity of the parameters, they have two virtues. First, an equation such as the one above forces us to think carefully about the processes involved in producing sediment. Secondly, assuming the model captures some aspects of reality, it assists managers in assessing the consequences of different practices within the watershed, and the possible effectiveness of proposed changes in land use.

A different approach is described by Grootjans *et al.* (1993). Here the objective was not the study of sediment yields, but rather the linking of hydrology to vegetation in small stream valleys in a catchment area of some 30 000 ha. However, since many species respond not to hydrology alone, but to fertility, Grootjans *et al.* concluded that a regional hydrological model was incapable of being used as a vegetation management tool. Instead, they advocate the use of indicator species or expert opinion to link vegetation to water management practices.

It is important to remember, at this point, that much of the literature on sediment, and most if not all of the models, implicitly rather than explicitly assume that sediment is undesirable. This is probably reasonable for heavily populated watersheds where humans have greatly increased rates of erosion by stripping forests and ploughing fields (e.g. Figure 9.6). This assumption, however, still has to be put into perspective. Recall that similar activities by humans can also increase flood peaks, and it is therefore frequently assumed (again implicitly) that it is always beneficial to reduce flooding. Chapter 4 showed, however, that flooding is absolutely essential to a wide array of wetlands, and further, that reduced flooding can have catastrophic consequences. The perspective of heavily populated watersheds can distract attention from the natural hydrology and erosional dynamics of watersheds. One wants to avoid the same mistake when dealing with sediment. Certainly, abnormally high levels of sediment are undesirable for vegetation types such as fens, or fish species such as salmon and trout. At the same time, fresh alluvial sediments are necessary for the establishment of tree species on floodplains, and therefore for all the plant and animal species (including fish, Welcomme (1976); Goulding (1980)) that require alluvial forests. Examples of this range from the arid American west (W. C. Johnson *et al.* 1976) to the boreal forests of Canada (Nanson and Beach 1977) to the Amazonian rain forests (Kalliola *et al.* 1991) to the heavily industrialized Rhine River valley (Schnitzler 1995). It is therefore entirely conceivable that damage can be done to alluvial wetlands by reducing sediment loading in the lower sections of rivers and in deltas. Large dams too, are known to hold back sediments that would tend to build deltas, thereby risking the actual loss of coastal wetlands. Having thought about the processes involved in building a simulation model of erosion within a watershed, let us return to sediment deposition as a natural feature of wetlands.

Importance of time scale: sedimentation, emergence and marsh dynamics

Burial, like fire, has two apparently contradictory effects. In the short term, it may cause immediate death. But, in the long term, it may create new habitat for the same organisms that were killed. In the short term, then, we are inclined to view sediment as a negative factor, more so when sedimentation rates are increased by human activities, and particularly when small wetlands are gradually filled in.

At the same time, we must accept that coastal wetlands are dynamic and dependent upon eroded sediment. Deltaic wetlands are self-evidently dependent upon sedimentation. The cutting of new channels and deposition of new sediments is part of the natural dynamics of deltaic wetlands (Figure 9.17). The succession of saltmarsh vegetation from low marsh to high marsh is caused by the accumulation of sediment as well as peat (Niering and Warren 1980; Frey and Basan 1978). Sediment accumulation can be very rapid in deltaic areas. Consider the example of the Yellow River (second only to the Ganges/Brahmaputra in sediment load (Figure 9.4)) which empties into the north end of the Yellow Sea between China and Korea. Sediment discharge reaches a maximum in the annual August floods, when more than 30% of the annual sediment load is discharged in a single month. In contrast, January accounts for less than 1% of the total. The modern delta has an area of 5400 km^2, and the shoreline is moving seawards at about 1.5 km yr^{-1}; since the estuary is already filled, most river-borne sediments are

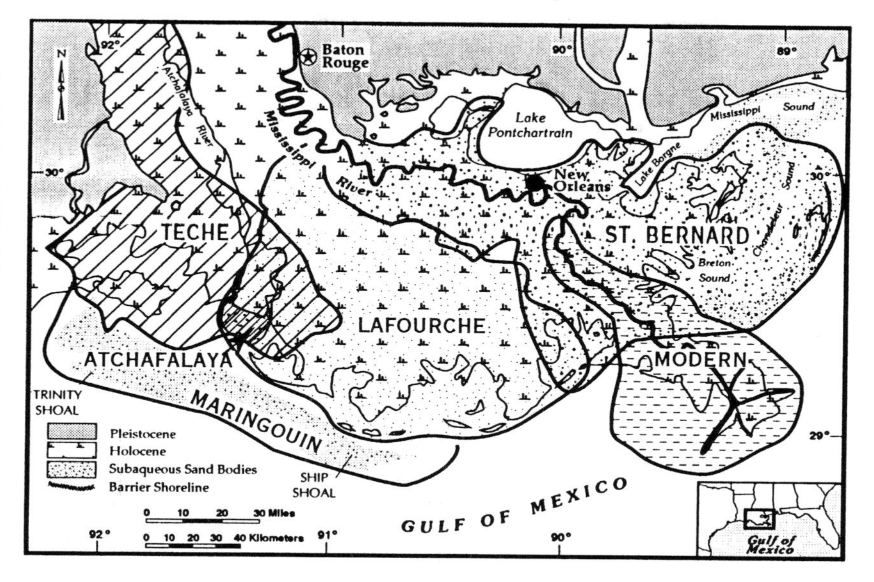

Figure 9.17 The Mississippi River delta is a composite of six distinct delta lobes produced by different courses of the Mississippi River over the past 7000 years. If the delta is to grow, rates of sedimentation must exceed the combined effects of subsidence and sea level rise (from Boesch *et al.* 1994).

now discharged directly into the sea (Schubel *et al.* 1986). Historical records emphasize the dynamic nature of these deposits (Chung 1982). In 1128 the Yellow River suddenly shifted its course southward, and from 1128 to 1855 the river mouth moved eastward by 90 km, adding an area of some 15 700 km^2. When, in 1855, the Yellow River again shifted northward, the absence of sediment and wave erosion allowed these older deposits to erode. About 1400 km^2 of land has now been reclaimed by the sea. Only a small portion of the tidal lands are vegetated with higher plants (e.g. *Sueda ussuriensis, Aeluropus littoralis, Phragmites communis* as well as some mangal); there are large expanses of 'barren flats' where *Spartina anglica* has been introduced to stabilize sediments, improve wildlife habitat and provide fodder and pasture (Chung 1982).

Where sea level is rising, the rates of sedimentation must exceed rates of sea level rise, or else, saltmarshes will disappear. For example, in Jiangsu Province, China, coastal recession of the southern Yellow River delta has occurred at 20 to 30 m annually and tidal land has sunk at rates of 5 to 10 cm/year over the past 50 years (Chung 1982). In Louisiana on the Gulf of Mexico, approximately 65 km^2 of wetlands are lost each year (Boesch *et al.* 1994). Direct destruction by draining or filling may account for less than a fifth of the Louisiana losses; indirect effects from alterations in natural hydrology and sedimentation may account for the rest. It appears that, in many Louisiana wetlands, rates of sedimentation can no longer keep pace with sea level rise. Reduced rates of sedimentation in coastal wetlands can be traced back to two main causes. First, the construction of levees restricts rivers from accumulating new sediment. Secondly, dams provide basins where sediment settles out and accumulates. As a result of these two activities, the suspended load in the Mississippi River decreased by about one half from 1963 to 1982 (Boesch *et al.* 1994). The remaining sediments appear insufficient to counteract the effects of coastal subsidence and rising sea levels. In addition, Shaffer *et al.* (1992) suggest that direct effects of grazing by *Myocaster coypus* (pp. 377–380), resulting in sensitivity to flooding, may be contributing to loss of vegetation cover.

Such deltaic situations stand in stark contrast to areas such as the Hudson Bay lowlands of Canada, where extensive areas of saltmarsh occur on a shoreline that is rising some 1.5 cm/yr (Glooschenko 1980) due to post-glacial rebound. The marshes here are similar in composition to those of Alaska and northern Europe (e.g. *Puccinellia phryganodes, Triglochin maritima*), but as the land rises, salinity falls and freshwater marsh species (e.g. *Carex palacea, Typha latifolia*) invade. Further inland are extensive bogs and fens interspersed with raised beach ridges. These wetlands are all very young, not because of newly deposited sediment, but because deglaciation occurred only some 8000 years ago, and new marshes continually form as land rises from beneath the sea. Emerging coastlines with saltmarsh vegetation are also found in other areas including Alaska, Scandinavia, Australia and South Africa (Stevenson *et al.* 1986).

A different shoreline zonation and sequence of events is associated with shorelines having rising sea levels. In north eastern North America, (Halifax, Nova Scotia to St John, New Brunswick) sea levels are rising in the range of 30 to 40 cm per century (Bégin *et al.* 1989), thereby causing a regression of forest, and sometimes also opening areas of wetland between the forest and the ocean. Along the coast of New Brunswick, for example (Bégin *et al.* 1989), sand dunes are moving upland with rising sea levels, and burying peatlands and spruce forests. Between the forest and dunes there are frequently areas of shallow water or marsh; this flooding reduces tree growth rates and regeneration. As sea levels continue to rise, these trees in turn are killed and buried by the migrating dunes. Depending upon topography and drainage, the area of marsh changes with time. A complex system of forest, peat bog, sand dunes and freshwater lagoons therefore occurs, with the dunes and lagoons revealing their origin through the still rooted stumps and standing dead trees emergent amidst them. Brinson *et al.* (1995) recognize four different zonation patterns that can arise as rising sea levels interact with the availability of sediment and the slope of the shoreline.

In contrast to well-studied areas in the temperate zone, rather less is known about sedimentation in tropical rivers. The entire Amazon lowlands appear, however, to be the result of three different kinds of large scale sedimentation processes (Irion *et al.* 1995). Sedimentation in the near Andean area is extremely high, reaching levels of almost 1000 tons km^{-2} yr^{-1}, leaving 100 m thick deposits downslope. Floodplains in the western Brazilian lowlands were largely ruled by sea levels, although studies are complicated by difficulties in dating sedimentary horizons and sequences. The main valley of the Rio Amazon has seen both periods of erosion during low sea levels, and deposition during periods of higher water. These rising (and falling) sea levels appear to have substantially influenced the entire development of the Amazon basin. Some 80 000 years BP, during the glacial maximum, sea levels may have fallen more than 100 m below recent levels (Irion *et al.* 1995), initiating a period of erosion that deepened the Amazon by some 20–25 m (Müller *et al.* 1995). After 15 000 BP, sea levels rose about 2 cm yr^{-1} and the Amazon valley was drowned because sedimentation rates were not high enough to balance rates of rising seas levels (Irion *et al.* 1995). During this period, a large freshwater lake about 1500 km long and up to 100 km wide may have extended from the mouth of the Amazon inland to about 65°W. The maximum size of this lake appears to have been reached around 6000 years BP. Sediment cores recovered from the deep-sea fan of the Amazon in the Atlantic Ocean suggest that, during this time, large quantities of continental detritus no longer reached the sea, being deposited instead in the sediment trap created by this lake. As sediments were deposited here, ridges, swales and levees would have formed in the middle Amazon area. Superimposed upon these large-scale processes are the ongoing processes of erosion and deposition (pp. 286–289) producing large meander complexes and the shallow lakes known as várzeas (Salo *et al.* 1986; Junk and Piedade 1997).

Over larger time scales of centuries and millennia, then, the balance among erosion, sedimentation, subsidence and emergence produces

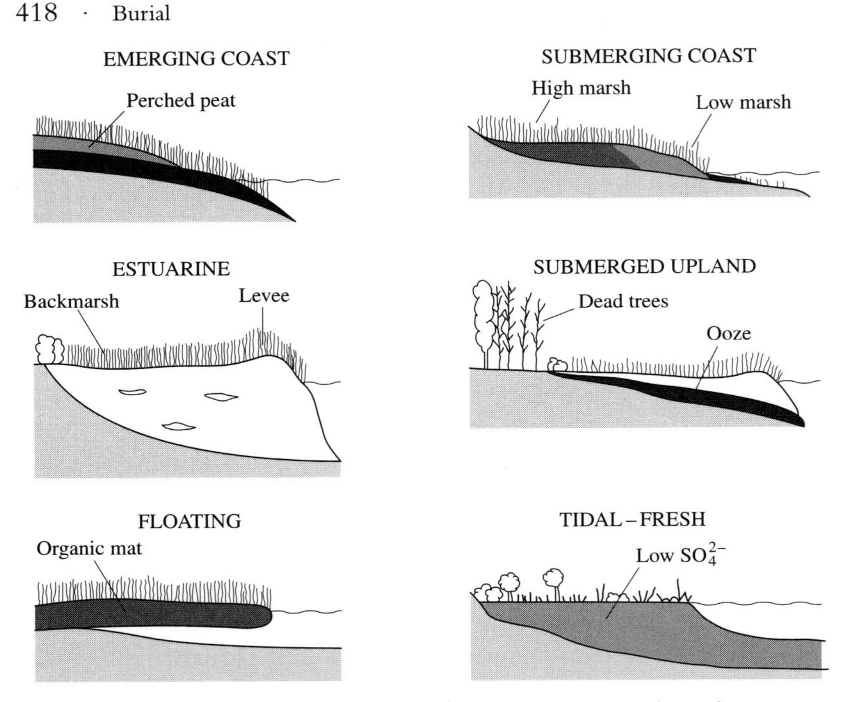

Figure 9.18 Six basic types of shoreline wetlands that can arise depending upon the type of sediment and the rate at which sea levels are rising or falling (from Stevenson *et al.* 1986).

much of the physiographic diversity of coastal wetlands (Figure 9.18). This physiographic diversity provides the habitat template that generates biological diversity.

Conclusions

The erosion of land generates sediment, and wetlands are the sites where this sediment accumulates. The accumulation of sediment alters species composition. Plants will shift towards species with deeply buried rhizomes and strong, pointed growing tips. The composition of invertebrates and fish will also change. Sedimentation rates can be related to land uses in the watershed. In the longer term, many coastal and alluvial wetlands are dependent upon sedimentation to create new habitat.

Part III
The path forward

10 · *Wetland restoration: assembly rules in the service of conservation*

Introduction

We have now finished our exploration of Principle 1: we have seen how the patterns and species in wetlands are produced by multiple environmental factors acting simultaneously. We could stop at this point and feel a certain amount of smug satisfaction that we know so much more about wetlands than other people. But this contentment could be a delusion. The real test of our understanding of ecology lies in (i) our ability to predict changes in communities and ecosystems, (ii) our ability to manipulate them to achieve certain goals, and (iii) our ability to choose those goals wisely. So, let us turn to management.

The diversity of wetland types, the vast numbers of species and complexity of environmental factors could appear daunting. It would be easy to put the book down, make some trite observation about the wonderful complexity of nature, and become deep ecologists. Or, we might throw up our hands with fear, and become autecologists. In either case, we could pick a philosophical perspective, or a species, and settle into a life-long sinecure, having concluded that ecosystem management is an impossible goal. It would then be necessary to emerge only to throw stones at those who persist, labelling them shallow and naive, or to defend our particular species when someone is challenged to manage the entire ecosystem. Before we join this reclusive assembly, we might recall that it *is* possible to simplify intelligently. First, we have seen that, although there are many patterns, and many species, these can be condensed into a smaller array of patterns (e.g. zonation, succession, boundary clustering) and a smaller array of properties of species (e.g. richness, dominant species, rare species). Secondly, although there are many interacting environmental factors to consider, only a few are likely to predominate in any situation (e.g. hydrology, fertility, disturbance, competition, grazing and burial), and many of these have relatively predictable effects.

So, if we now turn to Principle 2, the task for managers is to manipulate one or more of these factors in order to create the desired wetland. This can involve either the manipulation of existing wetlands, or the recreation of wetlands where none existed before, to achieve certain goals. In the first case, we may wish to restore water level fluctuations to a floodplain in order to change the composition of some valley wetlands. In the second case, we may have inherited a drained piece of landscape, and wish to recreate the wetlands that once occurred there. The former process is often termed ecosystem management, the principles of which have been laid out by Christensen *et al.* (1996). The latter process of recreating ecosystems or communities where none now exist is often called restoration (e.g. Lewis 1982; Jordan *et al.* 1987; Bonnicksen 1988). In practice, the term restoration is now used rather more widely, so that wise management of a damaged wetland is now frequently termed restoration as well. Thus restoration is part of the larger challenge posed by ecosystem management (Grumbine 1994; Christensen *et al.* 1996; Haeuber and Franklin 1996).

The scientific literature contains vast numbers of reports of restoration, whether narrowly or broadly defined (e.g. Cairns 1980, 1988; Lewis 1982; Jordan *et al.* 1987; Zelazny and Feierabend 1988; Galatowitsch and van der Valk 1994; Kusler and Kentula 1990; Pfadenhauer and Klotzli 1996; Kusler *et al.* 1995). In practice, many of the individual studies document a specific site, a particular manipulation, a selected set of introductions, and a restricted set of measurements to determine the relative degree of success (but see Larson 1988; Zedler 1988; Rosgen 1995; Brinson 1995; Galatowitsch and van der Valk 1994; Kershner 1997). This chapter could easily degrade into a set of case studies that illustrate what can be done on a case-by-case basis, without any common framework. Since there are already more than enough examples of this approach, I shall veer to the other end of the continuum, and review a conceptual framework that may unite all restoration studies. Further, I will draw upon an already established area of community ecology, assembly rules.

I will use an already established scientific framework for two reasons. First, I deny the all too convenient separation between science and management, or the alleged distinction between theory and application. All too often this alleged distinction is just an excuse. It allows theoreticians to both keep their feet dry and ignore societal implications of their work. Simultaneously, it provides managers with an excuse for being unfamiliar with general scientific principles and for continuing management practices that are years out of date. (No wonder this non-existent dichotomy is so persistent and widespread! Now that I have exposed the hidden motivations, perhaps it will be opined less frequently at symposia.) My second motivation is that, like General Patton (e.g. Essame 1974), I am of the opinion that it is a waste of resources to fight for the same ground twice. If we have a framework sitting on the scholarly shelf, let us pull it off and apply it to management. I risk annoying some applied ecologists, who will find that their particular restoration study is not included here, but I hope that they will appreciate the attempt to provide a context for all restoration studies, rather than a list of selected examples.

Some background to restoration

The real test of wetland ecology lies not in how much knowledge we feel we have, nor in how many papers appear in our learned journals, nor in how many species we can identify, nor in the mathematical sophistication of our models, but rather in our practical ability to protect and restore ecological communities. In order to do so efficiently, we must be able to predict the response of ecosystems to change. Change could include both natural processes (e.g. shifts in climate) and deliberate human manipulations (e.g. alterations to wetland hydrology). The most telling test of our knowledge arises when we set out to recreate an ecological community. The process of recreating an ecological community is termed 'ecosystem restoration' (e.g. Cairns 1980; Jordan *et al.* 1987; Bonnicksen 1988; Dahm *et al.* 1995). Ecosystem restoration is growing rapidly as an application of ecological practices; Bonnicksen (1988) refers to it as 'an emerging profession within the science of ecology'. Several goals are possible: to repair biotic communities, to re-establish them on the same sites if they are destroyed, or to construct synthetic communities if the originals are no longer possible.

We should not assume, however, that restoration is something dramatically new. Bonnicksen, like many other writers in this field, traces the roots of restoration ecology back only as far as Aldo Leopold in 1949. More than a decade earlier, however, Clements (1935) also wrote an essay appropriately titled 'Experimental ecology in the public service' in which he described the applications of ecology to a wide range of applied problems.

In this essay, Clements used the title of 'natural landscaping' (p. 359) to addressing the ecological challenges posed by the revegetation of cuts and fills along highways, but refers in the text to '*rehabilitation* on the basis of climax vegetation' (p. 352) and '*restorations* as are proposed for the floor of the Yosemite Valley, where the original flowery meadows have been entirely dispossessed by weedy grasses . . .' (p. 360) (italics mine). He explicitly lays down several 'canons' to guide natural landscaping.

> The chief of these is that nature is to be followed as closely as possible and hence native materials alone are to be employed, preferably from the outset but invariably in the final composition . . . The process of succession by which nature reclothes bare areas is to be utilized as the chief tool of landscaping, but the process is often to be hastened or telescoped to secure more rapid and varied results (p. 360)

On re-reading Clements, it seems that he had even foreseen the logical structure of this chapter, for under the topic of grazing lands, he adds 'A necessary adjunct is the use of *indicators* to record existing conditions and their gradual change into grazing communities of the desired composition and yield' (p. 353, italics mine).

In the previous century, 1883, Phipps wrote a book on the restoration of forests, and Larson (1996) has described what appears to be one of the earliest practical restoration projects in North America, the replanting of a forest in a gravel pit near the University of Guelph by Professor William Brown, an arboriculturalist from Scotland. In Beard's classic book on the vegetation of the Caribbean islands (Beard 1949), there is also a discussion of forest restoration activities in the early 1900s. In wetlands, the use of *Spartina anglica* for 'reclamation' of coastal mud flats was also of interest early in this century (Chung 1982). In the 1930s, the British blocked drainage ditches and used portable pumps to raise water levels in the Woodwalton Fen (Sheail and Wells 1983). Now there is even a separate scientific society (The Society for Ecological Restoration) and a journal (*Restoration Ecology*) devoted to the topic.

Restoration ecology has the potential to bring together a wide range of ecological activities. It challenges conservationists, applied ecologists, and theoreticians in different ways. Conservationists are challenged to shift some energy from protecting remnant fragments of habitat towards the longer term goal of restoring and re-connecting entire landscapes. The Wild Earth proposal for North America (Wild Earth 1992) is one example. Applied ecologists are challenged to move from traditional manipulation of single components of ecosystems, such as waterfowl or fish, towards the reconstruction of entire ecosystems so that they become self-regulating. Theoretical ecologists are challenged to develop practical tools to guide restoration and the indicators to monitor its success.

Let me re-assert that restoration is the ultimate test for ecologists. Restoration demands that we make clear predictions about the future state of communities, and take concrete steps to test those predictions by trying to create a specific ecological community. Large sums of money are already being spent. If our predictions are wrong, we will fail to create the desired community; we may not only waste the money invested in the program, but we may cause further environmental harm. In practice, however, one gains the impression that a good deal of practical restoration activity does not yet fully appreciate its historical roots, its scientific foundations, nor the availability of existing knowledge. There is therefore the risk that restoration ecology will be seen as yet another fragmented state within the increasingly Balkanized realm of ecology, and that the art and technique of restoration will be emphasized to the exclusion of the science. Certainly, a majority of the published papers on restoration largely ignore the relevant scientific literature, and tend to report on a single site and small group of species; such publications can be generated *ad inifinitum*, but it is difficult to extract general principles that have broader applicability. Further, the grave problems raised by medium number systems (see pp. 447–448) suggest that many such examples can be collected without any promise of finding principles that apply to a wide range of other sites. The problem of extracting general principles from large numbers of studies is well illustrated by an example from a different area of ecology, the voluminous studies, extensive reviews and ongoing divisive debates in the study of competition (e.g. Schoener 1985; Underwood 1986; Keddy 1989a, b; Grace 1990), debates that continue in spite of the increasing numbers of studies and the emerging potential of meta-analysis (Gurevitch *et al.* (1992). Apparently, volume does not guarantee clean answers to questions, nor the discovery of generally accepted principles.

I suggest that the unifying framework for restoration ecology already exists in the realm of community ecology. Restoration ecology really requires two tools: (i) the ability to manipulate ecosystems to recreate desired communities, and (ii) the ability to evaluate whether our manipulation has produced the desired change. All too often both of these are assumed without any scientific foundation. The first tool, step (i), is a rapidly developing field called assembly rules (Diamond 1975; Weiher and Keddy 1999), and I will first lay out this connection between assembly rules and restoration. The second tool, step (ii), evaluating our success at step (i), is rapidly developing in the realm of ecological indicators (e.g. Odum 1985; Rapport *et al.* 1985; Goldsmith 1991; McKenzie *et al.* 1992; Woodley *et al.* 1993). I will introduce wetland indicators later in this book (pp. 525–531), once the problems of wetland management have been explored in more detail.

Let us consider one example of restoration to illustrate the scale of the task at hand and the need for employing sound scientific principles from the start. An ambitious and large-scale restoration project is planned for the Kissimmee River in central Florida. This 166 km long river flows south into Lake Okeechobee; it initially consisted of a low-gradient, meandering channel with a wetland floodplain from 1.5 to 3 km wide. This floodplain supported more than 15 000 ha of wetlands. These were 56% broadleaf marsh (dominated by genera such as *Pontederia, Sagittaria* and *Leersia*), 14% wetland shrub swamps (dominated by *Cephalanthus* and *Salix*) and 26% wet prairies with coastal plain genera such as *Hydrocotyle, Cyperus* and *Fimbrystylis* (Toth *et al.* 1995). The pattern of water level fluctuations was considered to be more typical of tropical rivers (such as the Amazon or Niger) than those of other temperate zone rivers (Koebel 1995). There were diverse wading bird communities including White Ibis and Snowy Egret, as well as nearly 40 species of fish, including a nationally significant Largemouth Bass fishery.

Between 1962 and 1971, this river was transformed into a 90 km long, 10 metre deep, 100 metre wide canal. This eliminated historic water level fluctuations and resulted in the loss of more than 12 000 ha of wetland (Koebel 1995). It also resulted, as we would expect from Chapter 4, in losses of wet prairie and broadleaf marsh, but increases in wooded wetland (Toth *et al.* 1995). Wildlife populations were also adversely affected (Toth 1993; Dahm *et al.* 1995). Invertebrates such as prawns (*Palaemonetes paludosus*) and apple snails (*Pomacea paludosa*) declined; these are important food sources for wading birds such as Limpkins and Snail Kites, as well as alligators and snapping turtles. The Largemouth Bass fishery was eliminated, and other fish populations such as Bluegill and Black Crappie, declined. Two species of fish may have been extirpated.

In 1976, the Florida Legislature passed the Kissimmee River Restoration Act. The objective of this restoration project is to re-establish the structure and function of historic river channel and floodplain habitats. Current plans call for the back-filling of over 35 km of canal, recarving of 14 km of river channel and the removal of two water control structures (Koebel 1995). In order to monitor the success of this project, five groups of indicators will be used: river channel floodplain communities, wading birds, waterfowl, fishes and invertebrates. Echoing the theme in Chapter 4, Toth *et al.* (1995) emphasize that the ultimate challenge is to re-establish the habitat structure of the river. If this is done, it is believed that ecological integrity will follow.

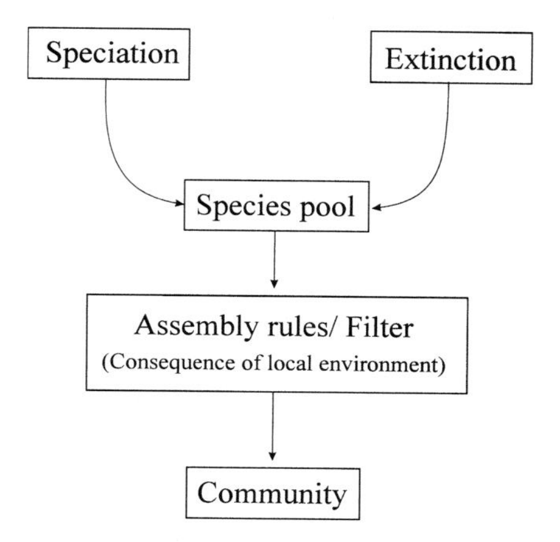

Figure 10.1 The local environment filters out species from the pool, thereby creating a community. This is the theoretical foundation for wetland restoration (after Wiens 1983).

The thesis of this chapter is that the intellectual framework of restoration ecology already exists in the realm of ecology. Restoration ecology really requires two tools: (i) the ability to manipulate ecosystems to recreate desired communities, and (ii) the ability to measure whether the manipulation has produced the desired change. The first is called assembly rules, and this chapter will focus on the connection between assembly rules and restoration. The second problem, that of monitoring, is rapidly developing in the realm of ecological indicators. The use of indicators in wetland monitoring will be reserved for the final chapter.

Let us turn to problem one. In any restoration project we have our intellectual raw materials: (i) starting conditions, (ii) a pool of species, and (iii) a list of key factors that we can manipulate. The practical problem is to identify which factors, applied to which set of species, will produce which communities.

Assembly rules as a foundation for restoration

The raw material for a restoration project is the pool of species available to colonize the site; the pool of species is the product of long-term processes of evolution and extinction (Figure 10.1). The study of these lies within the realm of evolutionary ecology and the time scales are generally sufficiently long that the pool can be regarded as a constant for any particular project. The objective of assembly rules, then, is to predict

Table 10.1. *The estimated relative importance of environmental factors that determine the properties of wetlands*

Environmental factor	Relative importance (%)
Hydrology	50
Fertility	15
Salinity	15
Disturbance	15
Competition	<5
Grazing	<5
Burial	<5

which subset of this species pool for a given region will occur in a specified habitat (Keddy 1992a; Weiher and Keddy 1999). It basically is a problem of deleting those species unsuited to a specified set of environmental conditions. A first objective would be simply to predict the presence or absence of species in a habitat. The second objective would be to predict abundance as well as presence.

The process of constructing communities from species pools is therefore in many ways analogous to the processes of evolution through natural selection. In natural selection, habitats serve as filters for genotypes, with the least suited genotypes being filtered out, and the best suited surviving to reproduce. In the case of assembly rules, habitats again serve as filters and eliminate those sets of traits which are unsuitable to that environment. The species which comprise the community are those with the traits that survive the filter. The filters that control wetland composition have already been systematically explored in this book: they are listed in Table 10.1 where I have also given a first guess as to their relative importance in controlling the community structure and composition of wetlands. This list (and the relevant chapters of the book) provide a shopping list for the factors that can be used to create filters for restoring wetlands.

For any particular wetland, the relative importances may vary. A restoration project in a delta may find that burial is relatively more important, whereas a restoration project in an agricultural landscape may find that fertility is relatively more important. In any case, hydrology requires first consideration and, if it is neglected, then other factors alone will probably be insufficient to create the desired wetland.

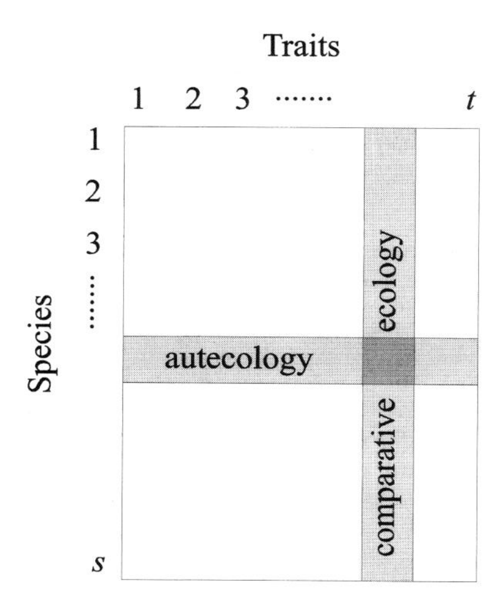

Figure 10.2 A trait matrix summarizes the properties of the species in the pool (from Keddy 1992b).

Given the list of environmental factors that act as filters in wetlands (Table 10.1), two biological data sets for ecological communities are needed: a species pool, and a matrix of the traits of species in this pool (Figure 10.2). 'Assembly rules' then specify which particular subset of traits (and therefore species possessing them) will be filtered out. More precisely, in the situation where we have knowledge of *n* traits for each species in the pool, we are looking for a procedure to specify whether or not certain traits (or sets of them) will permit a species to persist under a defined set of environmental conditions. The exact procedures for doing this most effectively need further work. Three examples illustrate some of the potential applications. These examples all involve the direct effects of abiotic factors upon living organisms. Biotic interactions do, of course, also deserve recognition. Predation can be a strong filter (Wilbur 1984; Carpenter *et al.* 1987) as can grazing (Jefferies 1988a, b; Shaffer *et al.* 1992; Smith and Kadlec 1985a, b) or competition (Snow and Vince 1984; Twolan-Strutt and Keddy 1996; Levine *et al.* 1998) and the presence of neighbours can also facilitate establishment or survival by reducing the constraining effects of physical factors (Castellanos *et al.* 1994; Bertness and Leonard 1997; Bertness and Hacker 1994). These biotic effects, sometimes termed examples of 'ecological engineering' (Jones *et al.* 1994) may be less obvious, but in principle, at least, can be treated in the same manner as the direct effects presented below in pp. 430–436.

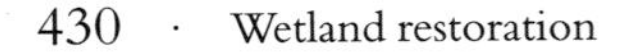

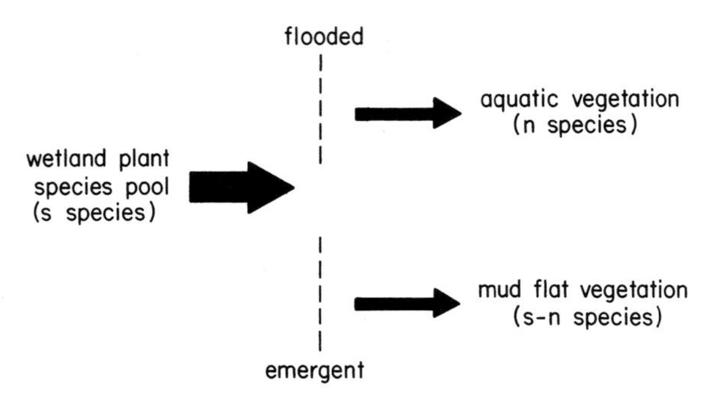

Figure 10.3 Flooding acts as a filter by controlling buried seed germination, thereby determining the composition of wetland plant communities (from Keddy 1992b).

Prairie wetlands

An early attempt to predict species composition in wetlands using filters can be found in van der Valk (1981). Species in prairie wetlands must periodically regenerate from buried seeds (Recall Table 4.4 and Figure 8.2). The problem is to predict species composition in these wetlands after a specified change in water level. A key element of van der Valk's model was the recognition that only one major trait was necessary to predict regeneration: whether or not a species could germinate under water. By measuring only this one trait on all species, one can predict which part of the species pool will occur under either set of conditions (Figure 10.3). The particular appeal of this model lies in the simplicity of the trait matrix.

In later work, van der Valk (1988) asked whether the densities of buried seeds of four emergent plants could predict the densities of adult shoots after a reduction in water level. Seedling densities were lower than expected, and rank order correlations between the abundance of buried seeds and seedling densities were variable (*Scolochloa festucacea*, 0.93; *Scirpus lacustris*, 0.76; *Typha glauca*, 0.44; and *Phragmites australis*, 0.03). In the case of *P. australis*, however, Dore and McNeill's (1980) monograph on grasses for adjoining Ontario expresses doubt as to whether viable seeds are formed, and suggests that reproduction may be almost entirely by vegetative means. The seed-seedling correlations are rather better for the other three species, although as van der Valk observes, 'there remain discrepancies between predicted and actual densities of seedlings during the drawdown'. This illustrates limitations on assembly rules. As species become increasingly similar in their traits, it is increasingly unlikely that simple relationships between traits and environment will predict abundance. When it comes to predicting individual species distributions, van der Valk suggests that '[R]eductionist models based on comparative life-history data wil provide the best possible foundation for vegetation management'. The point seems to be that the particular traits that must be selected will also depend upon the scale of the investigation and the kind of species involved. Equally, the use of such similar species to test assembly rules based upon traits perhaps illustrates the tendency of ecologists to think in terms of species rather than functional groups.

Fish in lakes

In flooded areas, access to oxygen is a critical problem for many living organisms (pp. 34–42). The Amazon basin has both the largest areas of flooded forest in the world and the largest number of freshwater fish species in the world (Goulding 1980; Lowe-McConnell 1987). Some 70 000 km^2 of forest are flooded, sometimes to depths of 15 metres, and for up to ten months of the year. Up to 3000 species of fish may inhabit this region; of the more than 1300 described to date, about 80% are either catfishes or characins (Figure 3.4). The annual cycle of tropical fish is closely tied to periods of inundation and periods of hypoxia particularly after flooding, when shallow pools in floodplains are left filled with debris and exposed to the warming effects of sunlight (recall pp. 187–190). The lowest concentration of oxygen in the water can then be considered to be a filter, which selectively removes different portions of the fish fauna. The ability to tolerate this filter can be determined for each fish species, whether by screening (*sensu* Grime and Hunt 1975) as in Junk (1984), or by reference to other traits.

Moving north to central North America, Magnuson and his co-workers (e.g. Tonn and Magnuson 1982; Magnuson *et al.* 1989) have studied the distributions of fish in lakes in the lake district of Wisconsin and Michigan. One county alone, Vilas County, has over 1300 lakes. Typical fish range from mud minnows (*Umbra limi*) and redbelly dace (*Chrosomus eos*) through to larger predators such as Northern Pike (*Esox lucius*) and Largemouth Bass (*Micropterus salmoides*). Many of the lakes adjoin peatlands and have emergent and aquatic plants. Rather than studying each lake in turn, some general rules would be useful.

The distribution of fish within 18 small lakes was closely connected to environmental factors thought to act as filters. The two principal fish community types were *Umbra*–cyprinid lakes and Centrarchid–*Esox* lakes. Low oxygen levels in winter and connectedness to larger bodies of

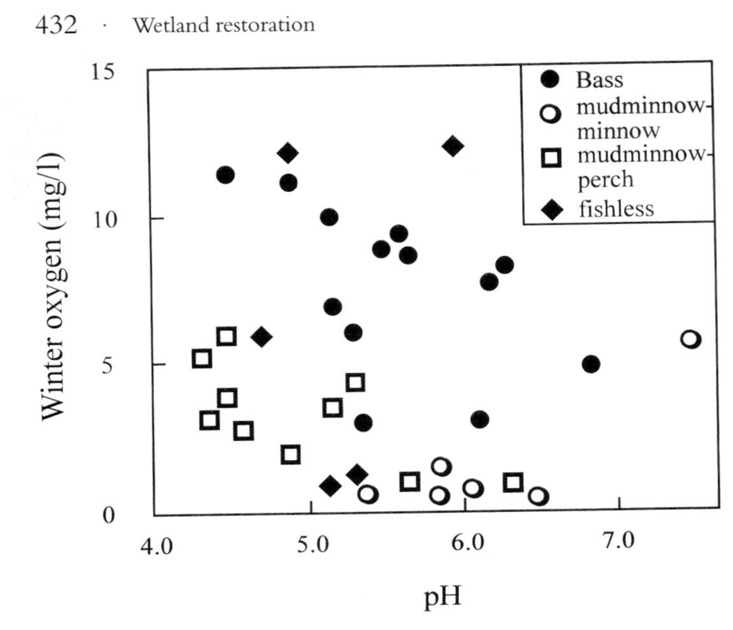

Figure 10.4 Winter oxygen levels and pH act as filters to create different fish communities (from Magnuson *et al.* 1989).

water which serve as refuges from low oxygen, were thought to be the main environmental factors producing these two community types (Tonn and Magnusson 1982). Further analysis added another lake community type, bass lakes, by splitting the centrarchid–*Esox* lakes into either pike lakes (*Esox lucius*) or bass lakes (*Micropterus salmoides, Ambloplites rupestris*) depending upon whether bass or pike was the dominant predator (Tonn *et al.* 1983).

Physical factors including lake area, maximum depth, watershed size and conductivity could be used to discriminate between the two main community types, as well as predict fish composition in a new sample of 11 lakes. Using a much larger sample, M. G. Johnson *et al.* (1977) classified 2496 Ontario lakes based on the occurrence of Walleye, Northern Pike, Lake Trout and Smallmouth Bass, and found these species were also related to seven limnological factors.

Magnuson *et al.* (1989) further separated the *Umbra*–cyprinid lakes into two groups: mudminnow–minnow lakes, and mudminnow – perch lakes. Winter oxygen levels and pH distinguished these later two groups. The lakes with only minnows had extremely low winter oxygen levels, with but one exception (Figure 10.4). If fish are screened for their

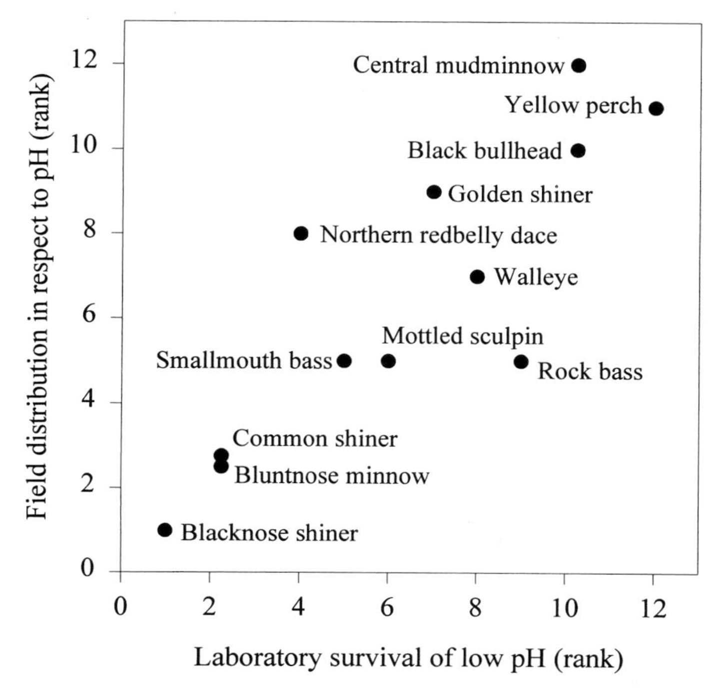

Figure 10.5 Laboratory measures of pH tolerance can predict the field distributions of fish (from Magnuson *et al.* 1989).

tolerance to pH, this trait is also a good predictor of their field distributions (Figure 10.5).

If low pH and low oxygen act as filters, this can explain why fish with high oxygen and high pH requirements do not inhabit shallow lakes with low pH. But it does not explain the reverse. Why do minnows and mudminnows not inhabit the lakes with higher oxygen and pH? Magnuson *et al.* (1989) conclude that the reason is predation; smaller fish are excluded to lakes from which predators are absent.

Sketching this as a series of filters (Figure 10.6), we begin with the pool of fish available to these lakes on the left. Physical factors and low pH eliminate centrarchid fish from small and shallow lakes. Predation eliminates minnows and mudminnows from the larger lakes. The overall structure of this community appears nested in a manner similar to that of zoned wetland plant communities. That is, these fish species appear to have inclusive niche structure, all being able to survive in the higher oxygen and higher pH lakes. In this case, however, it is predation that produces the

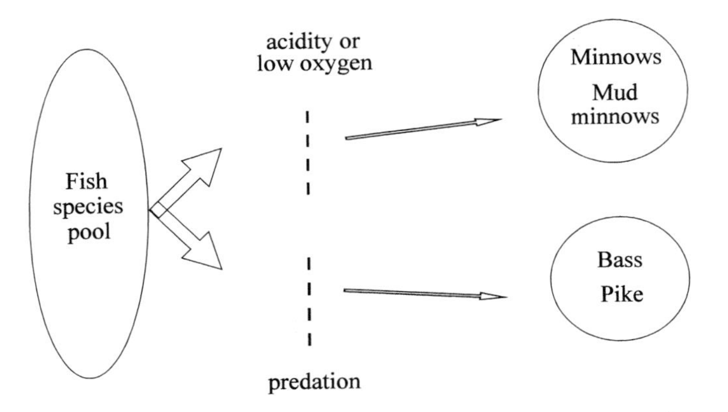

Figure 10.6 Different filters create different fish communities from a common pool.

different field distributions, whereas with plants it is a hierarchy of competitive ability.

Saltmarshes

These are marvellous systems for the application of assembly rules because they are so structured by a single factor: salinity. The practice of creating coastal marshes and mangal is already well established (Lewis 1982), although much of it emphasizes planting nursery-grown stock on mud flats. Let us consider salinity as a filter. A first task is to screen all common species for relative salinity tolerance. Partridge and Wilson (1987), for example, provide a fine example of this; species can be ranked in order of tolerance to salinity, and in general, they exhibit inclusive niche structure (Figure 10.7).

The vertical line in Figure 10.7 is sea water salinity. Only a small group of this pool can grow in sea water salinity. We can therefore regard this line as a filter. If the habitat in question has twice sea water salinity, the pool of 31 species is reduced to only one – *Suaeda nova-angliae*. With salinity at sea water concentrations, the subset of potential colonists is approximately ten. If we wanted to predict the effects of increasing salinity (say from increased tidal inundation associated with rising sea levels (Brewer and Grace 1990)) we could forecast which species might disappear first; conversely, if the site is to be impounded and flooded with fresh water, we could forecast which species might invade. We saw in Chapter

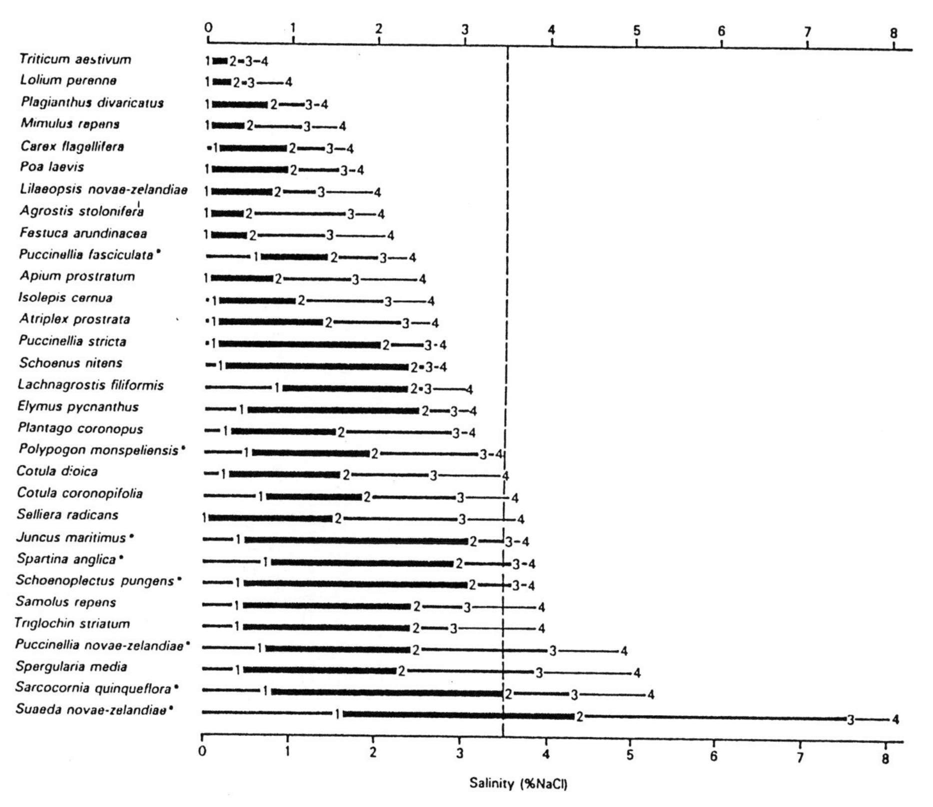

Figure 10.7 Saltmarsh plants can be ranked according to their tolerances for salinity. Vertical hatched line indicates ocean salinity. (From Partridge and Wilson 1987.)

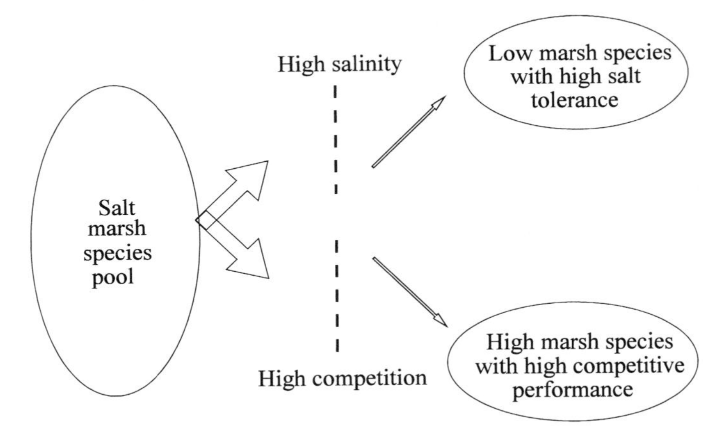

Figure 10.8 Salinity and competition can act as filters to create different marsh communities.

6 how fluctuations in salinity controlled regeneration of species in saltmarshes (Zedler and Beare 1986).

The experimental assembly of wetlands

If this approach to community assembly has validity, experimentally created filters should create different communities from one species pool. We have recently explored this process with a microcosm experiment (Weiher and Keddy 1995; Weiher *et al.* 1996). A pool of 20 wetland plant species was sown into 120 microcosm containers representing 24 wetland environments. The 20 species in the pool were selected from a group of 91 candidate species, based upon criteria including life history type, morphology and germinability. The environmental conditions manipulated included most of the major variables thought to influence wetlands: (i) water depth, (ii) timing and duration of flooding, (iii) leaf litter, (iv) soil surface texture, (v) sowing date and (vi) presence or absence of *Typha*. Each of these factors was repeated at high and low fertility. Species composition was then measured for five growing seasons.

Each environmental factor had a significant effect upon species composition (Table 10.2). Although 24 different treatment regimes were provided, Table 10.2 shows that water level and fertility were the most important filters. The subset of five water level regimes and two fertility levels (ten sets of conditions) produced only four major vegetation types (Figure 10.9). That is, different sets of environmental factors sometimes

Table 10.2. *Effects of different environmental factors (filters) on the species richness of 24 experimentally created wetland communities*

	Treatment	Fertility	Treatment × fertility	Year	Treatment × year	Fertility × year
Water depth	195.6***	47.48***	9.67***	153.66***	46.11***	9.36***
Water fluctuation	73.75***	46.76***	5.28*	99.13***	48.44***	5.41**
Litter	0.89	220.4***	8.65**	169.00***	4.98**	6.47**
Soil texture	0.50	77.63***	0.35	296.58***	1.88	12.14***
Sowing date	5.88**	56.96***	7.87**	268.52***	0.95	6.54**
Typha	0.05	81.87***	2.55	84.58***	3.52*	4.06*
Water Fluctuation	Treatment	220.4***	Treatment × fertility	296.58***	Treatment × year	Fertility × year

Notes:

F-values: *$P<0.05$, **$P<0.01$, ***$P<0.001$

Values in table are F values for repeated measures three-way ANOVA from a main effects and two-way interaction terms

Source: From Weiher and Keddy (1995).

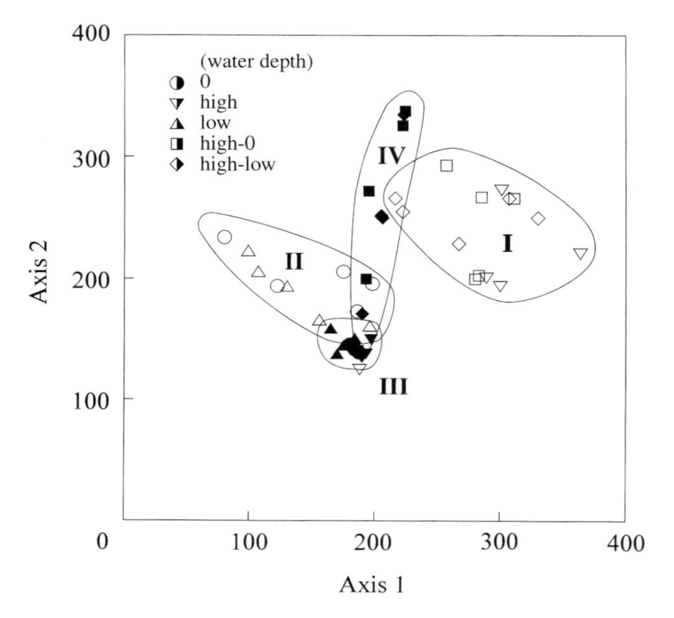

Figure 10.9 DCA ordination of microcosm plant community data. Symbols represent water depth/fluctuation treatments, filled symbols represent high fertility, open symbols represent low fertility. Note that the other treatments (litter, soil texture, starting date, and additional *Typha*) were not significantly different from communities III and IV (from Weiher and Keddy 1995).

produced convergent vegetation composition. For example, infertile treatments with high water or spring flooding (Type I, Figure 10.9) produced vegetation made up primarily of *Bidens cernua* and *Eleocharis smallii*. In contrast, infertile treatments with fluctuations in water depth were primarily composed of *Eleocharis smallii* or *Lythrum salicaria* (Type II, Figure 10.9). Community type III included fertile treatments with constant water levels, dominated by *Lythrum salicaria* with some *Scirpus acutus*. Finally, drier treatments (Type IV) were largely composed of *Lythrum salicaria* and *Carex crinita*. The interaction terms in Table 10.2 show that the arrangement of the communities did not follow a simple response to fertility on one axis and water level on the other.

Based upon these results, a few simple filters can be arranged to show how different environments sieve out sets of different communities from a species pool. If we take a community approach and consider relative abundance and position in DCA space, Figure 10.10 shows the result. First, there is one unknown filter that prevents establishment of one

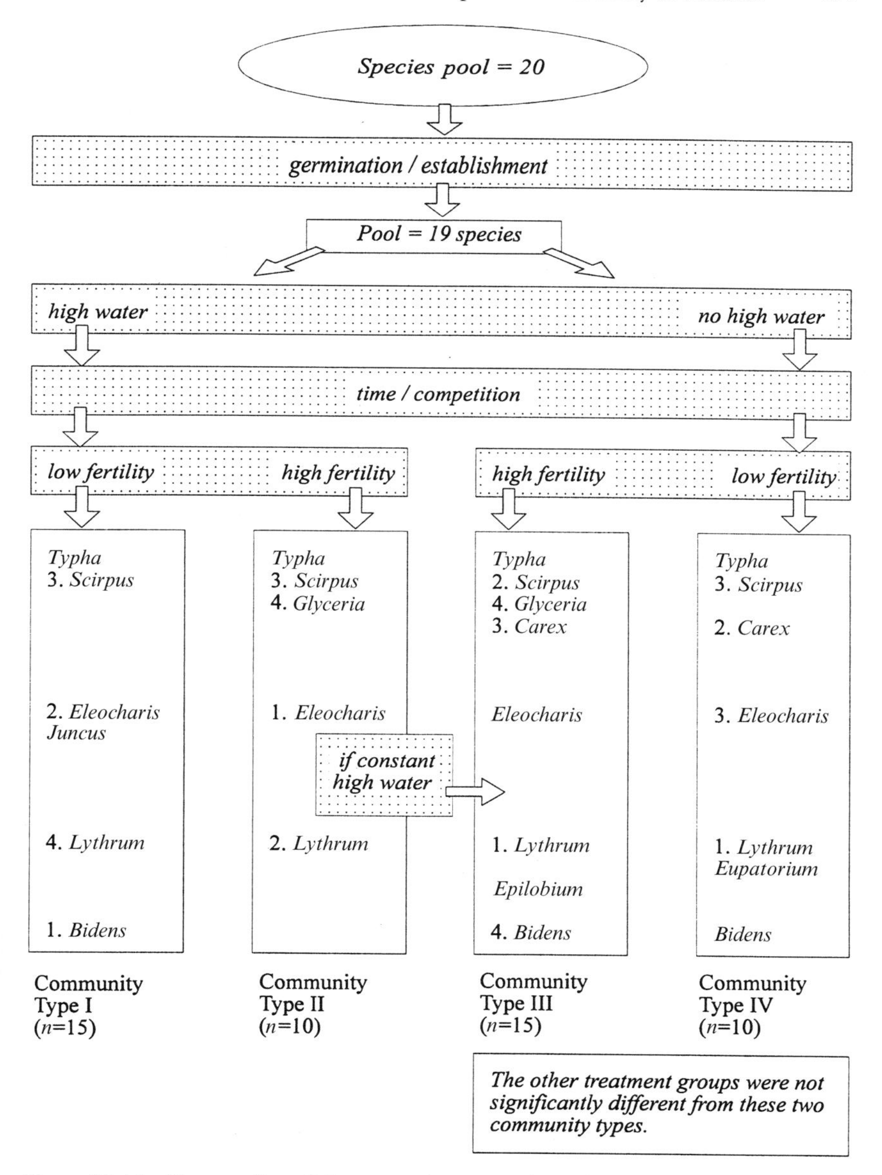

Figure 10.10 Twenty-four different environmental conditions, each replicated five times, produced four basic wetland vegetation types from a common species pool. It appears that five filters (stippled boxes) can account for the observed patterns. Numbers next to species names correspond to mean rank abundance (from Weiher and Keddy 1995).

species, *Panicum longifolium*. The second filter represents differences in establishment that were associated with water level. The next filter represents loss of species that did establish, but were unable to persist through time (e.g. *Xyris difformis*); presumably such species could not survive the effects of neighbours such as *Lythrum salicaria*. The remaining species were then sorted by response to high versus low fertility. In one exception, an added filter of constant high water transferred a community type from the high water (Type II) to the low water (Type III) group.

Response rules

In the case of assembly rules, the focus is upon the creation of new communities in previously unoccupied environments. While this may mimic some restoration problems, say, turning a gravel pit into a wetland, a majority of management problems begins with an established community which managers want to change in certain ways. Perhaps the goal is to eliminate exotic species, or increase rare species. Response rules specify how an initial vector of species composition will respond when an environmental factor is changed. Lewontin (1974) has called this 'transformation rules'. Other examples would include: how will prairie vegetation respond to fire or grazing? How will bird communities respond to forest clearance? How will stream invertebrates respond to siltation? There are two ways in which response rules would differ from assembly rules. First, one begins with a subset of species already present, and must predict how these will respond to the perturbation (deletion rules). Secondly, one must re-examine the species pool and trait matrix for species likely to replace those presently occurring (addition rules). The combination of deletion and addition rules is shown in Figure 10.11.

Some methodological considerations

Constructing trait matrices

The two essential data sets for 'assembly rules' are the species pool and the species traits. Species lists for habitats are relatively easy to come by compared with matrices of species traits. Therefore, let us explore only trait matrices here. Constructing a trait matrix is a large task. It requires careful consideration of the key species–environment interactions in any community. That is, it forces explicit consideration of (i) the filters operating and (ii) the key traits that organisms possess to protect themselves from these filters. Once this is done, the desired traits must be measured on all

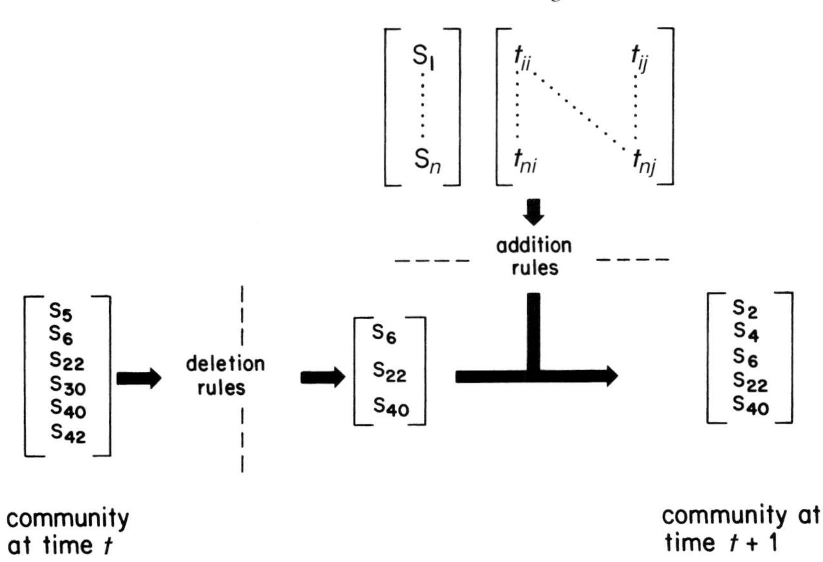

Figure 10.11 In order to predict change in an existing community ('response rules'), one needs to predict both deletions from, and additions to, the community (from Keddy 1992a).

members of the species pool. In the simplest case, this may require simply a morphological measure (e.g. bill length, seed size, gill structure). But in many cases, more complex functional traits will need to be measured such as relative growth rate, competitive performance or tolerance to specified perturbations. This will require developing a bioassay for these attributes so all species can be 'screened' for this property.

This approach was pioneered by Grime (e.g. Grime 1974, 1979) working on the Sheffield flora in England. It has since been applied to an increasing number of traits in wetland plants such as relative growth rate (Grime and Hunt 1975; Shipley and Peters 1990), fractional root porosity (Justin and Armstrong 1987), relative competitive ability (Gaudet and Keddy 1988, 1995), stress tolerance (Shipley and Keddy 1987), and palatability (Sheldon 1987; McCanny *et al.* 1990; Pennings *et al.* 1998). Let us consider flood tolerance – the obvious starting point for wetland plants. One approach would be to grow them all under different water levels, and then assign different degrees of flood tolerance. One could also try to measure traits which confer flood tolerance. Justin and Armstrong (1987) have screened for aerenchyma (Figure 10.12), Crawford and Tyler (1969) have screened for the presence of alcohol dehydrogenase (Figure 10.13)

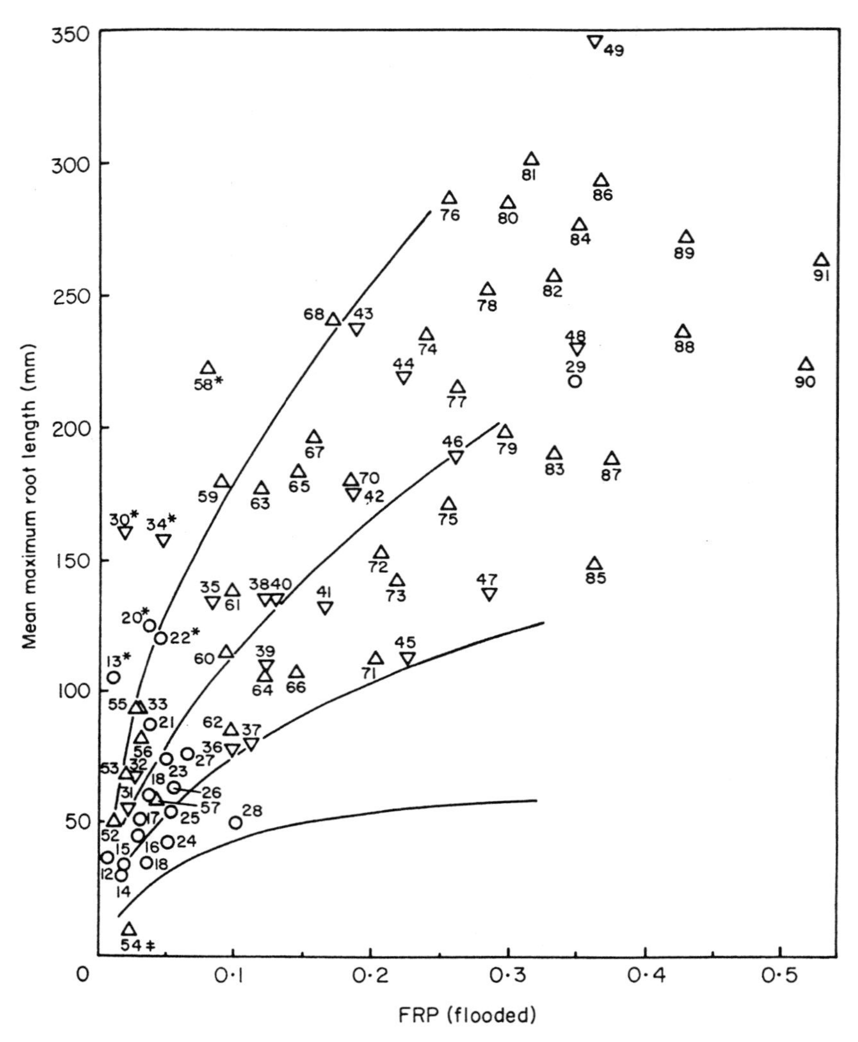

Figure 10.12 Ninety-one plant species were screened for fractional root porosity and mean maximum root length in flooded soils. O = non-wetland plant, ▽ = intermediate, and △ = wetland plant. Wetland plants scored high on both axes (from Justin and Armstrong 1987).

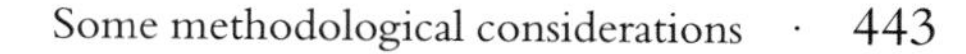

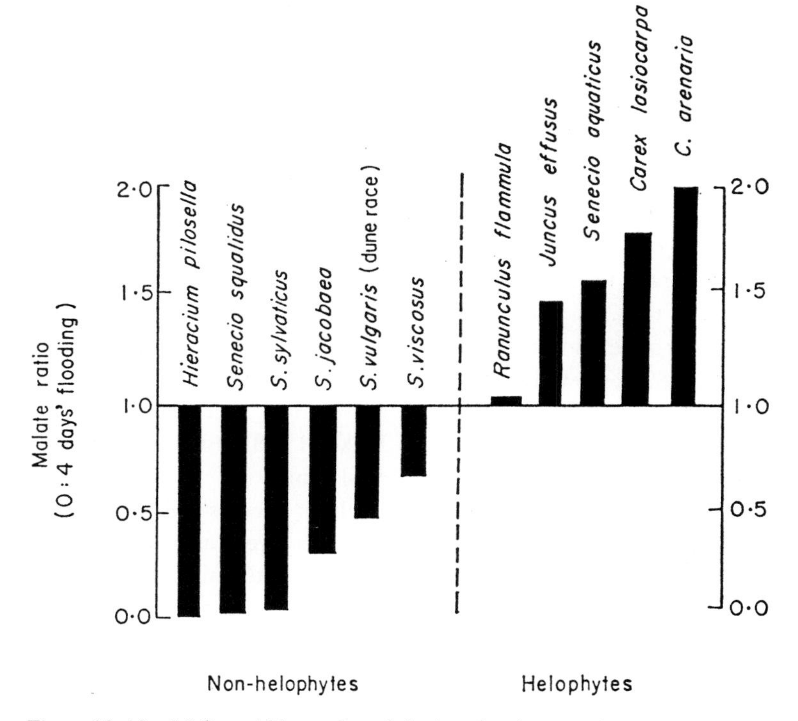

Figure 10.13 Malic acid is produced during flooding in plant species from wet habitats (from Crawford and Tyler 1969).

and Cavalieri and Huang (1979) have screened for the amino acid proline. In general, there are few complete lists of traits to accompany the list of species comprising the flora or fauna of an area. If such data were seen as a first step towards predictive ecology, more such matrices would undoubtedly be constructed.

Species or functional groups? A nagging question of scale

The preceding examples are based upon a species-by-species analysis. However, Chapter 9 argued that it is often preferable to aggregate species into groups sharing similar traits. Predicting which functional groups will be present will almost certainly be easier than predicting which species will represent in a particular functional group. We may need to begin with coarse scale prediction (among functional groups), allowing eventual refinement for fine scale prediction (within functional groups).

This distinction is illustrated by comparing Haefner's work on island birds (1978, 1981) with van der Valk's work on marsh plants (1981). Haefner's rules were much more complex than those of van der Valk, yet they dealt with fewer species. This probably results from Haefner working with a different scale of problem. In the van der Valk model, plants with strongly contrasting ecology were used, so that the model could deal with different functional groups or guilds. In contrast, Haefner selected a group of similar species where a greater degree of precision was needed to separate their habitat requirements. As a limiting case, we might imagine that, in some guilds, species are complete ecological equivalents (see, for example, Aarssen 1983, 1985; Agren and Fagerstrom 1984; Keddy 1989a), in which case no important functional traits would allow us to predict which species will represent a specified functional group. Perhaps this is partly why van der Valk's (1988) test was unsatisfactory, since all test species fit in one functional group.

It may also be the case that predicting which species will represent a particular functional group is qualitatively as well as quantitatively different. We may postulate that the traits which determine the presence or absence of specific guilds deal largely with traits directly related to the environment (e.g. aerenchyma to tolerate flooding, thick bark to tolerate fire). However, interactions within guilds may be among functional equivalents with nearly equal competitive abilities and resource requirements (Weiher *et al.* 1998). From the point of view of predicting function of vegetation, it may matter little whether a guild is represented by one or many species.

If our goal is to predict future states of communities for purposes of conservation of biological diversity, how many species and which ones represent a functional group may be critical. To illustrate this, consider the functional group of small evergreen rosette plants in wetlands, termed isoetids (Hutchinson 1975) or stress tolerators (Boston and Adams 1986). These species are characteristic of infertile, wave-washed shorelines, and their presence can be predicted from knowledge of substrate (Pearsall 1920), wave and ice scour (Keddy 1983) or vegetation biomass (Wisheu and Keddy 1989a). However, in some habitats this functional group is represented by the common and widespread species *Lobelia dortmanna* or *Eriocaulon septangulare*. In other cases, this group is represented by the nationally threatened and globally declining species *Sabatia kennedyana* (Keddy 1985a; Wisheu and Keddy 1989b). It is therefore critical for conservation that we are eventually able to predict at either the functional group level (presence or absence of evergreen rosette species) or at the species level (presence or absence of *Sabatia kennedyana*). However, given that these species share so many traits, the latter step may require very fine scale resolution. It is perhaps reassuring that the ability to simply predict the conditions for maintaining evergreen rosette species is an essential first step for conserving the threatened *Sabatia kennedyana*.

Monitoring success

Any restoration project must include an element of monitoring. From the point of view of basic science, the only way to develop workable assembly rules is to specify the combination of factors, traits and pool, and then test whether the community responds in the manner predicted. If not, refinements are necessary for progress to occur. From the strictly applied perspective, monitoring is necessary to determine whether the restoration is working. If management dollars are being spent, and no useful ecological responses are occurring, the programme is worthless. Monitoring requires us to select indicators for wetlands, and then to specify the target values for these indicators. Habits being what they are, it is always tempting to start counting selected species. In practice, there is a large body of science that deals with the selection of indicators and target values (e.g. Rapport *et al.* 1985; Odum 1985; Keddy 1991a; Keddy *et al.* 1993; Adamus 1992) . Any restoration project will need to draw upon this area of science as well. Selecting indicators for wetlands will be one of the main topics of the final chapter of this book.

Problems and prospects for restoration

Zedler (1988) describes some of the difficulties encountered in wetland restoration, particularly with replacing wetland functions. (i) Wetlands are complex and we have no blueprint for them. (ii) Wetlands species are mobile. This means that unwanted exotics can invade, or that valued species can disperse beyond the boundaries of the wetland. (iii) Wetlands are naturally dynamic. We may plan for one community type, and then find that changing water levels or soil fertility produces quite another. (iv) Persistence requires that the wetland contain all the components to achieve normal functions. Since we rarely know which species perform which functions, we may omit subcomponents that are necessary for persistence. (v) Unknown problems may arise during construction; landfill may be toxic, or insect eruptions may damage plantings. Such problems, she observes, illustrate why we need research involving cause and effect relationships.

These problems illustrate the value of dissecting wetlands in the manner used in this book. While we may lack blueprints, there are increasing numbers of quantitative relationships between observed patterns and ongoing processes in wetlands. The inherent variability of wetlands is a topic that has been repeatedly emphasized in the chapters on hydrology and natural disturbance. These processes require more attention in planning for restoration, and they illustrate the inherent risk of failure in small projects that fail to allow for natural processes. One of the reasons for studying functional groups of wetland organisms is to enable managers to connect sets of species with particular functions, and include necessary amounts of redundancy. In practice, there is still a great deal of work needed here, but in general, it is likely to be true that, the more diverse the wetland, the greater the probability that the desired functions will occur and the more redundancy there is likely to be. Similarly, a large pool of species will probably address problems that arise form changing environments or unexpected problems such as insect outbreaks. If we can predict such changes, managers can ensure that the species possessing appropriate traits are present. Natural systems are naturally resilient, or they would not have persisted into the present. Our challenge is to recreate these naturally resilient systems, and allow them to interact as much as possible with the natural processes in the landscape, from flooding to bank erosion to burial. The more we can allow natural dynamics, the greater the likelihood of success (Salo *et al.* 1986; White 1994; Rosgen 1995; Christensen *et al.* 1996). How to increase the odds of success is the topic of the next two chapters.

11 · *A functional approach*

Introduction

The basic functions of wetlands, such as oxygen production and nitrogen fixation (Chapter 1, pp. 55–77) are carried out by living organisms, and can be considered services performed for the biosphere and for humanity. The service provided by a wetland is the sum of the services provided by its individual components. What are these components? Individuals? Populations? Species? The process of studying and managing wetlands is hampered by the large number of components they possess. Rigler (1982) showed that it is impossible to use species-based models for understanding the behaviour of ecosystems:

> A temperate lake may support 1000 species. If each species interacted with every other species we would have $(1000 \times 999)/2$ or 0.5×10^6 potential interactions to investigate. Each potential interaction must be demonstrated to be insignificant or quantified. If we estimate one man–year per potential interaction it would take half a million years to gather the data required for one systems analysis model.

A complementary perspective on problems of modelling comes from the realm of general systems theory (Weinberg 1975), where one can recognize small, medium and large number systems. These have very different properties, and therefore generate different approaches to scientific investigation. *Small number systems* have very few components and few interactions, and these systems are amenable to precise mathematical description. Population ecology is an example, and while science has, in general, been successful with small number systems, in ecology these are usually inadequate because the small number systems can be created only by artificially removing populations from the many connections they have with other populations. At the other extreme are *large number systems* where there are so many components that the average behaviour becomes a useful description of the system. The ideal gas law provides one example; the position and velocity of a particular gas molecule are not of interest, but the properties of volume, temperature and pressure are. These systems too have been artificially analysed with some success.

The problem in ecology according to Lane (1985) is that ecosystems are neither large nor small number systems. These *medium number systems* contain too many components to be treated analytically, and too few for statistical analysis. Further, a rhinoceros and a grass plant cannot be averaged like an ideal gas molecule, nor are their behaviours and population dynamics equivalent to random events. As the number of components increases arithmetically, the number of interactions increases geometrically. Thus some method of simplification is necessary in order to solve problems involving medium number systems. This requires carefully preserving critical interactions and components while excising or ignoring others. The inherent difficulty in doing so wisely probably explains why, at present, medium number systems require modelling approaches that are as much an art as a science.

What then, are we to do?

Chapter 10 presented one alternative: the framework provided by assembly rules. Assembly rules first draw attention to a relatively small number of environmental factors that organize communities. Instead of an endless process of dissection, nature can perhaps be cleaved with a few sharp cuts into meaningful patterns. Hydrology, fertility, salinity and a few other factors provide the sword to cut apart the complexity of nature. Experiments can then manipulate these factors to sort out causal relationships, and managers can manipulate these same factors to produce the desired characteristics of wetlands. This approach is reductionist in the sense that it simplifies nature, but it is holist in that the pieces are being kept relatively large.

A complementary route to avoid Rigler's calculation is simplification. The focus is not upon environmental factors as filters, but upon the functions that organisms carry out in ecosystems. It may be possible to combine many individuals and species into one group with shared functions.

The strategy of simplification

The middle way

Taking the middle way between equally unpalatable extremes of holism and reductionism does not have a well-established name, but it can be

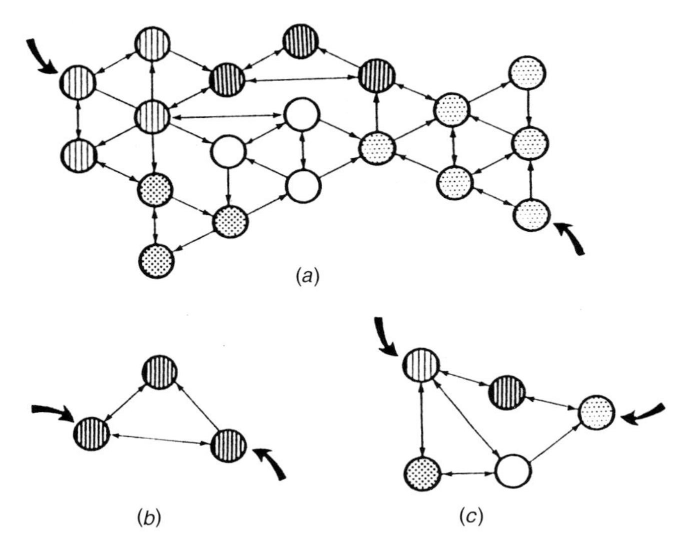

Figure 11.1 Representing an ecosystem at three different levels of resolution: (*a*) a detailed system model, (*b*) isolating a part of the system, (*c*) a less detailed ('lumped') system model (from Starfield and Bleloch 1991).

called *simplification* (Starfield and Bleloch 1991). Starfield and Bleloch say that learning to compromise is the first step towards building pragmatic models. Many people, they say, think of ecological modelling in terms of diagrams such as Figure 11.1(*a*). 'Their preconception is that ecosystems are made up of components that interact in a complex way and that models should be built to represent their complexity.' (p. 14) However, quite apart from Rigler's *reductio ad absurdum*, Starfield and Bleloch note that 'often the usefulness of such models, once they have been built, is disappointing'. Our first compromise, they conclude, is simplification. The way to accomplish this is to start with the management problem itself, rather than with a mental picture of the ecosystem. One then searches for the simplification that is most appropriate to solve the problem.

The most obvious approach is to excise one piece of the system (Figure 11.1(*b*)) and treat it in isolation. Yet the context of this piece cannot be ignored entirely. It may be possible to treat the rest of the system as an artificial driving force (thick arrows). Starfield and Bleloch consider some procedures for continuing with this approach, but in many ways it is likely to be less useful than the next alternative. Excising a few species is dangerous for another reason: it encourages ecologists to specialize upon a small group of organisms, and to view them out of context

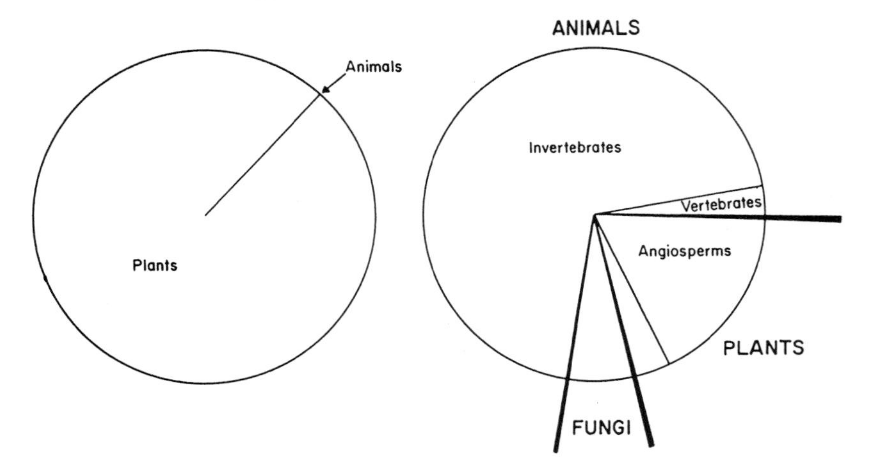

Figure 11.2 Plants and arthropods dominate life on Earth according to biomass (*left*) and numbers of species (*right*) (from Keddy 1989a).

from the rest of the system. This tendency is already so well established in wetland ecology that any procedure tending to reinforce it should be regarded with caution.

An alternative is to combine components that are similar to one another, as indicated by the similarities in shading in Figure 11.1. The most reasonable grounds for combining them would be similarities in function. One then ends up with a complete representation of the system, but one that has been simplified (Figure 11.1(*c*)). Choosing the appropriate level of simplification is thus 'a pragmatic compromise between the complexity of ecosystems on one hand, and the need to solve a problem, with limited data and in a reasonable amount of time on the other'. (p. 15)

A further advantage to simplification into functional groups is that it enhances communication among scientists and managers. Taxonomic classification exists to represent the evolutionary relationships of organisms, and the degree to which different kinds interbreed with each other. It did not originate as a tool to serve the needs of ecologists. When trying to communicate with scientists from other parts of the world, the nomenclature is often an obstacle to exchange of ideas; a different fauna or flora is like a different language. This problem is particularly severe for botanists and entomologists because of the large number of plant and insect species (Figure 11.2). One possible solution is to combine species not by their evolutionary lineage but by the similarity of the functions they perform.

Functional groups, therefore, have the dual benefit of providing a naturally simplified approach to wetlands and enhanced communication among ecologists. A third benefit may be that the emphasis upon function provides a natural bridge to those scientists, managers, administrators and politicians who think in terms of ecological services rather than wetland ecology.

Many of the examples in this chapter come from the realm of botany. I must emphasize that the principles, however, apply to any group of organisms. An introduction to the problem is provided by du Rietz (1931):

> That the plant-species may be classified not only according to their taxonomical relationship into genera, families etc., but also according to their . . . types, has been clear to botanists ever since the childhood of botanical science.

I have done du Rietz the courtesy of a direct quotation, but it should be obvious that to translate it into this chapter's context, we should restate it as:

> That species may be classified not only according to phylogeny, but also according to their types, has been clear to ecologists ever since the childhood of ecology.

How do we sort them into those types?

Functional classification for ecological prediction

Let us first remind ourselves of our goals: we want to be able to make predictions about the future states of ecosystems, and particularly to predict changes in function that may result from various human activities. This goal is widely accepted (e.g. Holling 1978; Rigler 1982; Starfield and Bleloch 1991). A major obstacle to being able to make such predictions is the large number of species which have to be included in community models. This can generate complexity which renders accurate prediction impracticable or, at best, very laborious. One route around this obstacle is to amalgamate groups of species into guilds or functionally similar groups of organisms which share ecological traits and which presumably play similar roles in ecological communities.

One of the major obstacles in creating these functional groups has nothing to do with the organisms themselves. Rather, it is the way we think about living systems. Classification could have two objectives: (i) forming groups with similar evolutionary histories in order to reconstruct phylogenies, or (ii) forming groups with similar ecological traits for predictive ecology. The former approach has had a major impact upon the historical development of ecology: many of the most high profile research questions in ecology dealing with diversity (e.g. Hutchinson 1959; May 1986; Connell 1987) can be traced back to the phylogenetic basis of species taxonomy. Such phylogenetic reconstructions are based upon analyses of traits, particularly on reproductive traits thought to reflect evolutionary trends in groups of organisms (Benson 1959). If we begin with phylogenetic species classifications, we naturally fall into a certain line of inquiry. The logic appears to go in the following manner. Since there is a large number of species (e.g. May 1988), how did so many species arise? Darwin provided an answer, and stimulated a century of research into the mechanisms and consequences of evolution through natural selection. This led to the second major question: how do all these species co-exist (May 1986)? The coexistence of many different species is the great question bequeathed by Darwin. Coexistence has been a central theme of ecology at least since Hutchinson's 1959 paper entitled 'Homage to Santa Rosalia' (Jackson 1981, May 1986), but it may have rather little to do with practical questions of ecosystem management.

If however, we begin with functional classifications, the path of inquiry has a different logic. While, on one hand, there is recognition of the overwhelming species diversity of the biosphere, there is also dramatic and obvious repetition of certain themes. The convergence in plant growth forms representing very different families in deserts is but one example: 'succulents' occur in groups including the Cactaceae, Liliaceae, Euphorbiaceae, Crassulaceae, and Asclepidaceae. At the opposite environmental extreme, in wetland habitats, 'mud flat annuals' are found in families including the Asteraceae, Cyperaceae, Juncaceae, Poaceae and Polygonaceae. From this viewpoint, it is the convergence in life history we emphasize. From this point of view, the important questions include (i) What are the major convergent groups and how many are there? (ii) How many do we need to recognize for a specific level of precision in our models? Growing out of this are other questions. What are the traits which they share? How do we use a knowledge of these traits to predict how a particular functional group will change after an external perturbation? How can we use a knowledge of these traits to predict the group of species that will be present in a specified environment?

The birds are perhaps the easiest to work with because food supply places strong selective pressure upon bill form, and provides a convenient means to sort species into basic feeding groups (Figure 11.3). At the finer

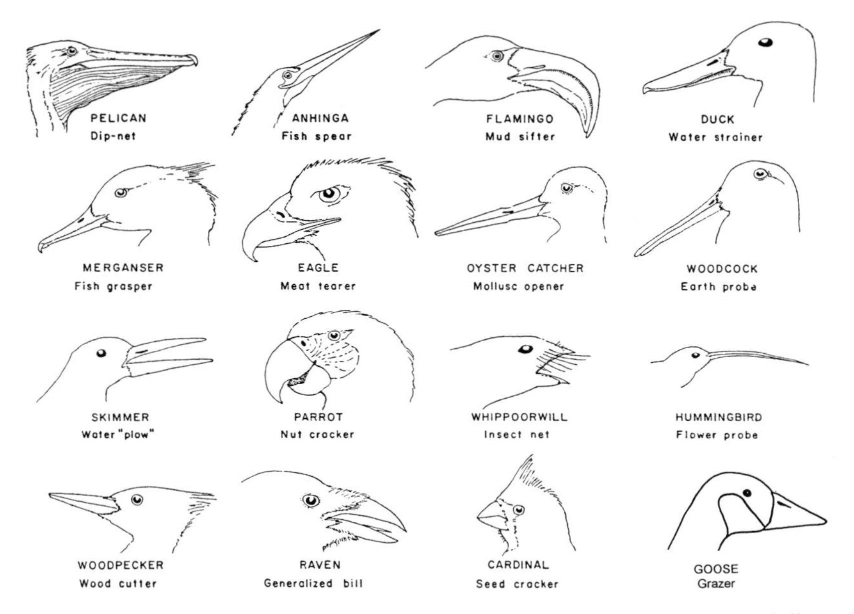

Figure 11.3 Birds can be arranged into functional groups based upon their bills, which in turn reflect their food sources (after Welty 1982).

scale, bills can vary in other attributes such as densities of comb-like lamellae used for filtering food particles from debris; in dabbling ducks, lamellar densities can range from 8–24 lamellae/cm (Nudds *et al.* 1994). Wider lamellar spacing (e.g. Mallards, *Anas platyrhynchos*) is associated with short, sparse vegetation typical of offshore habitats, whereas narrow spacing (e.g. Northern Shovelers, *A. clypeata*) is typical of ducks associated with the tall dense vegetation found in shallow water. Other attributes such as foraging habitat, nesting habitat, and migration can be used to recognize functional groups; with birds, these groups are usually called guilds (Root 1967).

Fish too can be classified by feeding strategy (Figure 11.4), with food type being reflected in the characteristics of the feeding apparatus. A still simpler classification, offered by Hoover and Killgore (1998), uses body shape to sort fish into one of four categories (accelerator, station holder, cruiser and manoeuvrer), along a morphological gradient running from fusiform and elongated (cigar-shaped) to broad and laterally compressed bodies. Other attributes such as foraging habitat, spawning habitat, and oxygen demands can be used to expand the classification. The need for functional feeding groups is illustrated by the great diversity of fishes in

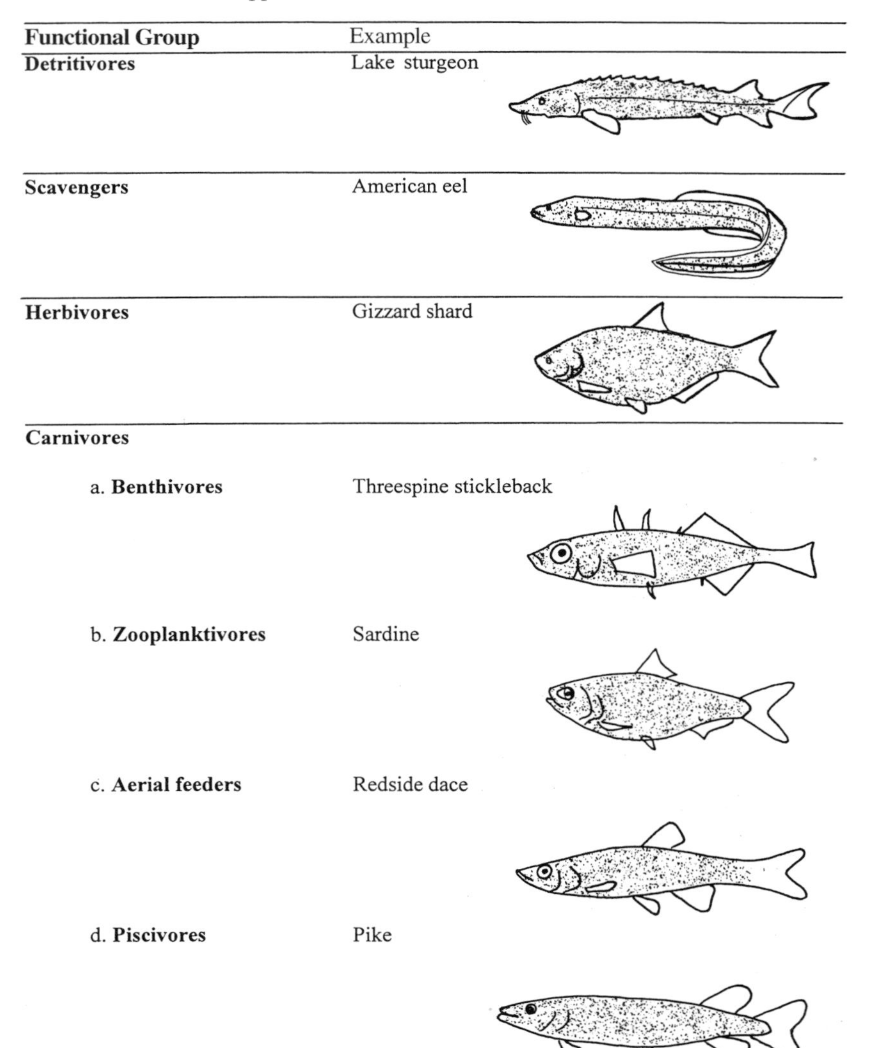

Figure 11.4 Fish can be divided into four main functional groups based upon their food sources (modified in part from Wootton 1990).

tropical rivers such as the Amazon (Chapter 3). Lowe-McConnell (1975) sorts these fish by three criteria: carnivore/herbivore, stenophage/euryphage, and non-specialists/specialists. Stenophages referred to fish that ate only plants or animals, euryphages ate both. Within the specialists she recognizes piscivores, insectivores, zooplankton feeders and mud eaters. Any of these attributes can be determined by examining stomach contents. However, in many cases it appears that, in spite of the very specialized teeth of groups such as the characoids, gut contents reveal rather little specialization. In the Parana/Pantanal system, diversity is again high: a single lagoon contained 52 species of fish representing more than three-quarters of a million individuals. The proportions of different trophic types were invertebrate feeders (62%), mud feeders (14%), herbivores (12%), detritus feeders (8%), omnivores (3%) and piscivores (0.5%). Wikramanayake (1990) sought correlations between morphological traits, feeding habitat and diet in tropical fishes from Sri Lanka. Morphological traits included length, body depth, body width, trunk shape, index of ventral flattening, pectoral fin length and eye diameter. Habitat variables included water velocity, depth, substrate type. Diets were estimated from gut contents. The first factor axis summarized size related traits, and the second described body shapes that ranged from deep-bodied laterally compressed species with large eyes and dorsally oriented mouths, to species with fusiform and elongated bodies with small eyes and ventrally oriented mouths. The first group was associated with the upper water column and slower velocities whereas the latter was found near the bottom in faster water. Overall, he concluded, there exists a close relationship among morphology, habitat use and diet. Winemiller (1991) has used the term 'ecomorphological convergence' to describe such functional groups arising in fish.

Insects similarly can be classified by their feeding system (Cummins 1973), considering both the dominant food type and the means by which they process it (Figure 11.5). Again, habitat, dispersal, life cycle and size can be used to expand the system. Cummins and Klug (1979) used the term 'functional feeding groups' for this classification system.

Finally, Severinghaus (1981) has re-emphasized the importance of functional classifications for simplification of complex natural systems. With respect to ecological assessment, he observes that particular guilds may have shared sensitivity to particular kinds of environmental stresses. 'Once the impact on any one species in a guild is determined, the impact on every other species in that guild is known. Furthermore, this information can be

Functional Group	Dominant Food	Feeding Mechanism	Example of Order
Shredders	Living tissue	Herbivore	Lepidoptera 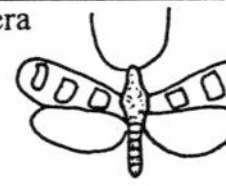
	Decomposing tissue	Detritivore	Plecoptera 
	Wood	Gouger	Coleoptera 
Collectors	Decomposing organic matter	Detritivore	Collembola 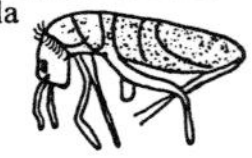
Scrapers	Periphyton	Herbivore	Coleoptera 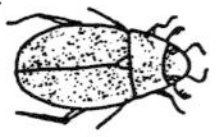
Macrophyte piercers	Living tissue	Herbivore	Neuroptera 
Predators	Living tissue	Engulfer	Megaloptera 
		Piercer	Neuroptera 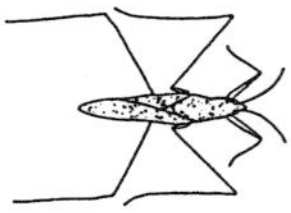
Parasites	Living tissue	Internal and external parasite	Hymenoptera 

Figure 11.5 Insects are divided into six functional groups based upon their dominant feeding mechanism (modified in part from Merritt and Cummins 1984).

applied to any ecosystem within which that guild is found. If an endangered species is contained in a guild, it is possible to predict the impact on that species without studying it specifically, which for most endangered species is virtually impossible to do anyway. Economically, the potential cost-savings are tremendous, since only a few species per guild need to be studied to establish the resulting impacts on all members of the guild.' While this may be somewhat of an overstatement in that it ignores possible differences among species within guilds, it none-the-less clearly states the

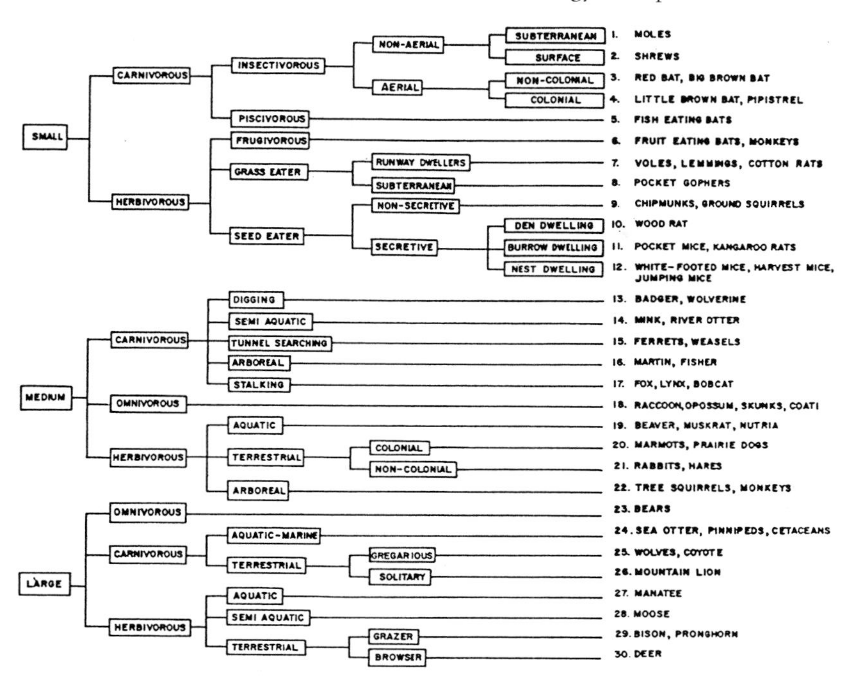

Figure 11.6 A functional classification for mammals of temperate regions based upon non-marine mammals inhabiting the continental United States of America (from Severinghaus 1981).

potential value of simplification. Severinghaus then produced a classification of mammals and terrestrial birds for the continental United States; Figure 11.6 shows his mammal classification because it may be of use in many other temperate areas, and might well be expanded to other biogeographic regions.

Although the names differ with both organism and investigator, the existence of such groups and their value to science and management is increasingly recognized (e.g. Southwood 1977; Severinghaus 1981; Terborgh and Robinson 1986; Simberloff and Dayan 1991).

The distinction between 'guilds' and 'functional groups'

In practice, there are important differences between the word 'guilds' and 'functional groups'. Most animal studies begin with food as the basic resource, and then group species into guilds that use a similar food. Terms such as 'water strainer' (Figure 11.3), 'zooplanktivore' (Figure 11.4), or 'macrophyte piercer' (Figure 11.5) clearly delineate groups using the type of resource being consumed. However, such studies retain a taxonomic bias, since the bird that is a 'water strainer', may, for example, be feeding on the same resource as a fish that is a 'zooplanktivore'. Moreover, while the term 'guild' has been used in many different senses (e.g. Terbough and Robinson 1986; Simberloff and Dayan 1991), the essential elements involve grouping of species 'that exploit the same class of resources in a similar way' (Root 1967). This intent seems clear in all of the animal examples (Figures 11.3 to 11.5). However, as Simberloff and Dayan remind us, this leaves two basic problems (i) what constitutes the same class of resources and (ii) just how similar should a similar way be? In practice, both of these questions may be answered by the scale of the inquiry; the finer the classification desired, the more groups there will be, and the more narrow the distinctions among resources will need to be. However, the idea of types of organisms is essentially Aristotelian and thus far more ancient than Root's work; for this reason alone I prefer the more generic term 'functional group'. (This may also have the salutory effect of reminding us that ecology did not suddenly originate in North America in the 1960s; Gorham 1953; Jackson 1981.)

Unlike guilds, functional groups are constructed based upon shared traits. While resource use may be one trait, the door remains open to include other important traits such as adaptations to extreme environments. Depending upon the management problem, the inclusion of other traits besides food will likely be of value.

Plants provide a far more thorny problem for such classification schemes. First, their resources are not as easily determined because resource uptake is not obvious upon inspection and cannot be determined easily from shape, size or gut contents. Secondly, plants use so few resources (largely CO_2, water, N, P, K) that it is hard to use these to erect a functional system that will work for large numbers of species. The following examples illustrate how ecologists have grappled with this problem.

Functional classification of wetland plants

Let us explore functional classification further. Plants will be the model because, even for zoologists, functional classifications of plants are necessary to describe the habitat their organisms occupy. Moreover, the process appears to be more difficult in plants, in part because they require similar resources and are not easily sorted into feeding groups.

One of the most ubiquitous methods of plant classification was proposed by Raunkaier (1937). His basic theme, paraphrased, is that life is

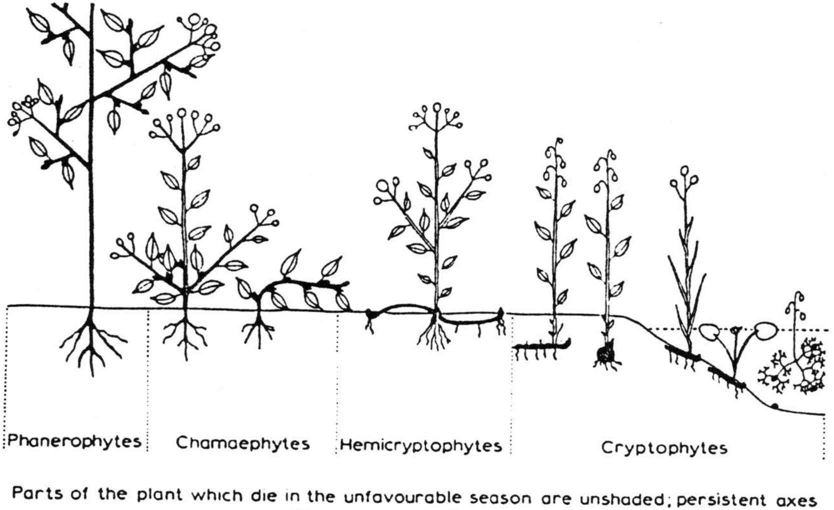

Figure 11.7 The Raunkaier system classifies plants based upon location and protection of their meristems (from Goldsmith and Harrison 1976).

not easy for plants. Unlike animals, they must remain rooted in a site as the environment changes around them. Raunkaier focused on the most important challenge faced by plants: protecting their meristems during unfavourable periods. For those less familiar with plants, it is worth emphasizing that here is one way in which vascular plants differ greatly from other living organisms. They have indeterminate growth, that is they continue growing throughout their lives. Moreover, their growth arises out of strictly defined areas of cell multiplication called the meristems. If these are killed, the plant can neither grow nor reproduce. Because of the many independent meristems, plants are modular organisms rather than single integrated units like most animals. Raunkaier focused our attention on how plants protect their meristems, and erected the categories shown in Figure 11.7.

Raunkaier's system is excellent for coarse-scale comparison. It is less useful for more fine-scale work. For example, in marshes, all of the plants may be either geophytes or therophytes. Managers may well need more accurate descriptions or predictions than simply whether cryptophytes or chamaephytes will be present. This requires a finer level of classification. We must understand that this does not mean there is something wrong with Raunkaier. Rather, the point is that we need different models to describe or predict at different scales. At the coarse scale, Raunkaier is excellent. At finer scales, we need more information about the organisms.

Dansereau (1959) developed a more complicated system that uses categories of traits to describe vegetation. His categories were life form, stratification (height), coverage, function (evergreenness), leaf shape and leaf texture (Table 11.1). Any vegetation type can be placed in a reduced number of functional groups such as 'tall erect woody plants with sclerophyllous needle leaves that are evergreen'; such a combination would describe many conifer stands around the world.

Here we encounter a slight problem. Once we drop to finer scale, the number of groups proliferates. At the same time, ecologists tend to restrict themselves to a smaller group of habitats. As a consequence, these finer scale classifications either have an unwieldy number of groups, or else they apply only to one habitat. Let us look at a few examples of these finer-scale classifications and use aquatic plants as a convenient example.

Aquatics are often treated separately, as their habitat would seem to present a fairly well-defined a natural group of organisms (e.g. Sculthorpe 1967; Hutchinson 1975). Dansereau (Table 11.2) erected a classification of eight life forms of aquatic plants. The principal traits were whether or not the plants were rooted, their relationship with the water surface and the nature of their leaves; Figure 11.8 presents a schematic version. This system still provides only eight groups, and in some circumstances it may be useful to have an even more fine-scaled functional classification. Table 11.3 gives Hutchinson's system, which he conveniently compares to both Dansereau and to a third system. Hutchinson also reviews some other systems for classifying aquatics, and presents an elaborate classification that stretches for eight pages (pp. 123–130) to which readers with a specific interest in aquatics are referred.

Finally, for completeness, wetland species can also be classified according to a completely different trait: method of dispersal. Propagules for dispersal are an important food source for many species of herbivores, and there may be some interesting possibilities here for those interested in herbivores in wetlands. Moreover, with increasing awareness of landscape ecology and large-scale patch dynamics there is again increasing emphasis upon dispersal. Long-distance dispersal has long been of interest to aquatic ecologists (e.g. Savile 1956; Sculthorpe 1967; van der Pijl 1972), but dispersal from one local patch to another will be increasingly important in a fragmented world. Table 11.4 illustrates the Dansereau scheme for classifying propagules.

Table 11.1. *A descriptive scheme involving six criteria to provide a structural description of vegetation types*

Criterion	Code	Description
1 Life-form		
	W	erect woody plants
	L	climbing or decumbent woody plants
	E	epiphytes
	H	herbs
	M	bryoids
2 Stratification		
	1	more than 25 metres
	2	10–25 metres
	3	8–10 metres
	4	2–8 metres
	5	0.5–2 metres
	6	0.1–0.5 metres
	7	0.0–0.1 metres
3 Coverage		
	b	barren or very sparse
	i	interrupted, discontinuous
	p	in patches, tufts, clumps
	c	continuous
4 Function		
	d	deciduous
	s	semideciduous
	e	evergreen
	j	evergreen–succulent; or evergreen–leafless
5 Leaf shape and size		
	n	needle or spine
	g	graminoid
	a	medium or small
	h	broad
	v	compound
	q	thalloid
6 Leaf texture		
	f	filmy
	z	membranous
	x	sclerophyll
	k	succulent; or fungoid

Source: Dansereau (1959).

Table 11.2. *Life-form of aquatic plants in relation to their substratum*

Relation to substratum	Type	Example	Type symbol
Free (not rooted or anchored)	natantia	*Lemna minor, Ceratophyllum demersum*	S
Rooted in soil emersed, at least in part			
• broadleaved	foliacea	*Sagittaria latifolia* *Pontederia cordata*	F
• narrow, tubular or linear-leaved	junciformia	*Scirpus validus* *Juncus nodosus*	J
• floating-leaved	nymphoidea	*Nymphaea odorata* *Nymphoides lacunosum*	N
submersed (at most a few floating leaves)			
• long, leafy stems and/or leaves ribbon-like	vittata	*Potamogeton richardsonii* *Vallisneria americana*	V
• leaves much reduced, crowded at base	rosulata	*Isoetes braunii* *Lobelia dortmanna*	R
• annuals	annua	*Najas flexilis* *Potamogeton pusillus*	T
Adnate or epiphytic	adnata	*Podstemon ceratophyllum* *Fontinalis* spp.	A

Source: Dansereau (1959).

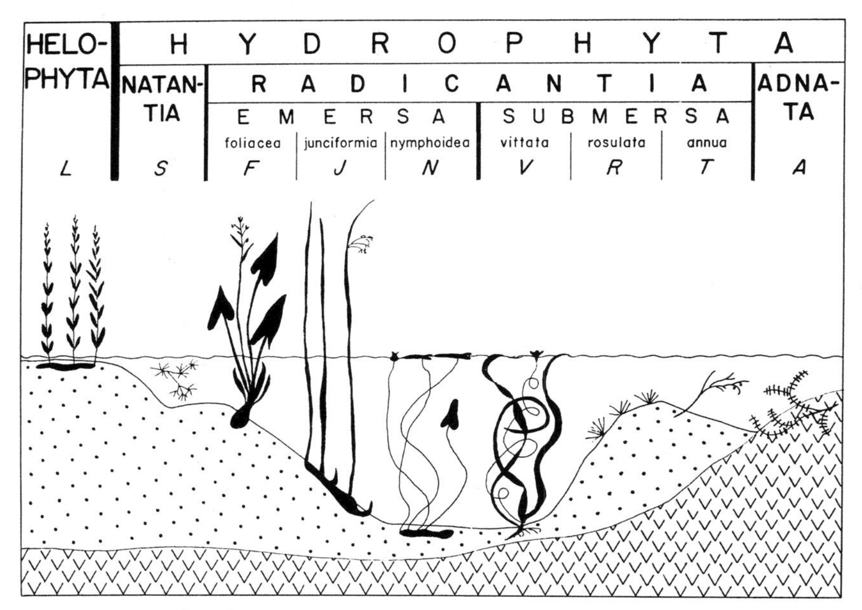

Figure 11.8 A classification of wetland plants based upon their growth form and habitat (from Dansereau 1959).

A general procedure for constructing functional groups

Figure 11.9 summarizes the process of constructing such functional groups. A key part is the trait matrix. Most traits considered in the above classifications are traits that can be determined by eye, such as life form, life span, method of vegetative propagation and position of overwintering shoots. However, if we consider the processes in vegetation such as nutrient uptake, competition, and interaction with agents of disturbance or stress, there are traits which may not be obvious upon inspection, but may be none the less closely related to the function of the plant in a community. We could begin by looking for traits that measure performance in these three areas: (i) ability to forage for essential elements in the absence of neighbours, (ii) traits associated with interaction with other plants and (iii) traits associated with ability to withstand agents such as fire, flooding, grazing, etc. These traits are fundamentally different from those used in most phylogenetic studies.

Since many of these traits are not obvious upon inspection, we need to apply the process of screening as developed by Grime and Hunt (1975) and Grime *et al.* (1981). The objective of screening is to develop a simple bioassay for a particular attribute, and then apply it systematically to an

Table 11.3. *A comparison of three schemes of ecological classification*

Hutchinson	Fassett-Wilson	Dansereau
A. Natant (Planophyta)		
I. At surface (Pleuston s.s. or Acropleustophyta)	Type 5	Natantia (S)
a. Lemnids		
b. Salviniids		
c. Hydrocharids		
d. Eichhorniids		
e. Stratiotids		
II. At mid-depth (Megaloplankton or Mesopleustophyta)	Type 5	Natantia (S)
a. Wolffiellids		
b. Utricularids		
c. Ceratophyllids		
B. Rooted in sediment (Rhizophyta)		
I. Part of vegetative structure above water (Hyperhydates)	Type 4	Junciformia (J)
a. Graminids		
b. Herbids		
c. Ipomeids		
d. Decodontids		
e. Aeschynomenids		
f. Sagittariids		Foliacea (F)
g. Nelumbids		Foliacea (F)

II. Leaves mostly floating, not regularly above surface (Ephydates)	Type 3	Nymphoidea (N)
a. Nymphaeids		
b. Natopotamids		
c. Marsileids		
d. Batrachids		
e. Trapids		
III. Leaves entirely submerged or almost so (Hyphydates)		
a. Vittate, with long stem	Type 1	Vittata (V)
1. Magnopotamids		
2. Parvopotamids		
3. Myriophyllids		
b. Rosulate, stem very short	Type 2	Rosulata (R)
1. Vallisneriids		
2. Otteliids		
3. Isoetids		

Source: From Hutchinson (1975).

Table 11.4. *An outline key of the major dispersal types in wetland plants*

Key	Dispersal type	No.
1. D. does not disarticulate from the parent plant before being	Auxochore	1
deposited at the site of further development. (Parent plant		
may be dead or alive at that time.)	Auxochore	1
1. D. disarticulates from the parent plant before the dispersal phase.		
2. D. very voluminous in relation to the actual reproductive part, consisting of a loose spherical framework.	Cyclochore	2
2. D. consisting for the main part of the reproductive organ.		
3. D. with scarious or stiff appendages.		
4. Appendages thin, light, often flexible.		
5. Appendages scarious, wing-like, or saccate	Pterochore	3
5. Appendages long, hairlike, or plumose	Pogonochore	4
4. Appendages short, stiff, spiny, or glandular, adhering to rough surfaces	Desmochore	5
3. D. without appendages (except for arils).		
6. D. with juicy or fleshy outer layers.	Sarcochore	6
6. D. with hard outer layer.		
7. D. small or light enough to be carried by breeze	Sporochore	7
7. D. too heavy to be carried by breeze.		
8. Parent, plant without mechanism of expulsion.		
9. D. light enough to be carried by wind	Sclerochore	8
9. D. very heavy	Barochore	9
8. Parent plant with mechanism of expulsion	Ballochore	10

Notes:
D. means diaspore, which is a seed and its attachments.
Source: From Dansereau (1959).

entire set of species. There are a few examples of this in wetlands. Shipley *et al.* (1989) created a matrix examining 7 juvenile traits and 13 adult traits. The objectives were to explore quantitative relationships among traits to (i) test whether juvenile and adult traits were independent and (ii) explore relationships among the traits. Figure 11.10 shows the first and second axes of variation in juvenile and adult traits. In juveniles, the important traits were variation in seed size, which was correlated with rapid germination in light (axis 1) and higher growth rates with reduced germination at constant temperatures (axis 2). More than half the variation in seedling life history traits was accounted for by these two axes

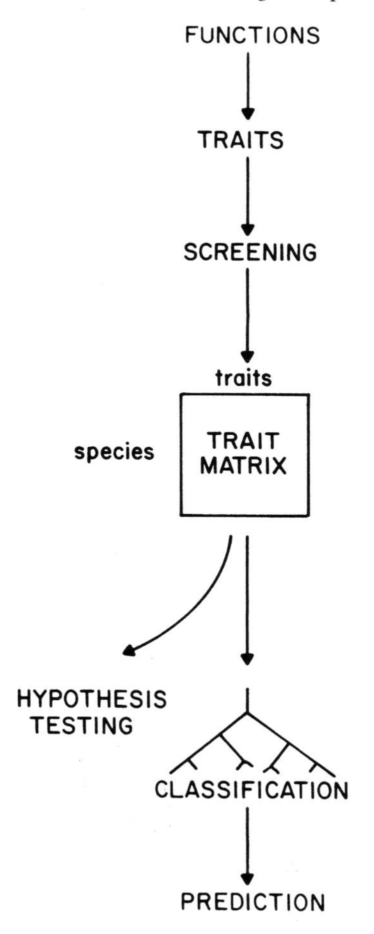

Figure 11.9 The process of classifying functional groups based upon a matrix of traits (from Boutin and Keddy 1993).

alone. Germination is known to be highly influenced by both light and fluctuating temperatures (Grime 1979; Grime *et al.* 1981) and it appears that wetland plants differ in their response to these two key environmental factors. Since the presence of established plants is likely to reduce the survival rates of seedlings, seedlings must either be able to escape adult plants by finding gaps, or else resist suppression. These first axes may be interpreted as two evolutionary solutions to this problem; it may be interpreted as having large, slow growing seeds at the left, and small rapidly growing seeds at the right. The two key axes in adult plants were the width of the

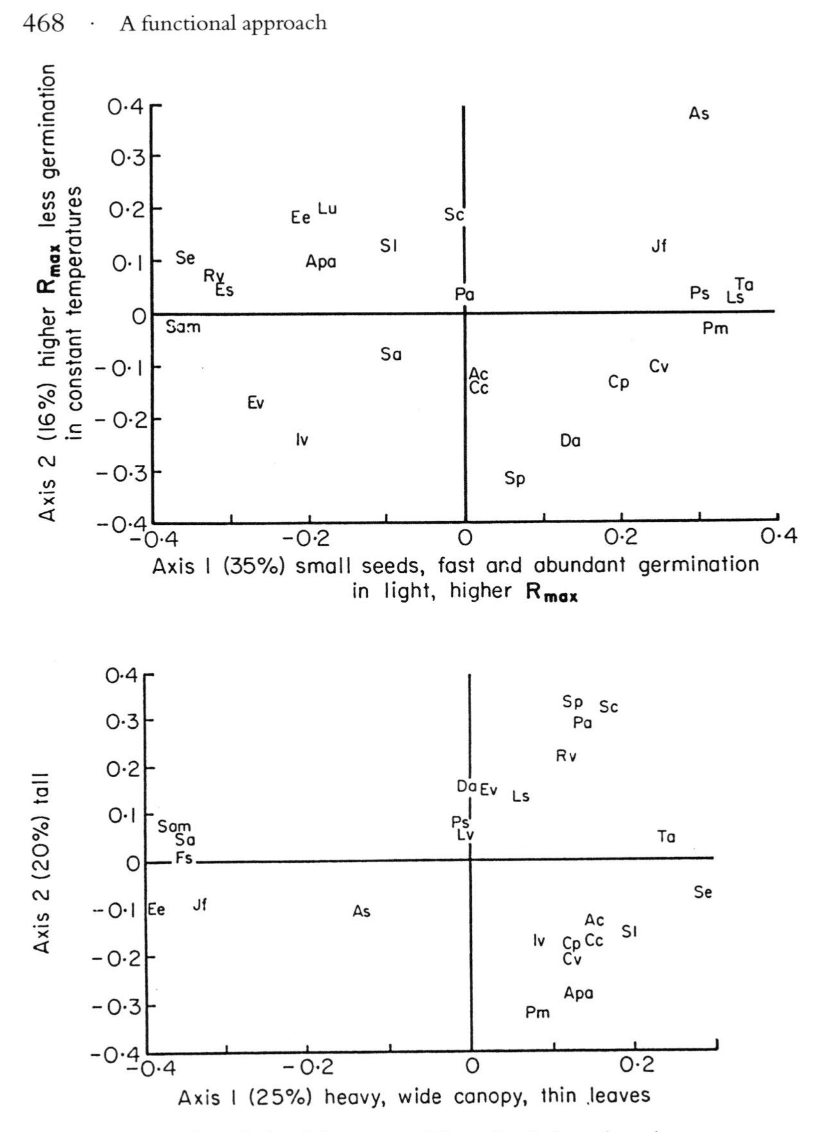

Figure 11.10 The relationships among 25 wetland plants based upon seven juvenile traits (*top*) and 13 adult traits (*bottom*). Ac = *Acorus calamus*; As = *Agrostis stolonifera*; Apa = *Alisma plantago-aquatica*; Cc = *Carex crinita*; Cp = *Carex projecta*; Cv = *Carex vulpinoidea*; Da = *Dulichium aurundinaceum*; Ee = *Eleocharis erythropoda*; Es = *Eleocharis smallii*; Ev = *Elymus virginicus*; Iv = *Iris versicolor*; Jf = *Juncus filiformis*; Lu = *Lycopus uniflorus*; Ls = *Lythrum salicaria*; Ps = *Penthorum sedoides*; Pa = *Phalaris arundinacea*; Pm = *Plantago major*; Rv = *Rumex verticillatus*; Sl = *Sagittaria latifolia*; Sam = *Scirpus americanus*; Sa = *Scirpus acutus*; Sc = *Scirpus cyperinus*; Se = *Sparganium eurycarpum*; Sp = *Spartina pectinata*; Ta = *Typha angustifolia* (from Shipley *et al.* 1989).

canopy (axis 1) and the height of the plants (axis 2). This can be interpreted as the importance of holding space and denying it to neighbours.

Perhaps the biggest surprise in the above work was the discovery that juvenile and adult traits were uncoupled. That is to say, the correlation matrices for adult traits showed no association with the correlation matrices for juvenile traits. Perhaps the traits required for regeneration in gaps are fundamentally different from the traits required to hold space as adults. This would mean that two categories, fugitive or stress-tolerant, could be constructed for each of two stages of life history (Figure 11.11). In turn, these four life history combinations can be related to three properties: frequency of gap formation, size of gaps and soil fertility.

Functional groups in marsh plants

The following example illustrates the progress to date in systematically applying the scheme in Figure 11.3 to the study of freshwater wetland plant communities. The classification is based upon a matrix of 43 species by 27 traits. The species were selected to represent wetland habitats and functional groups from across eastern North America. Species included rare or endangered taxa from infertile lake shores (*Coreopsis rosea*, *Panicum longifolium*), annuals typical of mud flats (*Bidens cernua*, *Cyperus aristatus*), large perennials (*Phalaris arundinacea*, *Typha glauca*), reeds from river banks (*Scirpus acutus*, *Eleocharis calva*), and an array of other species which represented other life forms and habitats. An emphasis upon traits associated with function would have been best, but owing to the magnitude of the project (coordinating seed collection, ensuring adequate germination of many species of contrasting ecology, growing a sufficient number of replicates), there were fewer functional measures than would be desirable. Traits included the following: (i) Relative growth rate (RGR), which is known to be correlated with rates of resource acquisition (e.g. Grime and Hunt 1975) and seedling stress tolerance (Shipley and Keddy 1987), (ii) height of juveniles, height of adults, and rates of shoot extension; height is associated with competitive ability for light (e.g. Givnish 1982; Gaudet and Keddy 1988), (iii) above and below ground biomass allocation, as well as photosynthetic area, are believed to be associated with foraging for different light to nutrient ratios (e.g. Tilman 1982, 1986), (iv) morphological traits such as shortest and longest distance between aerial shoots measured the way in which different species held space; such traits are important if, as seems to be the case, plant communities are largely under dominance control (Yodzis 1986).

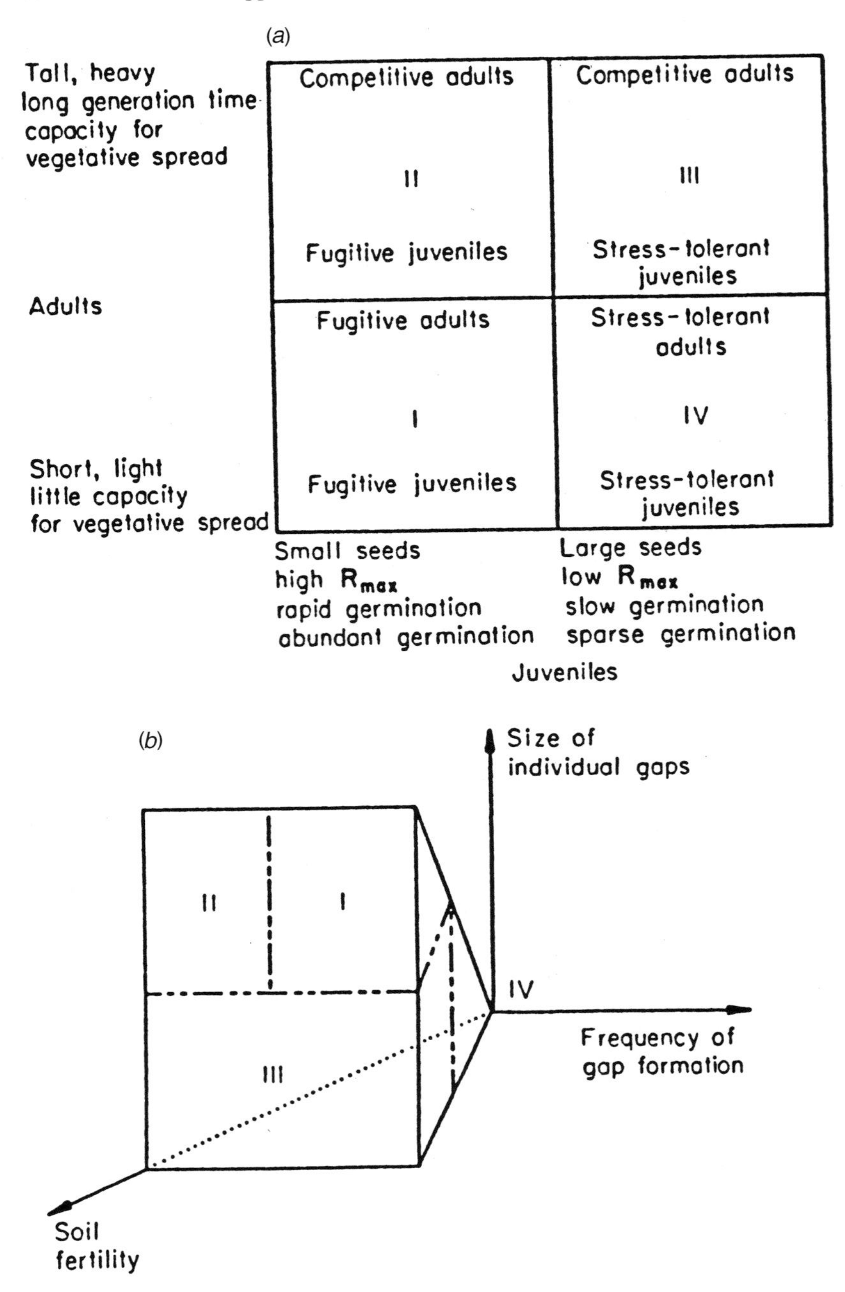

Figure 11.11 Four life history types can be recognized, based upon adult and juvenile life history types (from Shipley *et al.* 1989).

Figure 11.12 shows the dendrogram resulting from the cluster analysis. Long-lived clonal perennials are found near the top, whereas small annuals are found near the bottom. There were three main groups, each containing two or three subgroups, for a total of seven guilds or functional groups (provisional guild names were assigned on the left-hand side). Figure 11.13 summarizes these results, with the addition of key traits. One main group (left side) had a high percentage of individuals and species flowering in their first growing season, and no lateral spreading of their vegetative growth. They put emphasis on photosynthetic tissues with a high photosynthetic area and a low below-ground to above-ground biomass ratio. In contrast, the other group (right) did not flower much in their first year of growth but they expanded the vegetative parts, especially the below-ground system. These two groups apparently reflect the distinction between 'ruderal' (*sensu* Grime 1979) and 'perennial' strategies.

The 'ruderals' consisted of two further groups. While plants in both groups flowered in the first year, one subgroup died at the end of the growing season ('obligate annuals') whereas the other was perennial ('facultative annuals'). 'Obligate annuals' (e.g. *Bidens cernua*) also had numerous short stems, a large crown area and a tendency to direct more growth towards the above-ground structures relative to facultative annuals; they died at the end of the growing season. The 'facultative annuals' were tall with an erect, fast-growing main stem (e.g. *Verbena hastata*). Shoots emerged from the base of this stem the following year, again producing tall plants with a narrow, erect growth form topped by an inflorescence.

Within the 'perennials', there was a clear distinction between species which spread clonally and species with a more clumped growth form. The clonally spreading species had a deep and massive below-ground structure, which produced the high below-ground to above-ground biomass ratio. This also probably accounts for the high biomass to volume ratio of this group of species since roots and rhizomes are usually more dense than leaves and stems. The species with a clumped growth form were more compact in their growth with shallow and rather smaller below-ground structures. These two types can be considered 'matrix species' and 'interstitial' species (*sensu* Grubb 1986).

At a finer scale, the 'matrix species' were further composed of two groups. 'Clonal dominants' were tall and robust species with vigorous lateral spread that frequently produce large monospecific stands in fertile habitats (e.g. *Typha glauca*). 'Clonal stress tolerators' (*sensu* Grime 1977) had a similar growth pattern but were much smaller and consequently

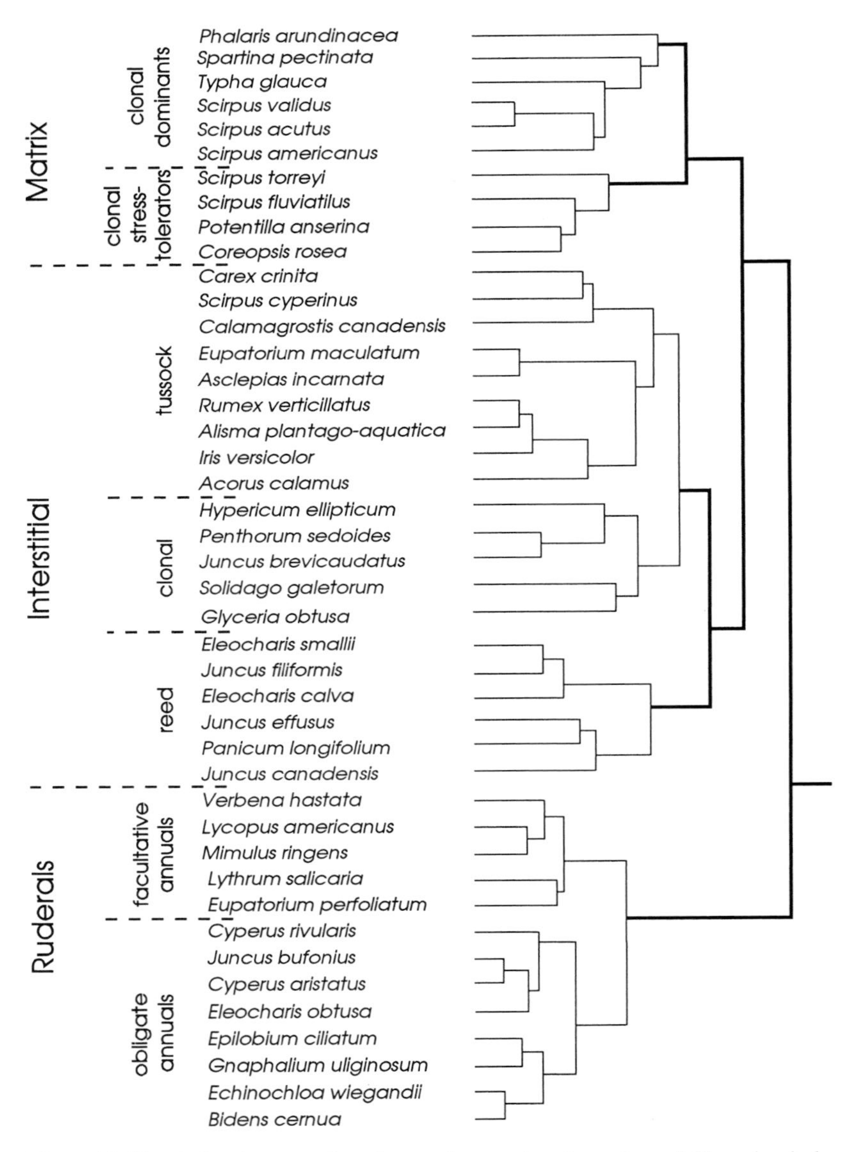

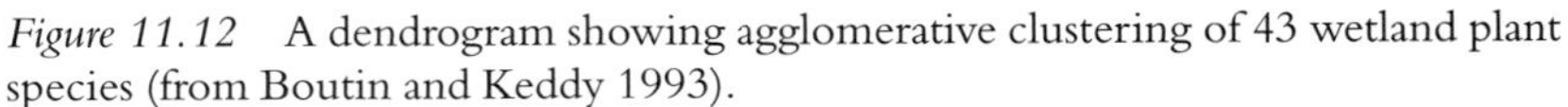
Figure 11.12 A dendrogram showing agglomerative clustering of 43 wetland plant species (from Boutin and Keddy 1993).

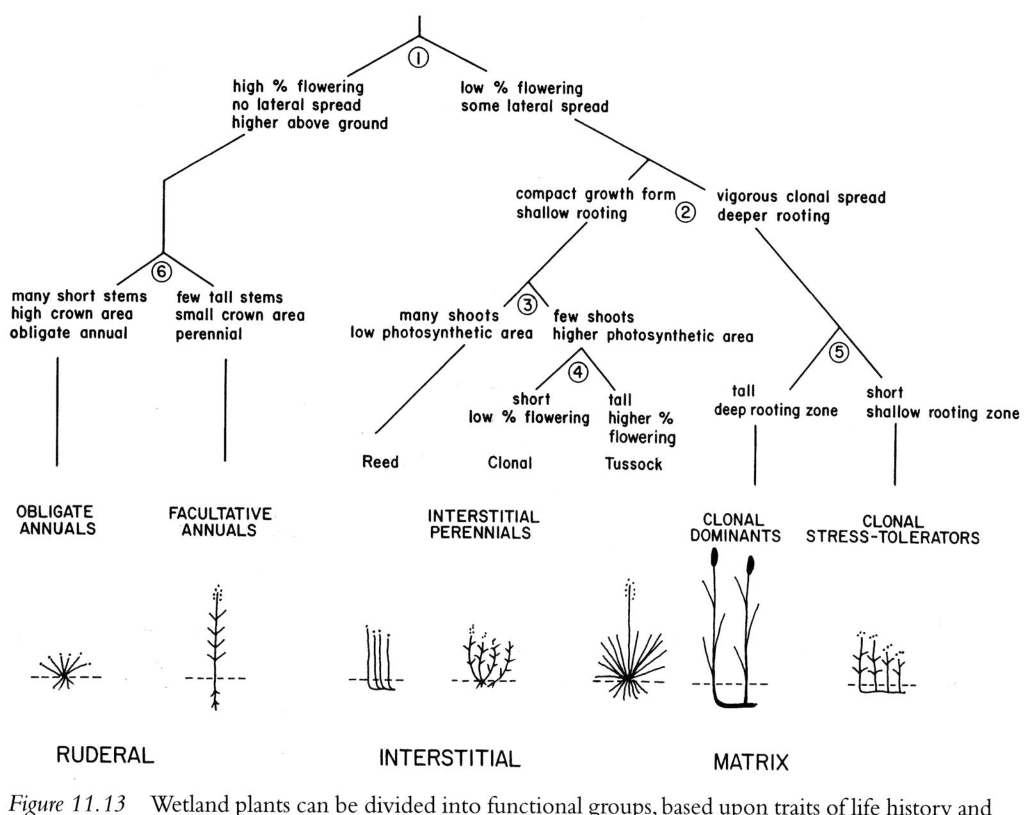

Figure 11.13 Wetland plants can be divided into functional groups, based upon traits of life history and morphology (from Boutin and Keddy 1993).

did not produce dense canopies; they were species more typical of infertile sand and gravel shorelines (e.g. *Scirpus torreyi*).

This approach shows some of the strengths and weaknesses of so-called objective approaches to functional groups. Weaknesses included the following. The selection of traits itself was somewhat arbitrary, although effort was made to include traits thought to be of importance in real communities. Many important traits were not included in the analysis; in some cases this was because there was no way that could be devised to screen for the traits; in other cases, a few species could not be included, and so that entire trait had to be removed from the analysis. The traits were based upon screening for only one growing season. Finally, the data set did not include distinctive groups such as evergreen rosette species (e.g. *Lobelia dortmanna*, *Eriocaulon septangulare*, *Isoetes* spp.), termed 'isoetids' (Hutchinson 1975), and woody plants. On the positive side, the number of components in the wetland vegetation was reduced from 43 species to seven functional groups. Of course, since these 43 species were selected from a much larger pool of wetland plants, the seven guilds actually represent much of the variation in this larger pool of hundreds of species. Also, the importance of traits such as ability to flower in the first year of growth was emphasized. If this classification has value, we would expect other traits to be predicted from knowledge of a species membership in these different functional types. Shipley and Parent (1991) tested this by examining three germination attributes: time to germination, maximum germination rate and proportion of seeds germinating for 64 wetland plant species. Dividing the species into three functional groups (annual, facultative annual and obligate perennial), they found that the obligate perennials took significantly longer to begin germination, and had significantly smaller maximum germination rates.

Functional types of wetlands

Using an analogous procedure, wetlands themselves can be divided into functional types. The objective again is to simplify from large numbers of entities with uncertain relationships to a small number with known properties and relationships. The procedure again uses a few traits that sort entities (the entities in this case are wetlands rather than species) into groups with similar characteristics. In the above work on organisms, the traits used were primarily those associated with the capture and processing of resources. To search for functional types in the case of entire wetlands, the traits might be associated with the capture and processing of water. If this seems to be pushing an analogy too far, it is equally possible to argue that any functional classification of wetlands should begin with water for three other reasons. First, water creates wetlands. Secondly, hydrology is the key factor that determines a vast array of wetland processes and properties (Chapter 4). Thirdly, the control of flood waters is an important service of wetlands.

Conveniently for us, Lent *et al.* (1997) have offered almost exactly such an approach to the classification of non-tidal wetlands. They use two sets of hydrologic indices to summarize the degree and nature of water exchange between a wetland and the adjacent landscape. The three main inputs of water to a wetland can be expressed as:

$$gwi + swi + pi = 100\%$$

where *gwi* is ground water input, *swi* is surface water input, and *pi* is direct precipitation input, each expressed as a fraction of total water input. In an identical way, outputs of water from a wetland scan be expressed as:

$$gwo + swo + eto = 100\%$$

where *gwo* is ground water output, *swo* is surface water output and *eto* is evapotranspiration as a fraction of total water output. In both cases, fractional indices are used to avoid sensitivity to differences among wetlands in water volume.

It is then possible to use each of these as axes along which different wetlands may be arrayed (Figure 11.14); the lower triangles represent the characteristics of inputs (left) and outputs (right). Three regions (I, V, VII) at the apices of each of these triangles represent wetlands dominated by a single input or output variable. The central quadrangle summarizes the importance of lateral water movement as opposed to atmospheric exchanges. Region I contains wetlands with strong lateral flows, as opposed to region IX with largely atmospheric flows.

Individual wetlands can then be located in these spaces using published data on wetland budgets (Figure 11.15). Thoreau's bog in Massachusetts (point 1) falls out in the region dominated by evapotranspiration and precipitation, whereas a *Typha* marsh in southern Ontario (point 17) falls out in the region dominated by lateral inflows and outflows. Cypress swamps (points 13 and 14) have an intermediate position. Prairie potholes (point 10) have significant lateral inflows, but largely atmospheric outflows. Widely different kinds of wetlands, then, can be represented on a single graph summarizing their hydrological properties.

Lent *et al.* also suggest that it is useful to calculate the degree to which the wetland interacts with the surrounding landscape, and the nature of

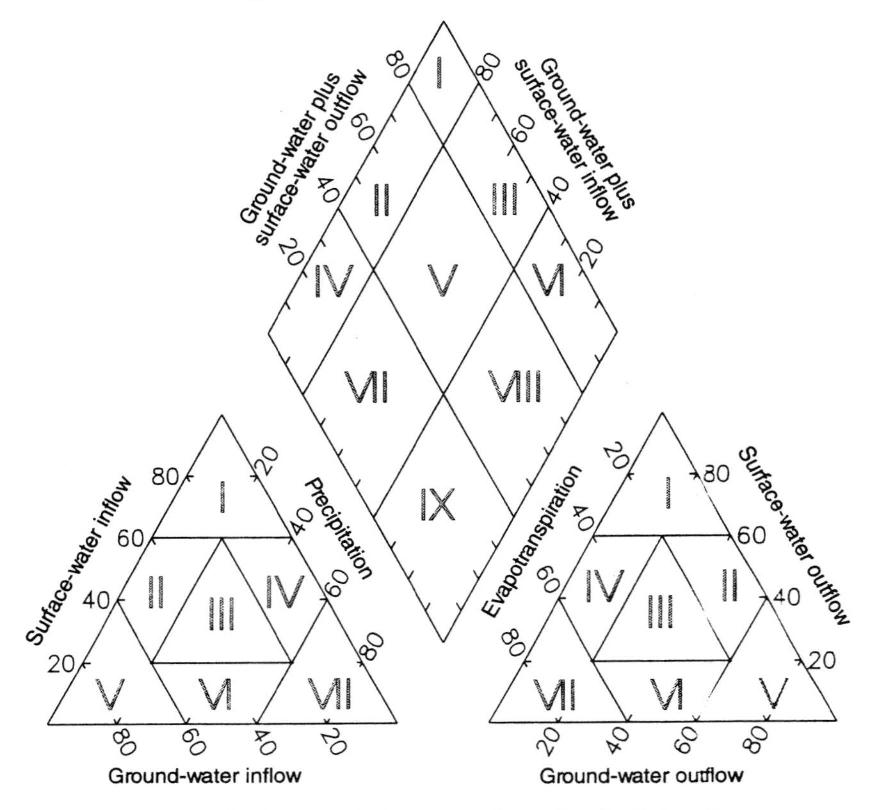

Figure 11.14 Axes for the quantitative comparison of wetlands, based upon water budget indices for lateral flows (ground water or surface water) and atmospheric flows (evapotranspiration and precipitation) (from Lent *et al.* 1997).

that interaction. The *degree of interaction* can be estimated from the amount by which evaporation and precipitation dominate hydrology

$$I = (2 - pi - eto) / 2$$

As values of I increase, the wetland is increasingly linked by surface and ground water flows to adjoining uplands.

The *nature of the interaction* quantifies the degree to which a wetland serves as a source as opposed to a sink for water. Again, this can be expressed in terms of precipitation and evapotranspiration alone, as:

$$S = pi - eto$$

Wetlands with S values less than -0.5 are strong sinks, whereas wetlands with S values greater than 0.5 are strong sources. Whereas I has only a

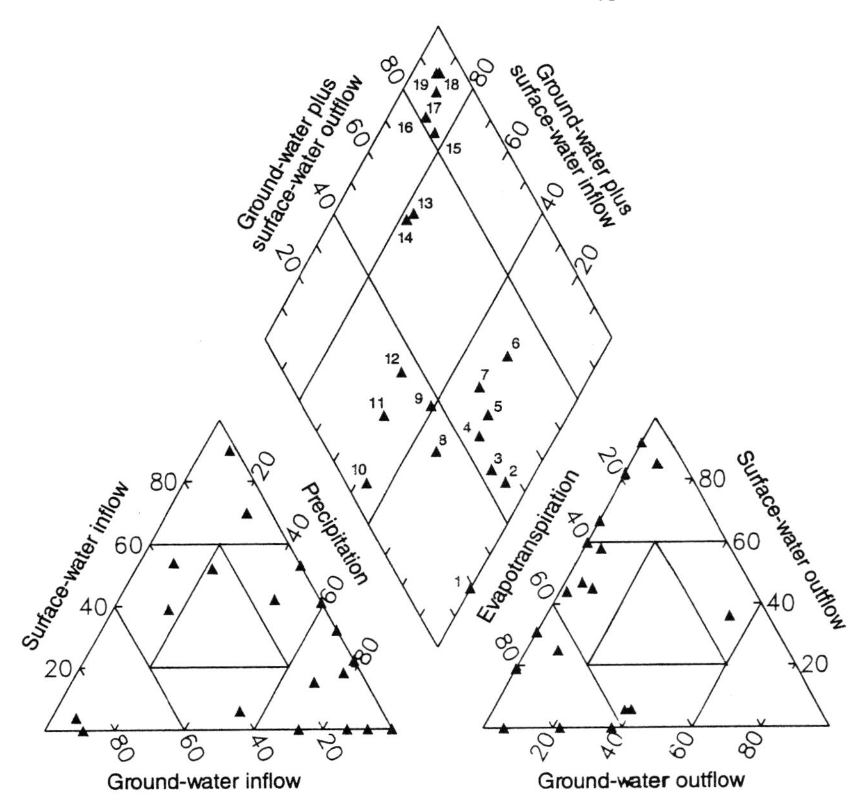

Figure 11.15 Nineteen wetlands plotted on the axes in Figure 11.14 based upon published sources. (1 is a bog, 10 is a prairie pothole, 6 is a fen and 17 is a *Typha* marsh; from Lent *et al.* 1997).

magnitude, *S* also has a sign to indicate net losses or gains. The indices *S* and *I* can be treated as two axes along which wetlands may be plotted to represent their interaction with the surrounding landscape, with nine different types if one uses the regions suggested by Lent *et al.* (1997).

Applying these indices to the same set of 19 published water budgets produced a wide array of *I* and *S* values. Wetlands near the top of the quadrangle in Figure 11.15 had high values for landscape interaction, with the five most interactive having *S* values very close to zero. These were weakly interactive with I values less than 0.5; the source/sink values ranged from $S = -0.41$ for the prairie pothole (point 10) to 0.40 for a rich fen in north Wales (point 6).

This procedure can therefore be used to graphically compare the hydrological functions of wetlands, and to divide them into seven input types or seven output types. The indices *I* and *S* provide a mean to summarize the nature of the interaction between a wetland and its surrounding watershed.

In using this procedure, many of the same issues will arise as with biological classifications. The degree of simplification required will probably determine the number of categories needed. If a landscape only contains a narrow range of wetland types, and further discrimination among them is required, it may be necessary to recognize further subdivisions of the categories proposed by Lent *et al.* In other cases it may only be of interest to find sets of wetlands in which one of the four main indices is predominant in order to quantify some other property (such as decomposition rates) thought to vary with hydrology; in such a case, one might deliberately search for wetlands near the corners of the quadrangle. As with biological classifications, one might also argue that other properties need to be included to clarify other aspects of wetland similarity. Whatever modifications might be required for future work, Lent *et al.* have provided a valuable method to visually and quantitatively compare the hydrological characteristics of wetlands.

Conclusions

Functional groups provide an important tool for simplifying the complexity of wetland communities. A major question would appear to be the relative value of good judgement vs. more rigorous methods of classifying functional groups. The examples above show some of the value of screening for traits and explicitly testing hypotheses about the relationships among traits. They also show a significant drawback: this work is laborious and we will, in any event, be dependent upon good judgement for some time yet. Additional effort is needed to collect trait information and tabulate it, and to relate traits to ecological functions.

12 · *Wetland conservation, management and research*

Let us now return to the ancient Epic of Gilgamesh with which we opened Chapter 4 (Sanders 1972). Those of us looking for an environmental message may find it noteworthy that, in an early part of the epic, Gilgamesh and his companion Enkidu travel to a mysterious cedar forest (which Sanders places in north Syria or south-west Persia) 'They gazed at the mountain of cedars, the dwelling-place of the gods . . . The hugeness of the cedar rose in front of the mountain, its shade was beautiful, full of comfort . . .' (p. 77). They encounter a monstrous guardian of the forest, Humbaba, whom they kill with their swords. 'They attacked the cedars, the seven splendours of Humbaba were extinguished.' (p. 83). The Epic of Gilgamesh therefore records an early episode of deforestation. Those of us familiar with the role of forests in wetlands will not consider it coincidental that four chapters later the gods, including Ninurta, the god of wells and canals, are 'cowering like curs' as a flood sweeps downstream.

Change in wetlands

The Tiber River

Moving into the realm of history as opposed to mythology, the Roman civilization originated with the Etruscans, who 'reclaimed Tuscany from forest and swamp' and built drainage tunnels to take the overflow from lakes (Durant 1944). The early history of Rome is little known, in part because the Gauls burned the city in 390 BC, presumably destroying most historical records. Although Rome was built on seven hills, it was not a healthy location, 'rains, floods and springs fed malarial marshes in the surrounding plain and even in the lower levels of the city'. (p. 12) but Etruscan engineers built walls and sewers for Rome, and 'turned it from a swamp into a protected and civilized capital'. One of the main sewers, the *Cloaca Maxima*, was large enough that wagons loaded with hay could pass beneath its arches; the city's refuse and rain water passed through openings in the streets into these drains and then into the Tiber, 'whose pollution was a lasting problem in Roman life' (Durant, 1944 p. 81). Meanwhile, deforestation occurred apace to provide building materials and fuel (Hughes and Thirgood 1982). It is unlikely to be a coincidence that the Tiber 'was perpetually silting its mouth and blocking Rome's port at Ostia; two hundred vessels foundered there in one gale . . . About 200 BC vessels began to put in at Puteoli, 150 miles south of Rome, and ship their goods overland to the capital' (Durant 1944, p. 78). Some 100 years later Julius Caesar had great plans 'to free Rome from malaria by draining Lake Fucinus and the Pontine marshes, and reclaiming these acres for tillage. He proposed to raise dykes to control the Tiber's floods; by diverting the course of that stream he hoped to improve the harbour at Ostia, periodically ruined by the river's silt' (p. 193). These plans were cut short when he was assassinated by a group of conspirators in 44 BC who saw, in these and other ambitions, the seeds of a potential monarch.

The Rhine and the low countries

The low countries of the Rhine delta also illustrate the long history of human interference with wetlands in Europe. The Netherlands are the delta of the Rhine river, which, like most European rivers, once had extensive floodplain forests dominated by woody species such as *Acer pseudoplatanus*, *Fraxinus excelsior*, *Populus alba* and *Quercus robur*. Altogether there may be some 40 tree species, depending upon flooding frequency and soil type (Wiegers 1990; Szczepanski 1990). Higher frequencies of flooding produce *Alnus* or *Salix* thickets. The long history of human activity such as agriculture, logging, drainage and dykes have eliminated most floodplain forests; in Poland, for example, only 1–2% of the landscape can be considered forested wetlands (Szczepanski 1990).

After the last Ice Age, levels of the North Sea rose rapidly, and about 6000 BP a system of barrier beaches formed along the coast (Figure 12.1). Sediments from the Rhine gradually filled the tidal basin behind the barrier coast, allowing marsh vegetation to develop. These marshes gradually changed into ombrotrophic raised bogs. In areas with higher tidal fluctuations, or less sediment, estuarine conditions persisted, and salt and brackish marshes formed, with fresh marshes and peat bogs developing landward. Closer to the Rhine and the Meuse, swamp forests and freshwater tidal areas formed (Wolff 1993).

Around 1500 years ago, colonists were attracted here to build dykes and polders and reclaim bogs. In the coastal region, dykes were built first to

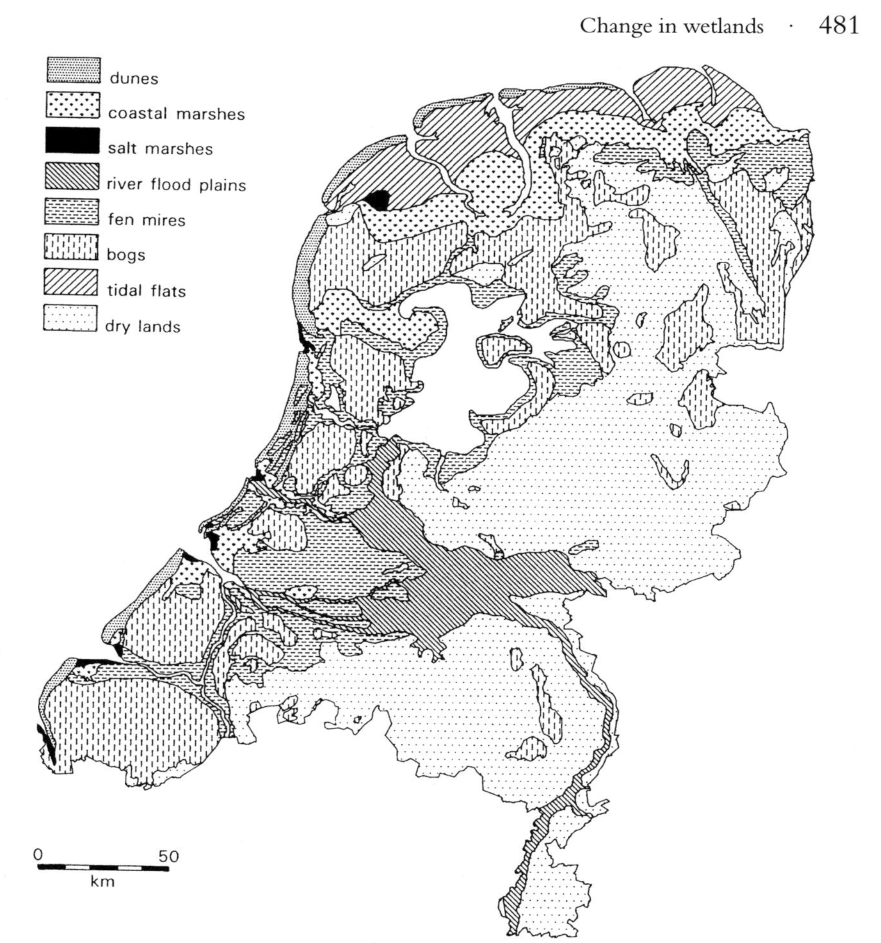

Figure 12.1 The vegetation of The Netherlands before significant human impact, around 100 AD (from Wolff 1993 after Zagwijn 1986).

defend farmland from flooding and then to extend the area of arable land. The Frisians in particular specialized in such work, followed by the Flemings and Hollanders, who extended their practices inland to the Elbe plain in Germany. The system consisted of digging drainage ditches to lower the water table, at first for cattle grazing and then for arable farming. Colonists were given permission to cut drainage ditches as far back from common watercourse as they wished. Thus by the twelfth and thirteenth centuries, a large area of peat bog plains was converted for agriculture. At the same time, water boards were established to co-ordinate building of dykes (van de Kieft, 1991).

Drainage of peatlands is followed by subsidence, particularly if the peatlands are also burned to provide extra nutrients for agriculture. This necessitated the construction of dams and dykes. Eventually, as sediment was deposited along water courses, and subsidence continued, the river channels increased in elevation relative to the land behind the embankments (Wolff 1993). In the 14th and 16th centuries, large areas of agricultural land were lost to flooding (e.g. the Dollard estuary, 150 km^2 inundated in the 14th to 15th century, the Biesbosch freshwater tidal area, 300 km^2 inundated in 1421; the Reimerswaal tidal flats in the Oosterschelde estuary, 100 km^2 in 1530). In total over 50% of the land area in the present Dutch province of Noord-Holland changed into lakes or disappeared back into the sea between the tenth and fourteenth century. This process was reversed again with technological improvements in the seventeenth century, and drainage of coastal areas reached a maximum in the twentieth century. The present landscape reflects these extensive changes in hydrology and vegetation. There are now several hundred polders along this coast. Many occur on peatlands along rivers and are drained by pumps; others are now raised above sea level by siltation and drained by sluice gates at ebb tide. The Zuiderzee, originally an estuary of the Rhine river, was divided in half in 1932 by a barrier dam, and the inner sections turned into four large polders fed with fresh water by the IJssel River.

During the First World War, Belgian engineers deliberately flooded parts of nearby Belgium in the Yser district by sea water in order to slow the advance of the German army. In his treatise, *Animal Ecology*, Elton (1927) summarizes work by Massart from *ca.* 1920:

> The sea-water killed off practically every single plant in this district, and all available places were very soon colonised by marine animals and plants . . . When the country was drained again at the end of the war, . . . the bare sea bottom was colonised by a flora of salt-marsh plants, but these gave way gradually to an almost normal vegetation until in many places the only traces of the advance and retreat of the sea were the skeletons of barnacles (*Balanus*) and mussels (*Mytilus*) on fences and notice-boards, and the presence of prawns (*Palaemonetes varians*) left behind in some of the shell holes. (p. 24–25)

In 1970 the Haringvliet area was separated from the North Sea by a dam with 17 sluices, and, as a consequence, the daily water table fluctuations declined from *ca.* 150 to 30 cm. Saltmarsh and brackish marsh communities disappeared within a few years. The current vegetation is largely controlled by grazing and flooding regimes, with *Phragmites* and *Scirpus* still present in wetter areas and *Agrostis stolonifera* in heavily grazed areas (van der Rijt *et al.* 1996).

The Wadden Sea, an estuarine environment that forms the northern coastline of The Netherlands is incompletely separated from the North Sea by barrier islands. Some 1200 km^2 was designated for nature protection in 1982. DeGroot (1992) has applied his system of functional values to the Wadden Sea, of which 45% is in this protected area. Regulation functions include moderating the climate and increasing precipitation on the adjoining land, primary production, and storage and recycling of nutrients. Production functions include the yields of crustaceans and shellfish for human consumption, as well as sand and shells for construction. Estimating (p. 215) that all functions together give goods and services in excess of US$ 6000 ha^{-1} y^{-1}, he concludes that many tidal areas in The Netherlands have been carelessly damaged, and

> Although the Dutch Wadden Sea, compared to other wetlands, is relatively well-protected and managed, it too is still threatened by many development plans and ongoing harmful human activities such as pollution and military training. (p. 218)

Overall, then, thousands of square km of peatlands, saltmarshes and shallow lakes have been lost. Further, nutrients and contaminants are carried into the area by the Rhine River. The remaining wetlands, however, occupy a key position on the West Palearctic flyway, and some 16% of The Netherlands is still classified as internationally important wetlands (Wolff 1993; Best *et al.* 1993). These reserves 'occur as small isolated patches in a matrix of agricultural land or as complexes formed by peat dredging, dyking of oxbow lakes, etc.' (Verhoeven *et al.* 1993, p. 33). In such small landscapes, hydrology is carefully controlled for the purpose of optimizing agricultural production in adjoining fields. Further, the remnant wetlands are being enriched with nutrients from four sources (Verhoeven *et al.* 1993): atmospheric deposition, surface water flow from heavily fertilized agricultural areas, inputs from eutrophic river water, and infiltration of contaminated ground water. The multiple factors of drainage, hydrological stabilization, eutrophication, grazing and pollution pose a major challenge to conservation managers.

When Goethe (1831) introduced Faust, the alchemist who sells his soul to the devil, he allows Faust to repent and aspire to carry out good works:

> Below the hills, a marshy plain
> infects what I so long have been retrieving:
> that stagnant pool likewise to drain
> were now my latest and my best achieving. (p. 222)

The Netherlands are of some interest because they illustrate land use changes in Europe as a whole, because they represent a delta of a major European river, and because they are well studied. At the other end of Europe, Greek wetlands face similar threats: 63% of wetlands have been lost and surveys report that more than half of all wetlands (and 100% of deltas) have experienced declines in water quality (Zalidis *et al.* 1997). Many others have been altered by changes in the water regime or loss of area. The extensive number of published papers on wetlands in the low countries (and even this section of the book) could be quite misleading: it should be borne in mind that The Netherlands comprise only 0.3% of the total area of Europe, and only 0.02% of the land area of the world (Wolff 1993). In spite of these, the number of papers on The Netherlands appears to outnumber the attention paid to the Pantanal, the Amazon and the Niger.

These few examples from the history of human impacts upon wetlands suggest that little has changed from the Tiber and Rhine of antiquity to the Parana (Chapter 1) and Amazon (Chapter 4) of modernity. But there are two possible sources of cautious optimism. First, while Europeans have badly damaged their own wetlands through several millennia of landscape modification, there is no essential reason to slavishly repeat these steps elsewhere. One can hope that other regions can learn from, rather than carelessly emulate, the European experience. Secondly, the scientific understanding of wetlands, while still incomplete, is vastly greater than it was in the days of the Etruscans. Whether human attitudes can change and science can advance to the point where we can avoid past mistakes is one of the unanswered questions of the coming millennium. It certainly extends well beyond the specific problems facing wetlands.

Two perspectives on conservation

There are two rather different perspectives upon the problems posed by wetland management and conservation. First, one could perceive wetlands primarily in terms of function, that is, as living machines that provide services to humans. So long as these services are performed, we have achieved our management goal. From this perspective, for example, the major groups of wetlands recognized in Chapter 1 (Figure 1.10, 1.11) can be relabelled as donors, receptors, or conveyors of water (Figure 12.2). The fact that they are bogs, fens and swamps, with different plant and animal species, and different rates of disturbance and fertility, may be unimportant to a hydrologist. Secondly, one could look at wetlands pri-

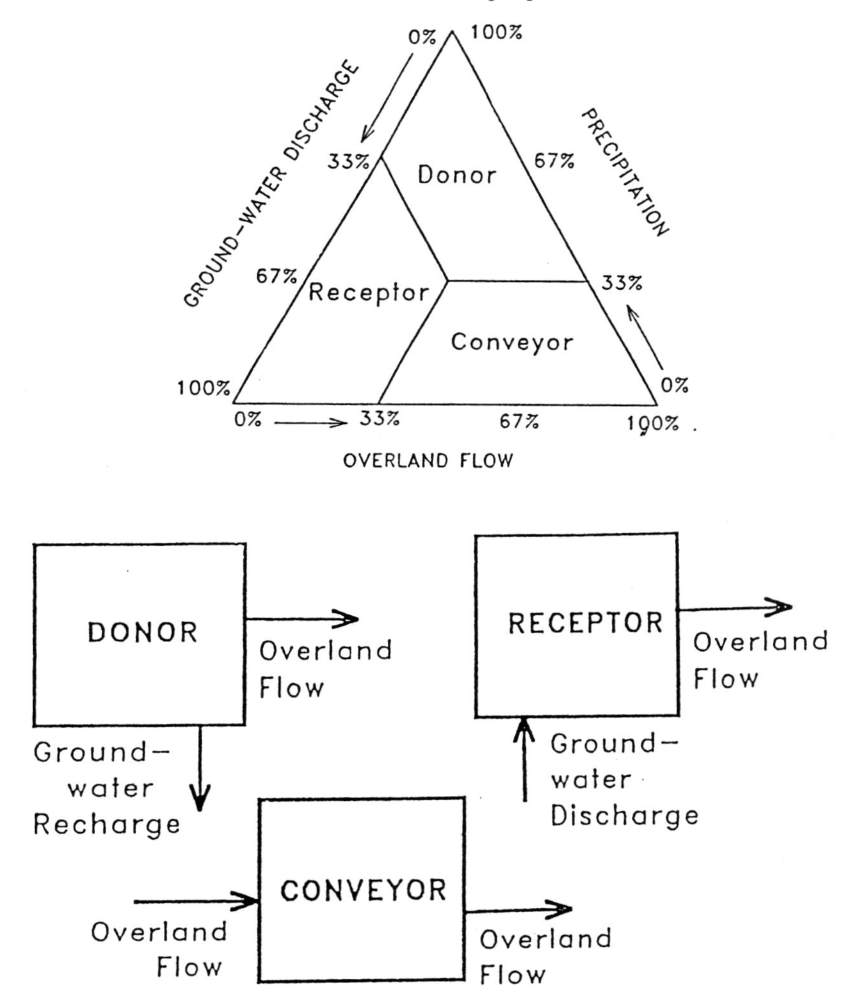

Figure 12.2 The three sources of water to wetlands corresponding to the functional classes of wetlands in *top* figure and the hydrological functions of each class (*bottom*) (after Brinson 1993a).

marily as natural communities of living organisms, and strive to maintain the patterns and processes within individual wetlands. So long as bogs, fens and swamps retain their normal complement of species and processes, we argue, one may also assume that they are providing the needed services.

Of course, these are not mutually exclusive perspectives, but they do naturally divide wetland ecologists, and their studies, into two rather different groups. Yet these perspectives are not as independent, nor as isolated, as they sometimes seem. For example, to return to Figure 12.2, if the primary function of a wetland is a conveyor of flood waters, then whether the wetland is a bottomland forest or wet prairie or sedge marsh may seem unimportant. Yet the particular vegetation types in a conveyor wetland will matter very much to humans concerned with fish that feed in alluvial forests, orchids that grow in wet prairies, and wrens that breed in sedge marshes. Further, if our primary concern is wetlands as sources of methane and its contribution to global warming (recall Chapter 1), our main goal may be to determine which of these wetland types produces which amounts of methane, and how different environmental factors regulate methane production within each wetland type. If, say, *Peltandra*, *Typha* and *Scheuzeria* all transmit equal amounts of methane to the atmosphere, we need not be concerned with which of these three species dominate the floodplain. On the other hand, if we discover that some species are much more efficient than others at emitting methane, it may be necessary to shift species composition to reduce methane production. Most wetlands, of course, have multiple functions: a single wetland will have a role in controlling hydrology, yielding wildlife, producing methane, fixing nitrogen, and supplying human recreation. Both the functional and community perspectives are necessary to understand the multiple consequences of different management actions. Since wetlands do perform multiple functions, one of the most thorny problems of management is ensuring that management for one function or goal does not cause loss of other equally important functions. Humans being what they are, it is all too easy to focus upon a single problem, a single function, or a single species, and ignore everything else.

These two perspectives overlap when one considers the function of maintaining biodiversity. First, there is a close connection between the global function of maintaining biodiversity, and the patterns and processes that occur within individual communities. Management for diversity was explored in Chapter 3. Secondly, functions are produced by the species that are present to perform that function; Chapter 10 explored some aspects of functional ecology of species. The very presence of specific plant and animal groups can be treated as an indicator that certain functions are occurring. The relative abundance of these functional groups is determined by changes in water level, nutrients, grazing and other factors. Thus, managing composition and structure will simultaneously manipulate function as well. There is an inherent linkage back to biological diversity of wetlands. Ehrlich and Ehrlich (1981) describe the loss of species from communities as being analogous to the loss of rivets from the wings of an aeroplane. A certain number of rivets can be lost without the wings falling off because there is some redundancy of function, but eventually if too many are lost, the functions decline. As a first approximation, we may assume that most functions in wetlands are carried out by more than one species; this is why species fall naturally into functional groups. If one species is lost, another may perform its function. But, if too many are lost, that function is no longer performed. The degree of redundancy, and hence the safety margin, is still an unknown.

At the risk of belabouring the obvious, let us look at one more example contrasting the perspectives of community ecology and ecosystem function: carbon storage. Those who consider wetlands as carbon pools for global climate regulation, may feel that far too much attention has been paid to local factors that influence species composition in wetlands. Yet, it is likely that species composition, hydrology, fertility and grazing can all affect rates of peat accumulation. There is still a need to understand these processes at the community level, even if one is ultimately concerned with the planetary scale. It would appear that the different groups of scientists and managers need to speak to one another to ensure that neither of these perspectives is neglected.

In this book, I have tried to emphasize this common ground between function and species composition, focusing upon wetlands as communities controlled by multiple factors acting simultaneously (Principle 1). The assumption is that, if we appropriately manage wetland communities to maintain their natural composition, their functions too will remain intact. That is, I have assumed, along with Holling (1978) and Christensen *et al.* (1996) that persistence of ecosystems and communities is important for ensuring the persistence of human civilization.

Persistence, however, is a dangerous word if it creates the impression that the future is static, and that dynamic changes in ecosystems have to be suppressed. In fact, as the book has illustrated, many kinds of dynamic changes, from water level fluctuations to river meanders to fire, are an essential part of maintaining and restoring the diversity and function of wetlands. Before we consider management further, however, let us digress for a broader look at the degree to which change is natural in wetlands.

The third principle of wetland ecology states *the multiple factors that produce a community or ecosystem will change through time.* Change in wetlands is nothing new. Many examples of change have been presented in this book, from peat bogs in Ireland changing with climate and human land use (Chapter 1) to Amazonian wetlands changing with river erosion and deposition (Chapter 6), from fire and water levels in the Florida everglades (Chapter 6) to flooding and salinity in Californian saltmarshes

(Chapter 6), from isostatic rebound around Hudson Bay (Chapter 9) to the dyking of the Rhine delta (Chapter 12). Accepting the dynamic character of ecosystems is the fifth principle of ecosystem management proposed by Christensen *et al.* (1996). A larger perspective on science and conservation in the light of ecosystem change has been discussed in Botkin's (1990) book, *Discordant Harmonies;* the basic conclusion is that there are no easy answers. Humans can damage ecosystems by suppressing natural dynamics, just as much as they can by causing change through damming rivers, dyking saltmarshes and draining peatlands. Of course, one of the difficult problems in conservation and management is to decide which changes are acceptable and which are not. Allowing meander systems to evolve in a floodplain would seem acceptable; allowing the proliferation of exotic species such as Purple Loosestrife, Water Hyacinth, or Nutria would seem unacceptable. One allows a natural process to continue, the other causes a rapid change that is not intrinsic to particular wetlands. Since the biota and the ecosystems of the Earth have changed steadily through time, management decisions require a broader appreciation of wetland dynamics. To conclude this discussion, let us look at four more examples of change over different time scales, from different causes, and with different amounts of human input.

Wetlands have existed for millions of years, but they have changed as the fauna and flora of Earth evolved. Coal swamps dominated by *Lepidodendron* trees no longer exist yet even they were subject to alternating wet and dry periods (Figure 12.3). We may recall Figure 1.20 and feel as estranged as a mire ecologist may feel encountering a tropical riparian woodland. But, looking more carefully, we would probably discover that the same processes we have seen in this book were occurring: water level fluctuations; fertility gradients; disturbance; herbivory; primary production; decomposition. In many ways coal swamps were very similar to swamps of today. We must learn to seek the similarities in process at the same time as we appreciate their differences; without this, we will slide into geographic, taxonomic and methodological Balkanization.

One need not look back millions of years to find the ebb and flow of waters. Over merely the last 30 000 years, the world has seen the formation of great pluvial lakes in Africa, south western North America and Australia (Figure 12.4). In Africa, these lakes reached their maximum extent around 9000 BP; in North America, between 12 000 and 24 000 BP and in Australia earlier still, perhaps 26 000 to 30 000 BP (Flint 1971; Street and Grove 1979). Imagine the extensive areas of wetlands, and the clouds of migratory waterfowl that must have once occupied areas of the Earth

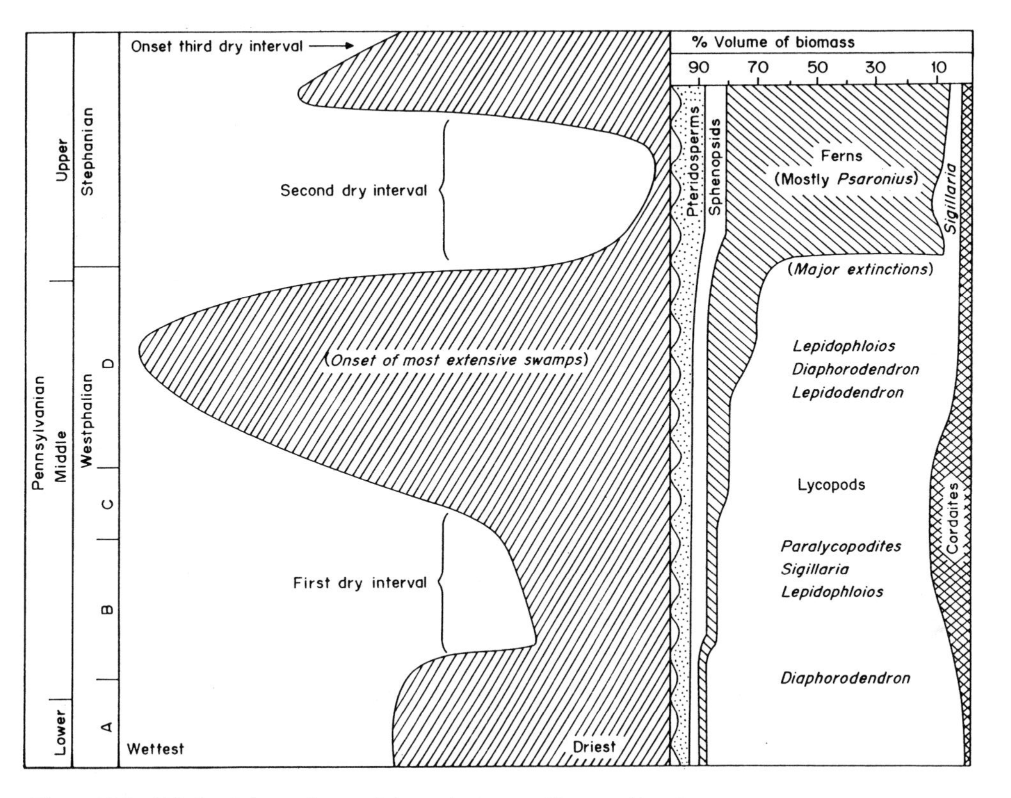

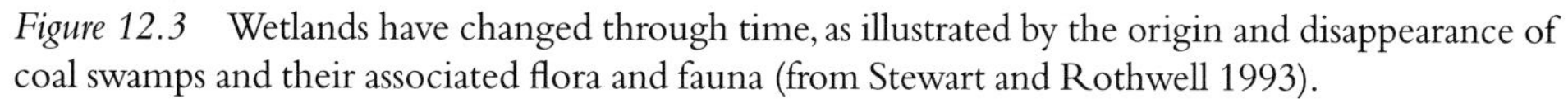

Figure 12.3 Wetlands have changed through time, as illustrated by the origin and disappearance of coal swamps and their associated flora and fauna (from Stewart and Rothwell 1993).

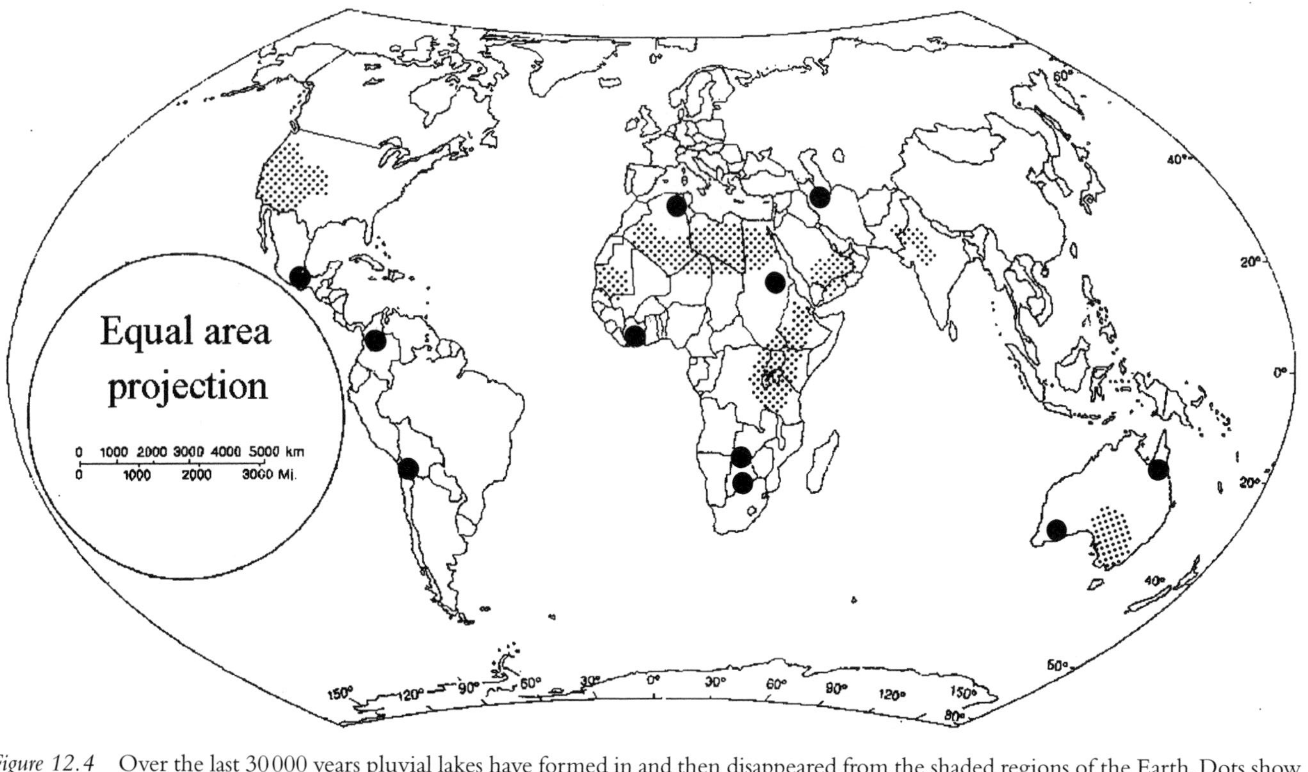

Figure 12.4 Over the last 30 000 years pluvial lakes have formed in and then disappeared from the shaded regions of the Earth. Dots show isolated lakes (after Street and Grove 1979).

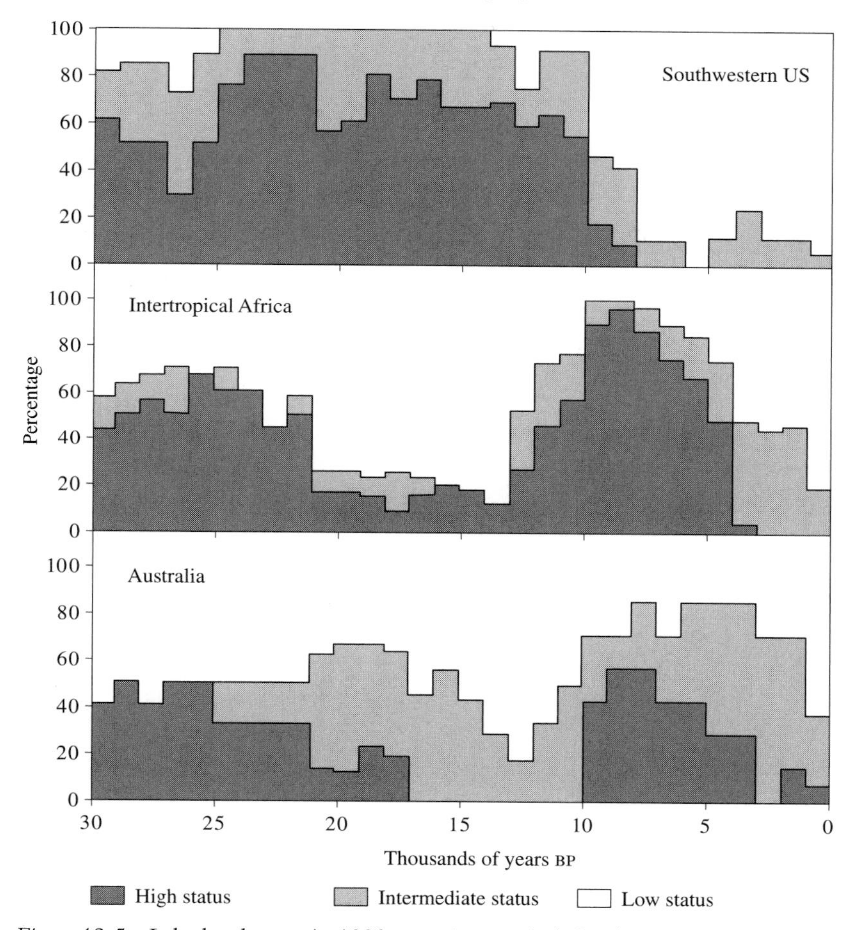

Figure 12.5 Lake level status in 1000-year time periods for the past 30 000 years in three parts of the world (after Street and Grove 1979).

that are now arid flats or remnant saline lakes. Our own millennium appears to be one of the most arid in the late Quaternary (Figure 12.5). If at times we despair about the impacts of our own species upon wetlands, perhaps Figures 12.4 and 12.5 can put it in a somewhat larger perspective.

Wetlands change at shorter time scales too, time scales at which evolution is probably comparatively unimportant, time scales which can be easily measured in human generations. Figure 1.21 showed the changes in a European mire as human civilization developed there. Figure 12.6 shows the estimated impacts of aboriginal civilizations in the Americas. Figure 12.7 shows us change on a shorter time scale still – the time over

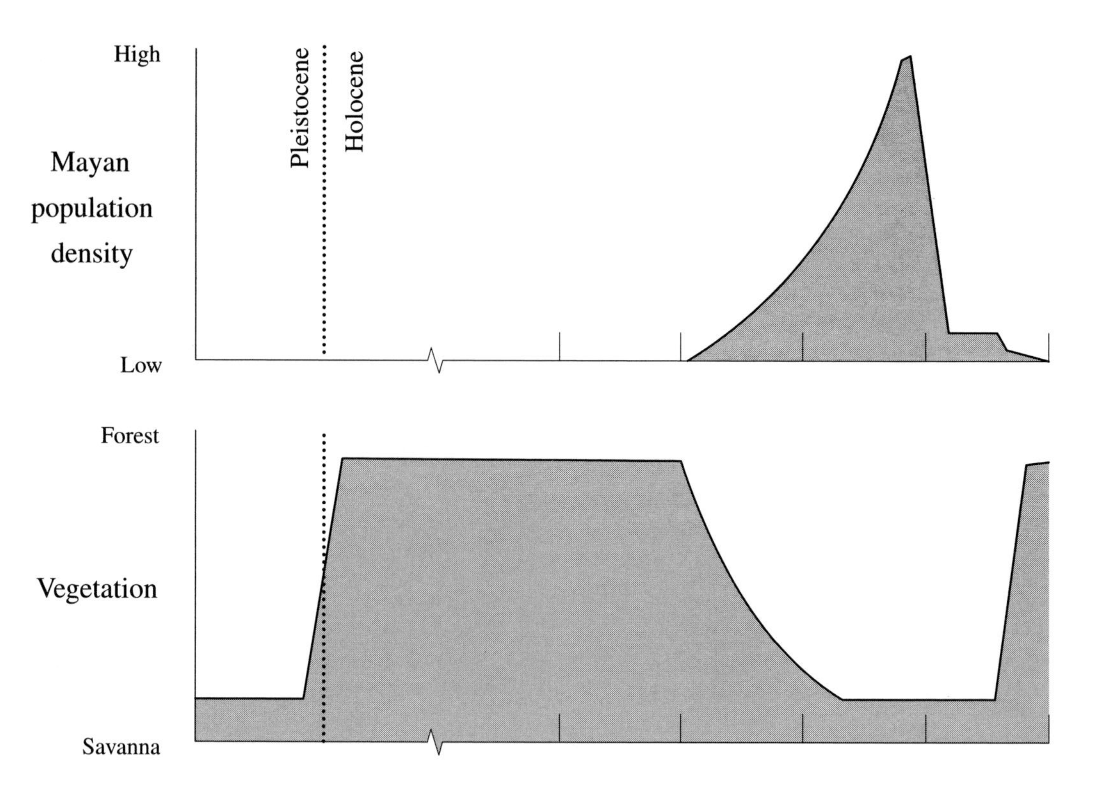
High
Mayan
population
density
Low
Pleistocene
Holocene
Forest
Vegetation
Savanna

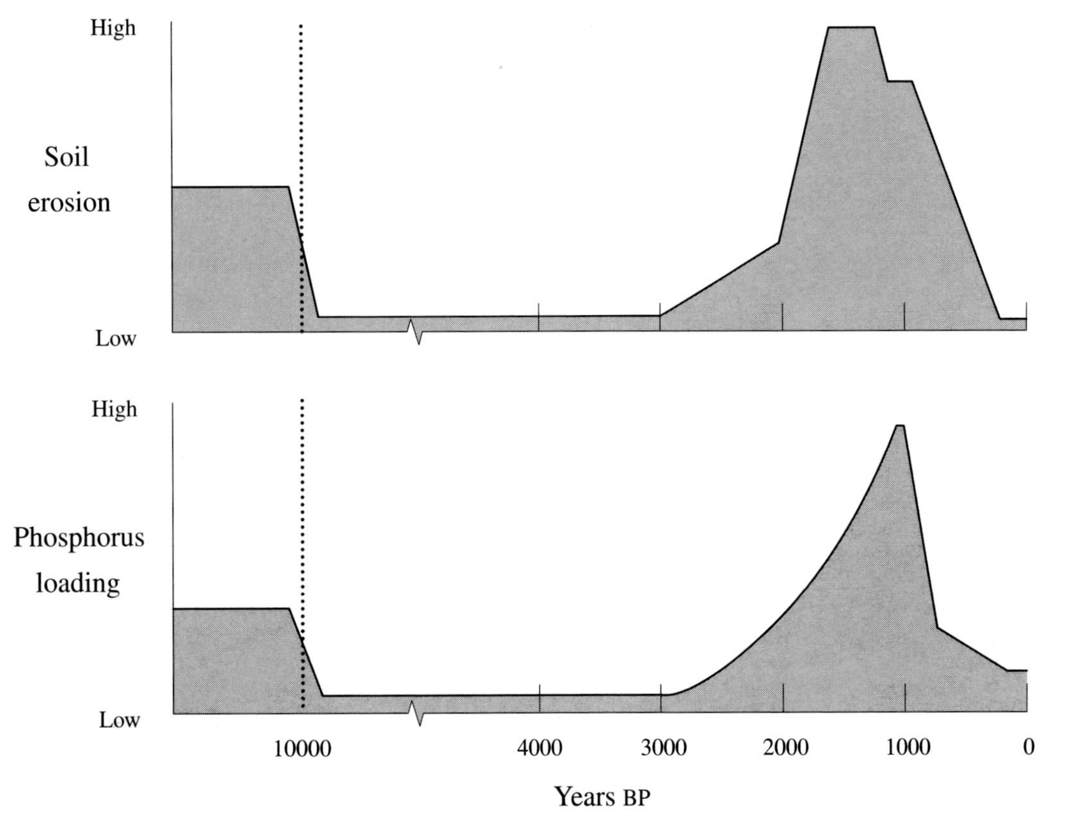

Figure 12.6 Aboriginal populations in America caused significant increases in erosion, as illustrated by impacts of the Mayans in Guatemala (after Binford *et al.* 1987).

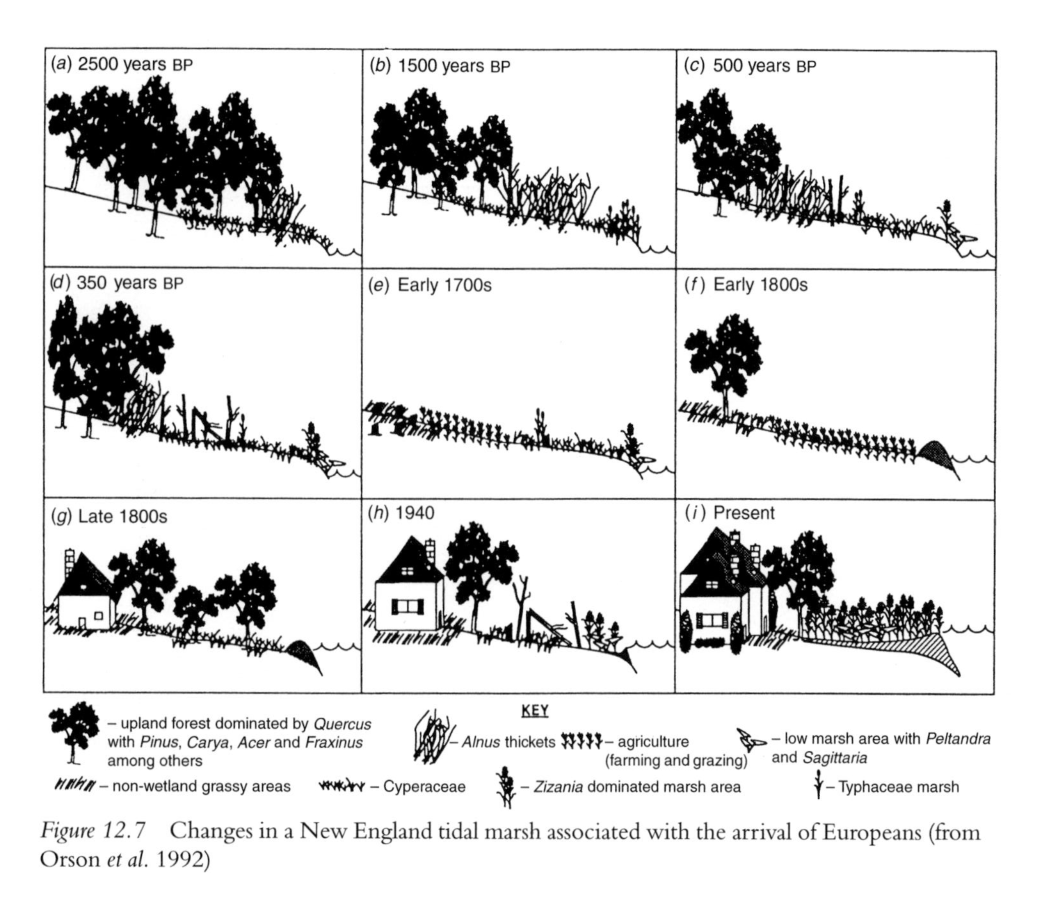

Figure 12.7 Changes in a New England tidal marsh associated with the arrival of Europeans (from Orson *et al.* 1992)

which Europeans arrived and modified wetlands in eastern North America.

Humans often fear change, and so we stabilize lake water levels, build dams to stop spring flooding, channelize rivers, and put rip rap along eroding river banks. As wetland ecologists we need to overcome these fears and learn to work with change. This does not mean that we must accept that all changes wrought by humans are desirable, or even acceptable. But, as Botkin (1990) reminds us, and as Figures 12.3–12.7 show, working with naturally dynamic systems is the situation with which we must contend.

Priorities for action

Accepting that wetlands are dynamic and that situations change does not mean that we can sit idly by while the world's wetlands are devastated and the world's biota destroyed. Although wetlands will always change with time, the *processes* that they perform, and the *functions* they provide, must be maintained. The immediate challenges facing wetland ecologists therefore fall into two areas: protecting representative ecosystems, and maintaining ecological function. In the longer term, we may turn our attention to the restoration of damaged wetlands. In all cases, we need to expand the scale at which we plan in order to put individual actions and decisions within a framework that includes ecosystems, landscapes and regions (Bedford and Preston 1988).

Protection

As the twentieth century has drawn to a close and the next century begins, our first challenge is to ensure that significant areas of wetland are protected from further degradation. Once these areas are protected, the next generation of managers will have to grapple with wise management of them; the first task in setting up reserve systems is to make management as easy as possible. The design of reserves, and of reserve systems, is a topic which itself deserves an entire book (e.g. Shrader-Frechette and McCoy 1993; Noss and Cooperrider 1994); a brief introduction has been provided by Noss (1995). Noss suggests a number of key steps in the design of reserve systems. The first two are (i) identification of core areas and buffer zones (ii) design of reserve networks to ensure that protected areas represent entire landscapes. This requires a number of simultaneous perspectives including local, regional, landscape and global.

Beginning with the core protected area, the *size* of each protected wetland should be large enough to retain the diversity of wetland types and full array of species present. The well-documented relationship between species richness and area (Chapter 3) shows that, in general, the bigger the site, the more species that are likely to be protected. Big areas have two particular merits. Big areas are important to maintain large predators that have large territories and are highly mobile (Weber and Rabinowitz 1996). Also, the bigger the site, the greater the possibility that natural processes can continue to generate habitat diversity. An alluvial wetland reserve, for example, ought in principle to be large enough to allow flooding and bank erosion to continue unabated. If these processes are missing, it may be impossible to retain the biological characteristics of the reserve, and it certainly will compound the difficulties and costs of management. Many other factors can be used to select core protected areas including naturalness, significance, rare species, ecological functions and value for research (Table 12.1). Each protected area needs to be surrounded by a *buffer zone*; in this zone, land use practices may have to be regulated to higher standards than elsewhere in order to ensure that nutrients, pollution or exotic species are not carried into the protected site from immediately adjoining sites. Biosphere reserves provide an example of such an arrangement with protected core areas such as a national park surrounded by larger landscapes in which human use includes consideration for the viability of the core area.

The reserves must be connected so that *dispersal* can occur from one reserve to the next. As reserves become increasingly smaller and more isolated from one anther, dispersal becomes increasingly constrained and species become increasingly broken into metapopulations with the dynamics typical of island species (MacArthur and Wilson 1967; Hanski and Gilpin 1991; Hanski 1994). While local extinction from small areas of wetland might be entirely normal given the natural dynamics of wetlands, once reserves become isolated fragments within a landscape, there may be no local populations available to recolonize the site. In the longer term, there will be constant selection for those species best able to colonize habitat islands; at the same time, species with limited dispersal such as flightless insects or ant-dispersed plants may be expected to slowly disappear from the entire reserve system. Large reserves help minimize this problem, but in other cases, reserves may need to be deliberately linked by corridors (Noss 1995).

On a broader scale, the protected wetland should *represent* a habitat type that is of significance at the local, regional or global scale. It may represent

Table 12.1. *Some important factors for selecting and prioritizing wetlands for conservation*

Factor	Comments
Area	All important ecological values and functions increase with area.
Naturalness	Minimal alteration to natural patterns and processes.
Representation	Serves as an example of one or more important ecosystem types.
Significance	Relative global importance: existing area of this habitat , rates of loss, percent of habitat type protected, better examples protected elsewhere?
Rare species	Globally and regionally significant species present.
Richness	Supports many species.
Productivity	Good production of commercial species (but high production may reduce rare species and diversity).
Hydrological functions	Flood reduction, ground water recharge.
Social functions	Education, tourism, recreation.
Carrier functions	Contribution to global life support system: oxygen production, nitrogen fixation, carbon storage.
Food production	Harvesting of species for human consumption.
Special functions	Spawning or nesting area, migratory stopover.
Potential	Potential for restoration to recover lost values and functions.
Prospects	Probability of long-term survival: future threats, buffer zones, possibilities for expansion, patrons, supporting organizations.
Corridors	Existing connections to other protected areas or site itself is a corridor.
Science function	Published work on site, existing use by scientists, existing research station, potential for future research.

Notes:
These are listed in approximate order of their importance.

a common wetland type, and serve as a protected example of a common landscape feature, or it may represent a rare ecosystem type; both representivity and protection of rare features are complementary objectives for setting up reserve systems. At this scale the site needs to be considered in the context of other protected wetland types: are there similar examples already protected? Are there more important vegetation types that are not yet protected? This process is often termed 'gap analysis', and algorithms now exist to evaluate different reserve scenarios in order to maximize the value of a reserve system (Pressey *et al.* 1993). The objective is to define the smallest number of sites needed to achieve certain goals, such as providing one, two, or three protected examples of each species, or each community type.

Most nations now have systems of protected areas such as wildlife areas, national parks and ecological reserves. Sites designated under Ramsar are of particular interest within the realm of wetlands (Figure 12.8). Even a cursory examination of this figure shows that wetlands in western Europe are heavily over-represented, yet this is the part of the world, where, in general, protected areas are both the smallest and most degraded by human activity. We need a shopping list to set future priorities, and where better to start than with the world's ten largest wetlands? There is bound to be some difference of opinion as to which wetlands should go on this list; for example, do prairie potholes with large numbers of ponds interspersed with prairies and farmland fit? Table 1.3, however, offered one such list. From the global perspective, we must be cautious about spending too much money on the precise management of tiny fragments of wetlands in heavily populated areas if this means that resources are being directed away from globally significant wetlands such as the Pantanal, the Sudd and the Amazon.

Setting up reserve systems literally is a race against time. There are growing reasons for optimism at the same time as one is discouraged by the ongoing rates of wetland destruction. Of course, it is possible to acquire degraded habitats and restore them, but this is a poor alternative to protecting areas that are pristine. Weber and Rabinowitz (1996) provide a few encouraging examples of progress in Central America and Africa, respectively.

In 1984 Belize created the Cockscomb Basin Jaguar Preserve (*ca.* 400 km^2), the first area in the world to be protected specially for the maintenance of jaguar populations. This was expanded by the creation of Chiquibul National Park (*ca.* 1000 km^2) also in Belize. This latter reserve has the potential to be linked to the Maya Biosphere Reserve (15 000 km^2).

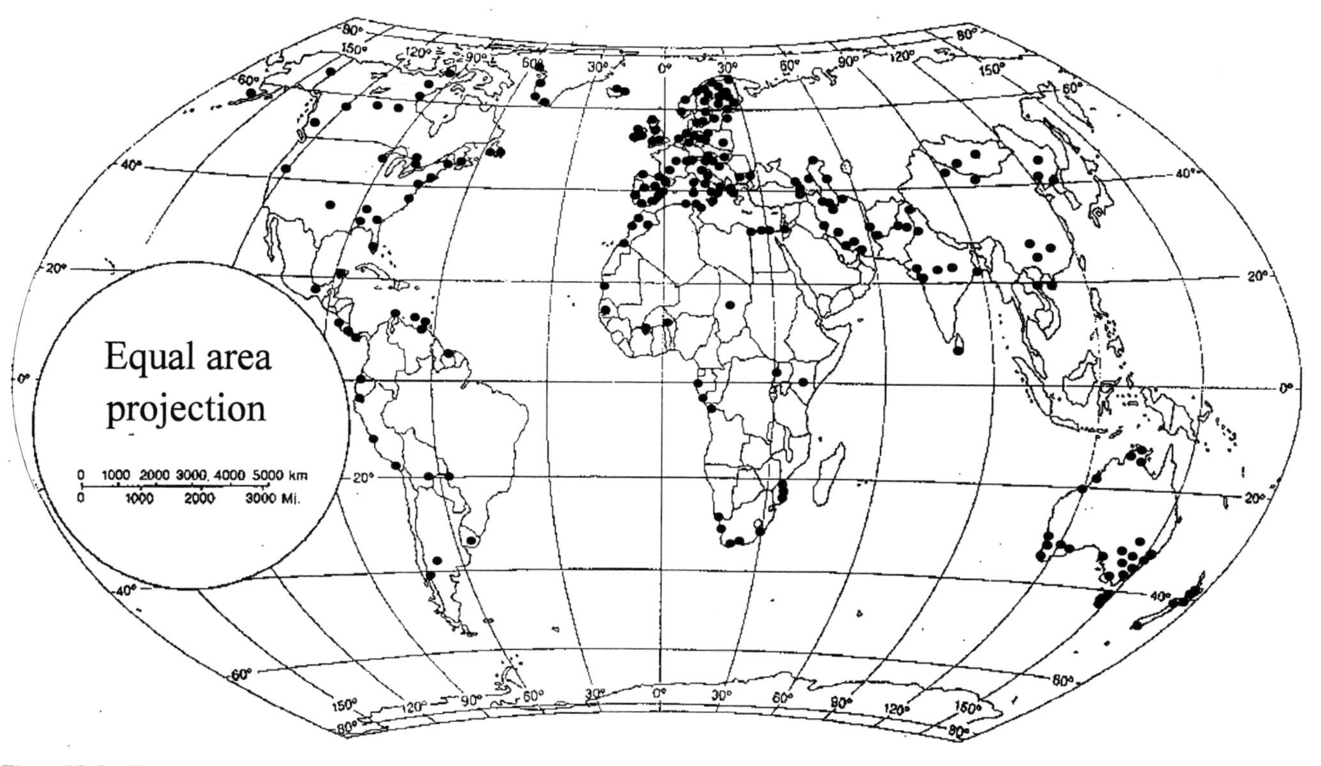

Figure 12.8 Ramsar sites designated as of 1993 (after Dugan 1993).

Similarly, Pasero Pantera (the path of the panther) is an international program designed to protect panther habitat along the entire Central American Isthmus. Costa Rica has already added lands to help contribute towards this international project.

Within Africa, there is now an international forest and wetland reserve of some 11 000 km^2 with species including gorillas and leopards, that includes protected lands in the Congo, the Central African Republic and Cameroon. The result is one large contiguous ecosystem protected across three national boundaries, so that should safeguards break down in any one country, the two adjoining areas could serve as reserves for extended periods. Designing reserve systems to buffer against political instability seems particularly wise in times of political uncertainty.

The problem of setting up a truly representative reserve system is illustrated by wetlands in the former USSR. There are ten different mire zones stretching from east European pine bogs and central Russian sedge fens in the west, to polygon mires and palsa mires in the far north, to maritime mires in Kamchatka and Sikhote-Alin (Botch and Masing 1983). In all, 43 zones and subzones can be delineated. Although each zone has certain characteristics set by climate and topography, other variation is superimposed depending upon whether the peatland is ombotrophic or minerotrophic, wooded or herbaceous, and depending upon the dominant species. In the north, extensive mires cover some 70% of the entire zone; in contrast, in the Caucasus and Carpathians, mires are smaller and isolated and serve as refugia for rare endemic plants. A proposed system of protected areas therefore must include a wide array of vegetation types and geographic regions in both Europe and Asia. Botch and Masing say that 310 mires are proposed for conservation, comprising some 1.4 million ha, less than 2% of the actual mire area.

One of the major obstacles to protection is the view that protecting ecosystems means withdrawing them from human use and thereby reducing human economic welfare. The functions performed by wetlands illustrate that this simplistic protect vs. use dichotomy does not exist; shrimp and fish production, are, for example, dependent upon saltmarshes and floodplains (Turner 1977; Welcomme 1976). For wetlands, the issue of function may provide economic arguments for preservation. Even if the obvious production functions are ignored, however, there is a further merit to protected areas; contrary to expectation, it appears that they actually stimulate economic activity (Rasker and Hackman 1996). Owing to the importance of this assertion, it is necessary to spend some time on this example.

Rasker and Hackman begin with the difficulty in protecting natural areas for large carnivores when there is a belief that a choice is required between environment and economy, '. . . a belief that, however, appealing, carnivore conservation is a luxury we cannot afford because the opportunity cost in terms of jobs and resources forgone is too high'. This is a commonly heard argument around the world; what may be surprising is the paucity of data for or against it. Rasker and Hackman set out to test this proposition by comparing economic indicators for two regions in north west Montana. Four counties with large protected areas (Flathead, Lewis and Clark, Teton, Powell) are compared with three resource extraction counties (Lincoln, Sanders, Mineral). The wilderness counties total some 3.4 million ha (839 000 protected), whereas the resource extractive counties totalled nearly 2 million ha (33 000 protected). The latter resource-extractive counties were chosen because the conflicts between jobs and environment are intense, and because timber harvesting and hard rock mining have traditionally played an important part in their economies. Although these are only counties, they are the size of nations in many other parts of the globe. If, indeed, 'locking up' land in reserves causes economic hardship, then the counties with protected areas should show reduced economic performance relative to the counties with few protected lands. Figure 12.9 shows the striking results. A range of economic indicators including unemployment rate (*bottom*) and personal income growth (*top*) were above the USA and Montana averages and even more above the means for the resource extractive counties. 'From 1969 to 1992 wilderness counties added new jobs and income in every non-agricultural sector of the economy. The resource-extractive counties lost more than 1300 jobs in the construction, transportation, and public utilities sectors.' The resource extractive counties also suffered from higher unemployment rates. Rasker and Hackman conclude:

> The bulk of growth in the Greater Yellowstone was in industries that do not rely on natural resources extracted from the ecosystem. From 1969 to 1992 more than 99% of all the new jobs and personal income (and 88% of existing jobs) came from industries other than mining, logging, and ranching or farming . . . Research on the economy of the Greater Yellowstone has uncovered a new paradigm for economic development in the West: protection of the wild and scenic character of the landscape and the quality of life in local communities serves as a magnet to retain local people and their businesses. These qualities are a vital part of the economic well-being of local residents . . .

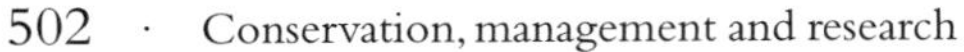

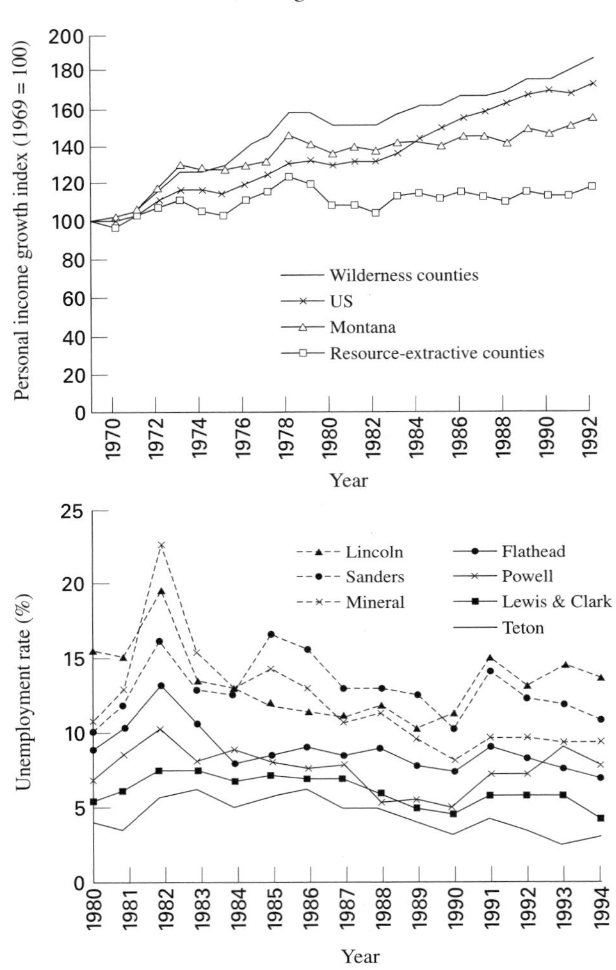

Figure 12.9 Personal income growth in four regions: wilderness counties, USA, Montana and resource extraction counties. Also, unemployment rates in resource extraction counties (----) contrasted with wilderness counties (——) (after Rasker and Hackman 1996).

While neither of these examples is exclusively wetland, they illustrate the possibilities of progress towards protecting large reserve systems that are more than a series of islands in an agricultural landscape. Even if we focus only upon wetlands, intact watersheds are essential to maintaining hydrology and water quality; in a sense then, any protected wetland really forces managers to focus upon the entire watershed with which the wetland interacts.

Once reserve systems have been organized (steps (i) and (ii) above), there are two further steps: (iii) management plans are needed for each site and for the system as a whole, and (iv) indicators are needed to provide a method for monitoring whether the management plans are achieving the goals set for the protected system (Noss 1995). In this book, I consider the next two sections, maintenance of function and restoration to fall under Noss's category three. Indicators are then a section unto themselves.

Maintaining the diversity and function of reserve systems

Representation and diversity

A reserve system is set up to protect the full array of ecosystems, communities and species that occur in a landscape. If there is systematic change within the reserve system, an entire section of the representivity may be lost. Exactly such a trend has been occurring in wetlands over the last century. Recall that hydrology and fertility are the two key factors that determine the kinds of wetlands that occur in a landscape. The variation in hydrology in wetlands has been steadily declining, from factors as diverse as drainage ditches that permanently lower the water table, to levees that prevent floodwaters from spilling onto alluvial marshes, to dams that hold back spring floods. The full array of hydrological regimes on Earth may therefore be converging upon increased stability and reduced variation. In an analogous way, there are systematic trends in fertility: a steady increase driven by eutrophication from sources including sewage from large cities, artificial production of fertilizers, burning of coal, run-off from agricultural landscapes, and atmospheric deposition. There is no need to repeat here the many examples we have seen of these processes, except to note that they are global in extent. Since the wetlands that arise in landscapes are produced by particular sets of hydrology and fertility, and since entire sets of conditions including high flooding levels and low fertility are vanishing from the landscape, we may assume that the corresponding wetland types are vanishing as well (Figure 12.10). That is to say, the array of wetland types within a landscape is being increasingly squeezed into the bottom right of the figure: eutrophic wetlands with relatively stable water levels. The invasion of woody plants into marshes, the disappearance of *Erica* heathlands, the replacement of wet meadows by *Typha* marshes and the replacement of native species by exotics in infertile wetlands, are all special cases of this widespread process of community convergence. Moreover, this change is being driven by two processes that

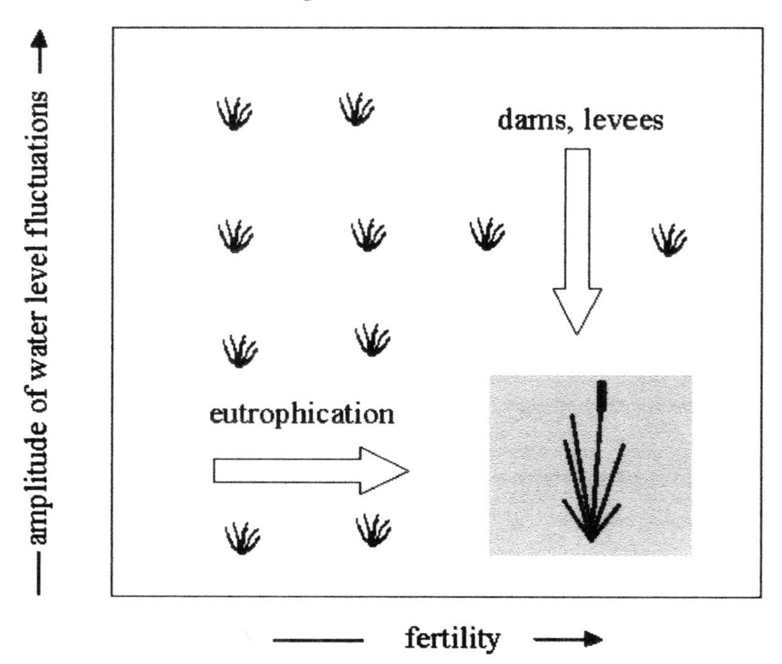

Figure 12.10 Human activities have compressed wetlands onto an increasingly narrow array of flooding and fertility regimes, leading to a loss of many wetland types.

continue largely unabated: the construction of dams, and the deposition of nutrients, as indicated by the large arrows, continue to squeeze wetlands into a progressively narrower region of possibilities. One of the high priorities for management is to reverse this process and re-establish sets of environmental conditions that represent the fuller array of possibilities in a landscape. All too often, each problem is seen as a special case and the overall significance is misunderstood. For example, in the 1265-page compendium *Freshwater Wetlands and Wildlife* (Sharitz and Gibbons 1989), there are no index entries under fertility, nutrients, eutrophication, nitrogen or phosphorus, as if the process of eutrophication is a phenomenon entirely unknown in the last 20 years of science. Similarly, as wetlands in the Athabasca floodplain are being lost to woody plant invasion, invasion by woody plants in eastern North America is being explained away as natural succession, with the suggestion that more wetlands are being lost to natural succession than to human causes (Larson *et al.* 1980; Golet and Parkhurst 1981).

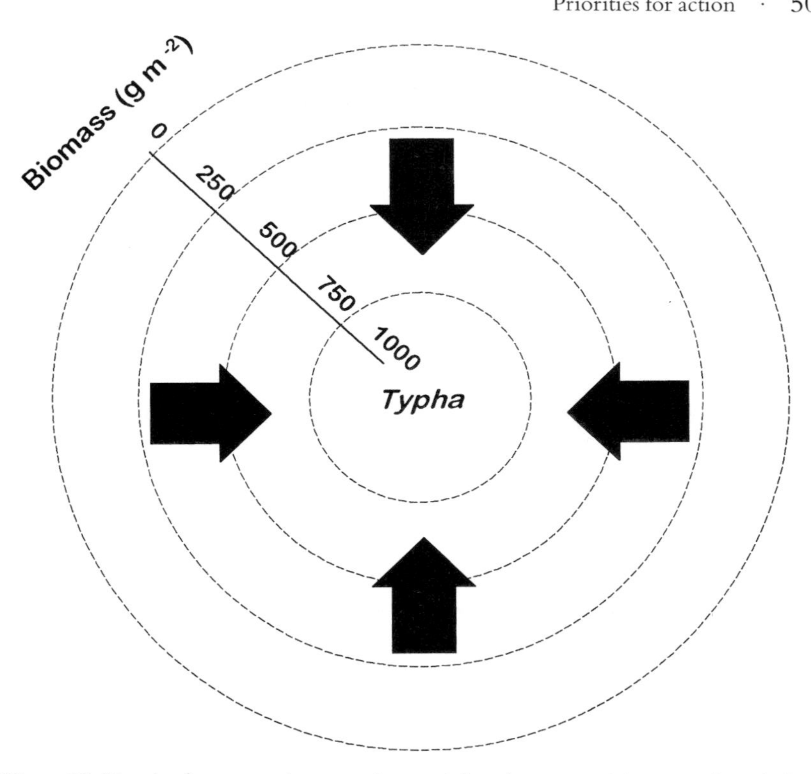

Figure 12.11 As the constraints creating peripheral communities are reduced, the vegetation converges upon stands of clonal perennials such as *Typha* spp., *Phalaris arundinacea*.

Another way of picturing this process is to return to the centrifugal organisation model in Figure 7.16. Recall that the diversity of vegetation types is produced by different constraint gradients, such as shortages of different nutrients, scouring by ice or waves, recurring fire, and other factors that limit primary production. The wide array of constrained sites produces the wide array of wetland types at the periphery of the model. As these constraints are reduced, the biomass increases, and the wetlands begin to converge (Figure 12.11). In the absence of strong hydrological constraints, and in the presence of high nutrient levels, cattails predominate. A reserve system that is a cattail monoculture is a reserve system that has lost most of its diversity, and a reserve system that no longer represents the majority of wetland types within a landscape.

Yet another way of viewing this problem is to envisage it as the removal of certain filters that once produced the structure in communities.

Consider the plants first. Removing long periods of flooding removes the filter that kept woody plants at bay, allowing them to invade herbaceous meadows. Removing the filter of infertility allows rapidly growing plants with dense canopies to invade herbaceous meadows. We can already predict endangered wetland groups of the future; rosette growth forms (e.g. *Parnassia, Saxifraga, Lobelia*), evergreen species (e.g. *Erica, Eriocaulon, Lilaeopsis*), carnivorous species (e.g. *Drosera, Dionaea, Utricularia*), species of infertile sands (e.g. *Castilleja, Cacalia, Gratiola*) or eroding shorelines (e.g. *Senecio, Pedicularis, Sabatia*) and species that require unusual nutrient ratios, recurring fire, intense flooding or high grazing pressure. What impact will this have upon animals? Presumably, species that forage in wet meadows, species that require the above plant guilds to complete their life cycles, insects specialized upon plants with low tissue nutrient levels, reptile species that nest in freshly deposited sands and silts, migratory birds that feed on mud flats around lakes and, in general, any functional group that tolerates extreme flooding and unusual fertility conditions, will be most at risk. Near where I write, the Spotted Turtle and Wood Turtle, which occupy shoreline fens and sandy floodplains, are more likely to be at risk, say than Red-winged Blackbirds that nest in cattail stands or Yellowthroats that nest in alders.

It is entirely possible that so much progressive wetland change has occurred that most of us have never seen the full array of wetland types that our landscape once possessed. That is to say, our frame of reference – the landscape we grew up with – may already be so altered that it is not a useful reference point for designing and managing reserve systems. This, of course, opens a broad spectrum of possibilities: just what is the array of wetland types we want to protect with a reserve system? Do we aim for the landscape of our childhood, the landscape of the mid-nineteenth century, or the landscape that may have occurred before humans appeared upon the scene?

There is no easy answer to such questions (Leopold 1949; Botkin 1990), but one approach might be to consider the array of environmental factors that would have occurred in the landscape before humans modified it. Hydrological and sedimentation models would allow us to determine a mean and standard deviation for both flooding regimes and fertility regimes in a landscape without human impacts. One could plot the original frequency distribution of habitats along the axes in Figure 12.10, and then overlay a current distribution, testing for a difference in the maxima or in standard deviations. Either could measure the degree of change in, or the loss, of wetland types within the landscape. Whether or not managers could ever recreate such landscapes, this would provide a realistic point of reference for more management. Imagine such a model for the Rhine River valley. What was the delta once like, and what has this to say about intensive management of remaining wetland fragments? To what extent could large floodplains be removed from human use so that natural flooding regimes could be allowed to re-establish? How would the apparent costs of this balance against the cost of building dams, maintaining levees and repairing the inevitable flood damage? The Rhine or the Mississippi may be too large to start with, but are there watersheds where this process could be started on a more regional scale?

Two challenges facing managers are therefore very clear. The first is to reduce the magnitude of the forces that are still driving wetlands to convergence, that is to (i) maintain the hydrological variation of wetlands and (ii) to reduce rates of eutrophication. The second is to reverse the process by re-establishing the full set of conditions shown in Figure 12.10; this will require re-establishing infertile conditions and high flood regimes. Re-establishing the full array of wetland types within a landscape leads naturally to the process of wetland restoration.

Function

If a landscape contains a full array of wetland types, from raised bog to floodplain forest, it is reasonable to assume, at least as a first approximation, that most functions are being performed. The actual rate of performance of each function could be calculated by determining function on a square metre basis, and then multiplying by the area of that wetland type in an area of landscape. This more utilitarian perspective concentrates upon the functions themselves rather than the processes within the community that carries out the function (e.g. Sather and Smith 1984; Adamus *et al.* 1987; Bardecki *et al.* 1989; Larson 1990; Denny 1995). Recall the 14 functions listed in Table 1.7. The magnitude of each of these functions will depend upon the particular environmental factors and ecological communities associated with each wetland. The first principle could therefore be rephrased to state 'The functions performed by any wetland are controlled by multiple environmental factors acting simultaneously.' Most such relationships are not yet well quantified, although exceptions include the relationship between wetland area and shrimp production (Turner 1977), biomass and plant diversity (Moore *et al.* 1989), or water table and methane flux (Gorham 1991). Determining such quantitative relationships is an important priority in wetland ecology; far too many studies report these functions for a single wetland rather than seeking general empirical relationships between basic properties and level of function.

Even when quantitative relationships are established, many other secondary factors can act upon them (Table 12.2), and in theory, each of these needs to be examined independently. In practice, only a few may be important in any situation. One starts with the problem of 18 functions each of which might be controlled by, say, four ecological properties (e.g. duration of flooding, rate of nitrogen loading, intensity of grazing and rate of burial). The 18 multiple regression equations linking functions to predictors would then have to be adjusted for the action of the attributes in Table 12.2. Although this prospect may seem daunting, it is still far simpler than the species-by-species, or wetland-by-wetland, approach that dominates so much current research activity. Further, returning to Figure 12.10, the impacts even of altered hydrology and increased fertility upon many basic functions are still unknown.

One way of summarizing human impacts on these functions of wetland ecosystems is to apply the framework of stressors and responses (e.g. Odum 1985; Freedman 1995). In the preceding chapters we have seen many environmental factors that can change the function or species composition of wetlands. These have ranged from alterations in hydrology through eutrophication to over-hunting of alligators. Each of these human alterations can be considered a stressor, that is, 'an environmental influence that causes measurable ecological detriment or change' (Freedman 1995). For each stressor (e.g. Table 12.3), we could list the expected changes in wetland function or structure. For example, increasing fertility will lead to increases in wetland production and biomass, but a probable reduction in species diversity. Comparisons of many ecosystems reveal a suite of responses that generally result from human stressors (Table 12.4).

In conclusion, even if we successfully designate important wetlands for legal protection, they must still be managed appropriately. This requires action by two quite different groups of people: regulators and managers. Regulators may fail to take the necessary steps to protect the functions of our designated wetlands. Maguire (1991) suggests that most scientists are 'too conservative in speech and action', leading to confusion among regulators about the true risks of action vs. non-action (Figure 12.12). Maguire challenges scientists to consider the risks associated with non-action. Managers may be even more dangerous than regulators; where regulators may fail to act, managers may fail to restrain their action. Over

Table 12.2. *Predictors of wetland functional value*

1. Contiguity
2. Constriction
3. Shape of basin
4. Fetch and exposure
5. Basin surface
6. Wetland surface area
7. Basin area/watershed area ratio
8. Basin area/subwatershed area ratio
9. Location in watershed
10. Stream order
11. Gradient of subwatershed
12. Gradient of tributaries
13. Gradient of basin
14. Perched condition
15. Land cover of subwatershed
16. Land cover trends
17. Soils of subwatershed
18. Lithologic diversity
19. Delta environment
20. Evaporation and precipitation balance
21. Wetland system
22. Vegetation form
23. Substrate type
24. Salinity and conductivity
25. pH
26. Hydroperiod
27. Flooding duration and extent
28. Artificial water level fluctuations
29. Natural water level fluctuations
30. Tidal range
31. Scouring
32. Flow velocity
33. Water depth (maximum)
34. Water depth (minimum)
35. Width
36. Oxygenation of sediments
37. Morphology of wetland
38. Flow blockage
39. Basin alterations
40. Pool/riffle ratio
41. Basin vegetation density
42. Wetland vegetation density
43. Sheet vs. Channel flow
44. Wetland water edge
45. Gradient of edge
46. Shoreline vegetation density
47. Shoreline soils
48. Disturbance
49. Plants: form richness
50. Plants: waterfowl value
51. Plants: anchoring value
52. Plants: productivity
53. Invertebrate density: freshwater
54. Invertebrate density: tidal flat
55. Shore erosion measurements
56. Ground water measurements
57. Suspended solids
58. Alkalinity
59. Eutrophic condition
60. Water quality correlates
61. Water quality anomalies
62. Water temperature anomalies
63. Bottom water temperature
64. Dissolved oxygen
65. Underlying strata
66. Discharge differential
67. Turbidity and suspended solids differential
68. Nutrient differential
69. Recharge effectiveness
70. Discharge effectiveness
71. Flood storage effectiveness
72. Shoreline anchoring opportunity
73. Shoreline anchoring effectiveness
74. Sediment trapping opportunity
75. Sediment trapping effectiveness

Source: From Adamus and Stockwell (1983) in Sather *et al.* (1990).

Table 12.3. *Stressors potentially occurring in wetlands*

Stressor
Enrichment/eutrophication
Organic loading and reduced dissolved oxygen
Contaminant toxicity
Acidification
Salinization
Sedimentation/burial
Turbidity/shade
Vegetation removal
Thermal alteration
Dehydration
Inundation
Fragmentation of habitat
Other human presence

Source: After Adamus (1992).

my short career as a biologist, I have seen fens dyked and flooded for enhancing waterfowl production, rare wetland plant communities flooded to maintain stocks of exotic sport fish and infertile watersheds fertilized in order to enhance waterfowl production. These are examples of the misapplication of ecological principles, and we shall have to remain on guard for them continually.

Problems and prospects of reserve systems

The importance of large areas, and interconnected reserve systems, is reinforced by the problems that face managers in trying to protect isolated fragments of habitat. These problems are particularly severe for Europeans, where there has been a long history of human modification of the landscape. Consider The Fens of Huntingdonshire in eastern England, adjacent to The Wash along the coast with the North Sea (Sheail and Wells 1983). These wetlands extend inland some 60 km from the ocean. The coastal areas are tidal marsh; in the more upland areas, the depth and character of the peat reflects differences in local drainage; in-between, along the River Nene, there are more alkaline conditions, and a series of lakes created by the meandering river. The largest lake is Whittlesea Mere, which in 1697, was said to be 3 miles broad and 6 miles long; most of this is less than 2 m deep. The number and area of lakes may have declined since the medieval period onwards, and in 1826,

Table 12.4. *Changes that may be expected in ecosystems upon an intensification of stress*

Energetics

1. Community respiration increases.
2. Production/respiration becomes unbalanced (i.e. P/R becomes greater than or less than 1).
3. Production/biomass and respiration/biomass (i.e. maintenance to biomass structure) ratios increase.
4. Importance of auxiliary energy increases.
5. Exported or unused primary production increases.

Nutrient cycling

6. Nutrient turnover increases.
7. Horizontal transport increases and vertical cycling of nutrients decreases.
8. Nutrient loss increases (system becomes more 'leaky').

Community structure

9. Proportion of *r*-strategists increases.
10. Size of organisms decreases.
11. Life spans of organisms or parts (leaves, for example) decrease.
12. Food chains shorten because of reduced energy flow at higher trophic levels and/or greater sensitivity of predators to stress.
13. Species diversity decreases and dominance increases; if original diversity is low, the reverse may occur.
14. General biotic impoverishment by extirpation of sensitive species and increased dominance by a few tolerant species.

General system-level trends

15. Ecosystem becomes more open (i.e. inputs and outputs become more important as internal cycling is reduced).
16. Successional trends reverse.
17. Efficiency of resource use decreases.
18. Parasitism and other negative interactions increase, and mutualism and other positive interactions decrease.
19. Functional properties (such as community metabolism) may be more robust (resistant to stressors) than species composition and other structural properties. In systems dominated by long-lived perennial plants (e.g. forests), the reverse may be true.

Source: After Freedman (1995).

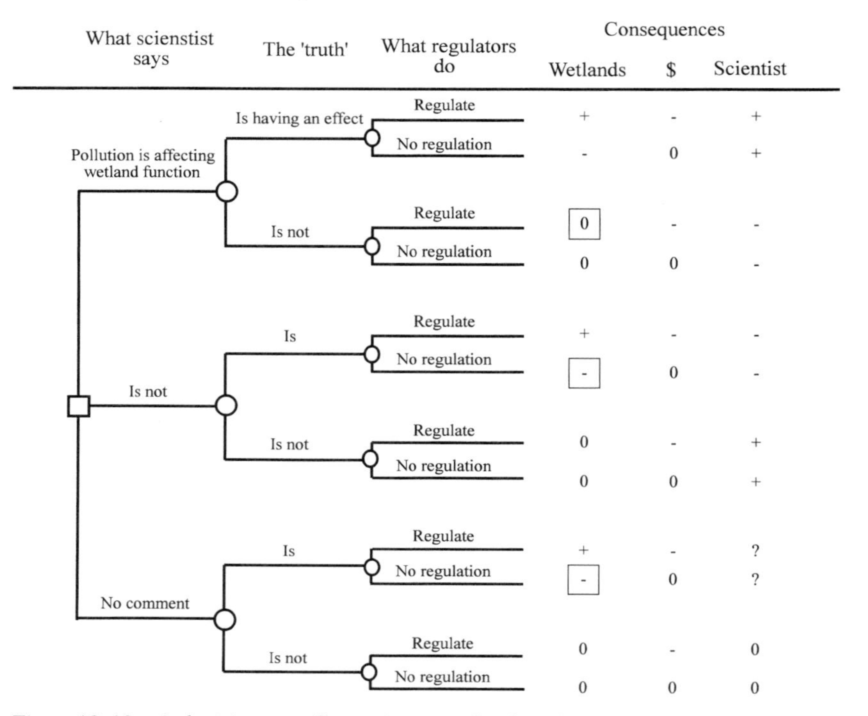

Figure 12.12 A decision tree illustrating a wetland ecologist's comments on pollution effects, possible mistakes, regulatory actions, environmental and personal consequences. + = beneficial change, − = detrimental change, 0 = status quo, boxes = consequences of regulators acting upon incorrect scientific opinion or lack thereof. If we say it is and it isn't, there are zero consequences for wetlands; in contrast, if we say it isn't and it is, there are negative consequences. The same is true if we say nothing (after Maquire 1991).

Whittlesea Mere dried up completely during one dry summer. The Domesday survey of 1086 outlined the various rights or privileges on the lakes, and later documents also drew attention to value for fish production and hunting for waterfowl. Settlement along the fen edge included both fenland and arable upland and pasture, and in the 1600s records from the manorial court at Upwood reveal attempts to regulate land use, including rights of grazing in the fen and of excavating turf for fuel. Farmers were forbidden from digging over '10000 cesses of turf from the fen in one year'.

By the 1700s the number and variety of species had begun to fall. Waterfowl were perhaps over-hunted, distinctive butterflies may have been over-collected, but habitat destruction was probably most important. In

1844 an Act of Parliament combined the drainage of the Huntingdonshire fen with the improvement of watercourses further downstream. It was not until 1850 that the last of the meres, Whittlesea Mere, could be drained. Both windmills and steam scoop wheels were used for further draining fens, and in 1851 it was the first site in England where a centrifugal pump from The Netherlands was used. In the 1890s, Pell remarked 'all is gone – reeds, sedges, the glittering water, the butterflies, the gypsies, the bitterns, the wild fowl, and its place . . . a dreary flat of black arable land, with hardly a jack snipe to give it a charm and characteristic attraction'.

The first attempt at preservation was made in 1910, with the purchase of 137 ha of the Woodwalton Fen. At the same time, the water table was falling, in part from peat cutting. In some cases, the peat cutting provided disturbed sites for unusual plants to colonize, but extensive cutting removed seed sources and also changed the nature of the peat. Woody plants began to invade the fen; some trees had established on the nature reserve as early as the 1860s, and by 1931 most of the reserve was covered by 'dense impenetrable thickets of sallow bushes'. It was recognized that this was the result not only of ending peat-cutting, but of the falling water table on the reserve. Drainage ditches were partially blocked to maintain water levels during times of drought, and in 1935 a portable pump was used to raise water from neighbouring drains into the reserve during dry weather. It would, of course, be possible to cut out the invading woody plants, but what would be the point if the fen was dried out? Drainage ditches were deepened further after the Second World War, and in 1972 a clay-cored bank was constructed on the northern and western perimeter of the reserve so as to reduce the amount of water percolating out of the reserve and into drainage ditches. A photograph in Sheail and Wells shows a small rectangular plot of land, largely wooded, forlornly surrounded by drainage ditches and agricultural land. The Holme Fen National Nature Reserve, 256 ha set aside in 1952, is some 3 km away. It has some species associated with undrained fenland such as *Calluna vulgaris*, *Erica tetralix* and *Cladium mariscus*, but it too is being invaded by scrub and trees as the water table falls.

Some 100 kilometres to the east, a similar discouraging history of habitat loss has been described for the Norfolk Broadlands (Moss 1983, 1984). The valleys here have a complex history, peat formation alternating with clay deposits from rising sea levels, with the latter influenced as well by a sand spit in Great Yarmouth intermittently serving as a barrier to the sea. In places, peat has accumulated up to 4 m in depth. These wetlands had *Phragmites* marsh alternating with woodland swamps dominated by alder and willow for several millennia. Peat accumulation in the valleys tended to more or less compensate for rising sea levels, although periods of flooding and periods of accretion can be detected in sediments. Some 46 shallow lakes, or broads, were created by peat cutting between the ninth and fourteenth centuries AD. Drainage by wind pumps in the late eighteenth and early nineteenth century, combined with intensification of agriculture and sewage disposal in the twentieth century, reduced wetland area and caused both the rapid growth of emergent macrophytes and the loss of aquatic plants. The Norfolk Broads developed some of the highest total phosphorus concentrations recorded for world freshwater lakes (Moss 1983). Further, the coypu (*Myocaster coypus*), a large South American rodent, was introduced for fur farming about 1929; some escaped and, by the 1960s, there were estimated to be 200 000 animals. The inevitable results have again been decreased numbers of species and habitats remaining in the landscape (Figure 12.13).

These are just two of many possible examples of isolated reserves in changing landscapes. Other examples from this book have included the drainage of prairie potholes combined with falling water tables from irrigation, the construction of large dams on rivers such as the Volta and Amazon, the impacts of grazing and canals in the Pantanal, phosphorus-laden water entering the Everglades, atmospheric deposition of nitrogen in western European heathlands, removal of annual flooding with levees along the Rhine and the Vistula, and possible changes in fire frequency in peatlands with global warming. Such examples serve to re-emphasize many of Noss's guidelines, including the need for large reserves, with buffer strips, as part of an interconnected system. Sinclair *et al.* (1995) point out, however, the risk posed by change within reserves themselves. They provide other examples to show that, while protection reduces the rates of decline in habitats, they none-the-less continue to decline after protection, albeit at a slower rate. Therefore, managers must place added emphasis upon habitat renewal within reserve systems. In an undisturbed landscape, natural habitat loss tends to be balanced by disturbance and succession that renews different habitat types (recall White 1979 and Chapter 6), but many of these dynamic processes are being negatively affected. Since the area of wildlife habitat is still in decline at the global scale (Figure 12.14), the challenge for conservationists and managers is not only to set up reserve systems, but to ensure that, within each system, natural habitats continue to be renewed. This requires sufficiently large reserves for natural dynamics to occur, or else increasingly expensive intervention by managers to attempt to simulate these processes. Such

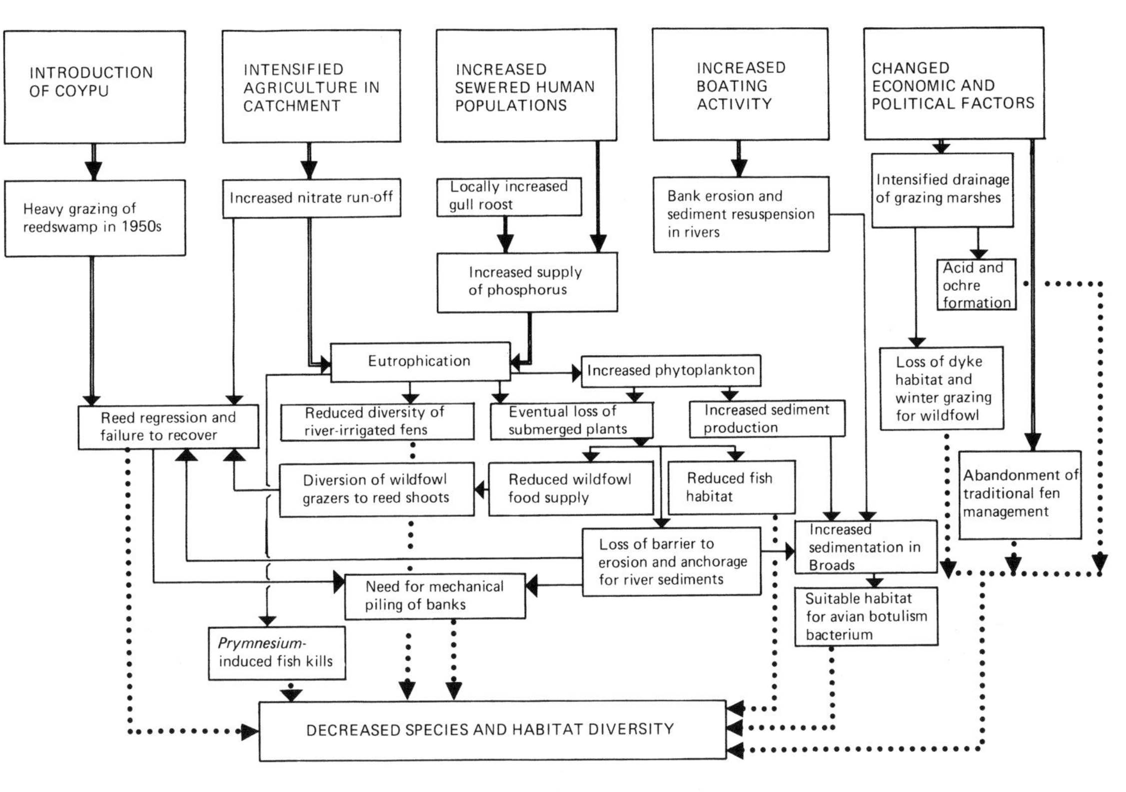

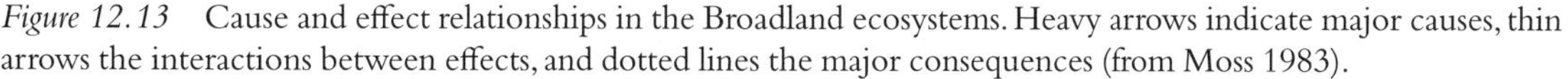
Figure 12.13 Cause and effect relationships in the Broadland ecosystems. Heavy arrows indicate major causes, thin arrows the interactions between effects, and dotted lines the major consequences (from Moss 1983).

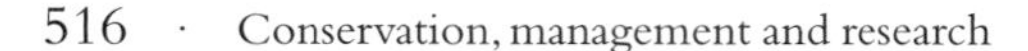

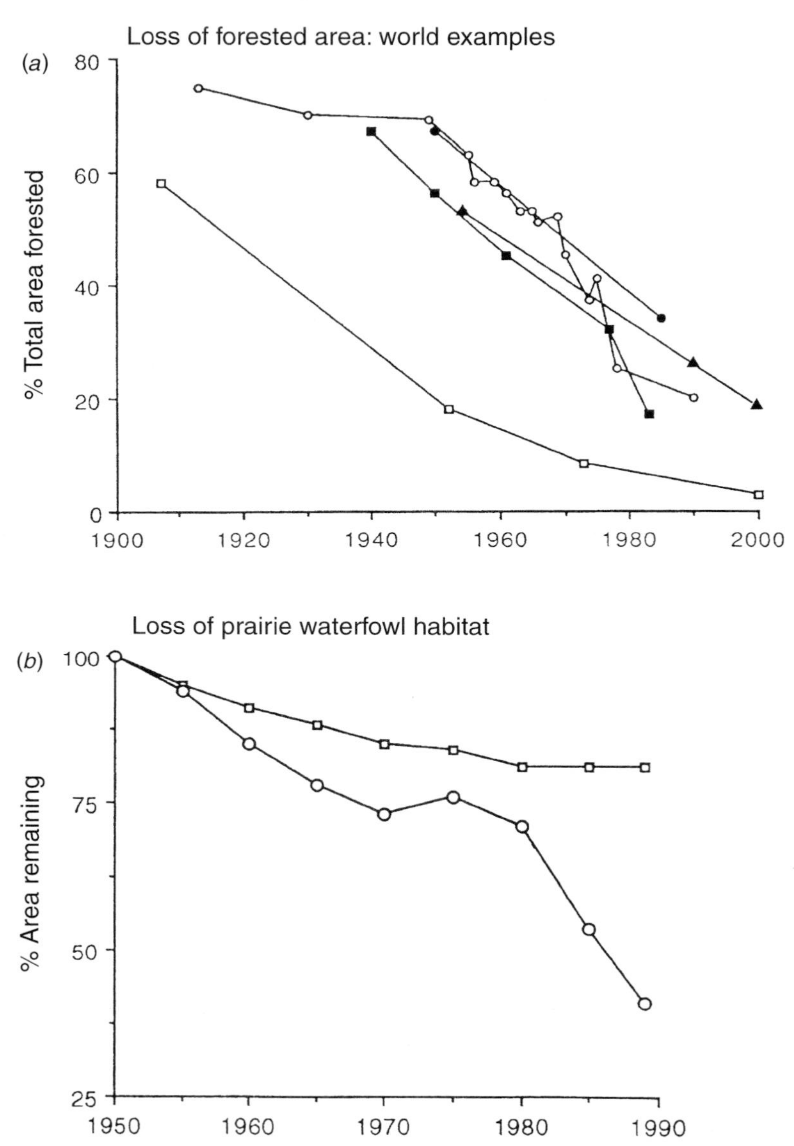

Figure 12.14 Examples of habitat loss. (*a*) Extent and loss of forested area shown as a proportion of total land area for each region. (□) Sao Paulo State, Brazil: total area: 229 450 km^2; (●) Madagascar: total area: 587 000 km^2; (○) Thailand: total area: 513 520 km^2; (■) Costa Rica: total area: 50 990 km^2; (▲) Vancouver Island, British Columbia: total area: 32 000 km^2. (*b*) Proportional loss of wetland habitat on Canadian prairies based on total wetland area in 1950: Manitoba (□) 1903 km^2; Alberta (○) 4064 km^2 (from Sinclair *et al.* 1995).

efforts might be possible with fire, or grazing, or even flooding, but certainly not for meander systems or deltas.

The commonly accepted goal of protecting 12% of a landscape in reserves (World Commission on Environment and Development 1987), is, Sinclair *et al.* (1995) observe, not based upon scientific criteria; it was an *ad hoc* calculation that was derived by assuming that since 4% of landscape was reserved at the time, a goal of three times this amount might be reasonable. 'There is a danger that such an *ad hoc* number will become a standard before we have any evidence that it is sufficient to protect biodiversity.' Noss (1995) therefore suggests that, after the first steps (mapping out a preliminary reserve network with core reserves, buffer zones and continuity), one should identify the species with the largest area requirements still extant in the region, and estimate the area needed to provide for both short-term and long-term viable populations of that species. A next step would be to identify the extirpated native species with the largest area requirements that could reasonably be reintroduced, and again estimate area for short-term and long-term survival. If the reserve systems are not sufficient to maintain long term viable populations of these species, plans must be made to enlarge the network or enhance connectivity within the system or to adjoining regions. Although Sinclair *et al.* use the term renewal to emphasize the balance between rates of loss and rates of creation of habitats, other words such as restoration, mitigation, and rehabilitation are already used to describe the process of recreating natural habitats. The need to recreate habitats, particularly when only isolated fragments remain, is the topic of the next section. The situation in the Huntingdonshire Fens (pp. 510–513), contrasted with guidelines provided by Sinclair *et al.* and Noss, illustrate the challenges to be faced in the coming decades.

Restoration

Designing a reserve system and managing it appropriately is a challenging mixture of basic and applied science. In Chapter 1, the second principle stated *to understand and manage wetlands we must determine the quantitative relationships between environmental factors and the properties of wetlands.* Since *wetlands are the product of many environmental factors acting simultaneously,* it follows that we manipulate wetlands by changing one or more of these factors, by changing flooding regimes, by reducing phosphorus in the water entering the wetlands, by reintroducing natural grazers, or by allowing fire. Each modification of an environmental factor is an act of management. Any management programme should only be undertaken with a specific goal in mind, and with an understanding of the known quantitative linkages that allow one to forecast the results of the manipulation. Noss (1995) emphasizes that all management should have a clearly articulated goal, because it is only when the goal is articulated that we can later determine whether or not the management has been successful. He begins with Leopold's 1949 essay on land ethics

> A thing is right when it tends to preserve the integrity, stability, and beauty of the biotic community. It is wrong when it tends otherwise.

Leopold did not explain what he meant by integrity, and although the word is increasingly used by managers, it is still poorly defined (e.g. Noss 1995; Woodley *et al.* 1993; Higgs 1997). Noss is of the opinion that the difficulty in defining integrity does not reduce its value; other terms like justice, freedom, love and democracy are also vague and slippery, and this has not kept scientists, philosophers and policy makers from thinking about them and being guided by their intent (Rolston 1994). Rather than enter this realm of discussion here, I will adopt the view that integrity has three essential components: (i) maintaining biological diversity, (ii) ensuring ecosystem persistence through time and (iii) maintaining performance of ecological functions. These are all relatively measurable and therefore operational, even if the term integrity is not. All three are also interrelated, in that if diversity declines, functions will naturally be impaired. Similarly, the continued performance of functions is probably essential for persistence. All of these aspects of integrity are equally assumed in the definition of 'ecosystem management' (e.g. Christensen *et al.* 1996; Grumbine 1997). The proliferation of terms for wise management should not distract us from setting clear goals and from ensuring that the best possible science is brought to bear for achievement of those goals.

Managers will rarely inherit a watershed with entirely intact and pristine wetland ecosystems; in most cases there will already have been considerable loss in wetland area, reductions in function and declines in biological diversity. Two of the principal challenges facing managers will therefore be (i) deciding to what degree it is possible to reverse these undesirable changes and (ii) implementing the programmes to make these changes. There is a variety of technical terms for the more specific options open to managers (Figure 12.15) depending upon the objectives of management for a particular site. Here I will mention only two: restoration and mitigation. The objective of restoration is the recreation of

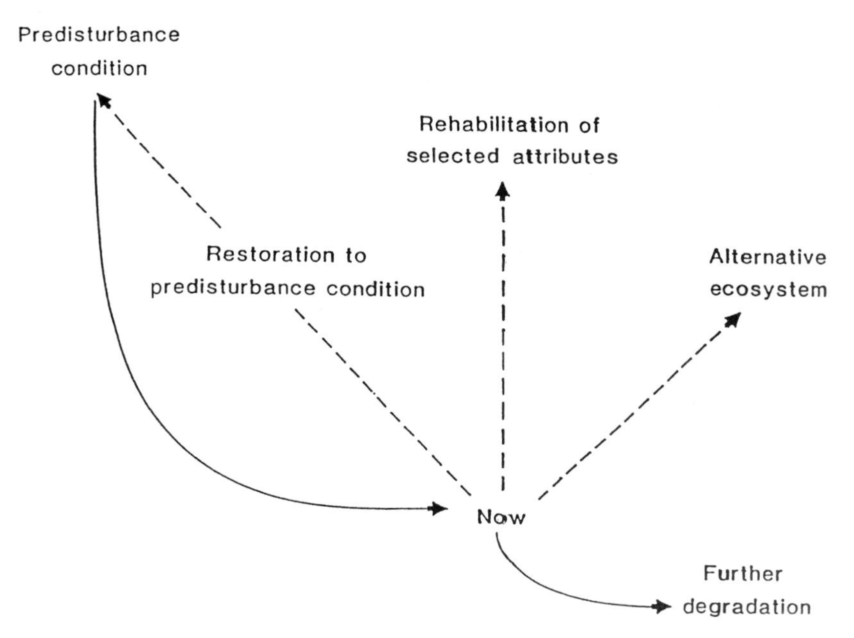

Figure 12.15 Possible goals for the management of disturbed ecosystems (from Cairns 1989 after Magnuson *et al.* 1980).

wetlands that have either been degraded or entirely lost from the landscape. Mitigation is a more specialized term, primarily American, for the construction of wetlands to replace the functions of wetlands that have been developed or damaged.

Ecosystem restoration (e.g. Lewis 1982; Jordan *et al.* 1987; Galatowitsch and van der Valk 1994) challenges managers to recreate ecosystems and communities that have been entirely lost or heavily degraded, and the field overlaps conceptually with 'ecosystem management' (e.g. Christensen *et al.* 1996; Grumbine 1997), the 'maintenance of ecological integrity' (e.g., Woodley *et al.* 1993; Noss 1995), 'rehabilitation' (Cairns 1988) and 'reconstruction' (Saunders *et al.* 1993). Whatever the label we use, the principles remain similar: the use of indigenous species and natural environmental factors to re-establish communities that can persist without continued human interference. Chapter 10 was entirely devoted to restoration principles, so here we will examine only a few more examples that follow naturally from the new material in Chapter 12.

Although the planting of mangrove and saltmarsh species to stabilize mudflats has been practised for over a century (Lewis 1982), the last few decades have seen a great deal more such activity, particularly in wetlands (Cairns 1980; Kusler *et al.* 1995). An entirely haphazard selection of recent examples include planting saltmarsh species to re-establish coastal marshes in China (Chung 1982), planting Baldcypress to recreate swamps in Louisiana (Myers 1995), mowing to re-establish species-rich meadows in Germany (Müller *et al.* 1992), peat cutting to maintain diversity in heathlands in Belgium (Sansen and Koedam 1996), and manipulation of flooding and grazing regimes in dyked coastal areas to maintain desired vegetation types (van de Rijt *et al.* 1996). An overview of wetland restoration experiments in central European lowlands (Pfadenhauer and Klotzli 1996) explored manipulations such as re-wetting, nutrient depletion of eutrophic sites and re-establishment of indigenous species. Their conclusion was 'Preliminary results indicate that the reconstruction of the former state (regeneration) is impossible within reasonable time spans' (note their use of the word regeneration rather than restoration). Their emphasis was upon fen and bog species, where the accumulation of peat is a crucial factor. Koerselman and Verhoeven (1995), however, suggest that one can create both N and P limitation in fens, thereby restoring their co-limitation by these nutrients; in order to accomplish this, however, one must reduce external nutrient inputs.

A principal distinction between North American and European perspectives on restoration and ecosystem management is their different biological reference points: there is a tendency for Europeans to set the goal of maintaining a familiar historical landscape created by humans (e.g. species rich pastures typical of the eighteenth and nineteenth century), whereas the North American tendency is to set the goal of recreating the ecosystems assumed to have been present before humans of European ancestry altered the landscape. Further, Europeans accept intensive management (e.g. cattle grazing, peat cutting, mowing), whereas North Americans tend to prefer natural controlling factors (erosion, fire, and flooding) One can hope for increasing overlap between these two views of management; in densely populated areas of Asia and North America, there may have to be increasing use of European management experience in order to maintain small examples of desired ecosystem types. Equally, Europeans may begin to value the possibility of managing larger areas of landscape for their original composition rather than for their cultural familiarity.

Consider one extreme case of the challenge posed by restoration: the re-establishment of prairie pothole vegetation in areas that have been drained and sown to row crops for at least 25 years. Galatowitsch and van der Valk (1996) examined the success of an entire set of such restoration projects. From a total of 62 restoration projects, ten were selected on hydric soils that had been tile drained and completely cultivated for corn and soybeans for 25–75 years. That is, in each of these sites, restoration involved recreation of a wetland where it had been absent for decades and where there was little reason to expect any residual seed bank. Given the importance of hydrology as a controlling factor or filter in the establishment of wetland communities, it might seem reasonable, at least as a first approximation, to assume that appropriate hydrology alone will re-establish wetlands. Perhaps steps as simple as plugging drainage ditches or removing tiles will suffice. In a comparison of ten restored wetlands with ten adjacent natural wetlands, the natural wetlands had a mean of 46 species compared with a mean of only 27 for the restored wetlands. Further, there were differences among functional groups; the restored sites had more species of submersed aquatics, but fewer species of sedge meadows. The seed banks of the communities also differed; natural sites had nearly twice as many species (15 vs. 8) and nearly twice as high a density of buried seeds (7300 vs. 3000 m^{-2}). Submersed aquatics, wet prairie and wet meadow species were all absent from the seed banks of the restored wetlands. Even the zonation patterns of individual species differed between the restored and natural sites. Galatowitsch and van der Valk propose the term 'efficient-community hypothesis' for the view that vegetation will re-establish itself rapidly after hydrology has been restored, and reject this hypothesis as a reasonable basis for restoration in prairie potholes. Part of the explanation may lie with rates of loss of wetland seeds; both seed densities and species richness decline with the duration a wetland has been drained. After 50 years, seed densities are <1000 m^{-2} (compared with 3000–7000 m^{-2} in natural sites) and richness is three species (compared with 12 in natural sites) (van der Valk *et al.* 1992).

Three years, is however, a relatively short period by which to judge restoration and species richness was increasing with time. Further, the restored sites they examined were an extreme situation; in many watersheds nuclei of remnant wetlands may remain as a source of propagules. Finally, extensive fields of soybeans and corn may be particularly effective barriers between wetlands, thereby minimizing natural dispersal from native sites. A strength of this study was the proper sampling design, reporting on two comparable sets of wetlands, rather than on the description of a single case. There is also evidence that aquatic invertebrates are under represented in restored wetlands (Galatowitsch and van der Valk

1994). Given the poor dispersal abilities of some species, Galatowitsch and van der Valk conclude that, at least in prairie potholes amidst farmland, species in guilds with poor dispersal capabilities will have to be reintroduced during restoration in order to re-establish the original ecological communities.

In a contrasting study in eastern North America, Padgett and Crow (1993) compared six natural wetlands with six created wetlands, and found a mean of 33 species per wetland, irrespective of the wetland history. Unlike the above study, however, several of the restored wetlands had been heavily planted, three of the sites having collectively received 44 500 tubers and 255 pounds of seeds of herbaceous species. Some species, such as *Alisma triviale*, *Carex scoparia*, *Eleocharis acicularis* and *Juncus effusus* seemed particularly common in restored wetlands, but overall both groups were quite heterogeneous. This work tends to suggest therefore, that restoration can be relatively successful if assisted by planting and seeding, although Padgett and Crow (1993, 1994) also warn about the need to carefully choose both the species used and their geographic origin. At present, simple criteria such as seed availability may play too large a role in determining the vegetation type that is created.

In general, apart from such specific examples as the two above, rather little is known about the success of wetland restoration projects. Even less is known about the cumulative effects – the degree to which sets of restored wetlands mimic the original communities and functions that were once present in landscapes. Increasing the area of wetlands being restored increases the risk that restoration may itself be changing the characteristics of wetlands of entire regions. The following example from the United States of America illustrates this potential problem.

In the United States of America there is a 'no-net-loss' policy for wetlands; damage to wetlands must be avoided, but if damage is necessary, it must be mitigated, which means that compensatory wetlands must be constructed to equal, or exceed, the functions that were performed by the damaged site. More precisely, mitigation is defined as 'the avoidance, minimization, rectification, and reduction or elimination of negative impacts or compensation by replacement or substitution' (Office of Technology Assessment, in Zedler 1996). Zedler simplifies this with the observation that successful mitigation means 'providing a habitat that is functionally equivalent to the one that will be lost', and assumes that ecosystems can be made to order. As a first step, Bedford (1996) observes the importance of ensuring that the replacement wetlands are hydrologically equivalent to the lost wetlands, since hydrology provides the template for the development of the wetland on a site. 'Any attempt to replace wetlands with ecologically or hydrologically equivalent types must be based on an understanding of the relationship of individual wetlands to the landscape.' The three key hydrological variables, she asserts are (i) relative importance of various water sources, (ii) mineral element and nutrient content and (iii) spatial and temporal dynamics. This comes close to the first three factors used in this book: hydrology, fertility and disturbance.

Mitigation, while well intended, may itself change the nature of wetlands in a landscape. Surveys of the kinds of wetlands being constructed for mitigation (Figure 12.16) suggest that shallow water wetlands along rivers are relatively easy to create, whereas wet meadows (lacustrine fringe, riverine fringe) are not. This problem is not necessarily restricted to mitigation: restoration could equally lead to such problems if the original distribution of wetland types and controlling factors in the landscape is not used as a reference point against which to judge the targets and results of individual projects.

The construction of dykes for restoration may also have unexpected negative consequences. In Ohio, for example, 90% of the Lake Erie wetlands have been drained, and 84% of the remaining ones are dyked and manipulated for waterfowl production. As with the construction of levees along rivers, dykes alter the hydrological regime of a wetland. In wetlands along lakes, water levels may fluctuate by many metres over several decades, producing a rich array of vegetation types and successional stages (Keddy and Reznicek 1986; Reznicek and Catling 1989; Charlton and Hilts 1989). Once such wetlands are dyked, and subjected to small but regular annual cycles, many of these vegetation types may be lost. Further, these wetlands are typically drained in the spring and refilled in the autumn (Johnson *et al.* 1977), which differs from natural patterns. Dykes on shoreline wetlands may prevent fish from migrating in and out of spawning or feeding areas. Johnson *et al.* (1977) sampled fish over one summer, and found that a dyked marsh had only 23 species compared with 40 in the undyked area. Further, some species in the undyked area had significantly higher body condition indices. They noted that when water levels are reduced in the spring, the wetland is dried at the very time when fish would benefit most from expanded areas for feeding and breeding.

Dykes are also commonly built along marine coasts; in Louisiana alone, some 2000 km^2 of wetland are manipulated by water control structures. Interference with tidal flows can cause marked changes in salinity (recall pp. 187–188). A common manipulation in Louisiana is a fixed-crest weir

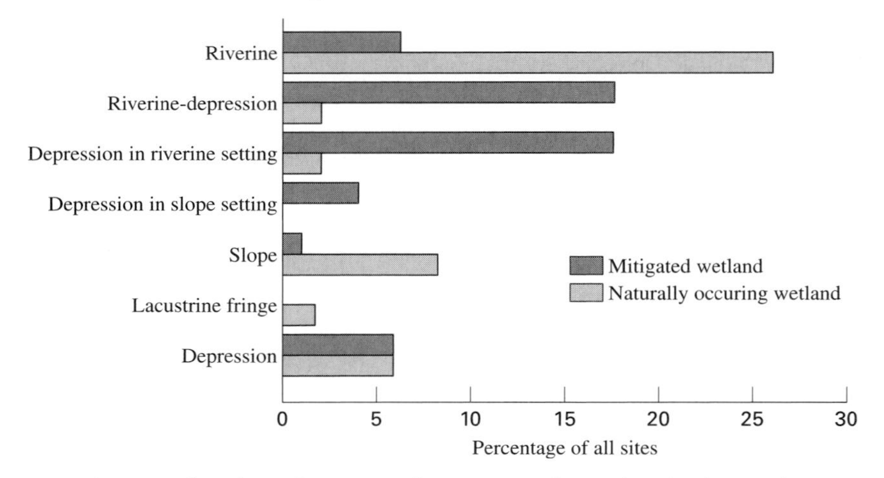

Figure 12.16 The relative frequency of seven types of natural wetland in north-western United States compared with their occurrence in created or managed wetlands. Note that the managed or created wetlands have an over-representation of two types of wetlands, 'riverine-depression' and 'depression in riverine setting'. Other wetland types, such as 'slope wetlands' and 'lacustrine fringe' wetlands, are rarely recreated (unpublished figure courtesy of Mary Kentula and United States EPA).

built on tidal inlets, the top being set at 15 cm below the soil level, thereby excluding all but the highest tides from reaching the marsh. Since drainage also builds up behind the weir, these marshes are termed semi-impounded. Rogers *et al.* (1992) found this kind of impoundment to be associated with decreased salinity, and significant changes in the abundance of aquatic animals. The unimpounded area had 32000 individuals from 32 taxa, whereas the area with a weir had more than 76000 individuals representing 26 taxa. Resident species such as shrimp increased to a total catch of more than 48000 individuals compared with just above 14000 in unimpounded areas; similarly Least Killifish, Western Mosquitofish and Golden Topminnows were all significantly more common. Three transient species that are typically marine, Gulf Menhaden, Blue Crab and Striped Mullet were significantly less common in the impoundment. Rogers *et al.* attribute these differences to two main causes: the weir being a barrier to the entry of marine species, and the submerged aquatic vegetation being more productive behind the weir.

Most published studies of impoundments, however, compare a managed with an unmanaged area, which does not provide proper replication (Hurlbert 1984). Future studies need to simultaneously compare a set of control sites with a set of treatment sites.

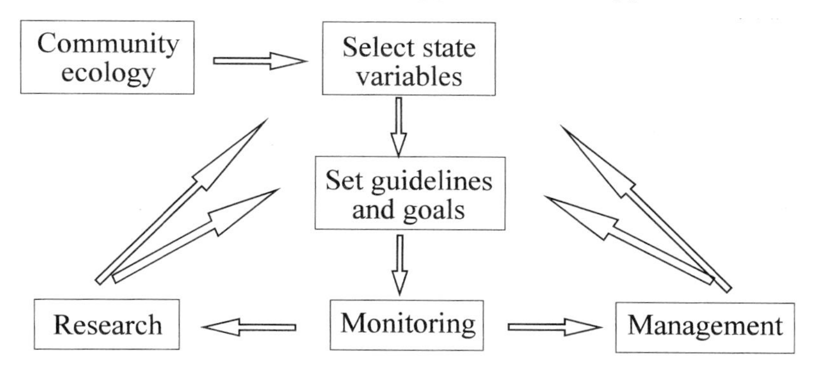

Figure 12.17 The only way to measure management success is to monitor a series of state variables (indicators) (Keddy *et al.* 1993).

Our task at hand is therefore clear: increased protection for wetland habitats around the world, better scientific management of them, and restoration of wetlands in areas where they have been lost. Two further tasks remain. The first is the selection of indicators to measure our performance, the second is the systematic application of scientific principles in order to solve practical problems. Let us consider them in turn.

Indicators: setting goals and measuring performance

In seeking to recreate, restore or simply manipulate natural wetlands, we need some procedure to measure success. This procedure must be based upon credible scientific criteria. The number of acres managed, or the amount of money spent, means nothing if the wetlands involved have been damaged by our management.

This is where indicators are helpful, indeed essential (e.g. Keddy 1991a; McKenzie *et al.* 1992; Adamus 1992, 1996; Woodley *et al.* 1993). Indicators provide an instrument panel for wetland management. As Tansley said in 1914 (long before the advent of computer controlled recording devices): 'The mere taking of an instrument in the field and recording of observations . . . is no guarantee of scientific results.' At present, we have difficulty in choosing indicators because ecology is not well enough developed as a science to tell us what the essential properties of wetlands are. We can, however, outline a process that begins with some fairly pragmatic decisions (Figure 12.17). It has three key steps: selecting the appropriate state variables, setting critical limits, and testing the indicators by following them through monitoring programs.

Selecting state variables

What properties of communities should we measure to guide our decision making? In the past, indicators have been developed haphazardly, often reflecting the interests of specific user groups and value systems, rather than according to more broad scale ecological criteria. This history is reflected in the kinds of databases we currently have. The following criteria might guide our efforts to select indicators (Keddy 1991a; Keddy *et al.* 1993).

(i) Ecologically meaningful: closely related to maintenance of essential environmental processes (e.g. water level fluctuations) and ecosystem functions (e.g. primary production).
(ii) Macro-scale: measuring the state of entire systems or key processes rather than small pieces or selected species.
(iii) Pragmatic: guided by measurable or empirical attributes of systems rather than conceptual or theoretical concepts and notions.
(iv) Sensitive: quick response to stresses and perturbations, to minimize lag and give maximum response times for decision makers.
(v) Simple: easy to measure, therefore inexpensive.

With these criteria in mind, there are at least three ways in which we might proceed with selecting indicators.

First, we might begin with measuring abiotic environmental factors that maintain and control the community type. Some of these key factors were listed above, and we have already examined how selected biota respond to filters such as dissolved oxygen, salinity and flooding. At one time, physical factors alone were monitored. Cairns *et al.* (1992) recalls that, in 1948, 'most pollution assessment was carried out by what were then called sanitary engineers (waste treatment specialists) and chemists. The accepted procedure was that, if certain limited chemical/physical conditions were met . . . there was little or no need to examine the biota' (p. ix).

More recently, biota have received increasing attention. The second approach looks at the consequences of the abiotic factors upon the biota, that is, the characteristics of the ecosystems that these environmental factors create. Biomass, diversity and life forms are all candidate ecosystem properties that have been examined in this book.

A third approach is to examine systems that have already been damaged or stressed to find indicators that appear sensitive to that damage or stress (Woodwell and Whittaker 1968; Rapport 1989; Rapport *et al.*

	Community types			
Indicators	Swamp	Marsh	Bog	Fen

Figure 12.18 We need a list of indicators characteristic of healthy wetland ecosystems in order to set goals for management or restoration (Keddy *et al.* 1993).

1985; Odum 1985; Schindler 1987; Freedman 1995). Ecosystems that are under stress display certain similar responses (Table 12.4); these include increased community respiration, increased nutrient loss and decreased diversity of organisms.

Setting critical limits

Once indicators are selected, the next step is to set acceptable and desirable levels for them. Each indicator would have a range of values specified, one limit being the tolerable level and the other being the desirable. If the system moved outside this specified range, managers would know that remedial action was needed to restore integrity. For example, one might set a goal of zero exotics as desirable for a rare wetland vegetation type, and two exotics as being tolerable. If more exotics than this invaded the site, one would investigate the reasons for the invasion, and then take the appropriate remedial action.

In the long run, managers need a handbook which would: (i) classify major ecosystem types, and (ii) specify for each wetland type the appropriate indicators with their desirable and acceptable levels. Figure 12.18 shows the structure such a guidebook might have, with a matrix of indicators by vegetation types. Some indicators (e.g. exotics) might have similar levels for all wetland types, whereas others (e.g. amphibian biomass) might have different critical limits for each wetland or habitat type.

Let us consider plant diversity as a possible indicator, and explore how critical limits might be set. Returning to Figure 3.13, one can see that eutrophication has a marked effect upon species richness within a quadrat, as well as the number of rare species. If the objective is to manage wetlands to protect nationally rare or endangered species, clearly there is a cut-off at 100 g/0.25m^2 above which many rare plants cannot occur. If the objective is to maintain local species richness, then biomass levels up to 200 g/0.25 m^2 are permissible. Thus, as a first approximation, we could set three ranges of biomass for protecting species richness: <100, 100–200 and >200 g/0.25 m^2. The critical limits could be set to maintain either the presence of specific endangered species (the lowest category) or mean diversity in quadrats (the middle category); in either case, the higher category is undesirable. In this example, biomass alone would suffice as a predictor. In terms of causation, however, increases in biomass are in many cases attributable to eutrophication, so it might be equally appropriate to monitor nutrient levels in the water. In this example, however, we assume that biomass is responding to nutrient levels, so biomass may be an inexpensive indicator of fertility. In practice, more environmental agencies are monitoring water quality than biomass, so there may be an element of historical momentum in the choice of indicator.

In their study of wetland plants in prairie potholes, Galtowitsch and van der Valk (1994) set critical limits on the number of species in six different species groups. Let us consider the category of wet meadows, 'perennial plants that do not occur in areas flooded more than one to two months in the spring. Most of these plants are most abundant in sedge meadows but are often common in wet prairies.' Three categories are recognized: depauperate (1–15), typical (16–30) and exceptional (31 +). In contrast, for emergent species that can withstand several years of flooding, the three categories are quite different: depauperate (1–2), typical (3–4) and exceptional (5 +). If values for many species groups fall in the depauperate category, Galatowitsch and van der Valk suggest that management action such as planting or seeding may be necessary. In this case, of course, the values are being set to measure the state of the vegetation. Ideally, in setting up a monitoring scheme, the vegetation might be used as an indicator of other ecosystem properties to achieve greater efficiency.

Monitoring

Selecting indicators and setting critical limits is obviously part of an evolutionary process. As scientific knowledge of community ecology and

Table 12.5. *Some candidate indicators for wetland restoration*

Production and persistence
Primary production
Decomposition
Nutrient loading
Sedimentation
Biological diversity and redundancy
Exotic species
Rare species
Functional groups
Production for humans
Abundance of commercial species
Harvest rates of commercial species

experience with ecosystem management increases, we need to remain open to changing both indicators and critical limits. Indicators would therefore evolve to reflect our constantly improving knowledge. It is therefore essential to monitor as projects occur, and then to use the information from monitoring to revise criteria for future projects (e.g. Holling 1978; Beanland and Duinker 1983; Keddy 1991a; Adamus 1992; Noss 1995; Rosenberg *et al.* 1995).

Although the blanks in Figure 12.18 show that we lack a list of standard wetland indicators, a candidate list would likely include three categories (Table 12.5). At present, we are often selecting indicators on a case-by-case basis. While there may indeed need to be a category for local, or site-specific, indicators, we should rapidly decide on a standard series. Not only would a standard list be more efficient, but it would allow us to compare and contrast among wetlands, and among management strategies.

In a frequently overlooked book chapter, Bloom (1980) has described a general procedure for setting critical limits and quantifying the recovery of communities after disturbance. He assumes that a preperturbation set of samples is available as a reference point. Based upon the species composition of these samples, one can define a 95% confidence interval envelope around the centroid of the original community (Figure 12.19). While Bloom used principal co-ordinate analysis to define the space occupied by these samples, any number of other

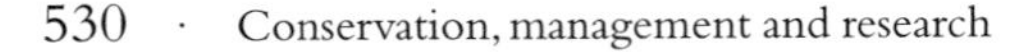

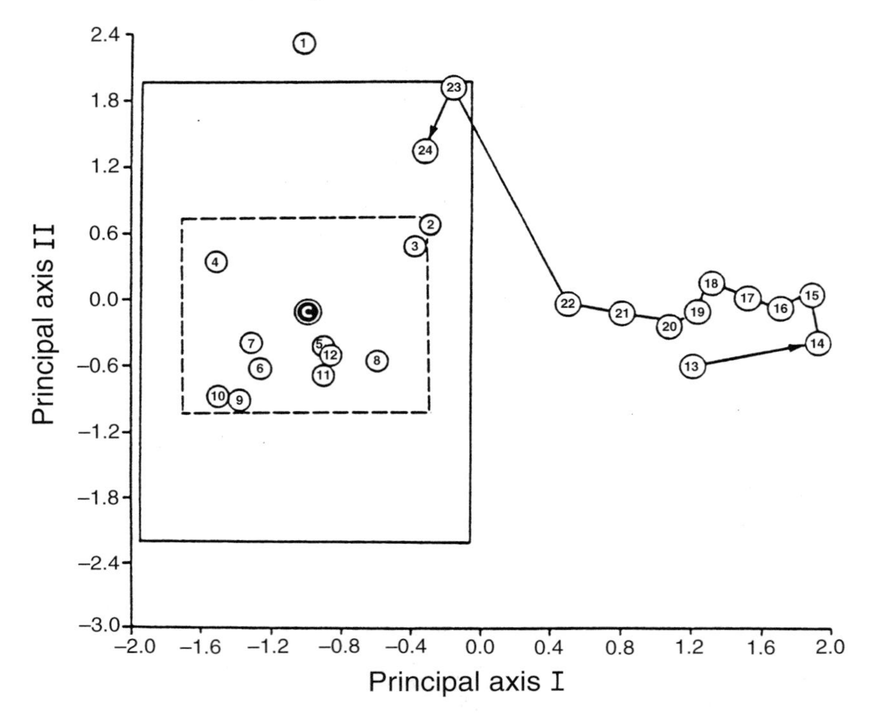

Figure 12.19 Defining and charting the recovery of a community (c) after perturbation. The data are from a benthic infaunal community, with sample units 1–12 before perturbation and sample units 13–24 after perturbation. The solid and dotted boxes represent two procedures for calculating a 95% rejection polygon using parametirc and non-parametic methods, respectively (from Bloom 1980).

possibilities exists; indeed, one might not even use species composition, but other properties such as functional composition, diversity or biomass, as suggested in Table 12.5. In any such case, the same principles would apply. A perturbation occurs when the community is pushed outside these 95% confidence limits. If the trajectory of the community is then monitored along the same axes, one can track both its degree of departure from the original composition, and its recovery; departures are defined as movement away from the centroid, and recovery movement towards it. Recovery is considered to have finished when the perturbed community returns to within the envelope defined by the original samples. The strengths of this approach are that one is provided with an objective space which defines the original state, as well as objective measures of perturbation and recovery.

In many cases, of course, the restoration ecologist inherits a perturbed site which may already be in the region of sample points 13 or 14 in Figure 12.19. In such a case, it is up to the recovery team to decide what the desired composition is. This could be based upon historical data from the site, published data from other sites that provide the desired end point, or new data from other less-perturbed sites. One might even choose a different region of the phase space if this region represents a community type that is disappearing from the landscape. Once that decision is made, and the appropriate recovery region is defined, the task of the restoration ecologist is to move the community from the right-hand part of the figure into the desired region on the left. The factors covered earlier in this book should provide the necessary shopping list.

Another perspective on ecosystem management

I have split the problem of conservation and management into two parts: priorities for action, and choosing indicators for those actions. In turn, I have broken each of these down into three goals. The priorities for action are protection, management and restoration. The priorities for indicators are selecting the appropriate factors to monitor, setting critical limits upon them, and monitoring to ensure that our management achieves its goals.

A more general framework is provided by Chistensen *et al.* (1996), and perhaps reflects the difficulties of achieving consensus among 13 authors who in turn are accountable to a scientific society. It provides many of the same components, but organizes them in slightly different fashion. The eight principles recognized by Christensen *et al.* are:

(i) *Sustainability.* The system is managed to ensure its persistence through time.
(ii) *Clear goals.* Goals must be explicitly stated in terms of desired future states for the system.
(iii) *Science.* Ecosystem management is based on sound ecological principles and emphasizes the role of processes and interconnections.
(iv) *Interconnectedness.* Managers must appreciate the vast array of interconnections that underlie ecosystem function.
(v) *Dynamic character of ecosystems.* Sustainability does not mean freezing systems at one point in time, but accepting that systems change over time within certain bounds.
(vi) *Scale.* Ecosystems operate over a wide range of temporal and spatial scales nested within one another.

(vii) *Humans.* Humans are part of ecosystems, and their behaviour must be included in models.

(viii) *Accountability.* The knowledge base is incomplete and subject to change. Monitoring is essential so that objective measures of results can be compared with specific expectations and forecasts.

One missing feature is the need for protected landscapes, which is on my list as the first priority for action. We have many years to manage ecosystems, but only a few years left in which to protect relatively undisturbed landscapes. I have also combined many of Christensen's categories into one, under the term 'maintenance of function'; the actions by which function is maintained will vary from situation to situation. Finally, I have placed greater emphasis upon indicators, in part, because I believe that this is a profound scientific challenge as well as a daunting management problem. The scientific challenge is to devise indicators that work, particularly large-scale indicators that apply to all of spaceship Earth. The management problem is to see that someone monitors these indicators, and more importantly, that action is taken to alter management practices when the indicators warn us of impending problems. For yet other perspectives upon the challenges posed by ecosystem management, Grumbine (1997) and Haeuber and Franklin (1996) can be consulted.

Goals, tactics and strategies for wetland scientists: a personal perspective

Goals

Where do we go from here? We must talk briefly about goals, tactics and strategies. Without these we are working blind. Imagine going to an airport and getting on a plane at random, hoping that by doing so enough times one will end up at one's destination. This is hardly effective or efficient. Instead, we decide where we want to go, and then a competent travel agent can suggest various routes that we might take to get there. Military planners have known this for years. It is often popular to criticize the military, but whether we like it or not, the military is a repository of a certain kind of wisdom. Sun Tzu was a military planner who lived millennia ago in a different culture, but he can still tell us about human nature, and he knew about attaining goals. I advise both graduate students and faculty to read and consider the wisdom in Sun Tzu's *The Art of War.* Further, some study of the Second World War might be appropriate, particularly with reference to the differing styles of the two great generals,

Montgomery and Patton (see Montgomery 1958 and Essame 1974, respectively).

I suggest that we wetland ecologists set a simple goal – the goal of being able to forecast the properties of communities that will arise under specified sets of environmental conditions. This goal will require all our energy as ecologists, and will very naturally integrate with the need for applications. To achieve this goal we need to be able to answer the following questions:

(i) What properties of ecosystems do we need to predict?
(ii) What environmental factors best predict the foregoing properties?
(iii) How do these properties and the predictors of them vary within and among community types?
(iv) How can we simplify the foregoing by reference to traits as opposed to species nomenclature?

I would call these four questions an avowedly pragmatic approach to community ecology. They require us to start where we are now, summarize the current state of our knowledge of these relationships, and then decide what further things we need to measure. We might add a fifth step, since wetlands also have effects in turn on other ecosystem processes such as light extinction, temperature, oxygen and nutrients. This would require us to erect a third class of variables, variables that measure the consequences of wetlands upon other parts of the biosphere. Carpenter and Lodge (1986) point out that, of 94 recent papers in *Aquatic Botany* that deal with interactions between aquatic macrophytes and the environment, 72 look at the effects of the environment upon macrophytes, whereas only 22 consider the reverse, the effects of macrophytes upon the environment.

Without some guidelines, we will clog the literature with all sorts of haphazard measurements on miscellaneous factors in wetlands scattered here and there with no discernible order. Some recent compendia illustrate this approach only too well. The mere fact that a trivial study is carried out in a wet place does not dignify it with the title of wetland ecology.

Six Tactical Guidelines

Money spent on the wrong kind of research is like money spent on buying the wrong kind of habitat – resources which could have been wisely allocated to conservation are lost. We must plan our research with the same dedication and effort as Montgomery or Patton planned their campaigns. Six guidelines may be of assistance.

Generality

First and foremost, scientific advice must be applicable to a variety of circumstances. Species and site-specific studies are not in themselves a viable approach to managing a global or national system of protected areas. For example, there are over 150 significant wetlands that have already been identified in just one part of Ontario (Ontario Ministry of Natural Resources (OMNR) 1988). Assume, as a first approximation, that each supports 1000 species. If we expect to understand how to manage such wetlands by studying each species, and allocate 1 year per species, we must allocate 1000 person–years per wetland. If we consider the interactions between the organisms (and we must), then 1000 species yields roughly half a million interactions, which translates into a half million person years per wetland (see also Rigler 1982). It is therefore not possible to study each interaction or even each species to provide management plans for natural areas.

The only way to manage a large collection of wetlands is to look for general principles that apply to numerous sites, or to combine species into groups with similar ecological properties. Such principles and general models can be applied to many specific sites or species, and refined if necessary (Peters 1980a, b). The continuum from general to site-specific models can be represented as a nested hierarchy of models, with the general principles at the top, and the specific site at the bottom (Figure 12.20). One can start at the top and work down to any site, but it is far more difficult to start at the bottom and then extrapolate to the rest of the world.

Explicit constraints

Now let us appear to contradict the first principle. When general principles are established and applied, it is necessary to be aware of constraints to the generality (e.g. Grubb 1985). Consider, for example, the seed bank model that was developed for prairie potholes (van der Valk 1981). It now seems that some managers believe that all wetlands must be managed by fluctuating water levels and allowing regeneration from buried seeds. Certainly, this model applies to lake shores (Keddy and Reznicek 1982, 1986) and some ponds (Salisbury 1970; McCarthy 1987). It may apply to many relatively fertile sites with a history of natural disturbance. Other wetland vegetation types, such as bogs and fens, do not rapidly regenerate

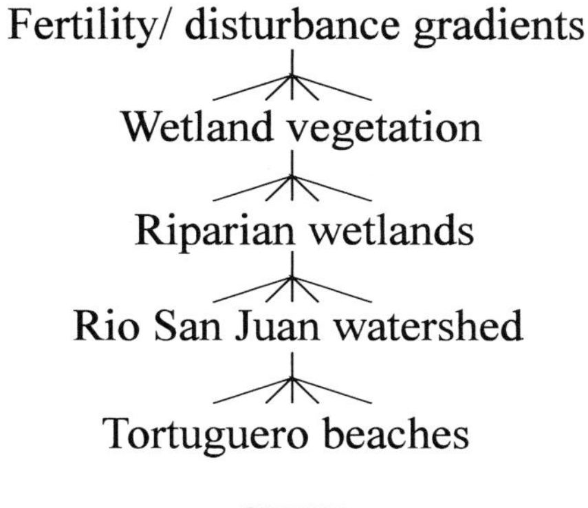

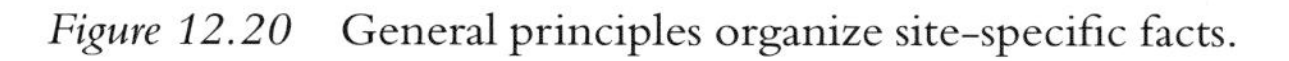

Figure 12.20 General principles organize site-specific facts.

from seed after periodic disturbance. Such vegetation types would be degraded or destroyed by application of the prairie pothole model.

We therefore need guidelines for determining which ecosystem types require which type of management. Every management model will have ecological constraints upon its application (Figure 12.21); lacking such constraints, a model is a dangerous weapon for managers to have in their arsenal.

First things first

When we build a house, we normally begin with the major features (foundation, walls) and only then work on the minor ones (door handles, light fixtures). Unfortunately, this perfectly common-sense approach to house building does not seem to carry over into ecology. A perusal of our journals would easily lead one to believe that many ecologists would pick out door knobs and then be puzzled that they have neither a door nor a house in which to install them. At the risk of restating the obvious, we should start with the most important factors and variables, and then and only then move to the finer ones. To start this discussion, I suggest that (at least in the areas I know) some 50% of the variation in wetland communities is attributable to hydrology. Fertility and salinity probably account for something like a further 20% each. All other factors (e.g. grazing, fire) address only residual variance (recall Table 10.1). We might therefore

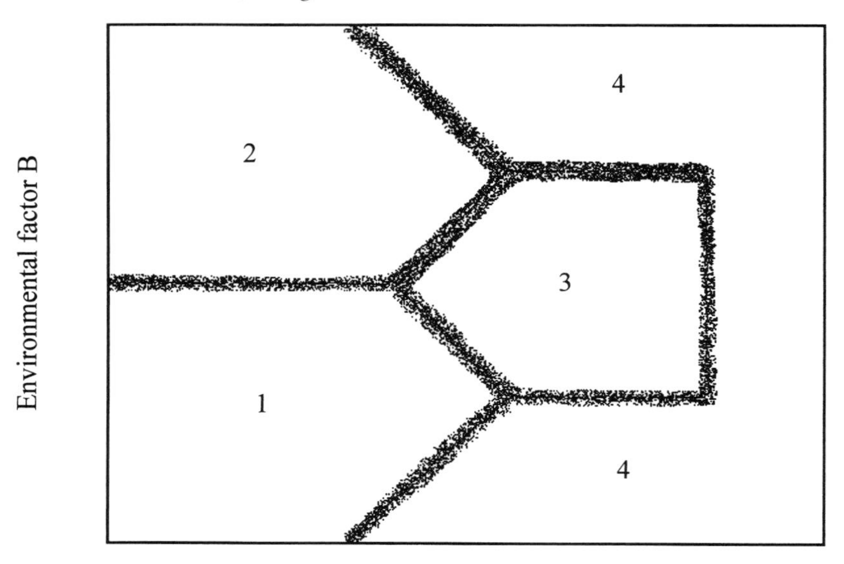

Figure 12.21 No single model will apply to all ecological conditions. In this case, four different models are needed to cover all possible subsets of conditions defined by environmental factors A and B. Any model should be presented with explicit details on its constraints.

anticipate that the Number 1 priority of wetland ecologists has been, and would be, the development of quantitative models linking wetland community structure to hydrological variables. We would be wrong.

Description and prediction

Protecting wetlands by zoning or acquisition is only the first step which prevents the obvious threats of filling and draining. Management plans are then required. Management plans require prediction – forecasting the possible effects of human impacts from surrounding activities in the landscape, as well as the consequences of different kinds of management. For example, a management plan will have to consider threats from eutrophication. Ehrenfeld (1983) studied the distinctive wetland vegetation in the infertile New Jersey pine barrens. She found that eutrophication led to the replacement of typical pine barrens plants by more widespread (and often exotic) species. Similarly, Morgan and Philipp (1986) studied aquatic macrophytes in the pine barrens and concluded that the primary

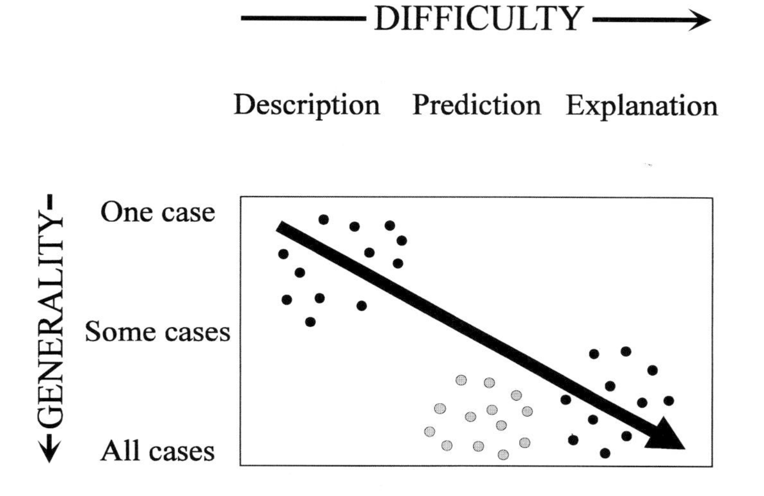

Figure 12.22 Descriptive studies of single cases are much easier to do than predictive or explanatory models applying to many cases or sites (after Leary 1985).

effect of nutrient pollution was to replace the distinctive pine barrens flora with common and exotic species.

Prediction is far more difficult than description, as Leary (1985) has recognized in trying to evaluate the costs and benefits of different kinds of government research (Figure 12.22). He concludes that generality and explanation (*lower right*) are the most difficult, even if they are the most important. They therefore require the most incentive, since all other things being equal, it is tempting to try to solve easy problems. In the case of conservation, it is probably true that we need prediction more than explanation (grey dots). A perusal of the kinds of grant applications submitted by biologists, as well as the kinds of projects funded by conservation agencies, suggests there is far more activity in description (Figure 12.22, *upper left*) than in prediction and understanding (Figure 12.22 *lower right*). This may be fine in the inventory stages of wetland protection, but long-term survival of natural systems requires a change in emphasis towards general predictive models and carefully designed field experiments to unravel the network of causation that produces ecological patterns.

Attitudinal inertia

Good research addresses important problems in new ways. Traditionally,

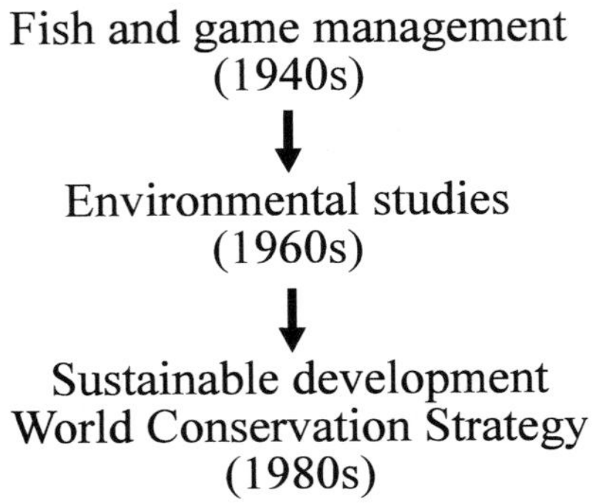

Fish and game management
(1940s)
↓
Environmental studies
(1960s)
↓
Sustainable development
World Conservation Strategy
(1980s)

Figure 12.23 Conservation paradigms can lag decades behind what the world needs (from Keddy and Wisheu 1989).

biologists have focused activities on selected species, particularly those big animals that are favoured by hunters. This 'moose–goose syndrome' (Keddy 1989a) still colours wetland research and conservation activities. Consider, for example, the effort put into mapping deer habitat as opposed to the effort in mapping turtle nesting beaches, or the number of biologists studying ducks as opposed to those studying invertebrates or plants. This produces inertia in the scientific response to conservation problems (Figure 12.23). This problem of attitudinal inertia is one of the most expensive and dangerous problems we currently face. Consider two personal examples; they are local issues but illustrate the principles that apply globally.

In spite of evidence of provincially and nationally rare plant communities on the shores of Matchedash Lake (e.g. Keddy and Reznicek 1982), the Ontario Ministry of Natural Resources flooded these wetlands in the unsubstantiated belief that it would increase the productivity of an introduced species of game fish. The final fate of these plant communities is still not known. This money could have been used for conservation if scientific advice had been sought or listened to in the first place. Desire to control the wetland for an exotic fish species outweighed protecting the rare native flora.

There is a wetland complex with relict sand dunes not far from where I write. The consultants who studied this area for the Ontario Ministry of Natural Resources were so impressed that they recommended provincial nature reserve status for the entire 350 ha. This shoreline habitat not only has unusual sand dune species, but uncommon wetland species and vege-

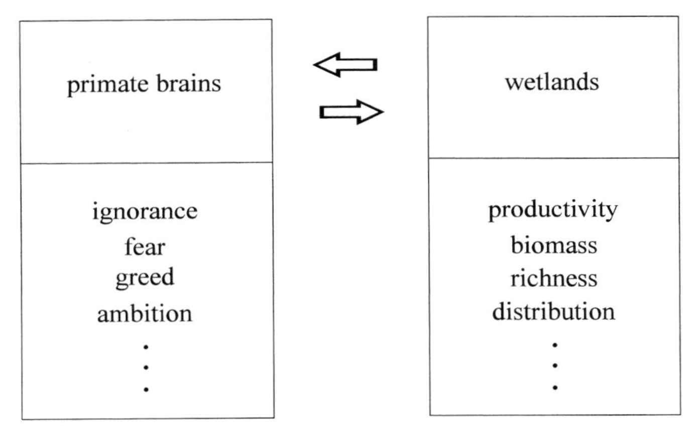

Figure 12.24 The two components of wetland conservation and management.

tation (e.g. Day *et al.* 1988; Moore *et al.* 1989). These are rapidly disappearing due to recreation on sandy shorelines. Instead, the Ministry proposed development of the site. This would entail over 100 campsites in the dunes, and the dunes and infertile wetlands would become a swimming beach. Their public information report noted 'the wet beach area possesses extensive weed growth . . . The wet beach requires clearance and maintenance should a swimming area be developed.'

Inner and outer obstacles

My two personal examples are typical of the sort of local horror stories that scientists can relate about natural areas that they know well. Such problems rarely arise because of scientific limitations (that is, from a lack of understanding of the external world). More often, the cause is human attitudes (that is, the inner realm of human thoughts and feelings). As scientists, we are trained to dissect and analyse living systems with exquisite care, but we often blunder into human interactions like drunken elephants in a minefield. It may well be that all young scientists should have some basic training in psychology and social science, particularly if they are planning on applied work. At the very least, let us remember that many of the obstacles to wetland research and conservation do not exist in the field where we can measure them with our instruments, but inside the heads of fellow citizens. Wetland management therefore has two separate components (Figure 12.24). If we ignore the left-hand one, we are like a general guilty of wishful thinking because he cannot admit that a minefield or mountain range is an obstacle to his campaign.

Summary

We are at a difficult point in human history. Our own civilization is increasingly threatened by the activities of our own species. Even if we doubt the likelihood of the most extreme scenarios, such as nuclear war and nuclear winter, or desertification and mass starvation, we cannot ignore the accumulating insidious effects of many less dramatic processes such as deforestation, soil erosion, drainage of wetlands, and rising rates of extinction (Figure 12.25) (Meadows *et al.* 1974). Ecologists are like the legendary thin red line of British soldiers; we stand between our civilization and the ecosystems upon which we depend. These ecosystems are mute. We alone provide them with voices. This is a heavy responsibility to bear, and one may wish that one had instead become a lawyer, a small town doctor or a store manager. One may wish that, instead of this book, one had read a murder mystery or a romance. But, given knowledge, we now have the duty to act. We could seek counsel from another professional organization where responsibility, duty, and the exercise of power are valued, the military.

> It is an honour to serve in the armed forces . . . It is also a duty of our citizens to serve in the armed forces, as volunteers or in accordance with our nation's laws, and to perform the military missions that this service may require. If the day should come when a large proportion of our citizens regard this service as less than an honour, and less than an obligation of citizenship, our proud nation will have begun the descent to lie beside other peoples who were unable or unwilling to fight for their principles or for the retention of their freedoms.
>
> (Crocker 1990 p. 31)

Action has several components. With respect to our own activities, there is the responsibility to work on significant problems rather than allowing our minds to flit about and occupy us with each autecological curiosity that catches our attention. We can avoid conducting research that is simply haphazardly selected problems in haphazardly selected sites with attractive species with no consideration of the literature outside one's own geographic region and taxonomic group. There is also the responsibility to speak clearly and act with integrity in defence of the world's ecosystems. To remain silent in the face of folly is irresponsible.

We have come so far from the first figure – wetlands as the cradle of human civilization. (Those with an ironic bent may observe that Figure

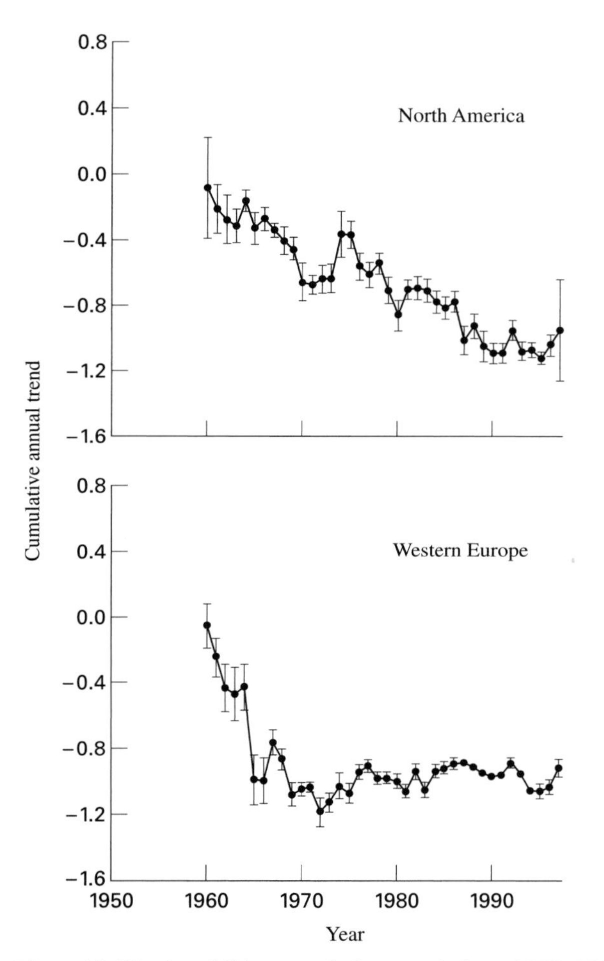

Figure 12.25 Amphibian population trends from 1960–1997 for North America and Western Europe (including the United Kingdom). (After Houlahan *et al.* 2000.)

1.1 also illustrates the terrain of one of our more recent wars.) Having reached the last page of the text, we are already preparing ourselves to put the book back on the shelf. Before doing so, we must ask ourselves where we go from here. Figure 12.26 reminds us that fundamentally our path forward is straightforward. We have a set of independent variables. These produce a series of measurable properties in wetland communities. In turn, wetland communities have certain functions beyond their own

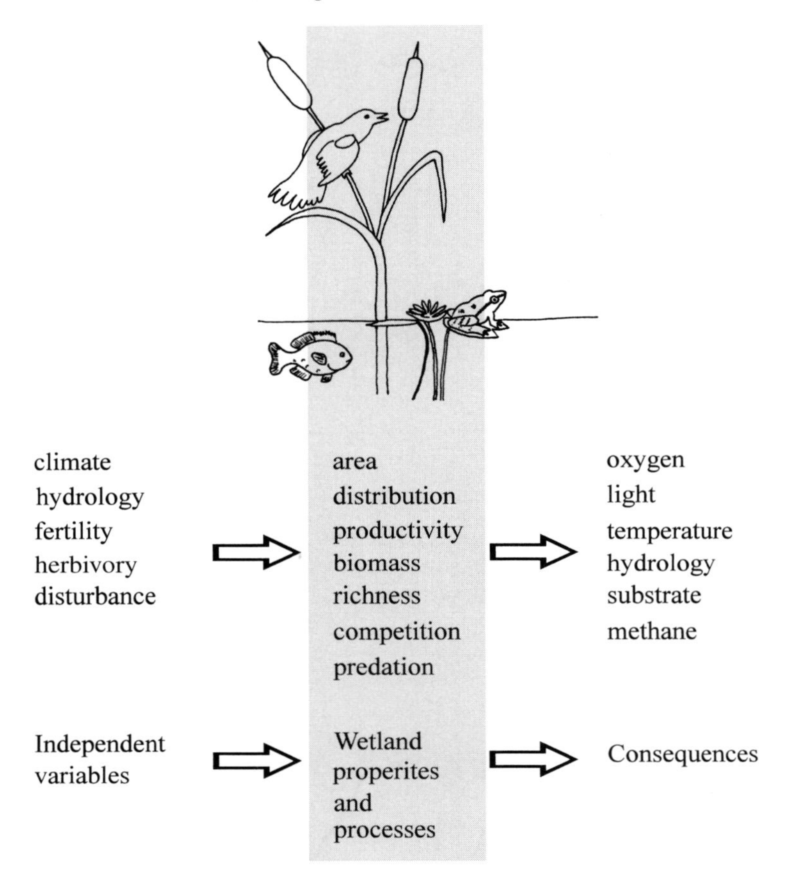

Figure 12.26 Wetland ecology is the study of how independent variables (*left*) determine wetland properties (*centre*) and the consequences of the foregoing linkages for wetlands and other ecosystems.

borders. We have the responsibility to sort out these relationships and convey them clearly and effectively to those around us. Within the simple framework of Figure 12.26, there is enough work for at least another generation of wetland ecologists.

References

Aarssen, L. (1983). Ecological combining ability and competitive combining ability in plants: towards a general evolutionary theory of coexistence in systems of competition. *The American Naturalist*, **122**, 707–31.

Aarssen, L. (1985). Interpretation of the evolutionary consequences of competition in plants: an experimental approach. *Oikos*, **45**, 99–100.

Adam, P. (1990). *Saltmarsh Ecology*. Cambridge: Cambridge University Press.

Adams, G. D. (1988). Wetlands of the prairies of Canada. In *Wetlands of Canada*. National Wetlands Working Group. Ecological Land Classification Series, No. 24. pp. 158–98. Sustainable Development Branch, Environment Canada, Ottawa, Ontario, and Polyscience Publications Inc., Montreal, Quebec.

Adamus, P. R. (1992). Choices in monitoring wetlands. In *Ecological Indicators*, ed. D. H. McKenzie, D. E. Hyatt and V. J. McDonald, pp. 571–92. London: Elsevier Applied Science.

Adamus, P. R. (1996). Bioindicators for assessing ecological integrity of prairie wetlands. *EPA/600/R-96/082*. Corvallis, OR: US Environmental Protection Agency, National Health and Environmental Effects Research Laboratory, Western Ecology Division.

Adamus, P. R. and Stockwell, L. T. (1983). *A Method for Wetland Functional Assessment: Volume I: Critical Review and Evaluation Concepts.* 176 pp. *Volume II: Federal Highway Administration Assessment Method.* 134 pp. Reports FHA-PI-82-23 and 24. Springfield VA: National Technical Information Service.

Adamus, P. R., ARA Inc., Clairtain, E. J. Smith, R. D. and Young, R. E. (1987). *Wetland Evaluation Technique (WET)*. Vol. II: *Methodology.* Vicksburg, Mississippi: Waterways Experiment Station, US Army Corps of Engineers.

Aerts, R. and Berendse, F. (1988). The effect of increased nutrient availability on vegetation dynamics in wet heathlands. *Vegetatio*, **76**, 63–9.

Agren, G. I. and Fagerstrom, T. (1984). Limiting dissimilarity in plants: randomness prevents exclusion of species with similar competitive abilities. *Oikos*, **43**, 369–75.

Alho, C. J. R., Lacher, Jr., T. E. and Gonçalves, H. C. (1988). Environmental degradation in the Pantanal ecosystem. *BioScience*, **38**, 164–71.

Allison, S. K. (1995). Recovery from small-scale anthropogenic disturbances by northern California salt marsh plant assemblages. *Ecological Applications*, **5**, 693–702.

Anderson, R. C., Liberta, A. E. and Dickman, L. A. (1984). Interaction of vascular plants and vesicular–arbuscular mycorrhizal fungi across a soil moisture–nutrient gradient. *Oecologia*, **64**, 111–17.

Anderson, R. C., Liberta, A. E., Dickman, L. A. and Katz, A. J. (1983). Spatial variation in vesicular-arbuscular mycorrhizal spore density. *Bulletin of the Torrey Botanical Club*, **110**, 519–25.

Archibold, O. W. (1995). *Ecology of World Vegetation*. London: Chapman and Hall.

Armentano, T. V. and Verhoeven, J. T. A. (1990). Biogeochemical cycles: global. In *Wetlands and Shallow Continental Water Bodies. Vol. 1. Natural and Human Relationships*, ed. B. C. Patten, pp. 281–311. The Hague, The Netherlands: SPB Academic Publishing.

Armstrong, W., Armstrong, J., Beckett, P. M. and Justin, S. H. F. W. (1991). Convective gas-flows in wetland plant aeration. *Plant Life Under Oxygen Deprivation*, ed. M. B. Jackson, D. D. Davies and H. Lambers, pp. 283–302. The Hague, The Netherlands, SPB Academic Publishing.

Arnold, S. J. (1972). Species densities of predators and their prey. *The American Naturalist*, **106**, 220–35.

Arnold, T. W. and Frytzell, E. K. (1990). Habitat use by male mink in relation to wetland characteristics and avian prey abundances. *Canadian Journal of Zoology*, **68**, 2205–8.

Arrhenius, O. (1921). Species and area. *Journal of Ecology*, **9**, 95–9.

Aselman, I. and Crutzen, P. J. (1989). Global distribution of natural freshwater wetlands and rice paddies, their net primary productivity, seasonality and possible methane emissions. *Journal of Atmospheric Chemistry*, **8**, 307–58.

Atwood, E. L. (1950). Life history studies of the nutria, or coypu, in coastal Louisiana. *Journal of Wildlife Management*, **14**, 249–65.

Auclair, A. N. D., Bouchard, A. and Pajaczkowski, J. (1976a). Plant standing crop and productivity relations in a *Scirpus–Equisetum* wetland. *Ecology*, **57**, 941–52.

Auclair, A. N. D., Bouchard, A. and Pajaczkowski, J. (1976b). Productivity relations in a *Carex*-dominated ecosystem. *Oecologia*, **26**, 9–31.

Austin, M. P. (1982). Use of a relative physiological performance value in the prediction of performance in multispecies mixtures from monoculture performance. *Journal of Ecology*, **70**. 559–70.

Austin, M. P., Pausas, J. G. and Nicholls, A. O. (1996). Patterns of tree species richness in relation to environment in southeastern New South Wales, Australia. *Australian Journal of Ecology*, **21**, 154–64.

Bacon, P. R. (1978). *Flora and Fauna of the Caribbean*. Trinidad: Key Caribbean Publications.

Bakker, J. P. (1985). The impact of grazing on plant communities, plant populations and soil conditions on salt marshes. *Vegetatio*, **62**, 391–8.

Bakker, S. A., Jasperse, C. and Verhoeven, J. T. A. (1997). Accumulation rates of organic matter associated with different successional stages from open water to carr forest in former turbaries. *Plant Ecology*, **129**, 113–20.

Ball, P. J. and Nudds, T. D. (1989). Mallard habitat selection: an experiment and implications for management. In *Freshwater Wetlands and Wildlife*. ed. R. R. Sharitz, and J. W. Gibbons, pp. 659–71. US Department of Energy. Proceedings of a Symposium held at Charleston, South Carolina, March 24–27, 1986.

Barbour, C. D. and Brown, J. H. (1974). Fish species diversity in lakes. *The American Naturalist*, **108**, 473–89.

Bardecki, M. J., Bond, W. K. and Manning, E. W. (1989). Assessing Greenock Swamp: functions, benefits and values, In *Wetlands. Inertia or Momentum?* pp. 235–44. Conference Proceedings Oct 21–22, Ontario: Federation of Ontario Naturalists.

Barko, J. W. and Smart, R. M. (1978). The growth and biomass distribution of two emergent freshwater plants, *Cyperus esculentus* and *Scirpus validus*, on different sediments. *Aquatic Botany*, **5**, 109–17.

Barko, J. W. and Smart, R. M. (1979). The nutritional ecology of *Cyperus esculentus*, an emergent aquatic plant, grown on different sediments. *Aquatic Botany*, **6**,13–28.

Barko, J. W. and Smart, R. M. (1980). Mobilization of sediment phosphorus by submersed freshwater macrophytes. *Freshwater Biology*, **10**, 229–38.

Barnard, J. R. (1978). Externalities from urban growth: the case of increased storm runoff and flooding. *Land Economics*, **54**, 298–315.

Barthélemy, A. (1874). De la respiration et de la circulation des gaz dans les végétaux. *Annales des Sciences Naturelles Botanique*, **19**, 131–75.

Batt, B. D. J, Anderson, M. G., Anderson, C. D. and Caswell, F. D. (1989). The use of prairie potholes by North American ducks. *Northern Prairie Wetlands*, ed. A. G. van der Valk, pp. 204–27. Ames, Iowa: Iowa State University Press.

Bauder, E. T. (1989). Drought stress and competition effects on the local distribution of *Pogogyne abramsii*. *Ecology*, **70**, 1083–9.

Bazely, D. R. and Jefferies, R. L. (1989). Lesser snow geese and the nitrogen economy of a grazed salt marsh. *Journal of Ecology*, **77**, 24–34.

Bazilevich, N. I., Rodin, L. Y. and Rozov, N. N. (1971). Geophysical aspects of biological productivity. *Soviet Geography. Review and Translations*, **12**, 293–317.

Beanland, G. E. and Duinker, P. N. (1983). *An Ecological Framework for Environmental Impact Assessment in Canada*. Halifax: Institute for Resource and Environmental Studies Dalhousie University and Federal Environmental Assessment Review Office.

Beard, J. S. (1949). *The Natural Vegetation of the Windward and Leeward Islands*. Oxford: Clarendon Press.

Bedford, B. L. (1996). The need to define hydrologic equivalence at the landscape scale for freshwater wetland mitigation. *Ecological Applications*, **6**, 57–68.

Bedford, B. L. and Preston, E. M. (1988). Developing the scientific basis for assessing cumulative effects of wetland loss and degradation on landscape functions: status, perspectives and prospects. *Environmental Management*, **12**, 751–71.

Beebee, T. J. C. (1996). *Ecology and Conservation of Amphibians*. London: Chapman and Hall.

Beeftink, W. G. (1977). The coastal salt marshes of western and northern Europe: an ecological and phytosociological approach. In *Wet Coastal Ecosystems*, ed. V. J. Chapman, pp. 109–55. Amsterdam: Elsevier.

Bégin, Y., Arseneault, S. and Lavoie, J. (1989). Dynamique d'une bordure forestière par suite de la hausse récente du niveau marin, rive sud-ouest du Golfe du Saint-Laurent, Nouveau-Brunswick. *Géographie physique et Quaternaire*, **43**, 355–366.

Belanger L. and Bedard, J. (1994). Role of ice scouring and goose grubbing in marsh plant dynamics. *Journal of Ecology*, **82**, 437–45.

Belkin, D. A. (1963). Anoxia: tolerance in reptiles. *Science*, **139**, 492–3.

Bender, E. A, Case, T. J. and Gilpin, M. E. (1984). Perturbation experiments in community ecology: theory and practice. *Ecology*, **65**, 1–13.

Benson, L. (1959). *Plant Classification*. Lexington: D. C. Heath and Company.

Berenbaum, M. R. (1991). Coumarins. In *Herbivores: Their Interactions with Secondary Plant Metabolites*, ed. G. A. Rosenthal and M. R. Berenbaum, pp. 221–49. San Diego: Academic Press.

Berendse, F. and Aerts, R. (1987). Nitrogen-use-efficiency: a biologically meaningful definition? *Functional Ecology*, **1**, 293–6.

Bernatowicz, S. and Zachwieja, J. (1966). Types of littoral found in the lakes of the Masurian and Suwalki Lakelands. *Komitet Ekolgiezny-Polska Akademia Nauk*, **XIV**, 519–45.

Bertness, M. D. (1991). Interspecific interactions among high marsh perennials in a New England salt marsh. *Ecology*, **72**, 125–37.

Bertness, M. D. and Ellison, A. E. (1987). Determinants of pattern in a New England salt marsh plant community. *Ecological Monographs*, **57**, 12–147.

Bertness, M. D. and Shumway, S. W. (1993). Competition and facilitation in marsh plants. *The American Naturalist*, **142**, 718–34.

Bertness, M. D. and Hacker, S. D. (1994). Physical stress and positive associations among marsh plants. *The American Naturalist*, **144**, 363–72.

Bertness, M. D. and Yeh, S. M. (1994). Cooperative and competitive interactions in the recruitment of marsh elders. *Ecology*, **75**, 2416–29.

Bertness, M. D. and Leonard, G. H. (1997). The role of positive interactions in communities: lessons from intertidal habitats. *Ecology*, **78**, 1976–89.

Bertness, M. D., Gough, L. and Shumway, S. W. (1992). Salt tolerances and the distribution of fugitive salt marsh plants. *Ecology*, **73**, 1842–51.

Bertness, M. D., Wikler, K. and Chatkupt, T. (1992). Flood tolerance and the distribution of *Iva frutescens* across New England salt marshes. *Oecologia*, **91**, 171–8.

Best, E. P. H., Verhoeven, J. T. A. and Wolff, W. J. (1993). The ecology of The Netherlands wetlands: characteristics, threats, prospects and perspectives for ecological research. *Hydrobiologica*, **265**, 305–20.

Bethke, R. W. and Nudds, T. D. (1993). Variation in the diversity of ducks along a gradient of environmental variability. *Oecologia*, **93**, 242–50.

Biesterfeldt, J. M., Petranka, J. W. and Sherbondy, S. (1993). Prevalence of chemical interference competition in natural populations of wood frogs, *Rana sylvatica*. *Copeia*, **3**, 688–95.

Binford, M. W., Brenner, M., Whitmore, T. J. Higuera-Gundy, A., Deevey, E. S. and Leyden, B. (1987). Ecosystems, paleoecology and human disturbance in subtropical and tropical America. *Quaternary Scientific Review*, **6**, 115–28.

Bliss, L. C. and Gold, W. G. (1994). The patterning of plant communities and edaphic factors along a high arctic coastline: implications for succession. *Canadian Journal of Botany*, **72**, 1095–107.

Blizard, D. (1993). *The Normandy Landings D-Day. The Invasion of Europe 6 June 1944*. London: Reed International Books.

Bloom, S. A. (1980). Multivariate quantification of community recovery. In *The Recovery Process in Damaged Ecosystems*, ed. J. Cairns, pp. 141–51. Ann Arbor, MI: Ann Arbor Science Publishers.

Boesch, D. F., Josselyn, M. N., Mehta, A. J., Morris, J. T., Nuttle, W. K., Simenstad, C. A. and Swift, D. P. J. (1994). Scientific assessment of coastal wetland loss, restoration and management in Louisiana. *Journal of Coastal Research*, Special Issue No. 20.

Bolen, E. G., Smith, L. M. and Schramm, H. L., Jr. (1989). Playa lakes: prairie wetlands of the southern High Plains. *BioScience*, **39**, 615–23.

Bond, G. (1963). *Plant Physiology*. In ed. F. B. Salisbury and C. W. Ross, 1985., 3rd edn, p. 254, Figure 13.3. Belmont, CA: Wadsworth Publishers.

Bonetto, A. A. (1986). The Paraná River system. In *The Ecology of River Systems*, ed. B. R. Davies and K. F. Walker, pp. 541–55. Dordrecht, The Netherlands: Dr W. Junk Publishers.

Bonnicksen, T. M. (1988). Restoration ecology: philosophy, goals and ethics. *The Environmental Professional*, **10**, 25–35.

Bonser, S. P. and Reader, R. J. (1995). Plant competition and herbivory in relation to vegetation biomass. *Ecology*, **76**, 2176–83.

Bormann, E. H. and Likens, G. E. (1981). *Patterns and Process in a Forested Ecosystem*. New York: Springer-Verlag.

Boston, H. L. (1986). A discussion of the adaptation for carbon acquisition in relation to the growth strategy of aquatic isoetids. *Aquatic Botany*, **26**, 259–70.

Boston, H. L. and Adams, M. S. (1986). The contribution of crassulacean acid metabolism to the annual productivity of two aquatic vascular plants. *Oecologia*, **68**, 615–22.

Botch, M. S. and Masing, V. V. (1983). Mire ecosystems in the USSR. In *Ecosystems of the World 4B*, ed. A. J. P. Gore, pp. 95–152. Amsterdam: Elsevier Scientific Publishing Company.

Botkin, D. B. (1990). *Discordant Harmonies. A New Ecology for the Twenty-first Century*. New York: Oxford University Press.

Boucher, D. H. (1985). *The Biology of Mutualism: Ecology and Evolution*. New York: Oxford University Press.

Boutin, C. and Keddy, P. A. (1993). A functional classification of wetland plants. *Journal of Vegetation Science*, **4**, 591–600.

Bowden, W. B. (1987). The biogeochemistry of nitrogen in freshwater wetlands. *Biogeochemistry*, **4**, 313–48.

Bowers, M. D. (1991). Iridoid glycosides. In *Herbivores: Their Interactions with Secondary Plant Metabolites*, ed. G. A. Rosenthal and M. R. Berenbaum, pp. 297–325. San Diego: Academic Press.

Boyd, C. E. (1978). Chemical composition of wetland plants. In *Freshwater Wetlands. Ecological Processes and Management Potential*, ed. R. E. Good, D. F. Whigham and R. L. Simpson, pp. 155–68. New York: Academic Press.

Bradley, C. E. and Smith, D. G. (1986). Plains cottonwood recruitment and survival on a prairie meandering river floodplain, Milk River, southern Alberta and northern Montana. *Canadian Journal of Botany*, **64**, 1433–42.

Brändle, R. A. (1991). Flooding resistance of rhizomatous amphibious plants. In *Plant Life Under Oxygen Deprivation*, ed. M. B. Jackson, D. D. Davis and H. Lambers, pp. 35–46. The Hague, The Netherlands: SPB Academic Publishing.

Brasher, S. and Perkins, D. F. (1978). The grazing intensity and productivity of sheep in the grassland ecosystem. In *Production Ecology of British Moors and Montane Grasslands*, ed. O. W. Heal and D. F. Perkins, pp. 354–74. Ecological Studies vol. 27. Berlin: Springer-Verlag.

Breman, H. and de Wit, C. T. (1983). Rangeland productivity and exploitation in the Sahel. *Science*, **221**, 1341–7.

Brewer, J. S. and Grace, J. B. (1990). Plant community structure in an oligohaline tidal marsh. *Vegetation*, **90**, 93–107.

Brinkman, R. and Van Diepen, C. A. (1990). Mineral soils. In *Wetlands and Shallow Continental Water Bodies. Vol. 1. Natural and Human Relationships*, ed. B. C. Patten, pp. 37–59. The Hague, The Netherlands: SPB Academic Publishing.

Brinson, M. M. (1993a). Changes in the functioning of wetlands along environmental gradients. *Wetlands*, **13**, 65–74.

Brinson, M. M. (1993b). A hydrogeomorphic classification for wetlands. Technical Report WRP-DE-4. US Army Corps of Engineers, Washington, DC.

Brinson, M. M. (1995). Functional classifications of wetlands to facilitate watershed planning. In *Wetlands and Watershed Management. Science Applications and Public Policy*, ed. J. A. Kusler, D. E, Willard, and H. C. Hull Jr., pp. 65–71. A collection of papers from a national symposium and several workshops at Tampa, FL, April 23–26. The Association of State Wetland Managers, NY.

Brinson, M. M., Christian, R. R. and Blum, L. K. (1995). Multiple states in the sea-level induced transition from terrestrial forest to estuary. *Estuaries*, **18**, 648–59.

Brinson, M. M., Lugo, A. E. and Brown, S. (1981). Primary productivity, decomposition and consumer activity in freshwater wetlands. *Annual Review of Ecology and Systematics*, **12**, 123–61.

Brönmark, C. (1985). Interactions between macrophytes, epiphytes and herbivores: an experimental approach. *Oikos*, **45**, 26–30.

Brönmark, C. (1990). How do herbivorous freshwater snails affect macrophytes? – a comment. *Ecology*, **71**, 1213–15.

Brown, J. F. (1997). Effects of experimental burial on survival, growth, and resource allocation of three species of dune plants. *Journal of Ecology*, **85**, 151–8.

Brown, S., Brinson, M. M. and Lugo, A. E. (1979). Structure and function of riparian wetlands. In *Strategies for Protection and Management of Floodplain Wetlands and other Riparian Ecosystems*, tech. coord. R. R. Johnson, J. F. McCormick, pp. 17–31. Washington DC: Gen. Tech. Rep. WO-12, Forest Serv., USDA.

Brunton, D. F. and Di Labio, B. M. (1989). Diversity and ecological characteristics of emergent beach flora along the Ottawa River in the Ottawa–Hull region, Quebec and Ontario. *Naturaliste Canadien*, **116**, 179–91.

Bubier, J. L. (1995). The relationship of vegetation to methane emission and hydrochemical gradients in northern peatlands. *Journal of Ecology*, **83**, 403–20.

Bucher, E. H., Bonetto, A., Boyle, T. P., Canevari, P., Castro, G., Huszar, P. and Stone, T. (1993). Hidrovia: An initial environmental examination of the Paraguay-Paraná waterway. Wetlands for the Americas, Manomet, Massachusetts, USA, and Buenos Aires, Argentina.

Burger, J., Shisler, J. and Lesser, F. H. (1982). Avian utilization on six salt marshes in New Jersey. *Biological Conservation*, **23**, 187–212.

Burnett, J. H. (1964). The study of Scottish vegetation. In *The Vegetation of Scotland*, ed. J. H. Burnett, pp. 1–11. Edinburgh: Oliver & Boyd.

Cairns, J. (ed.) (1980). *The Recovery Process in Damaged Ecosystems.* Ann Arbor, MI: Ann Arbor Science.

Cairns, J. (ed.) (1988). *Rehabilitating Damaged Ecosystems* Vols 1 and 2. Boca Raton, Florida: CRC Press, Inc.

Cairns, J. (1989). Restoring damaged ecosystems: is predisturbance condition a viable option? *The Environmental Professional*, **11**, 152–9.

Cairns, J., Jr., Niederlehner, B. R. and Orvos, D. R. (1992). In *Advances in Modern Environmental Toxicology*, ed. M. A. Mehlman, Vol. XX. Princeton Scientific Publishing Company, Princeton, NJ.

Callaway, R. M. and King, L. (1996). Temperature-driven variation in substrate oxygenation and the balance of competition and facilitation. *Ecology*, **77**, 1189–95.

Canada Committee on Ecological Land Classification. (1988). *Wetlands of Canada*. Ecological land classification series No. 24. National Wetlands Working Group, Environment Canada, Ottawa, Canada.

Canny, M. J. (1998). Transporting water in plants. *American Scientist*, **86**, 152–9.

Carignan, R. and Kalff, J. (1980). Phosphorus sources for aquatic weeds: water or sediments? *Science*, **207**, 987–9.

Carpenter, S. R., and Lodge, D. M. (1986). Effects of submersed macrophytes on ecosystem processes. *Aquatic Botany*, **26**, 341–70.

Carpenter, S. R. and Kitchell, J. F. (1988). Consumer control of lake productivity. *BioScience*, **38**, 764–9.

Carpenter, S. R., Chisholm, S. W., Krebs, C. J., Schindler, D. W. and Wright, R. F. (1995). Ecosystem experiments. *Science*, **269**, 324–7.

Carpenter, S. R., Kitchell, J. F., Hodgson, J. R., Cochran, P. A., Elser, J. J., Elser, M. M., Lodge, D. M., Kretchmer, D., He, X. and von Ende, C. N. (1987). Regulation of lake primary productivity by food web structure. *Ecology*, **68**, 1863–76.

Castellanos, E. M., Figueroa, M. E. and Davy, A. J. (1994). Nucleation and facilitation in saltmarsh succession: interactions between *Spartina maritima* and *Arthrocnemem perenne*. *Journal of Ecology*, **82**, 239–48.

Catling, P. M., Spicer, K. W. and Lefkovitch, L. P. (1988). Effects of the introduced floating vascular aquatic, *Hydrocharis morsus-ranae* (Hydrocharitaceae), on some North American aquatic macrophytes. *Naturaliste Canadien*, **115**, 131–7.

Cavalieri, A. J. and Huang, A. H. C. (1979). Evaluation of proline accumulation in the adaptation of diverse species of marsh halophytes to the saline environment. *American Journal of Botany*, **66**, 307–12.

Chaneton, E. J. and Facelli, J. M. (1991). Disturbance effects on plant community diversity: spatial scales and dominance hierarchies. *Vegetatio*, **93**, 143–56.

Chapin, F. S., III. (1980). The mineral nutrition of wild plants. *Annual Review of Ecology and Systematics*, **11**, 233–60.

Chapman, V. J. (1940). The functions of the pneumatophores of *Avicennia nitida* Jacq. *Proceedings of the Linnean Society of London*, **152**, 228–33.

Chapman, V. J. (1974). *Salt Marshes and Salt Deserts of the World*. Lehre: J. Cramer.

Chapman, V. J. (ed.) (1977). *Wet Coastal Ecosystems*. Amsterdam: Elsevier.

Charlton, D. L. and Hilts, S. (1989). Quantitative evaluation of fen ecosystems on the Bruce Peninsula. In *Ontario Wetlands: Inertia or Momentum*, ed. M. J. Bardecki and N. Patterson, pp. 339–54. Proceedings of Conference, Ryerson Polytechnical Institute, Toronto, Oct 21–22, 1988.

Chesson, P. L. and Warner, R. R. (1981). Environmental variability promotes coexistence in lottery competitive systems. *American Naturalist*, **117**, 923–43.

Christensen, N. L., Bartuska, A. M., Brown, J. H., Carpenter, S., D'Antonio, C., Francis, R., Franklin, J. F., MacMahon, J. A., Noss, R. F., Parsons, D. J., Peterson, C. H., Turner, M. G. and Woodmansee, R. G. (1996). The report of the Ecological Society of America Committee on the Scientific Basis for Ecosystem Management. *Ecological Applications*, **6**, 665–91.

Christensen, N. L., Burchell, R. B., Liggett, A. and Simms, E. L. (1981). The structure and development of pocosin vegetation. In *Pocosin Wetlands: An Integrated Analysis of Coastal Plain Freshwater Bogs in North Carolina*, ed. C. J. Richardson, pp. 43–61. Stroudsburg, PA: Hutchinson Ross Publishing Company.

Christie, W. J. (1974). Changes in the fish species composition of the Great Lakes. *Journal of the Fisheries Research Board of Canada*, **31**, 827–54.

Chung, C. (1982). Low marshes, China. In *Creation and Restoration of Coastal Plant Communities*, ed. R. R Lewis III, pp. 131–45. Boca Raton, FA: CRC Press.

Cicerone, R. J. and Ormland, R. S. (1988). Biogeochemical aspects of atmospheric methane. *Global Biogeochemical Cycles*, **2**, 299–327.

Clapham, W. B., Jr. (1973). *Natural Ecosystems.* New York: The Macmillan Company.

Clarke, L. D. and Hannon, N. J. (1967). The mangrove swamp and salt marsh communities of the Sydney district. I. Vegetation, soils and climate. *Journal of Ecology*, **55**, 753–71.

Clarke, L. D. and Hannon, N. J. (1969). The mangrove swamp and salt marsh communities of the Sydney district. II. The holocoenotic complex with particular reference to physiography. *Journal of Ecology*, **57**, 213–34.

Clements, F. E. (1916). Plant succession: an analysis of the development of vegetation. *Carnegie Institution*. 242.

Clements, F. E. (1935). Experimental ecology in the public service. *Ecology*, **16**, 342–63.

Clements, F. E. (1936). Nature and structure of climax. *Journal of Ecology*, **24**, 254–82.

Clements, F. E., Weaver, J. E. and Hanson, H. C. (1929). *Plant Competition.* Washington, DC: Carnegie Institution of Washington.

Clymo, R. S. and Hayward, P. M. (1982). The ecology of *Sphagnum*. In *Bryophyte Ecology*, ed. A. J. E. Smith, pp. 229–89. London: Chapman and Hall.

Clymo, R. S. and Duckett, J. G. (1986). Regeneration of *Sphagnum*. *New Phytologist*, **102**, 589–614.

Coles, B. and J. Coles. (1989). *People of the Wetlands. Bogs, Bodies and Lake-Dwellers.* Thames and Hudson.

Coley, P. D. (1983). Herbivory and defense characteristics of tree species in a lowland tropical forest. *Ecological Monographs*, **53**, 209–33.

Colinvaux, P. 1978. *Why Big Fierce Animals are Rare. An Ecologist's Perspective.* New Jersey: Princeton University Press.

Colwell, R. K. and Fuentes, E. R. (1975). Experimental studies of the niche. *The Annual Review of Ecology and Systematics*, **6**, 281–309.

Committee on Characterization of Wetlands. (1995). *Wetlands. Characteristics and Boundaries.* Washington, DC: National Academy Press.

Committee on Ecological Land Classification. (1988). *Wetlands of Canada.* Ecological land classification series No. 24. National Wetlands Working Group, Environment Canada, Ottawa, Canada.

Connell, J. H. (1978). Diversity in tropical rain forests and coral reefs. *Science*, **199**, 1302–10.

Connell, J. H. (1987). Maintenance of species diversity in biotic communities. In *Evolution and Coadaptation in Biotic Communities*, ed. S. Kavan, J. H. Connell, and T. Hidaka. Tokyo: University of Tokyo Press.

Connell, J. H. and Orias, E. (1964). The ecological regulation of species diversity. *The American Naturalist*, **98**, 399–414.

Connor, E. F. and McCoy, E. D. (1979). The statistics and biology of the species–area relationship. *The American Naturalist*, **113**, 791–833.

Connor, E. F. and Simberloff, D. (1979). The assembly of species communities: chance or competition? *Ecology*, **69**, 1132–40.

Cordone, A. J. and Kelley, D. W. (1961). The influences of inorganic sediment on the aquatic life of streams. *California Fish and Game*, **47**, 189–228.

Cowardin, L. M. and Golet, F. C. (1995). US Fish and Wildlife Service 1979 wetland classification: a review. *Vegetatio*, **118**, 139–52.

Cowardin, L. M., Carter, V. Golet, F. C. and LaRoe, E. T. (1979). *Classification of Wetlands and Deepwater Habitats of the United States.* USDI Fish and Wildlife Service. FWS/OBS-79/31. 103 pp.

Cowling, R. M., Rundel, P. W., Lamont, B. B., Arroyo, M. K. and Arianoutsou, M. (1996a). Plant diversity in mediterranean-climate regions. *Trends in Ecology and Evolution*, **11**, 362–6.

Cowling, R. M., MacDonald, I. A. W. and Simmons, M. T. (1996b). The Cape Peninsula, South Africa: Physiographical, biological and historical background to an extraordinary hot-spot of biodiversity. *Biodiversity and Conservation*, **5**, 527–550.

Craft, C. B., Vymazal, J. and Richardson, C. J. (1995). Response of everglades plant communities to nitrogen and phosphorus additions. *Wetlands*, **15**, 258–71.

Craighead, F. C., Sr. (1968). The role of the alligator in shaping plant communities and maintaining wildlife in the southern Everglades. *The Florida Naturalist*, **41**, 2–7, 69–74.

Crawford, R. M. M. (1982). Physiological response to flooding. In *Physiological Plant Ecology II. Encyclopedia of Plant Physiology*, ed. O. L. Large, P. S. Nobel, C. B. Osmond and H. Ziegler. Berlin: VRB Springer-Verlag.

Crawford, R. M. M. and Braendle, R. (1996). Oxygen deprivation stress in a changing environment. *Journal of Experimental Botany*, **47**, 145–59.

Crawford, R. M. M. and McManmon, M (1968). Inductive responses of alcohol and malic acid dehydrogenases in relation to flooding tolerance in roots. *Journal of Experimental Botany*, **19**, 435–41.

Crawford, R. M. M. and Tyler, P. D. (1969). Organic acid metabolism in relation to flooding tolerance in roots. *Journal of Ecology*, **57**, 235–44.

Crawley, M. J. (1983). *Herbivory. The Dynamics of Animal/Plant Interactions*. Oxford: Blackwell.

Crocker, L. P. (1990). *Army Officer's Guide. 45th edn.* Harrisburg, PA: Stackpole Books.

Crow, G. E. (1993). Species diversity in aquatic angiosperms: latitudinal patterns. *Aquatic Botany*, **44**, 229–58.

Crowder, A. A. and Bristow, J. M. (1988). Report: the future of waterfowl habitats in the Canadian lower Great Lakes wetlands. *The Journal of Great Lakes Research*, **14**, 115–27.

Cummins, K. W. (1973). Trophic relationships of aquatic insects. *Annual Review of Entomology*, **18**, 83–206.

Cummins, K. W. and Klug, M. J. (1979). Feeding ecology of stream invertebrates. *Annual Review of Ecology and Systematics*, **10**, 147–72.

Currie, D. J. (1991). Energy and large-scale patterns of animal- and plant-species richness. *The American Naturalist*, **137**, 27–49.

Cyr, H. and Pace, M. L. (1993). Magnitude and patterns of herbivory in aquatic and terrestrial ecosystems. *Nature*, **361**, 148–50.

Czaya, E. (1983). *Rivers of the World*. Cambridge: Cambridge University Press.

Dacey, J. W. H. (1980). Internal winds in water lillies: an adaptation for life in anaerobic sediments. *Science*, **210**, 1017–19.

Dacey, J. W. H. (1981). Pressurized ventilation in the yellow water lily. *Ecology*, **62**, 1137–47.

Dacey, J. W. H. (1988). In *Plant Physiology*, ed. F. B. Salisbury and C. W. Ross, 3rd edn, pp. 68–70. Belmont: Wadsworth Publishers.

Dahm, C. N., Cummins, K. W., Valett, H. M. and Coleman, R. L. (1995). An ecosystem view of the restoration of the Kissimmee River. *Restoration Ecology*, **3**, 225–38.

Damman, A. and Dowhan, J. (1981). Vegetation and habitat conditions in Western Head Bog, a southern Nova Scotian plateau bog. *Canadian Journal of Botany*, **59**, 1343–59.

Dansereau, P. (1959). Vascular aquatic plant communities of southern Quebec. A preliminary analysis. *Transactions of the Northeast Wildlife Conference*, vol. 10, pp. 27–54.

Dansereau, P. and Segadas-Vianna, F. (1952). Ecological study of the peat bogs of eastern North America. *Canadian Journal of Botany*, **30**, 490–520.

Darlington, P. J. (1957). *Zoogeography: The Geographical Distribution of Animals*. New York: John Wiley.

Day, R. T., Keddy, P. A., McNeill, J. and Carleton, T. (1988). Fertility and disturbance gradients: a summary model for riverine marsh vegetation. *Ecology*, **69**, 1044–1054.

Day, W. (1984). *Genesis on Planet Earth*. 2nd edn. New Haven: Yale University Press.

Dayton, P. K. (1979). Ecology: a science and a religion. In *Ecological Processes in Coastal and Marine Systems*, ed. R. J. Livingston, pp. 3-18. New York: Plenum Press.

DeBenedictis, P. A. (1974). Interspecific competition between tadpoles of *Rana pipiens* and *Rana sylvatica*: an experimental field study. *Ecological Monographs*, **44**, 129–51.

de Groot, R. S. (1992). *Functions of Nature*. The Netherlands: Wolters-Noordhoff.

Delcourt, H. R. and Delcourt, P. A. (1988). Quaternary landscape ecology: relevant scales in space and time. *Landscape Ecology*, **2**, 23–44.

Delcourt, H. R. and Delcourt, P. A. (1991). *Quaternary Ecology: A Paleoecological Perspective*, London: Chapman and Hall.

del Moral R., Titus, J. H. and Cook, A. M. (1995). Early primary succession on Mount St. Helens, Washington, USA. *Journal of Vegetation Science*, **6**, 107–20.

De Luc, J. A. (1810). Geologic travels, In Gorham, E. (1953). Some early ideas concerning the nature, origin and development of peat lands. *Journal of Ecology*, **41**, 257–74.

Denny, P. (1972). Sites of nutrient absorption in aquatic macrophytes. *Journal of Ecology*, **60**, 819–29.

Denny, P. (1985). *The Ecology and Management of African Wetland Vegetation*. Dordrecht: Junk.

Denny, P. (1993a). Wetlands of Africa: Introduction. In *Wetlands of the World 1*, ed. D. F. Whigham *et al.*, pp. 1–31. The Netherlands: Kluwer Academic Publishers.

Denny, P. (1993b). Eastern Africa. In *Wetlands of the World 1*, ed. D. F. Whigham *et al.*, pp. 32–46. The Netherlands: Kluwer Academic Publishers.

Denny, P. (1995). Benefits and priorities for wetland conservation: the case for national conservation strategies. In *Wetlands. Archaeology and Nature Conservation*, ed. M. Cox, V. Straker and D. Taylor, Proceedings of the International Conference, University of Bristol, April 1994. HMSO.

Desmukh, I. (1986). *Ecology and Tropical Biology*, Palo Alto, CA: Blackwell Scientific.

Diamond, J. M. (1975). Assembly of Species Communities. In *Ecology and Evolution of Communities*, ed. M. L. Cody and J. M. Diamond, pp. 342–444. Cambridge, MA: Belknap Press, Harvard University Press.

Diamond, J. M. (1983). Laboratory, field and natural experiments. *Nature*, **304**, 586–7.

Dickinson, C. H. (1983). Micro-organisms in peatlands. In *Ecosystems of the World 4A. Mires: Swamp, Bog, Fen and Moor*, ed. D. W. Goodall, pp. 225–45. Amsterdam: Elsevier Scientific Publishing Company.

Dinerstein, E. (1991). Seed dispersal by greater one-horned rhinoceros (*Rhinoceros unicornis*) and the flora of *Rhinoceros* latrines. *Mammalia*, **55**, 355–62.

Dinerstein, E. (1992). Effects of *Rhinoceros unicornis* on riverine forest structure in lowland Nepal. *Ecology*, **73**, 701–4.

Dittmar, L. A. and Neely, R. K. (1999). Wetland seed bank response to sedimentation varying in loading rate and texture. *Wetlands*, **19**, 341–51.

Dore, W. G. and McNeill, J. (1980). *Grasses of Ontario.* Monograph 26, Research Branch, Agriculture Canada, Ottawa.

Dugan, P. (ed.) (1993). *Wetlands in Danger.* New York City: Oxford University Press.

Dumortier, M., Verlinden, A., Beeckman H. and van der Mijnsbrugge, K. (1996). Effects of harvesting dates and frequencies on above and below-ground dynamics in Belgian wet grasslands. *Ecoscience*, **3**, 190–8.

Duncan, R. P. (1993). Flood disturbance and the coexistence of species in a lowland podocarp forest, south Westland, New Zealand. *Journal of Ecology*, **81**, 403–16.

Durant, W. (1944). *The Story of Civilization III. Caesar and Christ.* New York: Simon and Schuster.

du Rietz, G. E. (1931). *Life-forms of Terrestrial Flowering Plants.* Uppsala: Almqvist & Wiksells Boktryckeri A. B.

Dynesius, M. and Nilsson, C. (1994). Fragmentation and flow regulation of river systems in the northern third of the world. *Science*, **266**, 753–62.

Ehrenfeld, J. G. (1983). The effects of changes in land-use on swamps of the New Jersey pine barrens. *Biological Conservation*, **25**, 353–75.

Ehrlich, A. and Ehrlich, P. (1981). *Extinction. The Causes and Consequences of the Disappearance of Species.* New York: Random House.

Elakovich, S. D., and Wootten, J. W. (1989). Allelopathic potential of sixteen aquatic and wetland plants. *Journal of Aquatic Plant Management*, **27**, 78–84.

Ellenberg, H. (1985). Veränderungen der Flora Mitteleuropas unter dem Einfluß von Düngung und Immissionen. *Schweizerische Zeitschrift fuer Forstwesen*, **136**, 19–39.

Ellenberg, H. (1988). Floristic changes due to nitrogen deposition in central Europe. In *Critical Loads for Sulphur and Nitrogen*, ed. J. Nilsson, and P. Grennfelt. Report from a workshop held at Skokloster, Sweden, 19–24 March, 1988.

Ellenberg, H. (1989). Eutrophierung – das gravierendste Problem im Naturschutz? *Norddeutsche Naturschutzakademie*, **2**, 9–12.

Ellery, W. N., Ellery, K., Rogers, K. H., McCarthy, T. S. and Walker, B. H. (1993). Vegetation, hydrology and sedimentation processes as determinants of channel form and dynamics in the northeastern Okavango Delta, Botswana. *African Journal of Ecology*, **31**, 10–25.

Ellison, A. M. and Farnsworth, E. J. (1996). Spatial and temporal variability in growth of *Rhizophora mangle* saplings on coral cays: links with variation in insolation, herbivory, and local sedimentation rate. *Journal of Ecology*, **84**, 717–31.

Elton, C. (1927). *Animal Ecology*. London: Sidgwick and Jackson Ltd.

Elveland, J. (1978). Management of rich fens in northern Sweden. Studies of various factors influencing the vegetational dynamics. *Statens naturvardsverk PM 1007*. Solna, Sweden: Forskningsnamnden.

Elveland, J. (1979). Irrigated and naturally flooded hay-meadows in North Sweden – a nature conservancy problem. *Statens naturvardsverk PM 1174*. Solna, Sweden: Forskningssekretariatet.

Elveland, J. and Sjöberg, K. (1982). Some effects of scything and other management procedures on the plant and animal life of N. Swedish Wetlands formerly mown for hay. *Statens naturvardsverket PM 1516*. Solna, Sweden: Forskningssekretariatet.

Environment Canada (1976). Marine Environmental Data Service, Ocean and Aquatic Sciences. Monthly and yearly mean water levels. Vol 1 Inland. Department of Environment, Ottawa, Canada.

Eriksson, O. (1993). The species-pool hypothesis and plant community diversity. *Oikos*, **68**, 371–4.

Essame, H. (1974). *Patton. A Study in Command*. New York: Charles Scribner's Sons.

Facelli, J. M., Leon, R. J. C. and Deregibus, V. A. (1989). Community structure in grazed and ungrazed grassland sites in the flooding Pampa, Argentina. *American Midland Naturalist*, **121**, 125–33.

Faith D. P., Minchin, P. R. and Belbin, L. (1987). Compositional dissimilarity as a robust measure of ecological distance. *Vegetatio*, **69**, 57–68.

Farney, R. A., and Bookhout, T. A. (1982). Vegetation changes in a Lake Erie marsh (Winous Point, Ottawa County, Ohio) during high water years. *Ohio Journal of Science*, **82**, 103–7.

Faulkner, S. P. and Richardson, C. J. (1989). Physical and chemical characteristics of freshwater wetland soils. In *Constructed Wetlands for Wastewater Treatment*, ed. D. A. Hammer, pp. 41–72. Municipal, Industrial and Agricultural. Chelsea, Michigan: Lewis Publishers.

Fernández-Armesto, F. (1989). *The Spanish Armada. The Experience of War in 1588*. Oxford: Oxford University Press.

Field, R., Stuzeski, E. J., Masters, H. E. and Tafuri, A. N. (1974). Water pollution and associated effects from street salting. *Journal of Environmental Engineering Division*, **100**, 459–77.

Findlay, S. C. and Houlahan, J. (1997). Anthropogenic correlates of biodiversity in southeastern Ontario wetlands. *Conservation Biology*, **11**, 1000–9.

Finney, B. P. and Johnson, T. C. (1991). Sedimentation in Lake Malawi (East Africa) during the past 10,000 years: a continuous paleoclimatic record from the southern tropics. *Palaeogeography, Palaeoclimatology, Palaeoecology*, **85**, 351–66.

Flint, R. F. (1971). *Glacial and Quaternary Geology*. New York: John Wiley.

Forman, A. T. and L. E. Alexander. (1998). Roads and their ecological effects. *Annual Review of Ecology and Systematics*, **29**, 207–31.

Foster, D. R. and Glaser, P. H. (1986). The raised bogs of south-eastern Labrador,

Canada: classification, distribution, vegetation and recent dynamics. *Journal of Ecology*, **74**, 47–71.

Foster, D. R., King, G. A., Glaser, P. H. and Wright, Jr., H. E. (1983). Origin of string patterns in boreal peatlands. *Nature*, **306**, 256–7.

Foster, D. R. and Wright Jr., H. E. (1990). Role of ecosystem development and climate change in bog formation in central Sweden. *Ecology*, **71**, 450–63.

Fox, A. D. and Kahlert, J. (1999). Adjustments to nitrogen metabolism during wing moult in Greylag Geese, *Anser anser. Functional Ecology*, **13**, 661–9.

Fragoso, J. M. V. (1998). Home range and movement patterns of white-lipped Peccary (*Tayassu pecari*) herds in the Northern Brazilian Amazon. *Biotropica*, **30**, 458–69.

Freedman, B. (1995). *Environmental Ecology.* 2nd edn. San Diego: Academic Press.

Fremlin, G. (ed. in chief). (1974). *The National Atlas of Canada.* 4th edn, revised. Toronto: Macmillan.

Frenzel, B. (1983). Mires – repositories of climatic information or self-perpetuating ecosystems? In *Ecosystems of the World 4A. Mires: Swamp, Bog, Fen and Moor*, ed. D. W. Goodall, pp. 35–65. Amsterdam: Elsevier Scientific Publishing Company.

Fretwell, S. D. (1977). The regulation of plant communities by food chains exploiting them. *Perspectives in Biology and Medicine*, **20**, 169–85.

Frey, R. W. and Basan, P. B. (1978). Coastal salt marshes. In *Coastal Sedimentary Environments*, ed. R. A. Davis, pp. 101–69. New York: Springer-Verlag.

Fritzell, E. K. (1989). Mammals in prairie wetlands. In *Northern Prairie Wetlands*, ed. A. van der Valk, pp. 268–301. Ames, Iowa: Iowa State University Press.

Galatowitsch, S. M. and van der Valk, A. G. (1994). *Restoring Prairie Wetlands: An Ecological Approach.* Ames, IA: Iowa State University Press.

Galatowitsch, S. M. and van der Valk, A. G. (1996). The vegetation of restored and natural prairie wetlands. *Ecological Applications*, **6**, 102–12.

Galinato, M. and van der Valk, A. (1986). Seed germination of annuals and emergents recruited during drawdowns in the Delta Marsh, Manitoba, Canada. *Aquatic Botany*, **26**, 89–102.

García, L. V., Marañón, T., Moreno, A. and Clemente, L. (1993). Above-ground biomass and species richness in a Mediterranean salt marsh. *Journal of Vegetation Science*, **4**, 417–24.

Gaudet, C. L. (1993). Competition in shoreline plant communities: a comparative approach. PhD thesis. University of Ottawa, Ottawa, Ontario, Canada.

Gaudet, C. L. and Keddy, P. A. (1988). A comparative approach to predicting competitive ability from plant traits. *Nature*, **334**, 242–43.

Gaudet, C. L. and Keddy, P. A. (1995). Competitive performance and species distribution in shoreline plant communities: a comparative approach. *Ecology*, **76**, 280–91.

Geis, J. W. (1985). Environmental influences on the distribution and composition of wetlands in the Great Lakes basin. In *Coastal Wetlands*, ed. H. H. Prince and F. M. D'Itri, pp. 15–31. Chelsea: Lewis Publishers.

Gentry, A. H. (1988). Changes in plant community diversity and floristic composition on environmental and geographical gradients. *Annals of the Missouri Botanical Garden*, **75**, 1–34.

Gignac, L. D. and Vitt, D. H. (1990). Habitat limitations of *Sphagnum* along climatic, chemical, and physical gradients in mires of western Canada. *The Bryologist*, **93**, 7–22.

Gilbert, J. J. (1988). Suppression of rotifer populations by *Daphnia*: a review of the evidence, the mechanisms, and the effects on zooplankton community structure. *Limnology and Oceanography,* **33**, 1286–303.

Gilbert, J. J. (1990). Differential effects of *Ananbaena affinis* on cladoceran and rotifers: mechanisms and implications. *Ecology*, 71, 1727–40.

Giller, K. E. and Wheeler, B. D. (1986). Past peat cutting and present vegetation patterns in an undrained fen in the Norfolk Broadland. *Journal of Ecology*, **74**, 219–47.

Givnish, T. J. (1982). On the adaptive significance of leaf height in forest herbs. *The American Naturalist*, **120**, 353–81.

Givnish, T. J. (1988). Ecology and evolution of carnivorous plants. In *Plant–Animal Interactions*, ed. W. B. Abrahamson, pp. 243–90. New York: McGraw-Hill.

Glaser, P. H. (1992). Raised bogs in eastern North America – regional controls for species richness and floristic assemblages. *Journal of Ecology*, **80**, 535–54.

Glaser, P. H., Janssens, J. A. and Siegel, D. I. (1990). The response of vegetation to chemical and hydrological gradients in the Lost River peatland, northern Minnesota. *Journal of Ecology*, **78**, 1021–48.

Gleason, H. A. (1926). The individualistic concept of the plant association. *Bulletin of the Torrey Botanical Club*, **53**, 7–26.

Gleason, H. A. (1939). The individualistic concept of the plant association. *American Midland Naturalist*, **21**, 92–110.

Glob, P. V. (1969). *The Bog People. Iron-Age Man Preserved.* Translated from the Danish by R. Bruce-Mitford. Ithaca, NY: Cornell University Press.

Glooschenko, W. A. (1980). Coastal ecosystems of the James/Hudson Bay area of Ontario, Canada. *Zeitschrift fuer Geomorphologie NF*, **34**, 214–24.

Godwin, Sir H. (1981). *The Archives of the Peat Bogs.* Cambridge: Cambridge University Press.

Goethe, J. W. (1831). *Goethe's Faust, Part 2.* Trans. B. Taylor, revised and edited by S. Atkins, 1962. New York: Collier Books.

Goin, C. J. and Goin, O. B. (1971). *Introduction to Herpetology*. 2nd edn. San Francisco: W. H. Freeman.

Goldsmith, F. B. (1973). The vegetation of exposed sea cliffs at South Stack, Anglesey II. Experimental studies. *Journal of Ecology*, **61**, 819–29,

Goldsmith, F. B. (ed.) (1991). *Monitoring for Conservation and Ecology.* London: Chapman and Hall.

Goldsmith, F. B. and Harrison, C. M. 1976. Description and analysis of vegetation. In *Methods in Plant Ecology*, ed. S. B. Chapman. Oxford: Scientific Publishers.

Golet, F. C. and Parkhurst, J. A. (1981). Freshwater wetland dynamics in South Kingston, Rhode Island, 1939–1972. *Environmental Management*, **5**, 245–51.

Good, R. E., Whigham, D. F. and Simpson, R. L (eds.) (1978). *Freshwater Wetlands. Ecological Processes and Management Potential.* New York: Academic Press.

Goodall, D. W. (ed.) (1983). *Ecosystems of the World 4A. Mires, Swamp, Bog, Fen and Moor.* Amsterdam: Elsevier Scientific Publishing Company.

Gopal, B. (1990). Nutrient dynamics of aquatic plant communities. In *Ecology and Management of Aquatic Vegetation in the Indian Subcontinent*, ed. B. Gopal, pp. 177–97. The Netherlands: Kluwer Academic Publishers.

Gopal, B. and Goel, U. (1993). Competition and allelopathy in aquatic plant communities. *The Botanical Review*, **59**, 155–210.

Gopal, B., Kvet, J., Loffler, H., Masing, V. and Patten, B. (1990). Definition and classification. In *Wetlands and Shallow Continental Water Bodies. Vol. 1. Natural and Human Relationships*, ed. B. C. Patten, pp. 9–15. The Hague, The Netherlands: SPB Academic Publishing.

Gore, A. J. P. (1983). Introduction. In *Ecosystems of the World 4A. Mires: Swamp, Bog, Fen and Moor*, ed. A. J. P. Gore, Amsterdam: Elsevier Scientific Publishing Company.

Gorham, E. (1953). Some early ideas concerning the nature, origin and development of peat lands. *Journal of Ecology*, **41**, 257–74.

Gorham, E. (1957). The development of peatlands. *The Quarterly Review of Biology*, **32**, 145–66.

Gorham, E. (1961). Water, ash, nitrogen and acidity of some bog peats and other organic soils. *Journal of Ecology*, **49**, 103–6.

Gorham, E. (1990). Biotic impoverishment in northern peatlands. In *The Earth in Transition*, ed. G. M. Woodwell. Cambridge: Cambridge University Press.

Gorham, E. (1991). Northern peatlands role in the carbon cycle and probable responses to climatic warming. *Ecological Applications*, **1**, 182–95.

Gosselink, J. G. and Turner, R. E. (1978). The role of hydrology in freshwater wetland ecosystems. In *Freshwater Wetlands – Ecological Processes and Management Potential*, ed. R. E. Good, D. F. Whigham and R. L. Simpson, pp. 63–79. New York: Academic Press.

Gough, J. (1793). Reasons for supposing that lakes have been more numerous than they are at present; with an attempt to assign the causes whereby they have been defaced. *Memoirs of the Literary and Philosophical Society of Manchester*, **4**, 1–19. In Walker, D. (1970). *Direction and Rate in Some British Post-Glacial Hydroseres.* In *Studies in the Vegetational History of the British Isles*, ed. D. Walker and R. G. West, pp. 117–39. Cambridge: Cambridge University Press.

Gough, L. G., Grace, J. B. and Taylor, K. L. (1994). The relationship between species richness and community biomass: the importance of environmental variables. *Oikos*, **70**, 271–9.

Goulding, M. (1980). *The Fishes and the Forest: Explorations in Amazonian Natural History.* Berkley: University of California Press.

Grace, J. B. (1990). On the relationship between plant traits and competitive ability. In *Perspectives on Plant Competition*, ed. J. B. Grace and D. Tilman, pp. 51–65. San Diego, CA: Academic Press.

Grace, J. B. (1999). The factors controlling species density in herbaceous plant communities: an assessment. *Perspectives in Plant Ecology, Evolution and Systematics*, **2**, 1–28.

Grace, J. B. and Wetzel, R. G. (1981). Habitat partitioning and competitive displacement in cattails (*Typha*): Experimental field studies. *The American Naturalist*, **118**, 463–74.

Greening, H. (1995). Resource-based watershed management in Tampa Bay. In *Wetlands and Watershed Management. Science Applications and Public Policy*, ed. J. A. Kusler, D. E. Willard and H. C. Hull Jr., pp. 172–81. A collection of papers from a national symposium and several workshops at Tampa, FL, April 23–26. The Association of State Wetland Managers, NY.

Griffiths, R. A., Denton, J. and Wong, A. L. (1993). The effect of food level on competition in tadpoles: interference mediated by protothecan algae? *Journal of Animal Ecology*, **62**, 274–9.

Grime, J. P. (1973). Competitive exclusion in herbaceous vegetation. *Nature*, **242**, 344–7.

Grime, J. P. (1974). Vegetation classification by reference to strategies. *Nature*, **250**, 26–31.

Grime, J. P. (1977). Evidence for the existence of three primary strategies in plants and its relevance to ecological and evolutionary theory. *The American Naturalist*, **111**, 1169–94.

Grime, J. P. (1979). *Plant Strategies and Vegetation Processes.* Chichester: John Wiley and Sons.

Grime, J. P. and Hunt, R. (1975). Relative growth-rate: its range and adaptive significance in a local flora. *Journal of Ecology*, **63**, 393–422.

Grime, J. P., Mason, G., Curtis, A. V., Rodman, J., Band, S. R., Mowforth, M. A. G., Neal, A. M. and Shaw, S. (1981). A comparative study of germination characteristics in a local flora. *Journal of Ecology*, **69**, 1017–59.

Grishin, S. Y., del Moral, R., Krestov, P. V. and Verkholat, V. P. (1996). Succession following the catastrophic eruption of Ksudach volcano (Kamchatka, 1907). *Vegetatio*, **127**, 129–53.

Groombridge, B. (ed.) (1992). *Global biodiversity. Status of the Earth's living resources.* A report compiled by the World Conservation Monitoring Centre. London: Chapman and Hall.

Grootjans, A. P., van Diggelen, R., Everts, H. F., Schipper, P. C., Streefkerk, J., de Vries, N. P. and Wierda, A. (1993). Linking ecological patterns to hydrological conditions on various spatial scales: a case study of small stream valleys. In *Landscape Ecology of a Stressed Environment*, ed. C. C. Vos and P. Opdam, pp. 60–99. London: Chapman and Hall.

Grosse, W., Büchel, H. B. and Tiebel, H. (1991). Pressurized ventilation in wetland plants. *Aquatic Botany*, **39**, 89–98.

Grover, A. M. and Baldassarre, G. A. (1995). Bird species richness within beaver ponds in south-central New York. *Wetlands*, **15**, 108–18.

Grubb, P. J. (1977). The maintenance of species-richness in plant communities: the importance of the regeneration niche. *Biological Review*, **52**, 107–45.

Grubb, P. J. (1985). Plant populations and vegetation in relation to habitat disturbance and competition: problems of generalizations. In *The Population Structure of Vegetation*, ed. J. White, pp. 595–621. The Hague: Dr W. Junk,

Grubb, P. J. (1986). Problems posed by sparse and patchily distributed species in species-rich plant communities. In *Community Ecology*, ed. J. M. Diamond and T. J. Case, pp. 207–25. New York, NY: Harper and Row.

Grubb, P. J. (1987). Global trends in species-richness in terrestrial vegetation: a view from the northern hemisphere. In *Organization of Communities Past and Present*, ed. J. H. R. Gee and P. S. Giller, pp. 99–118. 27th symposium of The British Ecological Society, Aberystwyth. Oxford: Blackwell Scientific Publications.

Grumbine, R. E. (1994). What is ecosystem management? *Conservation Biology*, **8**, 27–38.

Grumbine, R. E. (1997). Reflections on 'What is ecosystem management?' *Conservation Biology,* **11**, 41–7.

Gurevitch, J., Morrow, L., Wallace, A. and Walsh, A. (1992). A meta-analysis of competition in field experiments. *The American Naturalist*, **140**, 539–72.

Guy, H. P. (1973). Sediment problems in urban areas. In *Focus on Environmental Geology*, ed. R. W. Tank. New York: Oxford University Press.

Guyer, C. and Bailey, M. A. (1993). Amphibians and reptiles of longleaf pine communities. In *Proceedings of the Tall Timbers Fire Ecology Conference*, No. 18, ed. S. M. Hermann, pp. 139–58. The Longleaf Pine Ecosystem: ecology, restoration and management. Tall Timbers Research Station, Tallahassee, Fl.

Hacker, S. D. and Bertness, M. D. (1999). Experimental evidence for factors maintaining plant species diversity in a new England salt marsh. *Ecology*, **80**, 2064–73.

Haefner, J. W. (1978). Ecosystem assembly grammars: generative capacity and empirical adequacy. *Journal of Theoretical Biology*, **73**, 293–318.

Haefner, J. W. (1981). Avian community assembly rules: the foliage-gleaning guild. *Oecologia*, **50**, 131–42.

Haeuber, R. and Franklin, J. (eds.) (1996). Perspectives on ecosystem management. *Ecological Applications*, **6**, 692–747.

Hairston, N. G., Smith, F. E. and Slobodkin, L. B. (1960). Community structure, population control, and competition. *The American Naturalist*, **XCIV**, 421–5.

Haith, D. A. and Shoemaker, L. L. (1987). Generalized watershed loading functions for stream-flow nutrients. *Water Resources Bulletin*, **23**, 471–8.

Hamilton, S. K., Sipel, S. J. and Melack, J. M. (1996). Inundation patterns in the Pantanal wetland of South America determined from passive microwave remote sensing. *Archiv fuer Hydrobiologie*, **137**, 1–23.

Hammer, D. A. (1969). Parameters of a marsh snapping turtle population Lacreek refuge, South Dakota. *Journal of Wildlife Management*, **33**, 995–1005.

Hammer, D. A. (ed.) (1989). *Constructed Wetlands for Wastewater Treatment. Municipal, Industrial and Agricultural.* Chelsea, MI: Lewis Publishers.

Hanski, I. (1994). Patch-occupancy dynamics in fragmented landscapes. *Trends in Ecology and Evolution*, **9**, 131–5.

Hanski, I. and Gilpin, M. (1991). Metapopulation dynamics: a brief history and conceptual domain. *Biological Journal of the Linnean Society*, **42**, 3–16.

Hardin, G. (1968). The tragedy of the commons. *Science*, **162**, 1243–8.

Hardin, G. and Baden, J. (1977). *Managing the Commons.* San Francisco: W. H. Freeman and Company.

Harper, J. L. (1977). *Population Biology of Plants.* London: Academic Press.

Harper, J. L. and Benton, R. A. (1966). The behavior of seeds in soil: II. The germination of seeds on the surface of a water supplying substrate. *Journal of Ecology*, **54**, 151–66.

Harper, J. L., Williams, J. T. and Sagar, G. R. (1965). The Behavior of seeds in soil: I. The heterogeneity of soil surfaces and its role in determining the establishment of plants from seed. *Journal of Ecology*, **53**, 273–86.

Harris, R. R., Fox, C. A. and Risser, R. (1987). Impact of hydroelectric development on riparian vegetation in the Sierra Nevada region, California, USA. *Environmental Management*, **11**, 519–27.

Harris, S. W. and Marshall, W. H. (1963). Ecology of water-level manipulations on a northern marsh. *Ecology*, **44**, 331–43.

Hart, D. D. (1983). The importance of competitive interactions within stream populations and communities. In *Stream Ecology. Application and Testing of General Ecological Theory*, ed. J. R. Barnes and G. W. Minshall, pp. 99–136. New York: Plenum Press.

Hartman, J. M. (1988). Recolonization of small disturbance patches in a New England salt marsh. *American Journal of Botany*, **75**, 1625–31.

Harvey, P. H., Colwell, R. K. Silvertown, J. W. and May, R. M. (1983). Null models in ecology. *Annual Review of Ecology and Systematics*, **14**, 189–211.

Haukos, D. A. and Smith, L. M. (1993). Seed-bank composition and predictive ability of field vegetation in playa lakes. *Wetlands*, **13**, 32–40.

Haukos, D. A. and Smith, L. M. (1994). Composition of seed banks along an elevational gradient in playa wetlands. *Wetlands*, **14**, 301–7.

Hayati, A. A. and Proctor, M. C. F. (1991). Limiting nutrients in acid-mire vegetation: peat and plant analyses and experiments on plant responses to added nutrients. *Journal of Ecology*, **79**, 75–95.

Heal, O. W., Latter, P. M. and Howson, G. (1978). A study of the rates of decomposition of organic matter. In *Production Ecology of British Moors and Montane Grasslands*, ed. O. W. Heal and D. F. Perkins, pp. 136–59. Ecological Studies vol. 27. Berlin: Springer-Verlag.

Hellquist, C. B. and Crow, G. E. (1984). *Aquatic Vascular Plants of New England: Part 7. Cabombaceae, Nymphaeaceae, Nelumbonaceae, and Ceratophyllaceae.* Station Bulletin 527 March 1984. Durham, New Hampshire: University of New Hampshire.

Hemphill, N. and Cooper, S. D. (1983). The effect of physical disturbance on the relative abundances of two filter-feeding insects in a small stream. *Oecologia*, **58**, 378–82.

Higgs, E. S. (1997). What is good ecological restoration? *Conservation Biology*, **11**, 338–48.

Hik, D. S., Jefferies, R. L. and Sinclair, A. R. E. (1992). Foraging by geese, isostatic uplift and asymmetry in the development of salt-marsh plant communities. *Journal of Ecology*, **80**, 395–406.

Hill, N. M. and Keddy, P. A. (1992). Prediction of rarities from habitat variables: coastal plain plants on Nova Scotian lakeshores. *Ecology*, **73**, 1852–9.

Hill, N. M., Keddy, P. A. and Wisheu, I. C. (1998). A hydrological model for predicting the effects of dams on the shoreline vegetation of lakes and reservoirs. *Environmental Management*, **22**, 723–36.

Hoagland, B. W. and Collins, S. L. (1997a). Gradient models, gradient analysis, and hierarchical structure in plant communities. *Oikos*, **78**, 23–30.

Hoagland, B. W. and Collins, S. L. (1997b). Heterogeneity in shortgrass prairie vegetation: the role of playa lakes. *Journal of Vegetation Science*, **8**, 277–86.

Hochachka, P. W., Fields, J. and Mustafa, T. (1973). Animal life without oxygen: Basic biochemical mechanisms. *American Zoology*, **13**, 543–55.

Hogenbirk, J. C. and Wein, R. W. (1991). Fire and drought experiments in northern wetlands: a climate change analogue. *Canadian Journal of Botany*, **69**, 1991–7.

Hogg, E. H., Lieffers, V. J. and Wein, R. W. (1992). Potential carbon losses from peat profiles: Effects of temperature, drought cycles, and fire. *Ecological Applications*, **2**, 298–306.

Holling, C. S. (ed.) (1978). *Adaptive environmental assessment and management.* Chichester: John Wiley

Hook, D. D. (1984). Adaptations to flooding with fresh water. In *Flooding and Plant Growth*, ed. T. T. Kozlowski, pp. 265–94. Orlando: Academic Press.

Hook, D. D., McKee, Jr. W. H., Smith, H. Gregory, J., Burrell, Jr. V. J., DeVoe, W. R., Sojka, R. E. Gilbert, S., Banks, R. Stolzy, L. H. Brooks, C., Matthews, T. D. and

Shear, T. H. (eds.) (1988). *The Ecology and Management of Wetlands Volume 1: Ecology of Wetlands*. Portland: Timber Press.

Hoover, J. J. and Killgore, K. J. (1998). Fish communities. In *Southern Forested Wetlands. Ecology and Management*, ed. M. G. Messina and W. H. Conner, pp. 237–60. Boca Raton, FL: Lewis Publishers.

Horn, H. (1976). Succession. In *Theoretical Ecology: Principles and Applications*, ed. R. M. May, pp. 187–204. Philadelphia: W. B. Saunders.

Hou, H-Y. (1983). Vegetation of China with reference to its geographical distribution. *Annals of the Missouri Botanical Garden*, **70**, 509–48.

Houlahan, J. E., Findlay, C. S., Schmidt, B. R., Meyer, A. H. and Kuzmin, S. L. (2000). Quantitative evidence for global amphibian population declines. *Nature*, **404**, 752–5.

Howard-Williams, C. and Thompson, K. (1985). The conservation and management of African wetlands. In *The Ecology and Management of African Wetland Vegetation*, ed. P. Denny, pp. 203–30. Dordrecht: Dr W. Junk Publishers.

Howarth, R. W., Fruci, J. R. and Sherman, D. (1991). Inputs of sediment and carbon to an estuarine ecosystem: Influence of land use. *Ecological Applications*, **1**, 27–39.

Hubbell, S. P. and Foster, R. B. (1986). Biology, chance, and the history and structure of tropical rain forest tree communities. In *Community Ecology*, ed. J. Diamond and T. J. Case, pp. 314–29. New York: Harper and Row.

Hughes, J. D. and Thirgood, J. V. (1982). Deforestation, erosion and forest management in ancient Greece and Rome. *Journal of Forestry*, **26**, 60–75.

Hunter, M. D. and Price, P. W. (1992). Playing chutes and ladders: heterogeneity and the relative roles of bottom-up and top-down forces in natural communities. *Ecology*, **73**, 724–32.

Hurlbert, S. H. (1984). Pseudoreplication and the design of ecological field experiments. *Ecological Monographs*, **54**, 187–211.

Hurlbert, S. H. (1990). Spatial distribution of the montane unicorn. *Oikos*, **58**, 257–271.

Huston, M. (1979). A general hypothesis of species diversity. *American Naturalist*, **113**, 81–101.

Huston, M. (1994). *Biological Diversity. The Coexistence of Species on Changing Landscapes*. Cambridge: Cambridge University Press.

Hutchinson, G. E. (1959). Homage to Santa Rosalia or why are there so many kinds of animals? *The American Naturalist*, **93**, 145–9.

Hutchinson, G. E. (1975). *A Treatise on Limnology. Volume 3. Limnological Botany.* New York: John Wiley.

Ingram, H. A. P. (1982). Size and shape in raised mire ecosystems: a geophysical model. *Nature*, **297**, 300–3.

Ingram, H. A. P. (1983). Hydrology. In *Ecosystems of the World 4A. Mires: Swamp, Bog, Fen and Moor*, ed. D. W. Goodall, pp. 67–158. Amsterdam: Elsevier Scientific Publishing Company.

International Joint Commission. (1980). *Pollution in the Great Lakes Basin from Land Use Activities.* International Joint Commission, pp 141.

Irion, G. M., Müller, J., de Mello, J. N. and Junk, W. J. (1995). Quaternary geology of the Amazon lowland. *Geo-Marine Letters*, **15**, 172–8.

Isabelle, P. S., Fooks, L. J., Keddy, P. A. and Wilson, S. D. (1987). Effects of roadside snowmelt on wetland vegetation: an experimental study. *Journal of Environmental Management*, **25**, 57–60.

Jackson, J. B. C. (1981). Interspecific competition and species distributions: the ghosts of theories and data past. *American Zoologist*, **21**, 889–901.

Jackson, M. B. and Drew, M. C. (1984). Effects of flooding on growth and metabolism of herbaceous plants. In *Flooding and Plant Growth*, ed. T. T. Kozlowski, pp. 47–128. Orlando: Academic Press.

Janzen, D. H. and Martin, P. S. (1982). Neotropical anachronisms: the fruits the gomphotheres ate. *Science*, **215**, 19–27.

Jean, M. and Bouchard, A. (1991). Temporal changes in wetland landscapes of a section of the St. Lawrence River, Canada. *Environmental Management*, **15**, 241–50.

Jefferies, R. L. (1977). The vegetation of salt marshes at some coastal sites in arctic North America. *Journal of Ecology*, **65**, 661–72.

Jefferies, R. L. (1988a). Pattern and process in Arctic coastal vegetation in response to foraging by lesser snow geese. In *Plant Form and Vegetation Structure*, ed. M. J. A. Werger, P. J. M. van der Aart, H. J. During and J. T. A. Verhoeven, pp. 281–300. The Hague, The Netherlands: SPB Academic Publishing.

Jefferies, R. L. (1988b). Vegetational mosaics, plant-animal interactions and resources for plant growth. In *Plant Evolutionary Biology*, ed. L. Gottlieb and S. K. Jain. London: Chapman and Hall.

Jeglum, J. K. and He, F. (1995). Pattern and vegetation–environment relationships in a boreal forested wetland in northeastern Ontario, **73**, 629–37.

Johnson, D. L., Lynch Jr., W. E. and Morrison, T. W. (1997). Fish communities in a diked Lake Erie wetland and an adjacent undiked area. *Wetlands*, **17**, 43–54.

Johnson, M. G., Leach, J. H., Minns, C. K. and Oliver, C. H. (1977). Limnological characteristics of Ontario lakes in relation to associations of walleye (*Stizostedion vitreum*), northern pike (*Esox lucius*), lake trout (*Salvelinus namaycush*) and smallmouth bass (*Micropterus dolomieui*). *Journal of the Fisheries Research Board of Canada*, **34**, 1592–601.

Johnson, W. C. (1994). Woodland expansion in the Platte River, Nebraska: patterns and causes. *Ecological Monographs*, **64**, 45–84.

Johnson, W. C., Burgess, R. L. and Keammerer, W. R. (1976). Forest overstory vegetation and environment on the Missouri River floodplain in North Dakota. *Ecological Monographs*, **46**, 59–84.

Johnston, C. A. and Naiman, R. J. (1990). Aquatic patch creation in relation to beaver population trends. *Ecology*, **71**, 1617–21.

Jones, C. G., Lawton, J. H. and Shachak, M. (1994). Organisms as ecosystem engineers. *Oikos*, **69**, 373–86.

Jones, R. H., Sharitz, R. R., Dixon, P. M., Segal, D. S. and Schneider, R. L. (1994). Woody plant regeneration in four floodplain forests. *Ecological Monographs*, **64**, 345–67.

Jordan, W. R. III, Gilpin, M. E. and Aber, J. D. (1987). *Restoration Ecology. Synthetic Approach to Ecological Research*. Cambridge: Cambridge University Press.

Judson, S. (1968). Erosion of the land, or what's happening to our continents? *American Scientist*, **56**, 356–74.

Junk, W. J. (1983). Ecology of swamps on the Middle Amazon. In *Ecosystems of the*

World 4B: Mires: Swamp, Bog, Fen and Moor, ed. D. W. Goodall. Amsterdam: Elsevier Science.

Junk, W. J. (1984). Ecology of the *várzea*, floodplain of Amazonian white-water rivers. In *The Amazon Limnology and Landscape Ecology of a Mighty Tropical River and its Basin*, ed. H. Sioli, pp. 215–43. The Netherlands: Junk Publishers.

Junk, W. J. (1986). Aquatic plants of the Amazon system. In *The Ecology of River Systems*, ed. B. R. Davies and K. F. Walker, pp. 319–37. Dordrecht, The Netherlands: Dr W. Junk Publishers.

Junk, W. J. (1993). Wetlands of tropical South America. In *Wetlands of the World I*, ed. D. F. Whigham, pp. 679–739. The Netherlands: Kluwer Academic Publishers.

Junk, W. J., and Welcomme, R. L. (1990). Floodplains In *Wetlands and Shallow Continental Water Bodies. Vol. 1. Natural and Human Relationships*, ed. B. C. Patten. pp. 491–524. The Hague, The Netherlands: SPB Academic Publishing.

Junk, W. J. and Piedade, M. T. F. (1994). Species diversity and distribution of herbaceous plants in the floodplain of the middle Amazon. *Ver. Internat. Verein. Limnol.*, **25**, 1862–5.

Junk, W. J. and Piedade, M. T. F. (1997). Plant life in the floodplain with special reference to herbaceous plants. In *The Central Amazon Floodplain*, ed. W. J. Junk, pp. 147–85. Berlin: Springer-Verlag.

Junk, W. J., Bayley, P. B. and Sparks, R. E. (1989). The flood pulse concept in river-floodplain systems. In *Proceedings of the International Large River Symposium*, ed. D. P. Dodge, pp. 110-27. Canadian Special Publication of Fisheries and Aquatic Sciences. 106.

Junk, W. J., Soares, M. G. M. and Saint-Paul, U. (1997). The fish. Pp. 385–408. In *The Central Amazon Floodplain Ecological Studies*, Vol. 126, pp. 385–408. Berlin: Springer-Verlag.

Jurik, T. M., Wand, S. and van der Valk, A. G. (1994). Effects of sediment load on seedling emergence from wetland seed banks. *Wetlands*, **14**, 159.

Justin, S. H. F. W. and Armstrong, W. (1987). The anatomical characteristics of roots and plant response to soil flooding. *New Phytologist*, **106**, 465–95.

Kajak, Z. (1993). The Vistula River and its riparian zones. *Hydrobiologia*, **251**, 149–57.

Kalamees, K. (1982). The composition and seasonal dynamics of fungal cover on peat soils. In *Peatland Ecosystems: Researches into the Plant Cover of Estonian Bogs and Their Productivity*, ed. V. Masing, pp. 12–29. Academy of Sciences of the Estonian S. S. R., Tallinn.

Kalliola, R., Salo, J., Puhakka, M. and Rajasilta, M. (1991). New site formation and colonizing vegetation in primary succession on the Western Amazon Floodplains. *Journal of Ecology*, **79**, 877–901.

Kaminski, R. M. and Prince, H. H. (1981). Dabbling duck and aquatic macroinvertebrate responses to manipulated wetland habitat. *Journal of Wildlife Management*, **45**, 1–15

Kaminski, R. M., Murkin, H. M. and Smith, C. E. (1985). Control of cattail and bulrush by cutting and flooding. In *Coastal Wetlands*, ed. H. H. Prince and F. M. D'Itri, pp. 253–62. Chelsea: Lewis Publishers.

Kantrud, H. A., Millar, J. B. and van der Valk, A. G. (1989). Vegetation of the wetlands of the prairie pothole region. In *Northern Prairie Wetlands*, ed. A. G. van der Valk, pp. 132–87. Ames, Iowa: Iowa State University Press.

Karrow, P. F. and P. E. Calkin (eds.) (1985). *Quaternary Evolution of the Great Lakes*. Geological Association of Canada. Special Paper 30. St John's, Newfoundland.

Keddy, C. J. and McCrae, T. (1989). *Environmental Databases for State of the Environment Reporting*. State of the Environment Reporting Branch, Environment Canada. Technical Report No. 19, 170 pp.

Keddy, P. A. (1976). Lakes as islands: the distributional ecology of two aquatic plants, *Lemna minor* L. and *L. trisulca* L. *Ecology*, **57**, 353–9.

Keddy, P. A. (1981). Vegetation with coastal plain affinities in Axe Lake, near Georgian Bay, Ontario. *Canadian Field Naturalist*, **95**, 241–8.

Keddy, P. A. (1982). Quantifying within lake gradients of wave energy, substrate particle size and shoreline plants in Axe Lake, Ontario. *Aquatic Botany*, **14**, 41–58.

Keddy, P. A. (1983). Shoreline vegetation in Axe Lake, Ontario: effects of exposure on zonation patterns. *Ecology*, **64**, 331–44.

Keddy, P. A. (1984). Plant zonation on lakeshores in Nova Scotia: a test of the resource specialization hypothesis. *The Journal of Ecology*, **72**, 797–808.

Keddy, P. A. (1985a). Lakeshores in the Tusket River Valley, Nova Scotia: distribution and status of some rare species, including *Coreopsis rosea* Nutt. and *Sabtia kennedyana* Fern. *Rhodora*, **87**, 309–20.

Keddy, P. A. (1985b). Wave disturbance on lakeshores and the within-lake distribution of Ontario's Atlantic coastal plain flora. *Canadian Journal of Botany*, **63**, 656–60.

Keddy, P. A. (1989a). *Competition*. London: Chapman and Hall.

Keddy, P. A. (1989b). Effects of competition from shrubs on herbaceous wetland plants: a 4-year field experiment. *Canadian Journal of Botany*, **67**, 708–16.

Keddy, P. A. (1990a). Competitive hierarchies and centrifugal organization in plant communities. In *Perspectives on Plant Competition*, ed. J. B. Grace and D. Tilman, pp. 265–90. San Diego: Academic Press.

Keddy, P. A. (1990b). Water level fluctuations and wetland conservation. In *Proceedings of an International Symposium Wetlands of the Great Lakes*. New York: Niagara Falls.

Keddy, P. A. (1990c). Is mutualism really irrelevant to ecology? *Bulletin of the Ecological Society of America*, **71**(2), 101–2.

Keddy, P. A. (1991a). Biological monitoring and ecological prediction: from nature reserve management to national state of environment indicators. In *Biological Monitoring for Conservation*, ed. F. B. Goldsmith. London: Chapman and Hall.

Keddy, P. A. (1991b). Water level fluctuations and wetland conservation. In *Wetlands of the Great Lakes*, ed. J. Kusler and R. Smardon, pp. 79–91. Proceedings of the Wetlands of the Great Lakes Symposium, Niagara, May 16 – May 18, 1990.

Keddy, P. A. (1991c). Reviewing a festschrift: what are we doing with our scientific lives? *Journal of Vegetation Science*, **2**, 419–24.

Keddy, P. A. (1992a). Assembly and response rules: two goals for predictive community ecology. *Journal of Vegetation Science*, **3**, 157–64.

Keddy, P. A. (1992b). A pragmatic approach to functional ecology. *Functional Ecology*, **6**, 621–6.

Keddy, P. A. (1994). Applications of the Hertzsprung–Russell star chart to ecology: reflections on the 21st birthday of Geographical Ecology. *Trends in Ecology and Evolution*, **9**, 231–4.

Keddy, P. A. and Reznicek, A. A. (1982). The role of seed banks in the persistence of Ontario's coastal plain flora. *American Journal of Botany*, **69**, 13–22.

Keddy, P. A. and Constabel, P. (1986). Germination of ten shoreline plants in relation to seed size, soil particle size and water level: an experimental study. *Journal of Ecology*, **74**, 122–41.

Keddy, P. A. and Reznicek, A. A. (1986). Great Lakes vegetation dynamics: the role of fluctuating water levels and buried seeds. *Journal of Great Lakes Research*, **12**, 25–36.

Keddy, P. A. and Shipley, B. (1989). Competitive hierarchies in herbaceous plant communities. *Oikos*, **54**, 234–41,

Keddy, P. A. and Wisheu, I. C. (1989). Ecology, biogeography, and conservation of coastal plain plants: some general principles from the study of Nova Scotian wetlands. *Rhodora*, **91**, 72–94.

Keddy, P. A. and MacLellan, P. (1990). Centrifugal organization in forests. *Oikos*, **59**, 75–84.

Keddy, P. A., Fraser, L. H. and Wisheu, I. C. (1998). A comparative approach to examine competitive responses of 48 wetland plant species. *Journal of Vegetation Science*, **9**, 777–86.

Keddy, P. A., Lee, H. T. and Wisheu, I. C. (1993). Choosing indicators of ecosystem integrity: wetlands as a model system. In *Ecological Integrity and the Management of Ecosystems*, ed. S. Woodley, J. Kay and G. Francis. Ottawa: St-Lucie Press.

Keddy, P. A., Twolan-Strutt, L. and Wisheu, I. C. (1994). Competitive effect and response rankings in 20 wetland plants: are they consistent across three environments? *Journal of Ecology*, **82**, 635–43.

Keeley, J. E., DeMason, D. A., Gonzalez, R. and Markham, K. R. (1994). Sediment-based carbon nutrition in tropical alpine *Isoetes*. In *Tropical Alpine Environments Plant Form and Function*, ed. P. W. Rundel, A. P. Smith and F. C. Meinzer, pp. 167–94. Cambridge: Cambridge University Press.

Kelly, K. (1975). The artificial drainage of land in nineteenth-century southern Ontario. *Canadian Geographer*, **4**, 279–98.

Kendall, R. L. (1969). An ecological history of the Lake Victoria Basin. *Ecological Monographs*, **39**, 121–76.

Keogh, T. M., Keddy, P. A. and Fraser, L. H. (1998). Patterns of tree species richness in forested wetlands. *Wetlands*, **19**, 639–47.

Kershaw, K. A. (1962). Quantitative ecological studies from Landmannahellir, Iceland. *Journal of Ecology*, **50**, 171–9.

Kershner, J. L. (1997). Setting riparian/aquatic restoration objectives within a watershed context. *Restoration Ecology*, **5**, 15–24.

Kirk, K. L. and Gilbert, J. J. (1990). Suspended clay and the population dynamics of planktonic rotifers and cladocerans. *Ecology*, **71**, 1741–55.

Klimas, C. V. (1988). River regulation effects on floodplain hydrology and ecology. In *The Ecology and Management of Wetlands*, Volume 1: *Ecology of Wetlands*, ed. D. D. Hook, W. H. McKee, Jr., H. K. Smith, J. Gregory, V. G. Burrell, Jr., M. R. DeVoe, R. E. Sojka, S. Gilbert, R. Banks, L. H. Stolzy, C. Brooks, T. D. Matthews and T. H. Shear, pp. 40-9. Portland: Timber Press.

Koebel, J. W., Jr. (1995). An historical perspective on the Kissimmee River restoration project. *Restoration Ecology*, **3**, 149–59.

Koerselman, W. and Verhoeven, J. T. A. (1995). In *Restoration of Temperate Wetlands*, ed. S. Wheeler, S. Shaw, W. Fojt and R. Robertson, pp. 91-112. Chichester, UK: John Wiley.

Koerselman, W. and Meuleman, F. M. (1996). The vegetation N:P ratio: a new tool to detect the nature of nutrient limitaton. *Journal of Applied Ecology*, **33**, 1441–50.

Kozlowski, T. T. (ed.) (1984a). *Flooding and Plant Growth.* Orlando: Academic Press.

Kozlowski, T. T. (1984b). Responses of woody plants to flooding. In *Flooding and Plant Growth*, ed. T. T. Kozlowski, pp. 129–63. Orlando: Academic Press.

Kozlowski, T. T. and Pallardy, S. G. (1984). Effect of flooding on water, carbohydrate, and mineral relations. In *Flooding and Plant Growth*, ed. T. T. Kozlowski, pp. 165–93. Orlando: Academic Press.

Kramer, D. L., Lindsay, C. C., Moodie, G. E. E. and Stevens, E. D. (1978). The fishes and the aquatic environment of the Central Amazon basin, with particular reference to respiratory patterns. *Canadian Journal of Zoology*, **56**, 717–29.

Kuhry, P. (1994). The role of fire in the development of *Sphagnum*-dominated peatlands in western boreal Canada. *Journal of Ecology*, **82**, 899–910.

Kuhry, P., Nicholson, B. J., Gignac, L. D., Vitt, D. H. and Bayley, S. E. (1993). Development of *Sphagnum*-dominated peatlands in boreal continental Canada. *Canadian Journal of Botany*, **71**, 10–22.

Kurihara, Y. and Kikkawa, J. (1986). Trophic relations of decomposers. In *Community Ecology: Pattern and Process*, ed. J. Kikkawa and D. J. Anderon, pp. 127–60. Melbourne, Blackwell Scientific Publications.

Kurimo, H. (1984). Simultaneous groundwater table fluctuation in different parts of the Virgin Pine Mires. *Silva Fennica*, **18**, 151–86.

Kusler, J. A. and Kentula, M. E. (eds.) (1990). *Wetland Creation and Restoration: Status of the Science.* Washington, DC: Island Press.

Kusler, J. A., Willard, D. E. and Hull Jr., H. C. (eds.) (1995). *Wetlands and Watershed Management. Science Applications and Public Policy.* A collection of papers from a national symposium and several workshops at Tampa, FL, April 23–26. New York: The Association of State Wetland Managers.

LaBaugh, J. W. (1989). Chemical characteristics of water in northern prairie wetlands. In *Northern Prairie Wetlands*, ed. A. G. van der Valk, pp. 56–90. Ames, Iowa: Iowa State University Press.

Laing, H. E. (1940). Respiration of the rhizomes of *Nuphar advenum* and other water plants. *American Journal of Botany*, **27**, 574–81.

Laing, H. E. (1941). Effect of concentration of oxygen and pressure of water upon growth of rhizomes of semi-submerged water plants. *Botanical Gazette*, **102**, 712–724.

Lane, P. A. (1985). A food web approach to mutualism in lake communities. In *The Biology of Mutualism. Ecology and Evolution*, ed. D. H. Boucher, pp. 344–74. Oxford University Press, NY.

Larcher, W. (1995). *Physiological Plant Ecology: Ecophysiology and Stress Physiology of Functional Groups.* 3rd edn. New York: Springer-Verlag.

Larson, D. W. (1996). Brown's Woods: an early gravel pit forest restoration project, Ontario, Canada. *Restoration Ecology*, **4**, 11–18.

Larson, J. S. (1988). Wetland creation and restoration: an outline of the scientific perspective. In *Increasing our Wetland Resources*, ed. J. Zelazny and J. S. Feierabend, pp.

73–9. Proceedings of a conference in Washington, DC October 4–7, 1987. National Wildlife Federation–Corporate Conservation Council.

Larson, J. S. (1990). Wetland value assessment. In *Wetlands and Shallow Continental Water Bodies. Vol. 1. Natural and Human Relationships*, ed. B. C. Patten, pp. 389–400. The Hague, The Netherlands: SPB Academic Publishing.

Larson, J. S., Mueller, A. J. and MacConnell, W. P. (1980). A model of natural and man-induced changes in open freshwater wetlands on the Massachusetts coastal plain. *Journal of Applied Ecology*, **17**, 667–73.

Latham, P. J., Pearlstine, L. G. and Kitchens, W. M. (1994). Species association changes across a gradient of freshwater, oligohaline, and mesohaline tidal marshes along the lower Savannah River. *Wetlands*, **14**, 174–83.

Latham, R. E. and Ricklefs, R. E. (1993). Continental comparisons of temperate-zone tree species diversity. In *Species Diversity in Ecological Communities: Historical and Geographical Perspectives*, ed. R. E. Ricklefs and D. Schluter, pp. 294–314. Chicago, IL, USA: The University of Chicago Press.

Laubhan, M. K. (1995). Effects of prescribed fire on moist-soil vegetation and soil macronutrients. *Wetlands*, **15**, 159–66.

Lavoisier (1789). *Elements of Chemistry*. In *Great Books of the Western World*, 2nd edn, ed. chief M. J. Adler, 1990, pp. 1–33 +XIII plates, Volume 42. Chicago: Encyclopaedia Britannica, Inc.

Lee, R. (1980). *Forest Hydrology*. New York: Columbia University Press.

Leary, R. A. (1985). A framework for assessing and rewarding a scientist's research productivity. *Scientometrics*, **7**, 29–38.

Leck, M. A. and Graveline, K. J. (1979). The seed bank of a freshwater tidal marsh. *American Journal of Botany*, **66**, 1006–15.

Leck, M. A., Parker, V. T. and Simpson, R. L. (eds.) (1989). *Ecology of Soil Seed Banks*. San Diego: Academic Press.

Legendre, L. and Legendre, P. (1983). *Numerical Ecology*. Amsterdam: Elsevier.

Leitch, J. A. (1989). Politicoeconomic overview of prairie potholes. In *Northern Prairie Wetlands*, ed. A. van der Valk, pp. 2-14. Ames: Iowa State University Press.

Leith, H. (1975). Historical survey of primary productivity research. In *Primary Productivity of the Biosphere*, ed. H. Leith and R. H. Whittaker, pp. 7-16. New York: Springer-Verlag.

Lemly, A. D. (1982). Modification of benthic insect communities in polluted streams: combined effects of sedimentation and nutrient enrichment. *Hydrobiologia*, **87**, 229–45.

Lent, R. M., Weiskel, P. K., Lyford, F. P. and Armstrong, D. S. (1997). Hydrologic indices for nontidal wetlands. *Wetlands*, **17**, 19–30.

Leopold, A. (1949). *A Sand County Almanac*. London: Oxford University Press.

Le Page, C. and Keddy, P. A. (1998). Reserves of buried seeds in beaver ponds. *Wetlands*, **18**, 242–8.

Levin, H. L. (1992). *The Earth Through Time*. 4th edn. Forth Worth: Saunders College Publishing, Harcourt Brace.

Levine, J., Brewer, J. S. and Bertness, M. D. (1998). Nutrients, competition and plant zonation in a New England salt marsh. *Journal of Ecology*, **86**, 285–92.

Levitt, J. (1977). The nature of stress injury and resistance. In *Responses of Plants to Environmental Stresses*, ed. J. Levitt, pp. 11-21. Academic Press, New York.

Levitt, J. (1980). *Responses of Plants to Environmental Stresses.* Vols. I and II. 2nd edn. New York: Academic Press.

Lewis, D. H. (1987). Evolutionary aspects of mutualistic associations between fungi and photosynthetic organisms. In *Evolutionary Biology of Fungi*, ed. A. D. M. Rayner, C. M. Brasier, and D. Moore, pp. 161–78. Cambridge: Cambridge University Press.

Lewis III, R. R (ed.) (1982). *Creation and Restoration of Coastal Plant Communities.* Boca Raton, FL: CRC Press.

Lewontin, R. C. (1974). The structure of evolutionary genetics. In *The Genetic Basis of Evolutionary Change.* New York: Columbia University Press.

Lieffers, V. J. (1984). Emergent plant communities of oxbow lakes in northeastern Alberta: salinity, water-level fluctuation, and succession. *Canadian Journal of Botany*, **62**, 310–16.

Llewellyn, D. W., Shaffer, G. P. Craig, N. J., Creasman, L., Pashley, D., Swan, M. and Brown, C. (1996). A decision-support system for prioritizing restoration sites on the Mississippi River alluvial plain. *Conservation Biology*, **10**, 1446–55.

Lodge, D. M. (1991). Herbivory on freshwater macrophytes. *Aquatic Botany*, **41**, 195–224.

Loffler, H. and Malkhazova, S. (1990). Impacts of wetlands on man. In *Wetlands and Shallow Continental Water Bodies, Vol. 1. Natural and Human Relationships*, ed. B. C. Patten, pp. 347–62. The Hague, The Netherlands: SPB Academic Publishing.

Louda, S. and Mole, S. (1991). Glucosinolates: chemistry and ecology. In *Herbivores: Their Interactions with Secondary Plant Metabolites*, ed. G. A. Rosenthal and M. R. Berenbaum, pp. 124–64. San Diego: Academic Press.

Louda, S. M., Keeler, K. H. and Holt, R. D. (1990). Herbivore influences on plant performance and competitive interactions. In *Perspectives in Plant Competition*, pp. 413–44. New York: Academic Press, Inc.

Loveless C. M. (1959). A study of the vegetation in the Florida everglades. *Ecology*, **40**, 1–9.

Lowe-McConnell, R. H. (1975). *Fish Communities in Tropical Freshwaters. Their Distribution, Ecology and Evolution.* London: Longman.

Lowe-McConnell, R. H. (1987). Fish of the Amazon System. Pp. 339–351. In *The Ecology of River Systems*, ed. B. R. Davies and K. F. Walker, pp. 339–51. Dordrecht, The Netherlands: Dr W. Junk Publishers.

Lu, J. (1995). Ecological significance and classification of Chinese wetlands. *Vegetatio*, **118**, 49–56.

Lugo, A. E and Brown, S. (1988). The wetlands of Caribbean islands. *Acta Cientifica*, **2**, 48–61.

Lugo, A. E. and Snedaker, S. C. (1974). The ecology of mangroves. *Annual Review of Ecology and Systematics*, **5**, 39–64.

Lugo, A. E., Brinson, M. and Brown, S. (eds.) (1990). *Forested Wetlands.* Amsterdam: Elsevier.

Lugo, A. E., Brown, S. and Brinson, M. M. (1988). Forested wetlands in freshwater and saltwater environments. *Limnology and Oceanography*, **33**, 849–909.

Lutman, J. (1978). The role of slugs in an *Agrostis-Festuca* grassland. In *Production Ecology of British Moors and Montane Grasslands*, ed. O. W. Heal and D. F. Perkins, pp. 332–47. Ecological Studies vol. 27. Berlin: Springer-Verlag.

Lynch, J. A., Grimm, J. W. and Bowersox, V. C. (1995). Trends in precipitation chemistry in the United States: A national perspective, 1980–1992. *Atmospheric Environment*, **29**, 1231–1246.

MacArthur, R. H. (1972). *Geographical Ecology.* New York: Harper and Row.

MacArthur, R. H. and Mac Arthur, J. (1961). On bird species diversity. *Ecology*, **42**, 594–8.

MacArthur, R. and Wilson, E. O. (1967). *The Theory of Island Biogeography.* Princeton: Princeton University Press.

Magnuson, J. J., Regier, H. A., Christie, W. J. and Sonzongi, W. C. (1980). To rehabilitate and restore Great Lake ecosystems. In *The Recovery Process in Damaged Ecosystems*, ed. J. Cairns Jr., pp. 95–112. Ann Arbor, MI: Ann Arbor Science Publishers.

Magnuson, J. J, Paszkowski, C. A., Rahel, F. J. and Tonn, W. M. (1989). Fish ecology in severe environments of small isolated lakes in northern Wisconsin. In *Freshwater Wetlands and Wildlife*, ed. R. Sharitz and J. W. Gibbons, pp. 487–515. Conf-8603101, DOE symposium Series No 61. Oak Ridge, TN: USDOE Office of Scientific and Technical Information.

Maguire, L. A. (1991). Risk analysis for conservation biologists. *Conservation Biology*, **5**, 123–5.

Malmer, N. (1986). Vegetational gradients in relation to environmental conditions in northwestern European mires. *Canadian Journal of Botany*, **64**, 375–83.

Maltby, E. and Turner, R. E. (1983). Wetlands of the world. *Geographical Magazine*, **55**, 12–17.

Maltby, E., Legg, C. J. and Proctor, C. F. (1990). The ecology of severe moorland fire on the North York Moors: effects of the 1976 fires, and subsequent surface and vegetation development. *Journal of Ecology*, **78**, 490–518.

Mandossian, A. and McIntosh, R. P. (1960). Vegetation zonation on the shore of a small lake. *The American Midland Naturalist*, **64**, 301–8.

Manfred, G. (1982). *World Energy Supply.* Berlin: Walter de Gruyter & Co.

Mark, A. F., Johnson, P. N., Dickinson, K. J. M. and McGlone, M. S. (1995). Southern hemisphere pattered mires, with emphasis on southern New Zealand. *Journal of the Royal Society of New Zealand*, **25**, 23–54.

Marquis, R. J. (1991). Evolution of resistance in plants to herbivores. *Evolutionary Trends in Plants*, **5**, 23–9.

Marschner, H. (1995). *Mineral Nutrition of Higher Plants.* 2nd edn. London: Academic Press.

Martin, P. S. and Klein, R. J. (1984). *Quaternary Extinctions: A Prehistoric Revolution.* Tucson, AZ: The University of Arizona Press.

Martini, I. P. (1982). Introduction. *Le Naturaliste Canadien*, **109**, 301–5.

Maseuth, J. D. (1995). *Botany: An Introduction to Plant Biology.* 2nd edn. PA: Saunders College Publishing.

Matthews, E. and Fung, I. (1987). Methane emission from natural wetlands: global distribution, area, and environmental characteristics of sources. *Global Biogeochemical Cycles*, **1**, 61–86.

Maun, M. A. and Lapierre, J. (1986). Effects of burial by sand on seed germination and seedling emergence of four dune species. *American Journal of Botany*, **73**, 450–5.

May, R. M. (1981). Patterns in multi-species communities. In *Theoretical Ecology*, ed. R. M. May, pp. 197–227. Oxford: Blackwell.

May, R. M. (1986). The search for patterns in the balance of nature: advances and retreats. *Ecology*, **67**, 1115–26.

May, R. M. (1988). How many species are there on Earth? *Science*, **241**, 1441–9.

Mayewski, P. A., Lyons, W. B., Spencer, M. J. Twickler, M. S., Buck, C. F. and Whitlow, S. (1990). An ice-core record of atmospheric response to anthropogenic sulphate and nitrate. *Nature*, **346**, 554–6.

Mayr, E. (1982). *The Growth of Biological Thought. Diversity, Evolution and Inheritance,* Cambridge, MA: Belknap Press of Harvard University Press.

McAuliffe, J. R. (1984). Competition for space, disturbance, and the structure of a benthic stream community. *Ecology*, **65**, 894–908.

McCanny, S. J., Keddy, P. A., Arnason, T. J., Gaudet, C. L., Moore, D. R. J. and Shipley, B. (1990). Fertility and the food quality of wetland plants: a test of the resource availability hypothesis. *Oikos*, **59**, 373–81.

McCarthy, K. A. (1987). Spatial and temporal distributions of species in two intermittent ponds in Atlantic County, NJ. MSc. Thesis, Rutgers University, NJ.

McClure, J. W. (1970). Secondary constituents of aquatic angiosperms. In *Phytochemical Phylogeny*, ed. J. B. Harborne, pp. 233–65. New York: Academic Press.

McCreary, N. J. (1991). Competition as a mechanism of submerged macrophyte community structure. *Aquatic Botany*, **41**, 177–93.

McCreary, N. J., Carpenter, S. R. and Chaney, J. E. (1983). Coexistence and interference in two submersed freshwater perennial plants. *Oecologia*, **59**, 393–6.

McIntosh, R. P. (1967). The continuum concept of vegetation. *Botanical Review*, **33**, 130–87.

McIntosh, R. P. (1985). *The Background of Ecology. Concept and Theory.* Cambridge: Cambridge University Press.

McKenzie, D. H., Hyatt, D. E. and McDonald, V. J. (1992). *Ecological Indicators*, Vols. 1 and 2. London: Elsevier.

McNaughton, S. J., Russ, R. W. and Seagle, S. W. (1988). Large mammals and process dynamics in African ecosystems. *Bioscience*, **38**, 794–800.

Meadows, D. H., Meadows, D. L., Randers, J. and Behrens III, W. W. (1974). *The Limits to Growth. A Report for the Club of Rome's Project on the Predicament of Mankind.* 2nd edn. New York: The New American Library.

Meave, J. and Kellman, M. (1994). Maintenance of rain forest diversity in riparian forests of tropical savannas: implications for species conservation during Pleistocene drought. *Journal of Biogeography*, **21**, 121–35.

Meave, J., Kellman, M., MacDougall, A. and Rosales, J. (1991). Riparian habitats as tropical refugia. *Global Ecology and Biogeography Letters*, **1**, 69–76.

Mendelssohn, I. A. and McKee, K. L. (1988). *Spartina alterniflora* die-back in Louisiana: time-course investigation of soil waterlogging effects. *Journal of Ecology*, **76**, 509–21.

Menges, E. S. and Gawler, S. C. (1986). Fourth-year changes in population size of the endemic Furbish's Lousewort: implications for endangerment and management. *Natural Areas Journal*, **6**, 6–17.

Merritt, R. W. and Cummins, K. W. (eds.). (1984). *An Introduction to the Aquatic Insects of North America.* 2nd edn. Iowa: Kendall/Hunt Publishing.

Messina, M. G. and Conner, W. H. (eds.) (1998). *Southern Forested Wetlands. Ecology and Management.* Boca Raton, FL: Lewis Publishers.

Michener, W. K. Blood, E. R., Bildstein, K. L., Brinson, M. M. and Gardner, L. R. (1997). Climate change, hurricanes and tropical storms, and rising sea level in coastal wetlands. *Ecological Applications*, **7**, 770–801.

Miller, G. R. and Watson, A. (1978). Heather productivity and its relevance to the regulation of red grouse populations. In *Production Ecology of British Moors and Montane Grasslands*, ed. O. W. Heal and D. F. Perkins, pp. 278–85. Ecological Studies vol. 27. Berlin: Springer-Verlag.

Miller, G. R. and Watson, A. (1983). Heather moorland in northern Britain. In *Conservation in Perspective*, ed. A. Warren and F. B. Goldsmith, pp. 101–17. Chichester: John Wiley.

Miller, M. W. and Nudds, T. D. (1996). Prairie landscape change and flooding in the Mississippi River valley. *Conservation Biology*, **10**, 847–53.

Miller, R. S. (1967). Pattern and process in competition. *Advances in Ecological Research*, **4**, 1–74.

Miller, R. S. (1968). Conditions of competition between redwings and yellow-headed blackbirds. *Journal of Animal Ecology*, **37**, 43–62.

Milliman, J. D. and Meade, R. H. (1983). World-wide delivery of river sediment to the oceans. *Journal of Geology*, **91**, 1–21.

Mitchell, G. F. (1965). Littleton Bog, Tipperary: an Irish vegetational record. *Geol. Soc. Am., Special Paper*, **84**, 1–16.

Mitsch, W. J. and Gosselink, J. G. (1986). *Wetlands*. New York: Van Nostrand Reinhold.

Mitsch, W. J. and Wu, X (1994). Wetlands and global change. In *Advances in Soil Science, Global Carbon Sequestration*, ed. B. A. Stewart, R. Lal, and J. M. Kimble. Chelsea, MI: Lewis Publishers.

Moeller, R. E. (1978). Carbon-uptake by the submerged hydrophyte *Utricularia purpurea*. *Aquatic Botany*, **5**, 209–16.

Monda, M. J., Ratti, J. T. and McCabe, T. R. (1994). Reproductive ecology of tundra swans on the arctic national wildlife refuge, Alaska. *Journal of Wildlife Management*, **58**, 757–73.

Montgomery, K. G. (1958). *The Memoirs of Field-marshal the Viscount Montgomery of Alamein*. St James' Place, London: K. G. Collins.

Moore, D. R. J. (1990). Pattern and process in wetlands of varying standing crop: the importance of scale. PhD thesis. University of Ottawa, Ottawa, Ontario, Canada.

Moore, D. R. J. (1998). The ecological component of ecological risk assessment: lessons from a field experiment. *Human and Ecological Risk Assessment*, **4**, 1103–1123.

Moore, D. R. J. and Wein, R. W. (1977). Viable seed populations by soil depth and potential site recolonization after disturbance. *Canadian Journal of Botany*, **55**, 2408–12.

Moore, D. R. J. and Keddy, P. A. (1989). The relationship between species richness and standing crop in wetlands: the importance of scale. *Vegetatio*, **79**, 99–106.

Moore, D. R. J., Keddy, P. A., Gaudet, C. L. and Wisheu, I. C. (1989). Conservation of wetlands: do infertile wetlands deserve a higher priority? *Biological Conservation*, **47**, 203–17.

Moore, P. D. (1973). The influence of prehistoric cultures upon the initiation and spread of blanket bog in upland Wales. *Nature*, **241**, 350–3.

Moorhead, K. K. and Reddy, K. R. (1988). Oxygen transport through selected aquatic macrophytes. *Journal of Environmental Quality*, **17**, 138–42.

Morgan, M. D. and Philipp, K. R. (1986). The effect of agricultural and residential development on aquatic macrophytes in the New Jersey Pine Barrens. *Biological Conservation*, **35**, 143–58.

Morowitz, H. J. (1968). *Energy Flow in Biology*. New York: Academic Press.

Morris, J. (1973). *Pax Britannica*. 3 Vols. Faber and Faber Ltd. Reprinted 1992 by Folio Society, London.

Moss, B. (1983). The Norfolk Broadland: experiments in the restoration of a complex wetland. *Biological Reviews of the Cambridge Philosophical Society*, **58**, 521–61.

Moss, B. (1984). Medieval man-made lakes: progeny and casualties of English social history, patients of twentieth century ecology. *Transactions of the Royal Society of South Africa*, **45**(2), 115–28.

Mountford, J. O., Lakhani, K. H. and Kirkham, F. W. (1993). Experimental assessment of the effects of nitrogen addition under hay-cutting and aftermath grazing on the vegetation of meadows on a Somerset peat moor. *Journal of Applied Ecology*, **30**, 321–32.

Mueller-Dombois, D. and Ellenberg, H. (1974). *Aims and Methods of Vegetation Ecology*. New York: Wiley.

Müller, J., Rosenthal, G. and Uchtmann, H. (1992). Vegetationsveränderungen und Ökologie nordwestdeutscher Feuchtgrünlandbrachen. *Tuexenia*, **12**, 223–44.

Müller, J., Irion, G., de Mello, J. N. and Junk, W. J. (1995). Hydrological changes of the Amazon during the last glacial-interglacial cycle in Central Amazonia (Brazil). *Naturwissenschaften*, **82**, 232–5.

Murkin, H. R. (1989). The basis for food chains in prairie wetlands. In *Northern Prairie Wetlands*, ed. A. G. van der Valk, pp. 316–38. Ames, Iowa: Iowa State University Press.

Myers, J. G. (1935). Zonation of vegetation along river courses. *Journal of Ecology*, **3**, 356–60.

Myers, R. S. (1995). Baldcypress (*Taxodium distichum* (L.) Rich.) restoration in southeastern Louisiana: the relative effects of herbivory, flooding, competition and macronutrients. *Wetlands*, **15**, 141–8.

Naiman, R. J., Johnston, C. A. and Kelley, J. C. (1988). Alteration of North American streams by beaver. *Bioscience*, **38**, 753–62.

Nanson, G. C. and Beach, H. F. (1977). Forest succession and sedimentation on a meandering-river floodplain, northeast British Columbia, Canada. *Journal of Biogeography*, **4**, 229–51.

Navid, D. (1988). Developments under the Ramsar convention. In *The Ecology and Management of Wetlands Volume 2: Management, Use and Value of Wetlands*, ed. D. D. Hook, W. H. McKee, Jr., H. K. Smith, J. Gregory, V. G. Burrell, Jr., M. R. DeVoe, R. E. Sojka, S. Gilbert, R. Banks, L. H. Stolzy, C. Brooks, T. D. Matthews and T. H. Shear, pp. 21–7. Portland: Timber Press.

Neiff, J. J. (1986). Aquatic plants of the Paraná system. In *The Ecology of River Systems*, ed. B. R. Davies and K. F. Walker pp. 557–71. Dordrecht, The Netherlands: Dr W. Junk Publishers.

Nicholson, A, and Keddy, P. A. (1983). The depth profile of a shoreline seed bank in Matchedash Lake, Ontario. *Canadian Journal of Botany*, **61**, 3293–6.

Niering, W. A. and Warren, R. S. (1980). Vegetation patterns and processes in New England salt marshes. *Bioscience*, **30**, 301–7.

Nilsson, C. (1981). Dynamics of the shore vegetation of a north Swedish hydro-electric reservoir during a 5-yr period. *Acta Phytogeographica Suecica*, **69**, 1–96.

Nilsson, C. and Jansson, R. (1995). Floristic differences between riparian corridors of regulated and free-flowing boreal rivers. *Regulated Rivers: Research and Management*, **11**, 55–66.

Nilsson, C. and Keddy, P. A. (1988). Predictability of change in shoreline vegetation in a hydroelectric reservoir, northern Sweden. *Canadian Journal of Fisheries and Aquatic Science*, **45**, 1896–904.

Nilsson, C., Grelsson, G., Johansson, M. and Sperens, U. (1989). Patterns of plant species richness along riverbanks. *Ecology*, **70**, 77–84.

Nilsson, C., Grelsson, G., Dynesius, M., Johansson, M. E. and Sperens, U. (1991). Small rivers behave like large rivers: effects of postglacial history on plant species richness along riverbanks. *Journal of Biogeography*, **18**, 533–41.

Noss, R. (1995). Maintaining ecological integrity in representative reserve networks. A World Wildlife Fund Canada/World Wildlife Fund United States Discussion Paper, WWF.

Noss, R. F. and Cooperrider, A. (1994). *Saving Nature's Legacy: Protecting and Restoring Biodiversity.* Washington, DC: Defenders of Wildlife and Island Press.

Novacek, J. M. (1989). The water and the wetland resources of the Nebraska sand-hills. In *Northern Prairie Wetlands*, ed. A. G. van der Valk, pp. 340–84. Ames, Iowa: Iowa State University Press.

Noy-Meir, I. (1975). Stability of grazing systems: an application of predator–prey graphs. *Journal of Ecology*, **63**, 459–81.

Nudds, T. D., Sjöberg, K. and Lundberg, P. (1994). Ecomorphological relationships among Palearctic dabbling ducks on Baltic coastal wetlands and a comparison with the Nearctic. *Oikos*, **69**, 295–303.

Odum, E. P. 1985. Trends expected in stressed ecosystems. *BioScience*, **35**, 419–22.

Oksanen, L. (1990). Predation, herbivory, and plant strategies along gradients of primary production. In *Perspectives on Plant Competition*, pp. 445–74. New York: Academic Press.

Oksanen, L., Fretwell, S. D., Arruda, J. and Niemelä, P. (1981). Exploitation ecosys-tems in gradients of primary productivity. *The American Naturalist*, **118**, 240–261.

OMNR. (1988). *Provincially and Regionally Significant Wetlands in Southern Ontario.* Interim Report – 1987. Toronto, Ontario: Wildlife Branch, Ontario Ministry of Natural Resources.

Oomes, M. J. M and Elberse, W. T. (1976). Germination of six grassland herbs in microsites with different water contents. *Journal of Ecology*, **64**, 745–55.

Orson, R. A., Simpson, R. L. and Good, R. E. (1990). Rates of sediment accumula-tion in a tidal freshwater marsh. *Journal of Sedimentary Petrology*, **60**, 859–69.

Orson, R. A., Simpson, R. L. and Good, R. E. (1992). The paleoecological develop-ment of a late Holocene, tidal freshwater marsh of the Upper Delaware River estuary. *Estuaries*, **15**, 130–46.

Osborne, P. L. and Polunin, N. V. C. (1986). From swamp to lake: recent changes in a lowland tropical swamp. *Journal of Ecology*, **74**, 197–210.

Ostrofsky, M. L. and Zettler, E. R. (1986). Chemical defenses in aquatic plants. *Journal of Ecology*, **74**, 279–87.

Padgett, D. J. and Crow, G. E. (1993). A comparison of floristic composition and species richness within and between created and natural wetlands of southeastern New Hampshire. In *Proceedings of the Twentieth Annual Conference on Wetlands Restoration and Creation*, ed. F. J. Webb, Jr., pp. 171–86. Tampa FL: Hillsborough Community College.

Padgett, D. J. and Crow, G. E. (1994). Foreign plant stock. Concerns for wetland mitigation. *Restoration and Management Notes*, **12**, 168–71.

Painter, T. J. (1991). Lindow Man, Tollund Man, and other peat-bog bodies: the preservative and antimicrobial action of sphagnan, a reactive glycuronoglycan with tanning and sequestering properties. *Carbohydrate Polymers*, **15**, 123–42.

Painter, S. and Keddy, P. A. (1992). *Effects of Water Level Regulation on Shoreline Marshes: A Predictive Model Applied to the Great Lakes.* Environment Canada: National Water Research Institute.

Palczynski, A. (1984). Natural differentiation of plant communities in relation to hydrological conditions of the Biebrza valley. *Polish Ecological Studies*, **10**, 347–85.

Partridge, T. R., and Wilson, J. B. (1987). Salt tolerance of salt marsh plants of Otago, New Zealand. *New Zealand Journal of Botany*, **25**, 559–66.

Patrick Jr., W. H. and Reddy, C. N. (1978). Chemical changes in rice soils. In *Soils and Rice*. International Rice Research Institute, Los Banos, Philippines. Brinkman and Van Diepen, 185 pp.

Patten, B. C. (ed.) (1990). *Wetlands and Shallow Continental Water Bodies. Vol. 1. Natural and Human Relationships.* The Hague, The Netherlands: SPB Academic Publishing.

Patten, D. T. (1998). Riparian ecosystems of semi-arid North America: diversity and human impacts. *Wetlands*, **18**, 498–512.

Peace–Athabasca Delta Implementation Committee. (1987). *Peace–Athabasca Delta Water Management Works Evaluation: Final Report*. Environment Canada, Alberta Environment, Saskatchewan Water Corporation. Canada.

Peace–Athabasca Delta Project Group. (1972). *The Peace–Athabasca Delta Summary Report, 1972.* Bulletin Commercial Printers, Ottawa: Department of the Environment.

Pearce, F. (1991). The rivers that won't be tamed. *New Scientist*, **1764**, 38–41.

Pearman, P. B. (1997). Correlates of amphibian diversity in an altered landscape of Amazonian Ecuador. *Conservation Biology*, **11**,1211–25.

Pearsall, W. H. (1920). The aquatic vegetation of the English Lakes. *Journal of Ecology*, **8**, 163–201.

Pearse, P. H., Bertrand, F. X. and MacLaren, J. W. (1985). Currents of change. Final report: inquiry on Federal Water Policy, Ottawa, Canada.

Peat, H. J. and Fitter, A. H. (1993). The distribution of arbuscular mycorrhizae in the British flora. *New Phytologist*, **125**, 845–54.

Pechmann, J. H. K., Scott, D. E., Gibbons, J. W. and Semlitsch, R. D. (1989). Influence of wetland hydroperiod on diversity and abundance of metamorphosing juvenile amphibians. *Wetlands Ecology and Management*, **1**, 3–11.

Pedersen, O., Sand-Jensen, K. and Revsbech, N. P. (1995). Diel pulses of O_2 and CO_2 in sandy lake sediments inhabited by *Lobelia dortmanna*. *Ecology*, **76**, 1536–45.

Peet, R. K. (1974). The measurement of species diversity. *Annual Review of Ecology and Systematics*, **5**, 285–307.

Peet, R. K. and Allard, D. J. (1993). Longleaf pine vegetation of the southern Atlantic and eastern Gulf Coast regions: A preliminary classification. In *Proceedings of the Tall Timbers Fire Ecology Conference*, ed. S. M. Hermann No. 18, pp. 45–81. The Longleaf Pine Ecosystem: Ecology, Restoration and Management. Tall Timbers Research Station, Tallahassee, FL.

Pehek, E. L. (1995). Competition, pH, and the ecology of larval *Hyla andersonii*. *Ecology*, **76**, 1786–93.

Pennings, S. C. and Callaway, R. M. (1992). Salt marsh zonation: the relative importance of competition and physical factors. *Ecology*, **73**, 681–90.

Pennings, S. C., Carefoot, T. H., Siska, E. L., Chase. M. E. and Page, T. A. (1998). Feeding preferences of a generalist salt-marsh crab: relative importance of multiple plant traits. *Ecology*, **79**, 1968–79.

Perkins, D. F. (1978). Snowdonia grassland: introduction, vegetation and climate. In *Production Ecology of British Moors and Montane Grasslands*, ed. O. W. Heal and D. F. Perkins, pp. 290–6. Ecological Studies vol. 27. Berlin: Springer-Verlag.

Peters, R. H. (1980a). From natural history to ecology. *Perspectives in Biology and Medicine*, **23**, 191–203.

Peters, R. H. (1980b). Useful concepts for predictive ecology. In *Conceptual Issues in Ecology*, ed. E. Saarinen. Dordrecht, Holland: D. Reidel Publishing Company.

Petr, T. (1986). The Volta River system. In *The Ecology of River Systems*, ed. B. R. Davies and K. F. Walker, pp. 163–83. Dordrecht, The Netherlands: Dr W. Junk Publishers.

Pfadenhauer, J. and Klotzli, F. (1996). Restoration experiments in middle European wet terrestrial ecosystems: an overview. *Vegetatio*, **126**, 101–15.

Phillips, G. L., Eminson, D. and Moss, B. (1978). A mechanism to account for macrophyte decline in progressively eutrophicated fresh-waters. *Aquatic Botany*, **4**, 103–26.

Phipps, R. W. (1883). *On the Necessity of Preserving and Replanting Forests*. Toronto, Canada: Blackett and Robinson.

Pianka, E. R. (1981). Competition and niche theory. In *Theoretical Ecology*, ed. R. M. May, pp. 114–41. Oxford: Blackwell.

Pickett, S. T. A. (1980). Non-equilibrium coexistence of plants. *Bulletin of the Torrey Botanical Club*, **107**, 238–48.

Pickett, S. T. A and White, P. S. (1985). *The Ecology of Natural Disturbance and Patch Dynamics*. Orlando: Academic Press.

Picman, J. (1984). Experimental study on the role of intra- and inter-specific behaviour in marsh wrens. *Canadian Journal of Zoology*, **62**, 2353–6.

Pieczynska, E. (1986). Littoral communities and lake eutrophication. In *Land Use Impacts on Aquatic Ecosystems*, ed. J. Lauga, H. Decamps and M. M. Holland. Proceedings of the Toulouse Workshop organized by MAB-UNESCO and PIREN-CNRS. April 1988.

Pielou, E. C. (1975). *Ecological Diversity*. New York: John Wiley.

Pielou, E. C. (1977). *Mathematical Ecology*. New York: John Wiley.

Pielou, E. C. and Routledge, R. D. (1976). Salt marsh vegetation: Latitudinal gradients in the zonation patterns. *Oecologia*, **24**, 311–21.

Pimental, D., Hurd, L. E., Bellotti, A. C., Forster, M. J., Oka, I., Sholes, O. D. and Whitman, W. J. (1973). Food production and the energy crisis. *Science*, **182**, 443–9.

Poiana, K. A. and Johnson, W. C. (1993). A spatial simulation model of hydrology and vegetation dynamics in semi-permanent prairie wetlands. *Ecological Applications*, **3**, 279–93.

Poljakoff-Mayber, A. and Gale, J. (eds.) (1975). *Plants in Saline Environments.* Berlin: Springer-Verlag.

Polunin, N. V. C. (1984). The decomposition of emergent macrophytes in fresh water. *Advances in Ecological Research*, **14**, 115–66.

Pomeroy, L. R. and Wiegert, R. J. (eds.) (1981). *The Ecology of a Salt Marsh.* Berlin: Springer-Verlag.

Ponnamperuma, F. N. (1972). The chemistry of submerged soils. *Advances in Agronomy*, **24**, 29–96.

Ponnamperuma, F. N. (1984). Effects of flooding on soils. In *Flooding and Plant Growth*, ed. T. Kozlowski, pp. 9–45. Orlando: Academic Press.

Poole, R. W. and Rathcke, B. J. (1979). Regularity, randomness, and aggregation in flowering phenologies. *Science*, **203**, 470–1.

Power, M. E. (1992). Top-down and bottom-up forces in food webs: Do plants have primacy? *Ecology*, **73**, 733–46.

Prance, G. T. and Schaller, J. B. (1982). Preliminary study of some vegetation types of the Pantanal, Mato Grosso, Brazil. *Brittonia*, **34**, 228–51.

Pressey, R. L., Humphries, C. J., Margules, C. R., Vane-Wright, R. I. and Williams, P. H. (1993). Beyond opportunism: key principles for systematic reserve selection. *Trends in Ecology and Evolution*, **8**, 124–8.

Preston, F. W. (1962a). The canonical distribution of commonness and rarity: Part I. *Ecology*, **43**, 185–215.

Preston, F. W. (1962b). The canonical distribution of commonness and rarity: Part II. *Ecology*, **43**, 410–32.

Price, M. V. (1980). On the significance of test form in benthic salt-marsh foraminifera. *Journal of Foraminiferal Research*, **10**, 129–35.

Prince, H. H. and D'Itri, F. M. (eds.) (1985). *Coastal Wetlands.* Chelsea: Lewis Publishers.

Prince, H. H. and Flegel, C. S. (1995). Breeding avifauna of Lake Huron. In *The Lake Huron Ecosytem: Ecology, Fisheries and Management*, ed. M. Munawar, T. Edsall and J. Leach. Ecovision World Monograph Series. Amsterdam, The Netherlands: SPB Academic Publishing.

Prince, H. H., Padding, P. I. and Knapton, R. W. (1992). Waterfowl use of the Laurentian Great Lakes. *Journal of Great Lakes Research*, **18**, 673–99.

Radford, A. E., Ahles H. E. and Bell, C. R. (1968). *Manual of the Vascular Flora of the Carolinas.* Chapel Hill, NC, USA: University of North Carolina Press.

Rapport, D. J. (1989). What constitutes ecosystem health? *Perspectives in Biology and Medicine*, **33**, 120–32.

Rapport, D. J., Thorpe, C. and Hutchinson, T. C. (1985). Ecosystem behaviour under stress. *American Naturalist*, **125**, 617–40.

Rasker, R. and Hackman, A. (1996). Economic development and the conservation of large carnivores. *Conservation Biology*, **10**, 991–1002.

Raunkaier, C. (1937). *Plant Life Forms.* Transl. by H. Gilbert-Cater. Oxford; Clarendon Press.

Raup, H. M. (1975). Species versatility in shore habitats. *Journal of the Arnold Arboretum*, **56**, 126–63.

Raven, P. H., Evert, R. F. and Eichhorn, S. E. (1992). *Biology of Plants*, 5th edn. New York: Worth Publishers.

Ravera, O. (1989). Lake ecosystem degradation and recovery studied by the enclosure method. In *Ecological Assessment of Environmental Degradation, Pollution and Recovery*, ed. O. Ravera. Amsterdam, The Netherlands: Elsevier Science Publishing.

Rawes, M. and Heal, O. W. (1978). The blanket bog as part of a Pennine moorland. In *Production Ecology of British Moors and Montane Grasslands*, ed. O. W. Heal and D. F. Perkins, pp. 224–43. Ecological Studies vol. 27. Berlin: Springer-Verlag.

Read, D. J., Koucheki, H. K. and Hodgson, J. (1976). Vesicular–arbuscular mycorrhizae in natural vegetation systems. 1. The occurrence of infection. *New Phytologist*, **77**, 641–53.

Read, D. J., Francis, R. and Finlay, R. D. (1985). Mycorrhizal mycelia and nutrient cycling in plant communities. In *Ecological Interactions in Soil*, ed. A. H. Fitter, pp. 193–217. Oxford: Blackwell Science.

Reddy, K. R. and Patrick, W. H. (1984). Nitrogen transformations and loss in flooded soils and sediments. *CRC Critical Reviews in Environmental Control*, **13**, 273–309.

Reid, D. M. and Bradford, K. J. (1984). Effect of flooding on hormone relations. In *Flooding and Plant Growth*, ed. T. W. Kozlowski, pp. 195–219. Orlando: Academic Press.

Reid, W. V., McNeely, J. A., Tunstall, J. B., Bryant, D. A. and Winograd, M. (1993). *Biodiversity Indicators for Policymakers*. World Resources Institute, Washington, DC.

Rejmankova, E., Pope, K. O., Pohl, M. D. and Rey-Benayas, J. M. (1995). Freshwater wetland plant communities of northern Belize: implications for paleoecological studies of Maya wetland agriculture. *Biotropica*, **27**, 28–36.

Reynoldson T. B. and Zarull, M. A. (1993). An approach to the development of biological sediment guidelines. In *Ecological Integrity and the Management of Ecosystems*, ed. S. Woodley, J. Kay and G. Francis, pp. 177–200. Ottawa, Canada: St Lucie Press.

Reznicek, A. A. and Catling, P. M. (1989). Flora of Long Point. *The Michigan Botanist*, **28**, 99–175.

Richardson, C. J. (ed.) (1981). *Pocosin Wetlands: An Integrated Analysis of Coastal Plain Freshwater Bogs in North Carolina*. Stroudsburg, PA: Hutchinson Ross Publishing Company.

Richardson, C. J. (1985). Mechanisms controlling phosphorus retention capacity in freshwater wetlands. *Science*, **228**, 1424–7.

Richardson, C. J. (1989). Freshwater wetlands: transformers, filters, or sinks? In *Freshwater Wetlands and Wildlife*, ed. R. R. Sharitz and J. W. Gibbons, pp. 25–46. Proceedings of a symposium held at Charleston, South Carolina, March 24–27, 1986. US Department of Energy.

Richardson, C. J. (1991). Pocosins: an ecological perspective. *Wetlands*, **11**, 335–54.

Richardson, C. J. (1995). Wetlands ecology. *Encyclopedia of Environmental Biology*, **3**, 535–50.

Richardson, C. J. and Gibbons, J. W. (1993). Pocosins, Carolina bays and mountain bogs. In *Biodiversity of the Southeastern United States*, ed. W. H. Martin, S. G. Boyce and A. C. Echternacht, pp. 257–310. New York: John Wiley.

Richter, K. O. and Azous, A. L. (1995). Amphibian occurrence and wetland characteristics in the Puget Sound Basin. *Wetlands*, **15**, 305–12.

Richter, B. D., Braun, D. P., Mendelson, M. A. and Master, L. L. (1997). Threats to imperiled freshwater fauna. *Conservation Biology*, **11**, 1081–93.

Rickerl, D. H., Sancho, F. O. and Ananth, S. (1994). Vesicular–arbuscular endomycorrhizal colonization of wetland plants. *Journal of Environmental Quality*, **23**, 913–16.

Ricklefs, R. E. (1987). Community diversity: relative roles of local and regional processes. *Science*, **235**, 167–71.

Rigler, F. H. (1982). Recognition of the possible: an advantage of empiricism in ecology. *Canadian Journal of Fisheries and Aquatic Sciences*, **39**, 1323–31.

Rigler, F. H. and Peters, R. H. (1995). *Science and Limnology*. Oldendorf/Lutie Germany: Ecology Institute.

Riley, J. L. (1982). Hudson Bay lowland floristic inventory, wetlands catalogue and conservation strategy. *Naturaliste canadien*, **109**, 543–55.

Riley, J. L. (1989). Southern Ontario bogs and fens off the Canadian Shield. In *Wetlands. Inertia or momentum?* Conference proceedings, Oct 21–22, pp. 355–67. Federation of Ontario Naturalists, Toronto, Ontario.

Riley, T. Z. and Bookhout, T. A. (1990). Responses of aquatic macroinvertebrates to early-spring drawdown in nodding smartweed marshes. *Wetlands*, **10**, 173–85.

Ritchie, J. C. (1987). *Postglacial Vegetation of Canada*. New York: Cambridge University Press.

Roberts, J. and Ludwig, J. A. (1991). Riparian vegetation along current-exposure gradients in floodplain wetlands of the River Murray, Australia. *Journal of Ecology*, **79**, 117–27.

Robertson, P. A., Weaver, G. T. and Cavanaugh, J. A. (1978). Vegetation and tree species patterns near the northern terminus of the southern floodplain forest. *Ecological Monographs*, **48**, 249–67.

Robertson, R. J. (1972). Optimal niche space of the redwinged blackbird (*Agelaius phoeniceus*). I. Nesting success in marsh and upland habitat. *Canadian Journal of Zoology*, **50**, 247–63.

Robinson, A. R. (1973). Sediment, our greatest pollutant? In *Focus on Environmental Geology*, ed. R. W. Tank. London: Oxford University Press.

Rogers, D. R., Rogers, B. D. and Herke, W. H. (1992). Effects of a marsh management plan on fishery communitires in coastal Louisiana. *Wetlands*, **12**, 53–62.

Rolston, H. (1994). Foreword. In *An Environmental Proposal for Ethics: The Principle of Integrity*, ed. L. Westra, pp. xi–xiii. Lanham, MD: Rowman & Littlefield. In Noss, R. (1995). *Maintaining Ecological Integrity in Representative Reserve Networks.* A World Wildlife Fund Canada/World Wildlife Fund United States Discussion Paper, WWF.

Rood, S. B. and Mahoney, J. M. (1990). Collapse of riparian poplar forests downstream from dams in western prairies: probable causes and prospects for mitigation. *Environmental Management*, **14**, 451–64.

Root, R. (1967). The niche exploitation pattern of the blue-grey gnatcatcher. *Ecological Monographs*, **37**, 317–50.

Rørslett, B. (1984). Environmental factors and aquatic macrophyte response in regulated lakes – a statistical approach. *Aquatic Botany*, **19**, 199–220.

Rørslett, B. (1985). Regulation impact on submerged macrophytes in the oligotrophic lakes of Setesdal, South Norway. *International Association for Theoretical and Applied Limnology*, **22**, 2927–36.

Rosen, B. H., Gray, S. and Flaig, E. (1995). Implementation of Lake Okeechobee watershed management strategies to control phosphorus load. In *Wetlands and Watershed Management. Science Applications and Public Policy*, ed. J. A. Kusler, D. E. Willard and H. C. Hull, Jr., pp. 199–207. A collection of papers from a national symposium and several workshops at Tampa, FL, April 23–26. NY: The Association of State Wetland Managers.

Rosenberg, D. M. and Barton, D. R. (1986). The Mackenzie river system. In *The Ecology of River Systems*, ed. B. R. Davies and K. F. Walker, pp. 425–33. Dordrecht, The Netherlands: Dr W. Junk Publishers.

Rosenberg, D. M., Bodaly, R. A. and Usher, P. J. (1995). Environmental and social impacts of large scale hydro-electric development: Who is listening? *Global Environmental Change*, **5**, 127–48.

Rosenthal G. A. and Berenbaum, M. R. (eds.). (1991). *Herbivores: Their Interactions with Secondary Plant Metabolites*. San Diego: Academic Press.

Rosgen, D. L. (1995). River restoration utilizing natural stability concepts. In *Wetlands and Watershed Management. Science Applications and Public Policy*, ed. J. A. Kusler, D. E. Willard and H. C. Hull, Jr., pp. 55–62. A collection of papers from a national symposium and several workshops at Tampa, FL, April 23–26. NY: The Association of State Wetland Managers.

Rosswall, T. (1983). The nitrogen cycle. In *The Major Biogeochemical Cycles and Their Interactions*, ed. B. Bolin and R. B. Cook, pp. 46–50. SCOPE Rep. No. 21. Chichester: John. Wiley.

Rothhaupt, K. O. (1990). Resource competition of herbivorous zooplankton: a review of approaches and perspectives. *Archives in Hydrobiology*, **118**, 1–29.

Rozan, T. F., Hunter, K. S. and Benoit, G. (1994). Industrialization as recorded in floodplain deposits of the Quinnipiac River. Connecticut. *Marine Pollution Bulletin*, **28**, 564–9.

Ryan, P. A. (1991). Environmental effects of sediment on New Zealand streams: a review. *New Zealand Journal of Marine and Freshwater Research*, **25**, 207–21.

Rybicki, N. B. and Carter, V. (1986). Effect of sediment depth and sediment type on the survival of *Vallisneria americana* Michx. grown from tubers. *Aquatic Botany*, **24**, 233–240.

Sakakura, Y. and Tsukamoto, K. (1998). Social rank in schools of juvenile yellowtail, *Seriola quinqueradiata*. *Journal of Applied Ichthyology*, **14**, 69–73.

Salisbury, F. B. and Ross, C. W. (1988). *Plant Physiology*. 3rd edn. Belmont: Wadsworth Publishers.

Salisbury, S. E. (1970). The pioneer vegetation of exposed muds and its biological features. *Royal Society of London, Philosophical Transactions, Series B*, **259**, 207–55.

Salo, J., Kalliola, R., Hakkinen, I., Makinen, Y., Niemela, P., Puhakka, M. and Coley, P. D. (1986). River dynamics and the diversity of Amazon lowland forest. *Nature*, **322**, 254–8.

Sanders, N. K. (1972). *The Epic of Gilgamesh*. An English version with an introduction by N. K. Sanders. Revised edition. London: Penguin Books.

Sand-Jensen, K. and Krause-Jensen, D. (1997). Broad-scale comparison of photosynthesis in terrestrial and aquatic plant communities. *Oikos*, **80**, 203–8.

Sansen, U. and Koedam, N. (1996). Use of sod cutting for restoration of wet heathlands: revegetation and establishment of typical species in relation to soil conditions. *Journal of Vegetation Science*, **7**, 483–6.

Santelmann, M. V. (1991). Influences on the distribution of *Carex exilis*: an experimental approach. *Ecology*, **72**, 2025–37.

Sather, J. H. and Smith, R. D. (1984). *An Overview of Major Wetland Functions.* US Fish and Wildlife Service. FWS/OBS-84/18.

Sather, J. H., Smith, R. D. and Larson, J. S. (1990). Natural values of wetlands In *Wetlands and Shallow Continental Water Bodies. Vol. 1. Natural and Human Relationships*, ed. B. C. Patten, pp. 373–87. The Hague, The Netherlands: SPB Academic Publishing.

Saunders, D. A., Hobbs, R. J. and Ehrlich, P. R. (eds.) (1993). *Nature Conservation 3: Reconstruction of Fragmented Ecosystems Global and Regional Perspectives*. Surrey: Beatty & Sons Pty Limited.

Savile, D. B. O. (1956). Known dispersal rates and migratory potentials as clues to the origin of the North American biota. *American Midland Naturalist*, **56**, 434–53.

Scagel, R. F., Bandoni, R. J., Rouse, G. E., Schofield, W. B., Stein, J. R. and Taylor, T. M. C. (1966). *Plant Diversity: An Evolutionary Approach.* Belmont, CA: Wadsworth.

Scharf, F. S., Juanes, F. and Sutherland, M. (1998). Inferring ecological relationships from the edges of scatter diagrams: comparison of regression techniques. *Ecology*, **79**, 448–60.

Schindler, D. W. (1977). Evolution of phosphorus limitation in lakes. *Science*, **195**, 260–2.

Schindler, D. W. (1987). Detecting ecosystem responses to anthropogenic stress. *Canadian Journal of Fisheries and Aquatic Sciences*, **44**, 6–25.

Schnitzler A. (1995). Successional status of trees in gallery forest along the river Rhine. *Journal of Vegetation Science*, **6**, 479–86.

Schoener, T. W. (1974). Resource partitioning in ecological communities. *Science*, **185**, 27–39.

Schoener, T. W. (1985). Some comments on Connell's and my reviews of field experiments on interspecific competition. *The American Naturalist*, **125**, 730–40.

Scholander P. F., Hammel, H. T., Bradstreet, E. D. and Hemmingsen, E. A. (1965). Sap pressure in vascular plants. *Science*, **148**, 339–46.

Schubel, J. R., Shen, H. and Park, M. (1986). Comparative analysis of estuaries bordering the Yellow Sea. In *Estuarine Variability*, ed. D. A. Wolfe, pp. 43–62. San Diego: Academic Press.

Scott, W. S. and Wylie, N. P. (1980). The environmental effects of snow dumping: a literature review. *Journal of Environmental Management*, **10**, 219–40.

Sculthorpe, C. D. (1967). *The Biology of Aquatic Vascular Plants*. Reprinted in 1985 London: by Edward Arnold.

Severinghaus, W. D. (1981). Guild theory development as a mechanism for assessing environmental impact. *Environmental Management*, **5**, 187–90.

Seward, A. C. (1931). *Plant Life Through the Ages.* London: Cambridge University Press.

Shaffer, G. P., Sasser, C. E., Gosselink, J. G. and Rejmanek, M. (1992). Vegetation dynamics in the emerging Atchafalaya Delta, Louisiana, USA. *Journal of Ecology*, **80**, 677–87.

Shannon, R. D., White, J. R., Lawson, J. E. and Gilmour, B. S. (1996). Methane efflux from emergent vegetation in peatlands. *Journal of Ecology*, **84**, 239–46.

Sharitz, R. R. and McCormick, J. F. (1973). Population dynamics of two competing annual plant species. *Ecology*, **54**, 723–40.

Sharitz, R. R. and Gibbons, J. W. (eds.) (1989). *Freshwater Wetlands and Wildlife.* US Department of Energy. Proceedings of a symposium held at Charleston, South Carolina, March 24–27, 1986.

Sharitz, R. R. and Mitsch, W. J. (1993). Southern floodplain forests. In *Biodiversity of the Southeast United States/Lowland Terrestrial Communities*, ed. W. H. Martin, S. G. Boyce and A. C. Echternacht, pp. 311-71. New York: John Wiley.

Sharp, M. J. and Keddy, P. A. (1985). Biomass accumulation by *Rhexia virginica* and *Triadenum fraseri* along two lakeshore gradients: a field experiment. *Canadian Journal of Botany*, **63**, 1806–10.

Shay, J. M. and Shay, C. T. (1986). Prairie marshes in western Canada, with specific reference to the ecology of five emergent macrophytes. *Canadian Journal of Botany*, **64**, 443–54.

Sheail, J. and Wells, T. C. E. (1983). The Fenlands of Huntingdonshire, England: a case study in catastrophic change. In *Ecosystems of the World 4B. Mires: Swamp, Bog, Fen and Moor*, ed. A. J. P. Gore, pp. 375–93. Amsterdam: Elsevier Scientific Publishing Company.

Sheldon, S. P. (1987). The effects of herbivorous snails on submerged macrophyte communities in Minnesota lakes. *Ecology*, **68**, 1920–31.

Sheldon, S. P. (1990). More on freshwater snail herbivory: a reply to Brönmark. *Ecology*, **71**, 1215–16.

Shimwell, D. W. (1971). *The Description and Classification of Vegetation.* Seattle: University of Washington Press.

Shipley, B. and Keddy, P. A. (1987). The individualistic and community-unit concepts as falsifiable hypotheses. *Vegetatio*, **69**, 47–55.

Shipley, B. and Peters, R. H. (1990). A test of the Tilman model of plant strategies: relative growth rate and biomass partitioning. *The American Naturalist*, **136**, 139–53.

Shipley, B. and Parent, M. (1991). Germination responses of 64 wetland species in relation to seed size, minimum time to reproduction and seedling relative growth rate. *Functional Ecology*, **5**, 111–18.

Shipley, B. and Keddy, P. A. (1994). Evaluating the evidence for competitive hierarchies in plant communities. *Oikos*, **69**, 340–5.

Shipley, B., Keddy, P. A., Gaudet, C. and Moore, D. R. J. (1991a). A model of species density in shoreline vegetation. *Ecology*, **72**, 1658–67.

Shipley, B., Keddy, P. A. and Lefkovitch, L. P. (1991b). Mechanisms producing plant zonation along a water depth gradient: a comparison with the exposure gradient. *Canadian Journal of Botany*, **69**, 1420–4.

Shipley, B., Keddy, P. A., Moore, D. R. J. and Lemky, K. (1989). Regeneration and establishment strategies of emergent macrophytes. *Journal of Ecology*, **77**, 1093–110.

Shrader-Frechette, K. S. and McCoy, E. D. (1993). *Methods in Ecology: Strategies for Conservation.* Cambridge, UK: Cambridge University Press.

Siegel, S. (1956). *Nonparametric Statistics for the Behavioral Sciences.* New York: McGraw-Hill.

Silander, J. A. and Antonovics, J. (1982). Analysis of interspecific interactions in a coastal plant community – a perturbation approach. *Nature*, **298**, 557–60.

Silvola, J., Alm, J., Ahlholm, U., Nykänen, H. and Martikainen, P. J. (1996). CO_2 fluxes from peat in boreal mires under varying temperature and moisture conditions. *Journal of Ecology*, **84**, 219–28.

Simberloff, D. and Dayan, T. (1991). The guild concept and the structure of ecological communities. *Annual Review of Ecology and Systematics*, **22**, 115–43.

Sinclair, A. R. E. (1983). The adaptations of African ungulates and their effects on community function. In *Tropical Savannas*, ed. F. Bouliere. Amsterdam: Elsevier.

Sinclair, A. R. E. and Fryxell, J. M. (1985). The Sahel of Africa: ecology of a disaster. *Canadian Journal of Zoology*, **63**, 987–94.

Sinclair, A. R. E, Hik, D. S., Schmitz, O. J., Scudder, G. G. E., Turpin, D. H. and Larter, N. C. (1995). Biodiversity and the need for habitat renewal. *Ecological Applications*, **5**, 579–87.

Sioli, H. (1964). General features of the limnology of Amazonia. *Verhandlungen/ Internationale Vereinigung fur theoretische und angewandte Limnologie*, **15**, 1053–8.

Sioli, H. (1986). Tropical continental aquatic habitats. In *Conservation Biology. The Science of Scarcity and Diversity*, ed. M. E. Soulé, pp. 383–93. Sunderland, Massachusetts: Sinauer Associates.

Sippel, S. J., Hamilton, S. K., Melack, J. M. and Novo, E. M. M. (1998). Passive microwave observations of inundation area and the area/stage relation in the Amazon River floodplain. *International Journal of Remote Sensing*, **19**, 3055–74.

Skellam, J. G. (1951). Random dispersal in theoretical populations. *Biometrika*, **38**, 196–218.

Slack, N. G., Vitt, D. H. and Horton, D. G. (1980). Vegetation gradients of minerotrophically rich fens in western Alberta. *Canadian Journal of Botany*, **58**, 330–50.

Small, E. (1972a). Water relations of plants in raised *Sphagnum* peat bogs. *Ecology*, **53**, 726–8.

Small, E. (1972b). Photosynthetic rates in relation to nitrogen recycling as an adaptation to nutrient deficiency in peat bog plants. *Canadian Journal of Botany*, **50**, 2227–33.

Smart, R. M. and Barko, J. W. (1978). Influence of sediment salinity and nutrients on the physiological ecology of selected salt marsh plants. *Estuarine and Coastal Marine Science*, **7**, 487–95.

Smith, D. C. and Douglas, A. E. (1987). *The Biology of Symbiosis*. London: Edward Arnold.

Smith, D. W. and Cooper, S. D. (1982). Competition among cladocera. *Ecology*, **63**, 1004–15.

Smith, L. M. and Kadlec, J. A. (1983). Seed banks and their role during the drawdown of a North American marsh. *Journal of Applied Ecology*, **20**, 673–84.

Smith, L. M. and Kadlec, J. A. (1985a). Fire and herbivory in a Great Salt Lake marsh. *Ecology*, **66**, 259–65.

Smith, L. M. and Kadlec, J. A. (1985b). Comparisons of prescribed burning and cutting of Utah marsh plants. *Great Basin Naturalist*, **45**, 463–6.

Smith, P. G. R., Glooschenko, V. and Hagen, D. A. (1991). Coastal wetlands of three Canadian Great Lakes: inventory, current conservation initiatives, and patterns of variation. *Canadian Journal of Fisheries and Aquatic Sciences*, **48**, 1581–94.

Smith, V. H. (1982). The nitrogen and phosphorus dependence of algal biomass in lakes: an empirical and theoretical analysis. *Limnology and Oceanography*, **27**, 1101–12.

Smith, V. H. (1983). Low nitrogen to phosphorus ratios favor dominance by blue-green algae in lake phytoplankton. *Science*, **221**, 669–71.

Snow, A. A. and Vince, S. W. (1984). Plant zonation in an Alaskan salt marsh II: an experimental study of the role of edaphic conditions. *Journal of Ecology*, **72**, 669–84.

Sousa, W. P. (1984). The role of disturbance in natural communities. *Annual Review of Ecology and Systematics*, **15**, 353–91.

Southwood, T. R. E. (1977). Habitat, the templet for ecological strategies? *Journal of Animal Ecology*, **46**, 337–365.

Southwood, T. R. E. (1988). Tactics, strategies, and templets. *Oikos*, **52**, 3–18.

Specht, A. and Specht, R. L. (1993). Species richness and canopy productivity of Australian plant communities. *Biodiversity and Conservation*, **2**, 152–67.

Spence, D. H. N. (1964). The macrophytic vegetation of freshwater lochs, swamps and associated fens. In *The Vegetation of Scotland*, ed. J. H. Burnett, pp. 306–425. Edinburgh: Oliver & Boyd .

Spence, D. H. N. (1982). The zonation of plants in freshwater lakes. *Advances in Ecological Research*, **12**, 37–125.

Spencer, D. F. and Ksander, G. G. (1997). Influence of anoxia on sprouting of vegetative propagules of three species of aquatic plants. *Wetlands*, **17**, 55–64.

Springuel, I. (1990). Riverain vegetation in the Nile valley in Upper Egypt. *Journal of Vegetation Science*, **1**, 595–8.

Starfield, A. M. and Bleloch, A. L. (1991). *Building Models for Conservation and Wildlife Management*. 2nd edn. Edina: MN Burgers International Group.

Stead, I. M., Bourke, J. B. and Brothwell, D. (1986). *Lindow Man: The Body in the Bog*. British Museum Publications.

Steedman, R. J. (1988). Modification and assessment of an index of biotic integrity to quantify stream quality in southern Ontario. *Canadian Journal of Fisheries and Aquatic Sciences*, **45**, 492–501.

Stevenson, J. C., Ward, L. G. and Kearney, M. S. (1986). Vertical accretion in marshes with varying rates of sea level rise. In *Estuarine Variability*, ed. D. A. Wolfe, pp. 241–59. San Diego: Academic Press.

Stewart, R. E. and Kantrud, H. A. (1971). *Classification of Natural Ponds and Lakes in the Glaciated Prairie Region*. US Fish and Wildlife Service, Resource Publication, 92.

Stewart, W. N and Rothwell, G. W. (1993). *Paleobotany and the Evolution of Plants*. 2nd edn. New York: Cambridge University Press.

Strahler, A. N. (1971). *The Earth Sciences*. 2nd edn. New York: Harper and Row.

Street, F. A. and Grove, A. T. (1979). Global maps of lake-level fluctuations since 30,000 yrs. B.P. *Quaternary Research*, **12**, 83–118.

Stuckey, R. L. (1975). A floristic analysis of the vascular plants of a marsh at Perry's Victory Monument, Lake Erie. *The Michigan Botanist*, **14**, 144–66.

Sutter, R. D. and Kral, R. (1994). The ecology, status, and conservation of two non-alluvial wetland communities in the south Atlantic and eastern Gulf coastal plain, USA. *Biological Conservation*, **68**, 235–43.

Szalay, F. A. de and Resh, V. H. (1997). Responses of wetland invertebrates and plants important in waterfowl diets to burning and mowing of emergent vegetation. *Wetlands*, **17**, 149–56.

Szczepanski, A. J. (1990). Forested wetlands of Poland. In *Forested Wetlands*, ed. A. E. Lugo, M. Brinson and S. Brown, pp. 437–46. Amsterdam: Elsevier.

Taiz, L. and Zeiger, E. (1991). *Plant Physiology.* California: Benjamin Cummings.

Talling, J. F. (1992). Environmental regulation in African shallow lakes and wetlands. *Revue D'hydrobiologie Tropicale*, **25**, 87–144.

Tallis, J. H. (1983). Changes in wetland communities. In *Ecosystems of the World 4A. Mires: Swamp, Bog, Fen and Moor*, ed. A. J. P. Gore, pp. 311–47. Amsterdam: Elsevier.

Tansley, A. G. (1939). *The British Islands and Their Vegetation.* Cambridge: Cambridge University Press.

Tansley, A. G. and Adamson, R. S. (1925). Studies of the vegetation of the English chalk. Part iii. The chalk grasslands of the Hampshire–Sussex border. *Journal of Ecology*, **XIII**, 177–223.

Taylor, D. R., Aarssen, L. W. and Loehle, C. (1990). On the relationship between *r*/*K* selection and environmental carrying capacity: a new habitat templet for plant life history strategies. *Oikos*, **58**, 239–50.

Taylor, J. A. (1983). The peatlands of Great Britain and Ireland. In *Ecosystems of the World 4B. Mires: Swamp, Bog, Fen and Moor*, ed. A. J. P. Gore, pp. 1–46. The Netherlands: Elsevier Scientific Publishing.

Taylor, K. L. and Grace, J. B. (1995). The effects of vertebrate herbivory on plant community structure in the coastal marshes of the Pearl River, Louisiana, USA. *Wetlands*, **15**, 68–73.

Teller, J. T. (1988). Lake Agassiz and its contribution to flow through the Ottawa–St. Lawrence system. In *The Late Quaternary Development of the Champlain Sea Basin*, ed. N. R. Gadd, pp. 281–9. Geological Association of Canada Special Paper 35. St. John's, Newfoundland.

Terborgh, J. and Robinson, S. (1986). Guilds and their utility in ecology. In *Community Ecology: Pattern and Process*, ed. J. Kikkawa and D. J. Anderson, pp. 65–90. Melbourne: Blackwell Scientific Publications.

Thibodeau, F. R. and Ostro, B. D. (1981). An economic analysis of wetland protection. *Journal of Environmental Management*, **12**, 19–30.

Thomas, J. D. (1982). Chemical ecology of the snail hosts of Schistosomiasis: snail–snail and snail–plant interactions. *Malacologia*, **22**, 81–91.

Thompson D. J. and Shay, J. M. (1988). First-year response of a *Phragmites* marsh community to seasonal burning. *Canadian Journal of Botany*, **67**, 1448–55.

Thompson, K. (1985). Emergent plants of the permanent and seasonally-flooded wetlands. In *The Ecology and Management of African Wetland Vegetation*, ed. P. Denny. Dordrecht: Dr W. Junk Publishers.

Thompson, K. and Hamilton, A. C. (1983). Peatlands and swamps of the African continent. In *Ecosystems of the World 4B. Mires: Swamp, Bog, Fen and Moor*, ed. A. J. P. Gore, pp. 331–73. Amsterdam: Elsevier Scientific Publishing Company.

Thoreau, H. D. (1854). Republished in 1965 as Walden and Civil Disobedience, New York: Airmont Pub.

Tilman, D. (1982). *Resource Competition and Community Structure.* Princeton, NJ: Princeton University Press.

Tilman, D. (1986). Evolution and differentiation in terrestrial plant communities:

The importance of the soil resource: light gradient. In *Community Ecology*, ed. J. Diamond and T. J. Case, p. 359–80. New York: Harper and Row.
Tilman, D. (1988). *Plant Strategies and the Dynamics and Structure of Plant Communities.* Princeton, NJ: Princeton University Press.
Todd, T. N. and Davis, B. M. (1995). Effects of fish density and relative abundance on competition between larval lake herring and lake whitefish for zooplankton. *Archiv fuer Hydrobiologie. Special Issues Advanced Limnology*, **46**, 163–71.
Tomlinson, P. B. (1986). *The Botany of Mangroves.* Cambridge: Cambridge University Press.
Toner, M. and Keddy, P. A. (1997). River hydrology and riparian wetlands: a predictive model for ecological assembly. *Ecological Applications*, **7**, 236–46.
Tonn, W. M., and Magnuson, J. J. (1982). Patterns in the species composition and richness of fish assemblages in northern Wisconsin lakes. *Ecology*, **63**, 1149–66.
Tonn, W. M., Magnuson, J. J. and Forbes, A. M. (1983). Community analysis in fishery management: an application with northern Wisconsin lakes. *Transactions of the American Fisheries Society*, **112**, 368–77.
Toth, L. A. (1993). The ecological basis of the Kissimmee River restoration plan. *Florida Scientist*, **1**, 25–51.
Toth, L. A., Arrington, D. A., Brady, M. A. and Muszick, D. A. (1995). Conceptual evaluation of factors potentially affecting restoration of habitat structure within the channelized Kissimmee River ecosystem. *Restoration Ecology*, **3**, 160–80.
Townsend, A. R., Braswell, B. H., Holland, E. A. and Penner, J. E. (1996). Spatial and temporal patterns in terrestrial carbon storage due to deposition of fossil fuel nitrogen. *Ecological Applications*, **6**, 806–14.
Townsend, G. H. (1984). *Simulating the Effect of Water Regime Restoration Measures on Wildlife Populations and Habitat within the Peace–Athabasca Delta.* Technical report No. 13, Western and Northern Region, Canadian Wildlife Service, Saskatoon, Canada.
Tsuyuzaki, S.,and Tsujii, T. (1990). Preliminary study on grassy marshland vegetation, western part of Sichuan Province, China, in relation to yak-grazing. *Ecological Research*, **5**, 271–6.
Tsuyuzaki, S., Urano, S. and Tsujii, T. (1990). Vegetation of alpine marshland and its neighboring areas, northern part of Sichuan Province, China. *Vegetatio*, **88**, 79–86.
Turner, R. E. (1977). Intertidal vegetation and commercial yields of penaeid shrimp. *Transactions of the American Fisheries Society*, **106**, 411–16.
Turner, R. E. (1982). Protein yields from wetlands. *Wetlands: Ecology and Management*, Proceedings of the First International Wetlands Conference. New Delhi, India, 10–17 Sept. 1980.
Turner, R. M. and Karpiscak, M. M. (1980). *Recent Vegetation Changes Along the Colorado River Between Glen Canyon Dam and Lake Mead, Arizona.* Geological Survey Professional Paper 1132. Washington: United States Government Printing Office.
Twolan-Strutt, L. and Keddy, P. A. (1996). Above- and below-ground competition intensity in two contrasting wetland plant communities. *Ecology*, **77**, 259–70.
Underwood, A. J. (1978). The detection of non-random patterns of distribution of species along a gradient. *Oecologia*, **36**, 317–26.
Underwood, A. J. (1986). The analysis of competition by field experiments. In

Community Ecology. Pattern and Process, ed. J. Kikkawa and D. J. Anderson, pp. 240–68. Melbourne: Blackwell.

Urban, D. L. and Shugart H. H. (1992). Individual based models of forest succession. In *Plant Succession*, ed. D. C. Glenn-Lewin, R. K. Peet, and T. T. Veblen, pp. 249–92. London: Chapman and Hall.

US Army Coastal Engineering Research Centre. (1977). *Shore Protection Manual*. Vol. 1. 3rd edn. Washington, DC: US Government Printing Office. 514 pp.

Valiela, I., Foreman, K., LaMontagne, M., Hersh, D., Costa, J., D'Avanzo, C., Babione, M., Sham, C., Brawley, J., Peckol, P., DeMeo-Anderson, B. and Lajtha, K. (1992). Couplings of watersheds and coastal waters: sources and consequences of nutrient enrichment in Waquoit Bay, Massachusetts. *Estuaries*, **15**, 443–57.

Vallentyne, J. R. (1974). *The Algal Bowl. Lakes and Man*. Miscellaneous Special Publication 22. Ottawa. Department of the Environment, Fisheries and Marine Service.

van Breeman, N. (1995). How *Sphagnum* bogs down [*sic*] other plants. *Trends in Ecology and Evolution*, **10**, 270–5

van de Kieft, C. (1991). The Low Countries. *The New Encyclopedia Britannica*, 15th edn. **23**, 314–25.

van de Rijt, C. W. C. J., Hazelhoff, L. and Blom, C. W. P. M. (1996). Vegetation zonation in a former tidal area: a vegetation-type response model based on DCA and logistic regression using GIS. *Journal of Vegetation Science*, **7**, 505–18.

van der Leeden, F. (ed.), Troise, F. and Tood, D. K. (1990). *The Water Encyclopedia*, 2nd edn. Chelsea: Lewis Publishers. Inc. pp. 808.

van der Pijl, L. (1972). *Principles of Dispersal in Higher Plants.* New York: Springer-Verlag.

van der Toorn, J., Verhoeven, J. T. A. and Simpson, R. L. (1990). Fresh water marshes. In *Wetlands and Shallow Continental Water Bodies, Volume 1*, ed. B. C. Patten, pp. 445–65. The Hague, The Netherlands: SPB Academic Publishing.

van der Valk, A. G. (1981). Succession in wetlands: a Gleasonian approach. *Ecology*, **62**, 688–96.

van der Valk, A. G. (1988). From community ecology to vegetation management: providing a scientific basis for management. In *Transactions of the Fifty-third North American Wildlife and Natural Resources Conference*, pp. 463–70. Washington, DC: Wildlife Management Institute.

van der Valk, A. G. (1989). *Northern Prairie Wetlands.* Ames, Iowa: Iowa State University Press.

van der Valk, A. G. and Davis, C. B. (1976). The seed banks of prairie glacial marshes. *Canadian Journal of Botany*, **54**, 1832–8.

van der Valk, A. G., and Davis, C. B. (1978). The role of seed banks in the vegetation dynamics of prairie glacial marshes. *Ecology*, **59**, 322–35.

van der Valk, A. G., Pederson, R. L. and Davis, C. B. (1992). Restoration and creation of freshwater wetlands using seed banks. *Wetlands Ecology and Management*, **1**, 191–7.

van der Valk, A. G., Swanson, S. D. and Nuss, R. F. (1983). The response of plant species to burial in three types of Alaskan wetlands. *Canadian Journal of Botany*, **61**, 1150–64.

Van Wijck, C. and de Groot, C. J. (1993). The impact of desiccation of a freshwater marsh (Garcines Nord, Camargue, France) on sediment–water–vegetation interactions. *Hydrobiologia*, **252**, 95–103.

Verhoeven, J. T. A and Liefveld, W. M. (1997). The ecological significance of organo-chemical compounds in *Sphagnum*. *Acta Botanica Neerl.*, **46**, 117–30.

Verhoeven, J. T. A., Kemmers, R. H. and Koerselman, W. (1993). Nutrient enrichment of freshwater wetlands. In *Landscape Ecology of a Stressed Environment*, ed. C. C. Vos and P. Opdam, pp. 33–59. London: Chapman and Hall.

Verhoeven, J. T. A., Koerselman, W. and Meuleman, A. F. M. (1996). Nitrogen- or phosphorus-limited growth in herbaceous, wet vegetation: relations with atmospheric inputs and management regimes. *Trends in Ecology and Evolution*, **11**, 493–7.

Vesey-FitzGerald, D. F. (1960). Grazing succession among East African game animals. *Journal of Mammology*, **41**, 161–72.

Vince, S. W. and Snow, A. A. (1984). Plant zonation in an Alaskan salt marsh I. Distribution, abundance, and environmental factors. *Journal of Ecology*, **72**, 651–67.

Vitousek, P. M. (1982). Nutrient cycling and nitrogen use efficiency. *The American Naturalist*, **119**, 553–72.

Vitousek, P. M., Aber, J., Howarth, R. W., Likens, G. E., Matson, P. A., Schindler, D. W., Schlesinger, W. H. and Tilman, G. D. (1997). Human alteration of global nitrogen cycle: causes and consequences. *Issues in Ecology*, Washington, DC: No. 1 Ecological Society of America.

Vitt, D. H. (1990). Growth and production dynamics of boreal mosses over climatic, chemical and topographic gradients. *Botanical Journal of the Linnean Society*, **104**, 35–59.

Vitt, D. H. (1994). An overview of factors that influence the development of Canadian peatlands. *Memoirs of the Entomological Society of Canada*, **169**, 7–20.

Vitt, D. H. and Slack, N. G. (1975). An analysis of the vegetation of *Sphagnum*-dominated kettle-hole bogs in relation to environmental gradients. *Canadian Journal of Botany*, **53**, 332–59.

Vitt, D. H., and Slack, N. G. (1984). Niche diversification of *Sphagnum* relative to environmental factors in northern Minnesota peatlands. *Canadian Journal of Botany*, **62**, 1409–30.

Vitt, D. H. and Chee, W. (1990). The relationships of vegetation to surface water chemistry and peat chemistry in fens of Alberta, Canada. *Vegetatio*, **89**, 87–106.

Vitt, D. H., Yenhung, L. and Belland, R. J. (1995). Patterns of bryophyte diversity in peatlands of continental western Canada. *The Bryologist*, **98**, 218–27.

Vivian-Smith, G. (1997). Microtopographic heterogeneity and floristic diversity in experimental wetland communities. *Journal of Ecology*, **85**, 71–82.

Vogl, R. (1969). One hundred and thirty years of plant succession in a southeastern Wisconsin lowland. *Ecology*, **50**, 248–55

Vogl R. (1973). Effects of fire on the plants and animals of a Florida wetland. *American Midland Naturalist*, **89**, 334–47.

Walker, B. H. and Wehrhahn, C. F. (1971). Relationships between derived vegetation gradients and measured environmental variables in Saskatchewan wetlands. *Ecology*, **52**, 85–95.

Walker, D. (1970). Direction and rate in some British post-glacial hydroseres. In *Studies in the Vegetational History of the British Isles*, ed. D. Walker and R. G. West, pp. 117–39. Cambridge: Cambridge University Press.

Wang, S., Jurik, T. M, and van der Valk, A. G. (1994). Effects of sediment load on various stages in the life and death of cattail (*Typha* × glauca). *Wetlands*, **14**, 166–73.

Wassen, M. J., Barendregt, A., Palczynski, A., de Smidt, J. T., and de Mars, H. (1990).

The relationship between fen vegetation gradients, groundwater flow and flooding in undrained valley mire at Biebrza, Poland. *Journal of Ecology*, **78**, 1106–22.

Waters, T. F. (1995). Sediment in streams. Sources, biological effects, and control. *American Fisheries Society Monograph* 7.

Watts, W. A. and Winter, T. C. (1966). Plant macrofossils from Kirchner Marsh, Minnesota – a paleoecological study. *Geological Society of America Bulletin*, **77**, 1339–60.

Weber, W. and Rabinowitz, A. (1996). A global perspective on large carnivore conservation. *Conservation Biology*, **10**, 1046–54.

Weiher, E. (1999). The combined effects of scale and productivity on species richness. *Journal of Ecology*, **87**, 1005–11.

Weiher, E. and Boylen, C. W. (1994). Patterns and prediction of *a* and *b* diversity of aquatic plants in Adirondack (New York) lakes. *Canadian Journal of Botany*, **72**, 1797–804.

Weiher, E. and Keddy, P. A. (1995). The assembly of experimental wetland plant communities. *Oikos*, **73**, 323–35.

Weiher, E. and Keddy, P. A. (1999). *Assembly Rules in Ecological Communities Perspectives, Advances, Retreats.* Cambridge: Cambridge University Press.

Weiher, E., Wisheu, I. C., Keddy, P. A. and Moore, D. R. J. (1996). Establishment, persistence, and management implications of experimental wetland plant communities. *Wetlands*, **16**, 208–18.

Weiher, E., Clarke, G. D. P. and Keddy, P. A. (1998). Community assembly rules, morphological dispersion, and the coexistence of plant species. *Oikos*, **81**, 309–22.

Wein, R. W. (1983). Fire behaviour and ecological effects in organic terrain. In *The Role of Fire in Northern Circumpolar Ecosystems*, ed. R. W. Wein and D. A. Maclean, pp. 81–95. New York; John Wiley,

Weinberg, G. M. (1975). *An Introduction of General Systems Thinking.* New York; John Wiley.

Weisner, S. E. B. (1990). *Emergent Vegetation in Eutrophic Lakes: Distributional Patterns and Ecophysiological Constraints.* Sweden: Grahns Boktryckeri.

Welcomme, R. L. (1976). Some general and theoretical considerations on the fish yield of African rivers. *Journal of Fish Biology*, **8**, 351–64.

Welcomme, R. L. (1979). *Fisheries Ecology of Floodplain Rivers.* London: Longman.

Welcomme, R. L. (1986). Fish of the Niger system. In *The Ecology of River Systems*, ed. B. R. Davies and K. F. Walker, pp. 25–48. Dordrecht, The Netherlands: Dr W. Junk Publishers.

Weller, M. W. (1978). Management of freshwater marshes for wildlife. In *Freshwater Wetlands. Ecological Processes and Management Potential*, ed. R. E. Good, D. F. Whigham and R. L. Simpson, pp. 267–84. New York: Academic Press.

Weller, M. W. (1994a). *Freshwater Marshes: Ecology and Wildlife Management.* 3rd edn. Minneapolis: University of Minnesota.

Weller, M. W. (1994b). Bird-habitat relationships in a Texas estuarine marsh during summer. *Wetlands*, **14**, 293–300.

Wells, H. G. (1949). *The Outline of History.* Garden City, NY: Garden City Books.

Welty, J. C. (1982). *The Life of Birds.* 3rd edn. New York: Saunders College Publishing.

Werner, E. E. (1984). The mechanisms of species interactions and community organization in fish. In *Ecological Communities. Conceptual Issues and the Evidence*, ed. D. R. Strong, Jr., D. Simberloff, L. G. Abele and A. B. Thistle, pp. 360–82. Princeton, NJ: Princeton University Press.

Werner, E. E. and Hall, D. J. (1976). Niche shifts in sunfishes: experimental evidence and significance. *Science*, **191**, 404–6.

Werner, E. E. and Hall, D. J. (1977). Competition and habitat shift in two sunfishes (Centrarchidae). *Ecology*, **58**, 869–76.

Werner, E. E. and Hall, D. J. (1979). Foraging efficiency and habitat switching in competing sunfishes. *Ecology*, **60**, 256–64.

Western, D. (1975). Water availability and its influence on the structure and dynamics of a savannah large mammal community. *African Wildlife Journal*, **13**, 265–86.

Westhoff, V. and Van der Maarel, E. (1973). The Braun–Blanquet approach. In *Ordination and Classification of Communities*, ed. R. H. Whittaker, pp. 617–726. The Hague: Junk.

Wetzel, R. G. (1975). *Limnology*. Philadelphia: W. B. Saunders Company.

Wetzel, R. G. (1989). Wetland and littoral interfaces of lakes: productivity and nutrient regulation in the Lawrence Lake ecosystem. In *Freshwater Wetlands and Wildlife*, ed. R. R. Sharitz and J. W. Gibbons, pp. 283–302. US Department of Energy (Proceedings of a Symposium held at Charleston, South Carolina, March 24–27, 1986).

Wheeler, B. D and Giller, K. E. (1982). Species richness of herbaceous fen vegetation in Broadland, Norfolk in relation to the quantity of above-ground plant material. *Journal of Ecology*, **70**, 179–200.

Wheeler, B. D. and Shaw, S. C. (1991). Above-ground crop mass and species richness of the principal types of herbaceous rich-fen vegetation of lowland England and Wales. *Journal of Ecology*, **79**, 285–301.

Whigham, D. F., Dykyjova, D. and Hejny, S. (eds.) (1992). *Wetlands of the World 1*. The Netherlands: Kluwer Academic Publishers.

Whillans, T. H. (1989). Wetlands and fisheries in central Ontario lakes. In *Ontario Wetlands: Inertia or Momentum*, ed. M. J. Bardecki and N. Patterson, Proceedings of Conference, Ryerson Polytechnical Institute, Toronto, Oct 21–22, 1988.

White, P. S. (1979). Pattern, process and natural disturbance in vegetation. *The Botanical Review*, **45**, 229–99.

White, P. S. (1994). Synthesis: vegetation pattern and process in the Everglades ecosystem. In *Everglades: The Ecosystem and its Restoration*, ed. S. Davis and J. Ogden, pp. 445–60. Del Ray Beach, FL: St. Lucia Press.

White T. C. R. (1993). *The Inadequate Environment*. Berlin: Springer-Verlag.

Whiting, G. J. and Chanton, J. P. (1993). Primary production control of methane emission from wetlands. *Nature*, **364**, 794–5.

Whitney, D. M., Chalmers, A. G., Haines, E. B., Hanson, R. B., Pomroy, L. R. and Sherr, B. (1981). The cycles of nitrogen and phosphorus. In *The Ecology of a Salt Marsh*, ed. L. R. Pomeroy and R. G Wiegert, pp. 161–78. Ecological Studies 38. New York: Springer-Verlag.

Whittaker, R. H. (1956). Vegetation of the Great Smoky Mountains. *Ecological Monographs*, **26**, 1–80.

Whittaker, R. H. (1962). Classification of natural communities. *Botanical Review*, **28**, 1–160.

Whittaker, R. H. (1967). Gradient analysis of vegetation. *Biological Reviews*, **42**, 207–64.

Whittaker, R. H. (1975). *Communities and Ecosystems*. New York: MacMillan Publishing.

Whittaker, R. H. and Likens, G. E. (1973). Carbon in the biota. In *Carbon in the Biosphere*, ed. G. M. Woodwell and E. R. Peacan, pp. 281–302. Springfield, VA: National Technical Information Services.

Wickware, G. M. and Rubec, C. D. A. (1989). *Ecoregions of Ontario*. Ecological Land Classification Series No. 26. Environment Canada: Sustainable Development Branch.

Wiegers, J. (1990). Forested wetlands in western Europe. In *Forested Wetlands*, ed. A. E. Lugo, M. Brinson and S. Brown, pp. 407–36. Amsterdam: Elsevier.

Wiegert, R. G., L. Pomeroy, R. and Wiebe, W. J. (1981). Ecology of salt marshes: an introduction. In *The Ecology of a Salt Marsh*, ed. L. R. Pomeroy and R. G. Wiegert. Ecological Studies 38. New York: Springer-Verlag.

Wiens, J. A. (1983). Avian community ecology: an iconoclastic view. In *Perspectives in Ornithology*, ed. A. H. Brush and G. A. Clark. Jr., pp. 355–403. Essays presented for the centennial of the American Ornithologists' Union. Cambridge: Cambridge University Press.

Wikramanayake, E. D. (1990). Ecomorphology and biogeography of a tropical stream fish assemblage: evolution of assemblage structure. *Ecology*, **71**, 1756–64.

Wilbur, H. M. (1972). Competition, predation and the structure of the *Ambystoma-Rana sylvatica* community. *Ecology*, **53**, 3–21.

Wilbur, H. M. (1984). Complex life cycles and community organization in amphibians. In *A New Ecology: Novel Approaches to Interactive Systems*, ed. P. W. Price, C. N. Slobodchikoff and W. S. Gaud, pp. 195–225. New York: John Wiley.

Wilcox, D. A. and Meeker, J. E. (1991). Disturbance effects on aquatic vegetation in regulated and unregulated lakes in northern Minnesota. *Canadian Journal of Botany*, **69**, 1542–51.

Wilcox, D. A. and Simonin, H. A. (1987). A chronosequence of aquatic macrophyte communities in dune ponds. *Aquatic Botany*, **28**, 227–42.

Wild Earth. (1992). *The Wildlands Project*. Special Issue. 88pp.

Williamson, G. B. (1990). Allelopathy, Koch's postulates and the neck riddle. In *Perspectives on Plant Competition*, ed. J. B. Grace and D. Tilman, pp. 143–62. San Diego, CA: Academic Press.

Willis, A. J. (1963). Braunton Burrows: the effects on the vegetation of the addition of mineral nutrients to the dune soils. *Journal of Ecology*, **51**, 353–74.

Wilson, E. O. and Bossert, W. H. (1971). *A Primer of Population Biology*. Sunderland, Massachusetts: Sinauer Associates, Inc. Publishers.

Wilson, J. A. (1972). *Principles of Animal Physiology*. New York: The Macmillan Company.

Wilson, S. D. and Keddy, P. A. (1985). Plant zonation on a shoreline gradient: physiological response curves of component species. *Journal of Ecology*, **73**, 851–60.

Wilson, S. D. and Keddy, P. A. (1986a). Species competitive ability and position along a natural stress/disturbance gradient. *Ecology*, **67**, 1236–42.

Wilson, S. D. and Keddy, P. A. (1986b). Measuring diffuse competition along an envi-

ronmental gradient: results from a shoreline plant community. *The American Naturalist*, **127**, 862–9.
Wilson, S. D. and Keddy, P. A. (1988). Species richness, survivorship, and biomass accumulation along an environmental gradient. *Oikos*, **53**, 375–80.
Wilson, S. D. and Keddy, P. A. (1991). Competition, survivorship and growth in macrophyte communities. *Freshwater Biology*, **25**, 331–7.
Winemiller, K. O. (1991). Ecomorphological diversification in lowland freshwater fish assemblages from five biotic regions. *Ecological Monographs*, **61**, 343–65.
Winter, T. C. and Rosenberry, D. O. (1995). The interaction of ground water with prairie pothole wetlands in the Cottonwood Lake area, east-central North Dakota, 1979–1990. *Wetlands*, **15**, 193–211.
Wisheu, I. C. (1996). Shared preference niche organization: implications for community organization and diversity. PhD Thesis. University of Ottawa, Ottawa, Canada.
Wisheu, I. C. (1998). How organisms partition habitats: different types of community organization can produce identical patterns. *Oikos*, **83**, 246–58.
Wisheu, I. C. and Keddy, P. A. (1989a). Species richness – standing crop relationships along four lakeshore gradients: constraints on the general model. *Canadian Journal of Botany*, **67**, 1609–17.
Wisheu, I. C. and Keddy, P. A. (1989b). The conservation and management of a threatened coastal plain plant community in eastern North America (Nova Scotia, Canada). *Biological Conservation*, **48**, 229–38.
Wisheu, I. C. and Keddy, P. A. (1991). Seed banks of a rare wetland plant community: distribution patterns and effects of human-induced disturbance. *Journal of Vegetation Science*, **2**, 181–8.
Wisheu, I. C. and Keddy, P. A. (1992). Competition and centrifugal organization of plant communities: theory and tests. *Journal of Vegetation Science*, **3**, 147–56.
Wisheu, I. C. and Keddy, P. A. (1996). Three competing models for predicting the size of species pools: a test using eastern North American wetlands. *Oikos*, **76**, 253–8.
Wisheu, I. C., Keddy, P. A., Moore, D. J., McCanny, S. J. and Gaudet, C. L. (1990). Effects of eutrophication on wetland vegetation. In *Proceedings of an International Symposium Wetlands of the Great Lakes.* Niagara Falls, NY.
Wium-Anderson, S. (1971). Photosynthetic uptake of free CO_2 by the roots of *Lobelia dortmanna*. *Plantarum*, **25**, 245–8.
Wolff, W. J. (1993). Netherlands Wetlands. *Hydrobiologia*, **265**, 1–14.
Woo, M., Rowsell, R. D. and Clark, R. G. (1993). Hydrological classification of Canadian prairie wetlands and prediction of wetland inundation in response to climatic variability. Occasional paper No.79. Canadian Wildlife Service.
Woodley, S., Kay, J. and Francis, G. (eds) (1993). *Ecological Integrity and the Management of Ecosystems*. Delray Beach, FL: St-Lucie Press.
Woodward, F. I. and Kelly, C. K. (1997). Plant functional types: towards a definition by environmental constraints. In *Plant Functional Types*, ed. T. M. Smith, H. H. Shugart, and F. I. Woodward, pp. 47–65. Cambridge: Cambridge University Press.
Woodwell, G. M. and Whittaker, R. H. (1968). Effects of chronic gamma radiation on plant communities. *Quarterly Review of Biology*, **43**, 42–55.

Woodwell, G. M., Mackenzie, F. T., Houghton, R. A., Apps, A. J., Gorham, E. and Davidson, E. A. (1995). Will the warming speed the warming? In *Biotic Feedbacks in the Global Climatic System*, ed. G. M. Woodwell and F. T. Mackenzie. New York: Oxford University Press.

Wootton, R. J. (1990). Biotic interaction II: Competition and mutualism. In *Ecology of Teleost Fishes*, ed. R. J. Wootton, pp. 216–37. London: Chapman and Hall.

World Commission on Environment and Development. (1987). *Our Common Future*. Oxford, UK: Oxford University Press.

World Conservation Monitoring Centre. (1992). *Global Biodiversity: Status of the Earth's Living Resources.* London: Chapman and Hall.

World Resources Institute. (1992). World Resources 1992–1993. Oxford: Oxford University Press.

Wright, D. H. and Reeves, J. H. (1992). On the meaning and measurement of nestedness of species assemblages. *Oecologia*, **92**, 416–28.

Wright, H. E. and Bent, A. M. (1968). Vegetation bands around Dead Man Lake, Chuska Mountain, New Mexico. *American Midland Naturalist*, **79**, 8–30.

Yabe, K. (1993). Wetlands of Hokkaido. In *Biodiversity and Ecology in the Northernmost Japan*, ed. S. Higashi, A. Osawa and K. Kanagawa, pp. 38–49. Hokkaido: Hokkaido University Press.

Yabe, K. and Numata, M. (1984). Ecological studies of the Mobawa–Yatsumi marsh. Main physical and chemical factors controlling the marsh ecosystem. *Japanese Journal of Ecology*, **34**, 173–86.

Yabe, K. and Onimaru, K. (1997). Key variables controlling the vegetation of a cool–temperate mire in northern Japan. *Journal of Vegetation Science*, **8**, 29–36.

Yodzis, P. (1986). Competition, mortality, and community structure. In *Community Ecology*, ed. J. Diamond and T. J. Case, pp. 480–92. New York: Harper and Row.

Yodzis, P. (1989). *Introduction to Theoretical Ecology.* New York: Harper and Row.

Yu, Z., McAndrews, J. H. and Siddiqi, D. (1996). Influences of Holocene climate and water levels on vegetation dynamics of a lakeside wetland. *Canadian Journal of Botany*, **74**, 1602–15.

Zagwijn, W. H. (1986). Nederland in het Holoceen. Geologie van Nederland, deel 1. Haarlem/Staatssuitgeverij's – Gravenhage: Rijks Geologische Dienst. 46pp.

Zalidis, G. C., Mantzavelas, A. L. and Gourvelou, E. (1997). Environmental impacts on Greek wetlands. *Wetlands*, **17**, 339–45.

Zedler, J. B. (1988). Why it's so difficult to replace wetland functions. In *Increasing our Wetland Resources*, ed. J. Zelazny and J. S. Feierabend. Proceedings of a conference in Washington, DC October 4–7, 1987. National Wildlife Federation–Corporate Conservation Council.

Zedler, J. B. (1996). Ecological issues in wetland mitigation: an introduction to the forum. *Ecological Applications*, **6**, 33–7.

Zedler, J. B. and Onuf, C. P. (1984). Biological and physical filtering in arid-region estuaries: seasonality, extreme events, and effects of watershed modification. In *The Estuary as a Filter*, ed. V. S. Kennedy, pp. 415–32. New York: Academic Press.

Zedler, J. B. and Beare, P. A. (1986). Temporal variability of salt marsh vegetation: the role of low-salinity gaps and environmental stress. In *Estuarine Variability*, ed. D. A. Wolfe, pp. 295–306. San Diego: Academic Press.

Zedler, J. B., Paling, E. and McComb, A. (1990). Differential responses to salinity help